Spectral Properties of Banded Toeplitz Matrices

Spectral Properties of Banded Toeplitz Matrices

Albrecht Böttcher
Chemnitz University of Technology
Chemnitz, Germany

Sergei M. Grudsky
CINVESTAV del I. P. N.
Mexico City, Mexico
and
Rostov-on-Don State University
Rostov-on-Don, Russia

siam Society for Industrial and Applied Mathematics
Philadelphia

10 9 8 7 6 5 4 3 2 1

Library of Congress Cataloging-in-Publication Data

Böttcher, Albrecht.
Spectral properties of banded Toeplitz matrices / Albrecht Böttcher, Sergei M. Grudsky.
p. cm.
Includes bibliographical references and index.
ISBN 0-89871-599-7 (pbk.)
Toeplitz matrices. I. Grudsky, Sergei M., 1955- II. Title.

QA188.B674 2005
512.9'434–dc22

2005051608

Partial royalties from the sale of this book are placed in a fund to help students attend SIAM meetings and other SIAM-related activities. This fund is administered by SIAM, and qualified individuals are encouraged to write directly to SIAM for guidelines.

Contents

Preface

Toeplitz matrices emerge in plenty of applications and have been extensively studied for about a century. The literature on them is immense and ranges from thousands of articles in periodicals to huge monographs. This does not imply that there is nothing left to say on the topic. To the contrary, Toeplitz matrices are an active field of research with many facets, and the amount of material gathered only in the last decade would easily fill several volumes.

The present book lives within its limitations: to *banded* Toeplitz matrices on the one hand and to the *spectral properties* of such matrices on the other. As a third limitation, we consider large matrices only, and most of the results are actually *asymptotics*.

When speaking of banded Toeplitz matrices, we have in mind an $n \times n$ Toeplitz matrix of bandwidth $2r + 1$, and we silently assume that n is large in comparison with $2r + 1$. A Toeplitz matrix is completely specified by the (complex) numbers that constitute its first row and its first column. The function on the complex unit circle whose Fourier coefficients are just these numbers is referred to as the symbol of the matrix. In the case of Toeplitz band matrices, the symbol is a Laurent polynomial. Thus, we need not struggle with piecewise continuous or oscillating symbols, which arise in many applications, but "only" with Laurent polynomials. This circumstance simplifies part of the investigation. On the other hand, Laurent polynomials cause questions that are different from those one encounters in connection with more general symbols. Eventually, Toeplitz band matrices form their own realm in the world of Toeplitz matrices.

We understand spectral properties in a broad sense. Of course, we study such problems as the evolution of the eigenvalues of banded $n \times n$ Toeplitz matrices as n goes to infinity. The pioneering result in this direction was already proved by Schmidt and Spitzer in 1960, and every worker in the field has a personal copy of the Schmidt/Spitzer paper. Here we cite a full proof of this result for the first time in the monographical literature. This proof is Schmidt and Spitzer's original proof with several simplifications and improvements introduced by Hirschman and Widom.

We regard the singular values of a matrix as its most important spectral characteristics after the eigenvalues and pseudoeigenvalues; hence, we pay due attention to the asymptotic behavior of the singular values as the matrix dimension increases to infinity. Clearly, questions about the norm, the norm of the inverse, and the condition numbers of a matrix are questions about the extreme singular values.

Normal Toeplitz matrices raise specific problems, and these will be discussed. However, typically a Toeplitz matrix is nonnormal; hence, pseudospectra tell us more about Toeplitz matrices than spectra. Accordingly, we embark on pseudospectra of Toeplitz matrices and on related issues, such as the transient behavior of powers of large Toeplitz matrices.

Finally, the book contains some very recent results on the spectral behavior of Toeplitz matrices under certain structured perturbations. These results are far from what one wants to know about Toeplitz matrices with randomly perturbed main diagonal, but they are beautiful, they point in a good direction, and they have the potential to stimulate further research.

As already stated, the majority of the results describe the asymptotic behavior as the matrix dimension n goes to infinity. Many questions considered here can be easily answered by a few MATLAB commands if the matrix dimension is moderate, say in the low hundreds. We try to deliver answers in the case where n is really large and the computer quits. Part of the results are equipped with estimates of the convergence speed, which provides the user at least with a vague feeling for as to whether one can invoke the result for n in the hundreds. And, most importantly, several problems of this book are motivated by applications in statistical physics, where n is around 10^8, the cube root of the Avogadro number 10^{23}, and, for such astronomic values of n, asymptotic formulas are the only chance to describe and to understand something.

In summary, the book provides several pieces of information about the eigenvalues, singular values, determinants, norms, norms of inverses, (unstructured and structured) condition numbers, (unstructured and structured) pseudospectra, transient behavior, eigenvectors and pseudomodes, and spectral phenomena caused by perturbations of large Toeplitz band matrices. The selection of the material represents our taste and is to some extent determined by subjects we have worked on ourselves, and we think we can tell the community something about. Naturally, numerous problems are left open. Moreover, various important topics, such as fast inversion of Toeplitz matrices or fast solution of Toeplitz systems, are not touched at all. These topics are the business of other books (see, for example, [157] and [177]). However, the material of the present book is certainly useful and in many cases even indispensable when dealing with such practical problems as the effective solution of a large banded Toeplitz system.

The book is intended as an introductory text to some advanced topics. We assume that the reader is familiar with the basics of real and complex analysis, linear algebra, and functional analysis. Almost all results are accompanied by full proofs.

A baby version of this book was published under the title *Toeplitz Matrices, Asymptotic Linear Algebra, and Functional Analysis* by Hindustan Book Agency, New Delhi, and Birkhäuser, Basel, in 2000.

S. M. Grudsky thankfully acknowledges financial support by CONACYT grant N 40564-F (México).

We sincerely thank our wives, Sylvia Böttcher and Olga Grudskaya, for their usual patient and excellent work on the LaTeX masters and on part of the illustrations. We are also greatly indebted to Mark Embree for valuable remarks on a draft of this book and to Linda Thiel and the staff of SIAM for their help with publishing the book.

Tragically, Olga Grudskaya died in a car accident in February 2004. We have lost an exceptional woman, a wonderful friend, and an irreplaceable colleague. Her early death leaves an emptiness that can never be filled. In late 2003, she began working on the illustrations for this book with great enthusiasm. She could not accomplish her visions. We were left with the drafts of her illustrations and included some of them. They provide us with an idea of the beauty that would have emerged if she would have been able to complete her work. May this book keep the memory of our irretrievable Olga.

Chemnitz and Mexico City, spring 2005 The authors

Chapter 1

Infinite Matrices

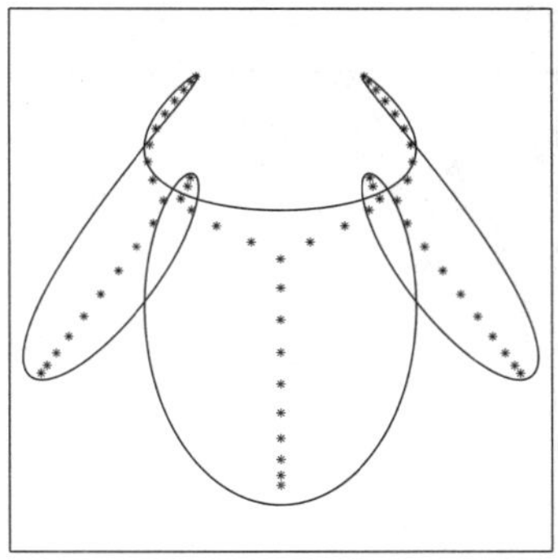

When studying large finite matrices, it is natural to look also at their infinite counterparts. The spectral phenomena of the latter are sometimes easier to understand than those of the former. The question whether properties of infinite Toeplitz matrices mimic the corresponding properties of their large finite sections is very delicate and is, in a sense, the topic of this book.

We regard infinite Toeplitz matrices as operators on ℓ^p. This chapter is concerned with some basic properties of these operators, including boundedness, norms, invertibility and inverses, spectrum, eigenvalues, and eigenvectors. Wiener-Hopf factorization provides us with a fairly effective tool for the inversion of infinite (but not of finite) Toeplitz matrices. We also embark on some of the problems that are specific for selfadjoint operators.

1.1 Toeplitz and Hankel Matrices

An *infinite Toeplitz matrix* is a matrix of the form

$$(a_{j-k})_{j,k=0}^{\infty} = \begin{pmatrix} a_0 & a_{-1} & a_{-2} & \dots \\ a_1 & a_0 & a_{-1} & \dots \\ a_2 & a_1 & a_0 & \dots \\ \dots & \dots & \dots & \dots \end{pmatrix}. \tag{1.1}$$

Such matrices are characterized by the property of being constant along the parallels to the main diagonal. Clearly, the matrix (1.1) is completely determined by its entries in the first row and first column, that is, by the sequence

$$\{a_k\}_{k=-\infty}^{\infty} = \{\dots, a_{-2}, a_{-1}, a_0, a_1, a_2, \dots\}. \tag{1.2}$$

Throughout this book we assume that the a_k's are complex numbers.

The matrix (1.1) is a band matrix if and only if at most finitely many of the numbers in (1.2) are nonzero. Although our subject is Toeplitz band matrices, it is also necessary to study Toeplitz matrices which are not band matrices. For example, the inverse of the band matrix

$$\begin{pmatrix} 1 & -\frac{1}{2} & 0 & 0 & \dots \\ 0 & 1 & -\frac{1}{2} & 0 & \dots \\ 0 & 0 & 1 & -\frac{1}{2} & \dots \\ 0 & 0 & 0 & 1 & \dots \\ \dots & \dots & \dots & \dots & \dots \end{pmatrix}$$

is the Toeplitz matrix

$$\begin{pmatrix} 1 & \frac{1}{2} & \frac{1}{2^2} & \frac{1}{2^3} & \dots \\ 0 & 1 & \frac{1}{2} & \frac{1}{2^2} & \dots \\ 0 & 0 & 1 & \frac{1}{2} & \dots \\ 0 & 0 & 0 & 1 & \dots \\ \dots & \dots & \dots & \dots & \dots \end{pmatrix},$$

and this is not a band matrix.

There is another type of matrix that arises when working with Toeplitz matrices. These are the Hankel matrices. An *infinite Hankel matrix* has the form

$$(a_{j+k+1})_{j,k=0}^{\infty} = \begin{pmatrix} a_1 & a_2 & a_3 & \dots \\ a_2 & a_3 & \dots & \dots \\ a_3 & \dots & \dots & \dots \\ \dots & \dots & \dots & \dots \end{pmatrix}. \tag{1.3}$$

Notice that (1.3) is completely given by only the numbers with positive indices in (1.2). Obviously, if the sequence (1.2) has finite support, then the matrix (1.3) contains only finitely many nonzero entries.

1.2 Boundedness

The Wiener algebra. Let $\mathbf{T} := \{t \in \mathbf{C} : |t| = 1\}$ be the complex unit circle. The *Wiener algebra* W is defined as the set of all functions $a : \mathbf{T} \to \mathbf{C}$ with absolutely convergent Fourier series - that is, as the collection of all functions $a : \mathbf{T} \to \mathbf{C}$ which can be represented in the form

$$a(t) = \sum_{n=-\infty}^{\infty} a_n t^n \ (t \in \mathbf{T}) \quad \text{with} \quad \|a\|_W := \sum_{n=-\infty}^{\infty} |a_n| < \infty. \tag{1.4}$$

Notice that instead of (1.4) we could also write

$$a(e^{i\theta}) = \sum_{n=-\infty}^{\infty} a_n e^{in\theta} \ (e^{i\theta} \in \mathbf{T}) \quad \text{with} \quad \|a\|_W := \sum_{n=-\infty}^{\infty} |a_n| < \infty. \tag{1.5}$$

The numbers a_n are the *Fourier coefficients* of a, and they can be computed by the formula

$$a_n = \frac{1}{2\pi} \int_0^{2\pi} a(e^{i\theta}) e^{-in\theta} \, d\theta. \tag{1.6}$$

Sometimes it will be convenient to identify a function $a : \mathbf{T} \to \mathbf{C}$ with the function $\theta \mapsto a(e^{i\theta})$; the latter function may be thought of as being given on $[0, 2\pi)$, $(-\pi, \pi]$, or even on all of the real line $\mathbf{R}$. Clearly, functions in W are continuous on $\mathbf{T}$ and, when regarded as functions on $\mathbf{R}$, they are 2π-periodic continuous functions.

Now let $a \in W$ and let $\{a_n\}_{n=-\infty}^{\infty}$ be the sequence of the Fourier coefficients of a. We denote by $T(a)$ and $H(a)$ the matrices (1.1) and (1.3), respectively:

$$T(a) := \begin{pmatrix} a_0 & a_{-1} & a_{-2} & \dots \\ a_1 & a_0 & a_{-1} & \dots \\ a_2 & a_1 & a_0 & \dots \\ \dots & \dots & \dots & \dots \end{pmatrix}, \quad H(a) := \begin{pmatrix} a_1 & a_2 & a_3 & \dots \\ a_2 & a_3 & \dots & \dots \\ a_3 & \dots & \dots & \dots \\ \dots & \dots & \dots & \dots \end{pmatrix}.$$

The matrix $T(a)$ is called the infinite Toeplitz matrix generated by a, while a is referred to as the *symbol* of the matrix $T(a)$. Note that if $\sum_{n=-\infty}^{\infty} |a_n| < \infty$, then there is exactly one $a \in W$ such that (1.6) holds for all n. On the other hand, although $H(a)$ is uniquely determined by a, it is only the numbers a_n with $n \geq 1$ that can be recovered from the matrix $H(a)$. In other words: $H(a) = H(b)$ if and only if $a_n = b_n$ for all $n \geq 1$.

Infinite matrices as operators. We let $\ell^p := \ell^p(\mathbf{Z}_+)$ $(1 \leq p \leq \infty)$ stand for the usual Banach spaces of complex-valued sequences $\{x_n\}_{n=0}^{\infty}$: for $1 \leq p < \infty$,

$$x = \{x_n\}_{n=0}^{\infty} \in \ell^p \iff \|x\|_p^p := \sum_{n=0}^{\infty} |x_n|^p < \infty,$$

and for $p = \infty$,

$$x = \{x_n\}_{n=0}^{\infty} \in \ell^\infty \iff \|x\|_\infty := \sup_{n \geq 0} |x_n| < \infty.$$

An infinite matrix $A = (a_{jk})_{j,k=0}^{\infty}$ is said to *induce a bounded operator* on ℓ^p if there is a constant $M \in (0, \infty)$ such that for every $x = \{x_n\}_{n=0}^{\infty} \in \ell^p$ the inequality

$$\sum_{j=0}^{\infty} \left| \sum_{k=0}^{\infty} a_{jk} x_k \right|^p \leq M^p \sum_{k=0}^{\infty} |x_k|^p \tag{1.7}$$

holds; we remark that (1.7) includes the requirement that the series

$$y_j = \sum_{k=0}^{\infty} a_{jk} x_k \ (j \geq 0) \quad \text{and} \quad \sum_{j=0}^{\infty} |y_j|^p$$

are convergent. If $A = (a_{jk})_{j,k=0}^{\infty}$ induces a bounded operator on ℓ^p, we can simply think of A as *being* a bounded operator on ℓ^p which, after writing the elements of ℓ^p as column vectors, acts by the rule

$$y = Ax \quad \text{with} \quad \begin{pmatrix} y_0 \\ y_1 \\ y_2 \\ \vdots \end{pmatrix} = \begin{pmatrix} a_{00} & a_{01} & a_{02} & \dots \\ a_{10} & a_{11} & a_{12} & \dots \\ a_{20} & a_{21} & a_{22} & \dots \\ \vdots & \vdots & \vdots & \end{pmatrix} \begin{pmatrix} x_0 \\ x_1 \\ x_2 \\ \vdots \end{pmatrix}.$$

If A induces a bounded operator on ℓ^p, then there is a smallest M for which (1.7) is true for all $x \in \ell^p$. This number M is the norm of A, and it is denoted by $\|A\|_p$:

$$\|A\|_p = \sup_{x \neq 0} \frac{\|Ax\|_p}{\|x\|_p} = \sup_{\|x\|_p=1} \|Ax\|_p.$$

If A does not induce a bounded operator on ℓ^p, we put $\|A\|_p = \infty$.

Let $\mathbf{Z}$ be the set of all integers. For $n \in \mathbf{Z}$, define $\chi_n \in W$ by

$$\chi_n(t) = t^n \quad (t \in \mathbf{T}).$$

The matrix $T(\chi_n)$ has units on a single parallel to the main diagonal and zeros elsewhere. Obviously, for $n \geq 0$,

$$T(\chi_n)x = \{\underbrace{0, \ldots, 0}_{n}, x_0, x_1, \ldots\}, \quad T(\chi_{-n})x = \{x_n, x_{n+1}, \ldots\}. \tag{1.8}$$

Similarly, $H(\chi_n)$ is the zero matrix for $n \leq 0$ and is a matrix with units on a single "antidiagonal" and zeros elsewhere for $n \geq 1$:

$$H(\chi_n)x = \{x_{n-1}, x_{n-2}, \ldots, x_0, 0, 0, \ldots\} \quad \text{for} \quad n \geq 1. \tag{1.9}$$

Proposition 1.1. *If $a \in W$, then $T(a)$ and $H(a)$ induce bounded operators on the space ℓ^p $(1 \leq p \leq \infty)$ and*

$$\|T(a)\|_p \leq \|a\|_W, \quad \|H(a)\|_p \leq \|a\|_W.$$

Proof. If a is given by (1.4), then

$$T(a) = \sum_{n=-\infty}^{\infty} a_n T(\chi_n), \quad H(a) = \sum_{n=1}^{\infty} a_n H(\chi_n),$$

and from (1.8) and (1.9) we infer that $\|T(\chi_n)\|_p = 1$ for all n and $\|H(\chi_n)\|_p = 1$ for $n \geq 1$, whence

$$\|T(a)\|_p \leq \sum_{n=-\infty}^{\infty} |a_n|, \quad \|H(a)\|_p \leq \sum_{n=1}^{\infty} |a_n|. \qquad \square$$

By virtue of Proposition 1.1, we can regard $T(a)$ and $H(a)$ as bounded linear operators on ℓ^p. For Hankel operators, we can say even much more.

Proposition 1.2. *If $a \in W$, then $H(a)$ is compact on ℓ^p $(1 \leq p \leq \infty)$.*

Proof. Write a in the form (1.4) and put

$$(S_N a)(t) := \sum_{n=-N}^{N} a_n t^n \quad (t \in \mathbf{T}).$$

The operator $H(S_N a)$ is given by the matrix

$$H(S_N a) = \begin{pmatrix} a_1 & \dots & a_N & 0 & \dots \\ \vdots & & \vdots & \vdots & \\ a_N & \dots & 0 & 0 & \dots \\ 0 & \dots & 0 & 0 & \dots \\ \vdots & & \vdots & \vdots & \end{pmatrix}$$

and is therefore a finite rank operator. From Proposition 1.1 we infer that

$$\begin{aligned} \|H(a) - H(S_N a)\|_p &= \|H(a - S_N a)\|_p \\ &\le \|a - S_N a\|_W = \sum_{|n| \ge N} |a_j| = o(1) \quad \text{as} \quad N \to \infty. \end{aligned}$$

Therefore, $H(a)$ is a uniform limit of finite rank operators. This implies that $H(a)$ is compact. □

1.3 Products

It is easily seen that W is a Banach algebra with pointwise algebraic operations and the norm $\|\cdot\|_W$, i.e., $(W, \|\cdot\|_W)$ is a Banach space, and if $a, b \in W$, then $ab \in W$ and $\|ab\|_W \le \|a\|_W \|b\|_W$.

Given $a \in W$, we define the function $\tilde{a}$ by $\tilde{a}(t) := a(1/t)$ $(t \in \mathbf{T})$. Clearly, $\tilde{a}$ also belongs to W. Since

$$a(t) = \sum a_n t^n \quad \Longrightarrow \quad \tilde{a}(t) = \sum a_{-n} t^n,$$

we see that $T(\tilde{a})$ and $H(\tilde{a})$ are the matrices

$$T(\tilde{a}) = \begin{pmatrix} a_0 & a_1 & a_2 & \dots \\ a_{-1} & a_0 & a_1 & \dots \\ a_{-2} & a_{-1} & a_0 & \dots \\ \dots & \dots & \dots & \dots \end{pmatrix}, \quad H(\tilde{a}) = \begin{pmatrix} a_{-1} & a_{-2} & a_{-3} & \dots \\ a_{-2} & a_{-3} & \dots & \dots \\ a_{-3} & \dots & \dots & \dots \\ \dots & \dots & \dots & \dots \end{pmatrix}.$$

Thus, $T(\tilde{a})$ is simply the transpose of $T(a)$, but $H(\tilde{a})$ has nothing to do with $H(a)$.

Proposition 1.3. *If $a, b \in W$ then $T(ab) = T(a)T(b) + H(a)H(\tilde{b})$.*

Proof. The jk entry of $T(ab)$ is

$$(ab)_{j-k} = \sum_{m-n=j-k} a_m b_n = \sum_{\ell=-\infty}^{\infty} a_{j+\ell} b_{-k-\ell},$$

the jk entry of $T(a)T(b)$ equals

$$(a_j \, a_{j-1} \dots) \begin{pmatrix} b_{-k} \\ b_{-k+1} \\ \vdots \end{pmatrix} = \sum_{\ell=-\infty}^{0} a_{j+\ell} b_{-k-\ell},$$

and the jk entry of $H(a)H(\tilde{b})$ is equal to

$$(a_{j+1}\, a_{j+2} \dots) \begin{pmatrix} b_{-k-1} \\ b_{-k-2} \\ \vdots \end{pmatrix} = \sum_{\ell=1}^{\infty} a_{j+\ell}\, b_{-k-\ell}. \qquad \square$$

Moral: The product of two infinite Toeplitz matrices is in general not a Toeplitz matrix, but it is always a Toeplitz matrix minus the product of two Hankel matrices. The previous proposition indicates the role played by Hankel matrices in the theory of Toeplitz matrices.

We now introduce two important subalgebras W_+ and W_- of W. Let W_+ and W_- stand for the set of all functions $a \in W$ which are of the form

$$a(t) = \sum_{n=0}^{\infty} a_n t \ (t \in \mathbf{T}) \quad \text{and} \quad a(t) = \sum_{n=-\infty}^{0} a_n t^n \ (t \in \mathbf{T}),$$

respectively. Equivalently, for $a \in W$ we have

$$\begin{aligned} a \in W_+ &\Longleftrightarrow H(\tilde{a}) = 0 \Longleftrightarrow T(a) \quad \text{is lower-triangular,} \\ a \in W_- &\Longleftrightarrow H(a) = 0 \Longleftrightarrow T(a) \quad \text{is upper-triangular.} \end{aligned}$$

Proposition 1.4. *If $a_- \in W_-$, $b \in W$, $a_+ \in W_+$, then*

$$T(a_- b a_+) = T(a_-)T(b)T(a_+).$$

Proof. Since $H(a_-) = H(\tilde{a}_+) = 0$, we deduce from Propostion 1.3 that

$$\begin{aligned} T(a_- b a_+) &= T(a_-)T(ba_+) + H(a_-)H(\widetilde{b a_+}) \\ &= T(a_-)T(ba_+) = T(a_-)T(b)T(a_+) + T(a_-)H(b)H(\tilde{a}_+) \\ &= T(a_-)T(b)T(a_+). \qquad \square \end{aligned}$$

1.4 Wiener-Hopf Factorization

In what follows, we have to work with a few important subsets of the Wiener algebra: GW, $\exp W$, $GW_\pm$, $\exp W_\pm$.

Wiener's theorem. We let GW stand for the group of the invertible elements of the algebra W. Thus, $a \in GW$ if and only if $a \in W$ and if there is a $b \in W$ such that $a(t)b(t) = 1$ for all $t \in \mathbf{T}$. Clearly, a function $a \in GW$ cannot have zeros on $\mathbf{T}$. The following famous theorem by Wiener says that the converse is also true.

Theorem 1.5. $GW = \{a \in W : a(t) \neq 0$ *for all* $t \in \mathbf{T}\}$.

The winding number. The set $\exp W$ is defined as the collection of all $a \in W$ which have a logarithm in W, that is, which are of the form $a = e^b$ with $b \in W$. To characterize $\exp W$, we need the notion of the winding number. If $a : \mathbf{T} \to \mathbf{C} \setminus \{0\}$ is a continuous function, then $a(t)$ traces out a continuous and closed curve in $\mathbf{C} \setminus \{0\}$ as t moves once around the counterclockwise oriented unit circle. The number of times this curve surrounds the origin counterclockwise is called the *winding number* of a and is denoted by wind a. Another (equivalent) definition is as follows. Every continuous function $a : \mathbf{T} \to \mathbf{C} \setminus \{0\}$ can be written in the form $a(e^{i\theta}) = |a(e^{i\theta})|e^{ic(\theta)}$ ($e^{i\theta} \in \mathbf{T}$), where $c : [0, 2\pi) \to \mathbf{R}$ is a continuous function. The number

$$\frac{1}{2\pi}\Big(c(2\pi - 0) - c(0 + 0)\Big)$$

is an integer which is independent of the particular choice of c. This integer is wind a.

Theorem 1.6. $\exp W = \{a \in GW : \text{ wind } a = 0\}$.

Analytic Wiener functions. In Section 1.3, we introduced the algebras $W_\pm$. We denote by $GW_\pm$ the functions $a_\pm \in W_\pm$ for which there exist $b_\pm \in W_\pm$ such that $a_\pm(t)b_\pm(t) = 1$ for all $t \in \mathbf{T}$, and we let $\exp W_\pm$ stand for the functions $a_\pm \in GW_\pm$ which can be represented in the form $a_\pm = e^{b_\pm}$ with $b_\pm \in W_\pm$. Notice that $GW_\pm$ is a proper subset of $W_\pm \cap GW$: for example, if $a_+ \in W_+$ is given by $a_+(t) = t$, then $1/a_+(t) = t^{-1}$ is a function in W_-.

Let $\mathbf{D} := \{z \in \mathbf{C} : |z| < 1\}$ be the open unit disk. Every function $a_+ \in W_+$ can be extended to an analytic function in $\mathbf{D}$ by the formula

$$a_+(z) = \sum_{n=0}^{\infty} a_n z^n \quad (z \in \mathbf{D}),$$

where $\{a_n\}_{n=0}^{\infty}$ is the sequence of the Fourier coefficients of a. Analogously, every function $a_- \in W_-$ admits analytic continuation to $\{z \in \mathbf{C} : |z| > 1\} \cup \{\infty\}$ via

$$a_-(z) = \sum_{n=0}^{\infty} a_{-n} z^{-n} \quad (1 < |z| \le \infty).$$

Theorem 1.7. *We have*

$$\begin{aligned} &GW_+ = \{a \in W : a(z) \neq 0 \textit{ for all } |z| \le 1\}, \\ &GW_- = \{a \in W : a(z) \neq 0 \textit{ for all } |z| \ge 1 \textit{ and for } z = \infty\}, \\ &\exp W_+ = GW_+, \quad \exp W_- = GW_-. \end{aligned}$$

Theorems 1.5 to 1.7 are standard results of the theory of commutative Banach algebras and are essentially equivalent to the facts that the maximal ideal spaces of W, W_+, W_- are $\mathbf{T}$, $\mathbf{D} \cup \mathbf{T}$, $(\mathbf{C} \cup \{\infty\}) \setminus \mathbf{D}$, respectively.

Theorem 1.8 (Wiener-Hopf factorization). *Let $a \in W$ and suppose that $a(t) \neq 0$ for all $t \in \mathbf{T}$ and that* wind $a = m$. *Then a can be written in the form*

$$a(t) = a_-(t)t^m a_+(t) \quad (t \in \mathbf{T}) \quad \textit{with} \quad a_\pm \in GW_\pm.$$

Proof. Recall that $\chi_m(t) = t^m$. We have wind $(a\chi_{-m})$ = wind a+wind $\chi_{-m} = m - m = 0$, whence $a\chi_{-m} = e^b$ with some $b \in W$ by Theorems 1.5 and 1.6. Let

$$b(t) = \sum_{n=-\infty}^{\infty} b_n t^n \quad (t \in \mathbf{T})$$

and put

$$b_-(t) = \sum_{n=-\infty}^{-1} b_n t^n, \quad b_+(t) = \sum_{n=0}^{\infty} b_n t^n.$$

It is obvious that $e^{b_\pm} \in GW_\pm$. The representation $a = e^{b_-}\chi_m e^{b_+}$ is the desired factorization. □

Laurent polynomials. These are the functions in the Wiener algebra with only finitely many nonzero Fourier coefficients. Thus, $b : \mathbf{T} \to \mathbf{C}$ is a *Laurent polynomial* if and only if b is of the form

$$b(t) = \sum_{j=-r}^{s} b_j t^j \quad (t \in \mathbf{T}), \tag{1.10}$$

where r and s are integers and $-r \le s$. We denote the set of all Laurent polynomials by $\mathcal{P}$ and we write $\mathcal{P}_{r,s}$ for the Laurent polynomials of the form (1.10). We also put $\mathcal{P}_s := \mathcal{P}_{s,s}$. Finally, we let $\mathcal{P}_s^+ := \mathcal{P}_{0,s-1}$ stand for the analytic polynomials of degree at most $s-1$ and we set $\mathcal{P}^+ = \cup_{s\ge 1} \mathcal{P}_s^+$.

Let us assume that $b \in \mathcal{P}_{r,s}$ is not identically zero and that $b_{-r} \ne 0$ and $b_s \ne 0$. We can write $b(t) = t^{-r}(b_{-r} + b_{-r+1}t + \cdots + b_s t^{r+s})$. If $b(t) \ne 0$ for $t \in \mathbf{T}$, we further have

$$b(t) = t^{-r} b_s \prod_{j=1}^{J} (t - \delta_j) \prod_{k=1}^{K} (t - \mu_k), \tag{1.11}$$

where $|\delta_j| < 1$ for all j and $|\mu_k| > 1$ for all k. Obviously, wind $(t - \delta_j) = 1$ and wind $(t - \mu_k) = 0$, whence

$$\text{wind}\, b = J - r; \tag{1.12}$$

that is, wind b is the number of zeros of b in $\mathbf{D}$ minus the number of poles of b in $\mathbf{D}$ (all counted according to their multiplicity). The factorization

$$b(t) = b_s \prod_{j=1}^{J} \left(1 - \frac{\delta_j}{t}\right) t^{J-r} \prod_{k=1}^{K} (t - \mu_k) \tag{1.13}$$

is a Wiener-Hopf factorization; notice that

$$\left(1 - \frac{\delta_j}{t}\right)^{-1} = 1 + \frac{\delta_j}{t} + \frac{\delta_j^2}{t^2} + \cdots \quad (t \in \mathbf{T}) \tag{1.14}$$

and

$$(t - \mu_k)^{-1} = -\frac{1}{\mu_k}\left(1 + \frac{t}{\mu_k} + \frac{t^2}{\mu_k^2} + \cdots\right) \quad (t \in \mathbf{T}) \tag{1.15}$$

are functions in W_- and W_+, respectively.

1.5 Spectra

Fredholm operators. Let X be a Banach space. We denote by $\mathcal{B}(X)$ and $\mathcal{K}(X)$ the bounded and compact linear operators on X, respectively. The *spectrum* of an operator $A \in \mathcal{B}(X)$ is the set

$$\text{sp}\, A = \{\lambda \in \mathbf{C} : \ A - \lambda I \quad \text{is not invertible}\}.$$

The operator valued function $\mathbf{C} \setminus \text{sp}\, A \to \mathcal{B}(H)$, $\lambda \mapsto (A - \lambda I)^{-1}$ is well defined and analytic. It is called the *resolvent* of A. An operator $A \in \mathcal{B}(X)$ is said to be *Fredholm* if it is invertible modulo compact operators, that is, if there is an operator $B \in \mathcal{B}(X)$ such that $AB - I$ and $BA - I$ are compact. We define the *essential spectrum* of $A \in \mathcal{B}(X)$ as the set

$$\text{sp}_{\text{ess}}\, A = \{\lambda \in \mathbf{C} : \ A - \lambda I \quad \text{is not Fredholm}\}.$$

Clearly, $\text{sp}_{\text{ess}}\, A \subset \text{sp}\, A$ and $\text{sp}_{\text{ess}}\, A$ is invariant under compact perturbations.

The *kernel* and the *image* (= range) of $A \in \mathcal{B}(X)$ are defined as usual:

$$\text{Ker}\, A = \{x \in X : \ Ax = 0\}, \quad \text{Im}\, A := A(X).$$

An operator $A \in \mathcal{B}(X)$ is said to be *normally solvable* if $\text{Im}\, A$ is a closed subspace of X. In that case the *cokernel* of A is

$$\text{Coker}\, A = X/\text{Im}\, A.$$

One can show that $A \in \mathcal{B}(X)$ is Fredholm if and only if A is normally solvable and both $\text{Ker}\, A$ and $\text{Coker}\, A$ have finite dimensions. The *index* of a Fredholm operator $A \in \mathcal{B}(X)$ is the integer

$$\text{Ind}\, A := \dim \text{Ker}\, A - \dim \text{Coker}\, A.$$

Theorem 1.9. *Let $a \in W$. The operator $T(a)$ is Fredholm on ℓ^p $(1 \le p \le \infty)$ if and only if $a(t) \neq 0$ for all $t \in \mathbf{T}$. In that case* $\text{Ind}\, T(a) = -\text{wind}\, a$.

Proof. If a has no zeros on $\mathbf{T}$ and if the winding number of a is m, then $a = a_- \chi_m a_+$ with $a_\pm \in GW_\pm$ by virtue of Theorem 1.5. From Proposition 1.4 we infer that

$$T(a) = T(a_-)T(\chi_m)T(a_+),$$

and the same proposition tells us that $T(a_\pm)$ are invertible, the inverses being $T(a_\pm^{-1})$. From (1.8) we see that $T(\chi_m)$ has closed range and that

$$\dim \text{Ker}\, T(\chi_m) = \begin{cases} 0 & \text{if} \quad m \ge 0, \\ |m| & \text{if} \quad m < 0, \end{cases} \qquad \dim \text{Coker}\, T(\chi_m) = \begin{cases} m & \text{if} \quad m \ge 0, \\ 0 & \text{if} \quad m < 0, \end{cases}$$

which implies that $T(\chi_m)$ is Fredholm of index $-m$. Consequently, $T(a)$ is also Fredholm of index $-m$.

Conversely, suppose now that $T(a)$ is Fredholm and let m be the index. Contrary to what we want, let us assume that $a(t_0) = 0$ for some $t_0 \in \mathbf{T}$. We can then find $b, c \in GW$

such that $\|a-b\|_W$ and $\|a-c\|_W$ are as small as desired and such that $|\text{wind } b - \text{wind } c| = 1$. Since Fredholmness and index are stable under small perturbations, it follows that $T(b)$ and $T(c)$ are Fredholm and that $\text{Ind } T(b) = \text{Ind } T(c) = m$. However, from what was proved in the preceding paragraph and from the equality $|\text{wind } b - \text{wind } c| = 1$ we know that $|\text{Ind } T(b) - \text{Ind } T(c)| = 1$. This contradiction shows that a cannot have zeros on $\mathbf{T}$. □

Corollary 1.10. *If $a \in W$, then* $\text{sp}_{\text{ess}}\, T(a) = a(\mathbf{T})$.

Proof. Apply Theorem 1.9 to $a - \lambda$. □

Corollary 1.11. *Let $a \in W$. The operator $T(a)$ is invertible on ℓ^p* $(1 \le p \le \infty)$ *if and only if $a(t) \neq 0$ for all $t \in \mathbf{T}$ and* $\text{wind } a = 0$.

Proof. If $T(a)$ is invertible, then $T(a)$ is Fredholm of index zero and Theorem 1.9 shows that a has no zeros on $\mathbf{T}$ and that $\text{wind } a = 0$. If $a(t) \neq 0$ for $t \in \mathbf{T}$ and $\text{wind } a = 0$, then $a = a_- a_+$ with $a_\pm \in GW_\pm$ due to Theorem 1.5. From Proposition 1.4 we deduce that $T(a_+^{-1})T(a_-^{-1})$ is the inverse of the operator $T(a) = T(a_-)T(a_+)$. □

The following beautiful purely geometric description of the spectrum of a Toeplitz operator is illustrated by Figure 1.1.

Corollary 1.12. *If $a \in W$, then*

$$\text{sp } T(a) = a(\mathbf{T}) \cup \big\{\lambda \in \mathbf{C} \setminus a(\mathbf{T}) : \text{ wind } (a - \lambda) \neq 0\big\}.$$

Proof. This is Corollary 1.11 with a replaced by $a - \lambda$. □

In Section 1.2, we observed that $H(a)$ is compact for every $a \in W$. The following result shows that the zero operator is the only compact Toeplitz operator.

Corollary 1.13. *If $a \in W$ and $T(a)$ is compact on ℓ^p* $(1 \le p \le \infty)$, *then a vanishes identically.*

Proof. If $T(a)$ is compact, then $\text{sp}_{\text{ess}}\, T(a) = \{0\}$, and Corollary 1.10 tells us that this can only happen if $a(\mathbf{T}) = \{0\}$. □

1.6 Norms

The cases $p = 1$ and $p = \infty$. It is well known that an infinite matrix $A = (a_{jk})_{j,k=0}^{\infty}$ induces a bounded operator on ℓ^1 and ℓ^∞, respectively, if and only if

$$\sup_{k \ge 1} \sum_{j=1}^{\infty} |a_{jk}| < \infty \quad \text{and} \quad \sup_{j \ge 1} \sum_{k=1}^{\infty} |a_{jk}| < \infty,$$

in which case

$$\|A\|_1 = \sup_{k \ge 1} \sum_{j=1}^{\infty} |a_{jk}| \quad \text{and} \quad \|A\|_\infty = \sup_{j \ge 1} \sum_{k=1}^{\infty} |a_{jk}|. \tag{1.16}$$

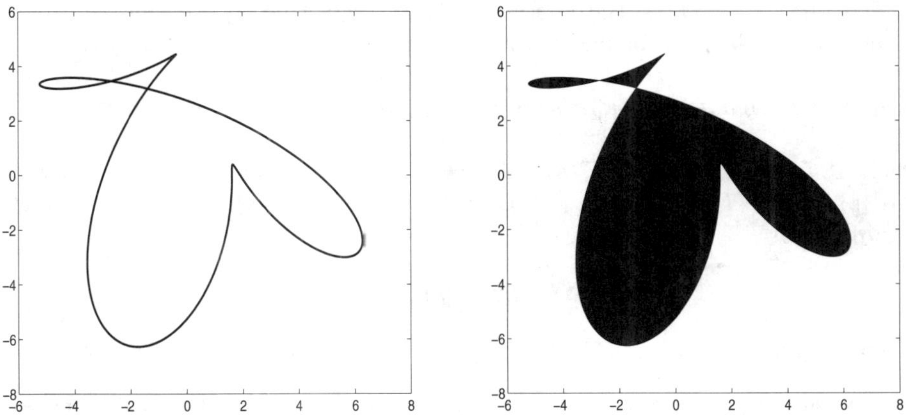

Figure 1.1. *The essential spectrum* $\mathrm{sp}_{\mathrm{ess}}\, T(a) = a(\mathbf{T})$ *on the left and the spectrum* $\mathrm{sp}\, T(a)$ *on the right.*

This easily implies the following.

Theorem 1.14. *If* $a \in W$ *then* $\|T(a)\|_1 = \|T(a)\|_\infty = \|a\|_W$. □

The case $p = 2$. Let $L^2 := L^2(\mathbf{T})$ be the usual Lebesgue space of complex-valued functions on $\mathbf{T}$ with the norm

$$\|f\|_2 := \left(\int_{\mathbf{T}} |f(t)|^2 \frac{|dt|}{2\pi} \right)^{1/2} = \left(\int_0^{2\pi} |f(e^{i\theta})|^2 \frac{d\theta}{2\pi} \right)^{1/2}.$$

The set $H^2 := H^2(\mathbf{T}) := \{f \in L^2 : f_n = 0 \text{ for } n < 0\}$ is a closed subspace of L^2 and is referred to as the *Hardy space* of L^2. Let $P : L^2 \to H^2$ be the orthogonal projection. Thus, if $f \in L^2$ is given by

$$f(t) = \sum_{n=-\infty}^{\infty} f_n t^n \quad (t \in \mathbf{T}),$$

then

$$(Pf)(t) = \sum_{n=0}^{\infty} f_n t^n \quad (t \in \mathbf{T}).$$

The map

$$\Phi : H^2 \to \ell^2, \quad f \mapsto \{f_n\}_{n=0}^{\infty} \tag{1.17}$$

is a unitary operator of H^2 onto ℓ^2. It is not difficult to check that if $a \in W$, then $\Phi^{-1}T(a)\Phi$ is the operator

$$\Phi^{-1}T(a)\Phi : H^2 \to H^2, \quad f \mapsto P(af), \tag{1.18}$$

where $(af)(t) := a(t)f(t)$. The observation (1.18) is in fact the key to the theory of Toeplitz operators on ℓ^2. We here confine ourselves to the following consequence of (1.18).

Theorem 1.15. *If $a \in W$ then $\|T(a)\|_2 = \|a\|_\infty$, where $\|a\|_\infty := \max_{t\in\mathbf{T}} |a(t)|$.*

Proof. If $f \in H^2$, then $\|\Phi^{-1}T(a)\Phi f\|_2 = \|P(af)\|_2 \le \|af\|_2 \le \|a\|_\infty \|f\|_2$, whence $\|T(a)\|_2 = \|\Phi^{-1}T(a)\Phi\|_2 \le \|a\|_\infty$. On the other hand, Corollary 1.12 implies that the *spectral radius*

$$\operatorname{rad} T(a) = \max\left\{|\lambda| : \lambda \in \operatorname{sp} T(a)\right\}$$

is equal to $\|a\|_\infty$. Because $\operatorname{rad} T(a) \le \|T(a)\|_2$, it follows that $\|a\|_\infty \le \|T(a)\|_2$. □

Other values of p. This case is more delicate, but one has at least two-sided estimates.

Proposition 1.16. *If $a \in W$ and $1 \le p \le \infty$, then $\|a\|_\infty \le \|T(a)\|_p \le \|a\|_W$.*

Proof. The inequality $\|T(a)\|_p \le \|a\|_W$ results from Proposition 1.1, and the inequality $\|a\|_\infty \le \|T(a)\|_p$ is a consequence of Corollary 1.12 together with the estimate $\|a\|_\infty = \operatorname{rad} T(a) \le \|T(a)\|_p$. □

1.7 Inverses

Let b be a Laurent polynomial of the form (1.10). Suppose $b(t) \ne 0$ for $t \in \mathbf{T}$ and $\operatorname{wind} b = 0$. From Section 1.4 we know that b can be written in the form $b = b_- b_+$ with

$$b_-(t) = \prod_{j=1}^{r}\left(1 - \frac{\delta_j}{t}\right), \quad b_+(t) = b_s \prod_{k=1}^{s}(t - \mu_k), \tag{1.19}$$

where $\delta := \max(|\delta_1|, \ldots, |\delta_r|) < 1$ and $\mu := \min(|\mu_1|, \ldots, |\mu_s|) > 1$. When proving Corollary 1.11, we observed that

$$T^{-1}(b) = T(b_+^{-1})T(b_-^{-1}). \tag{1.20}$$

From (1.19) we see that

$$b_-^{-1}(t) = \prod_{j=1}^{r}\left(1 + \frac{\delta_j}{t} + \frac{\delta_j^2}{t^2} + \cdots\right) =: \sum_{m=0}^{\infty} \frac{c_m}{t^m},$$

$$b_+^{-1}(t) = \frac{1}{b_s}\prod_{k=1}^{s}\left(-\frac{1}{\mu_k}\right)\prod_{k=1}^{s}\left(1 + \frac{t}{\mu_k} + \frac{t^2}{\mu_k^2} + \cdots\right) =: \sum_{m=0}^{\infty} d_m t^m.$$

With the coefficients c_m and d_m, formula (1.20) takes the form

$$T^{-1}(b) = \begin{pmatrix} d_0 & & & \\ d_1 & d_0 & & \\ d_2 & d_1 & d_0 & \\ \ldots & \ldots & \ldots & \ldots \end{pmatrix} \begin{pmatrix} c_0 & c_1 & c_2 & \ldots \\ & c_0 & c_1 & \ldots \\ & & c_0 & \ldots \\ & & & \ldots \end{pmatrix}. \tag{1.21}$$

Proposition 1.3 and (1.20) imply that

$$T^{-1}(b) = T(b^{-1}) - H(b_+^{-1})H(\widetilde{b_-^{-1}}). \tag{1.22}$$

Representation (1.20) gives us $T^{-1}(b)$ as the product of the lower triangular matrix $T(b_+^{-1})$ and the upper triangular matrix $T(b_-^{-1})$, while (1.22) shows that $T^{-1}(b)$ is the difference of the (in general full) Toeplitz matrix $T(b^{-1})$ and the product $H(b_+^{-1})H(\widetilde{b_-^{-1}})$ of two Hankel matrices.

Let α be any number satisfying

$$0 < \alpha < \min\left(\log\frac{1}{\delta}, \log\mu\right). \tag{1.23}$$

Lemma 1.17. *For every $n \geq 0$,*

$$|\left(b_-^{-1}\right)_{-n}| \leq \left(\min_{|z|=\delta+\varepsilon} |b_-(z)|\right)^{-1} (\delta+\varepsilon)^n \quad (\varepsilon > 0), \tag{1.24}$$

$$|\left(b_+^{-1}\right)_{n}| \leq \left(\min_{|z|=\mu-\varepsilon} |b_+(z)|\right)^{-1} (\mu-\varepsilon)^n \quad (0 < \varepsilon < \mu). \tag{1.25}$$

Consequently, $(b_-^{-1})_{-n}$ and $(b_+^{-1})_n$ are $O(e^{-\alpha n})$ as $n \to \infty$.

Proof. Since $1/b_-(z)$ is analytic for $|z| > \delta$, we get

$$(b_-^{-1})_{-n} = \frac{1}{2\pi i}\int_{|z|=1} \frac{z^{n-1}dz}{b_-(z)} = \frac{1}{2\pi i}\int_{|z|=\delta+\varepsilon} \frac{z^{n-1}dz}{b_-(z)}$$

and hence

$$|(b_-^{-1})_{-n}| \leq \frac{1}{2\pi}\left(\min_{|z|=\delta+\varepsilon} |b_-(z)|\right)^{-1} (\delta+\varepsilon)^{n-1} 2\pi(\delta+\varepsilon),$$

which is (1.24). Estimate (1.25) can be verified analogously. □

Proposition 1.18. *For the j, k entry of $T^{-1}(b)$ we have the estimate*

$$\left[T^{-1}(b)\right]_{j,k} = \left(b^{-1}\right)_{j-k} + O\left(e^{-\alpha(j+k)}\right).$$

Proof. From (1.22) we see that

$$\left|\left[T^{-1}(b)\right]_{j,k} - (b^{-1})_{j-k}\right| = \left|\sum_{\ell=1}^{\infty} (b_+^{-1})_{j+\ell}(b_-^{-1})_{-k-\ell}\right|$$

$$\leq \left(\sum_{\ell=1}^{\infty} |(b_+^{-1})_{j+\ell}|^2 \sum_{\ell=1}^{\infty} |(b_-^{-1})_{-k-\ell}|^2\right)^{1/2}, \tag{1.26}$$

and Lemma 1.17 implies that (1.26) is

$$O\left(\left(\sum_{\ell=1}^{\infty} e^{-2\alpha(j+\ell)}\right)^{1/2}\right) O\left(\left(\sum_{\ell=1}^{\infty} e^{-2\alpha(k+\ell)}\right)^{1/2}\right) = O(e^{-\alpha j})\, O(e^{-\alpha k}). \quad \square$$

Given two sequences $x = \{x_k\}$ and $y = \{y_k\}$, we set $(x, y) = \sum x_k \overline{y}_k$. The j, k entry of $T^{-1}(b)$ is just $(T^{-1}(b)x, y)$ for $x = e_k$ and $y = e_j$, where $\{e_n\}$ is the standard basis of ℓ^2. The following useful observation evaluates $(T^{-1}(b)x, y)$ at another interesting pair (x, y). For $z \in \mathbf{D}$, define $w_z \in \ell^2$ by $(w_z)_n = z^n$ $(n \geq 0)$.

Proposition 1.19. *Let $b = b_- b_+$ with $b_\pm$ given by* (1.19). *Then for $\alpha, \beta \in \mathbf{D}$,*

$$(T^{-1}(b)w_\alpha, w_\beta) = \frac{1}{b_s} \frac{1}{1 - \alpha\overline{\beta}} \prod_{j=1}^{r} \frac{1}{1 - \delta_j \alpha} \prod_{k=1}^{s} \frac{1}{\overline{\beta} - \mu_k} \tag{1.27}$$

$$= \frac{1}{1 - \alpha\overline{\beta}} \frac{1}{b_-(1/\alpha) b_+(\overline{\beta})}. \tag{1.28}$$

Proof. We have

$$(T^{-1}(b)w_\alpha, w_\beta) = (T(b_+^{-1})T(b_-^{-1})w_\alpha, w_\beta) = (T(b_-^{-1})w_\alpha, T(\overline{b_+^{-1}})w_\beta).$$

Define $\chi_n(t) := t^n$ $(t \in \mathbf{T})$. It is easily seen that, for $|\delta| < 1$,

$$T^{-1}(1 - \delta\chi_{-1})w_\alpha = T(1 + \delta\chi_{-1} + \delta^2\chi_{-2} + \cdots)w_\alpha = \frac{1}{1 - \delta\alpha} w_\alpha,$$

whence

$$T(b_-^{-1})w_\alpha = \left(\prod_{j=1}^{r} \frac{1}{1 - \delta_j \alpha} \right) w_\alpha.$$

Analogously, if $|\mu| > 1$,

$$T^{-1}(\chi_{-1} - \overline{\mu})w_\beta = -\frac{1}{\overline{\mu}} T^{-1} \left(1 - \frac{1}{\overline{\mu}} \chi_{-1} \right) w_\beta = -\frac{1}{\overline{\mu}} \frac{1}{1 - \overline{\mu}^{-1}\beta} w_\beta = \frac{1}{\beta - \overline{\mu}} w_\beta,$$

which implies that

$$T(\overline{b_+^{-1}})w_\beta = \overline{b}_s^{-1} \left(\prod_{k=1}^{s} \frac{1}{\beta - \overline{\mu_k}} \right) w_\beta.$$

Consequently,

$$(T(b_-^{-1})w_\alpha, T(\overline{b_+^{-1}})w_\beta) = \frac{1}{b_s} \prod_{j=1}^{r} \frac{1}{1 - \delta_j \alpha} \prod_{k=1}^{s} \frac{1}{\overline{\beta} - \mu_k} (w_\alpha, w_\beta).$$

As $(w_\alpha, w_\beta) = 1/(1 - \alpha\overline{\beta})$, we arrive at (1.27). Clearly, (1.28) is nothing but another way of writing (1.27). □

1.8 Eigenvalues and Eigenvectors

Let b be a Laurent polynomial. In this section we study the problem of finding the $\lambda \in \operatorname{sp} T(b)$ for which there exist nonzero $x \in \ell^p$ such that $T(b)x = \lambda x$. These λ are called eigenvalues of $T(b)$ on ℓ^p and the corresponding x's are referred to as eigenelements or eigenvectors (which sounds much better). Since $T(b) - \lambda I = T(b - \lambda)$, our problem is equivalent to the question of when a Toeplitz operator has a nontrivial kernel. Throughout this section we assume that b is not constant on the unit circle $\mathbf{T}$.

Outside the essential spectrum. For a point $\lambda \in \mathbf{C} \setminus b(\mathbf{T})$, we denote by $\operatorname{wind}(b, \lambda)$ the winding number of b about λ, that is, $\operatorname{wind}(b, \lambda) := \operatorname{wind}(b - \lambda)$. A sequence $\{x_n\}_{n=0}^\infty$ is said to be *exponentially decaying* if there are $C \in (0, \infty)$ and $\gamma \in (0, \infty)$ such that $|x_n| \le Ce^{-\gamma n}$ for all $n \ge 0$.

Proposition 1.20. *Let $1 \le p \le \infty$. A point $\lambda \notin b(\mathbf{T})$ is an eigenvalue of $T(b)$ as an operator on ℓ^p if and only if* $\operatorname{wind}(b, \lambda) = -m < 0$, *in which case* $\operatorname{Ker}(T(b) - \lambda I)$ *has the dimension m and each eigenvector is exponentially decaying.*

Proof. From Theorem 1.8 (or simply from (1.13)) we get a Wiener-Hopf factorization $b(t) - \lambda = b_-(t)t^{-m}b_+(t)$. Proposition 1.4 implies that $T(b - \lambda)$ decomposes into the product $T(b_-)T(\chi_{-m})T(b_+)$ and that the operators $T(b_\pm)$ are invertible, the inverses being $T(b_\pm^{-1})$. Thus, $x \in \operatorname{Ker} T(b - \lambda)$ if and only if $T(\chi_{-m})T(b_+)x = 0$. If $m \le 0$, this is equivalent to the equation $T(b_+)x = 0$ and hence to the equality $x = 0$. So let $m > 0$. We denote by $e_j \in \ell^p$ the sequence given by $(e_j)_k = 1$ for $k = j$ and $(e_j)_k = 0$ for $k \neq j$. Clearly, $T(\chi_{-m})T(b_+)x = 0$ if and only if $T(b_+)x$ belongs to the linear hull $\operatorname{lin}\{e_0, \ldots, e_{m-1}\}$ of $e_0, \ldots, e_{m-1}$. Consequently,

$$\operatorname{Ker} T(b - \lambda) = \operatorname{lin}\{T(b_+^{-1})e_0, \ldots, T(b_+^{-1})e_{m-1}\}.$$

This shows that $\dim \operatorname{Ker} T(b-\lambda) = m$, and from Lemma 1.17 we deduce that the sequences in $\operatorname{Ker} T(b - \lambda)$ are exponentially decaying. □

Inside the essential spectrum. Things are a little bit more complicated for points $\lambda \in b(\mathbf{T})$. In that case $b - \lambda$ has zeros on $\mathbf{T}$. For $\tau \in \mathbf{T}$, we define the function ξ_τ by

$$\xi_\tau(t) = 1 - \frac{\tau}{t} \quad (t \in \mathbf{T}).$$

Notice that ξ_τ has a single zero on $\mathbf{T}$ and that $T(\xi_\tau)$ is the upper triangular matrix

$$T(\xi_\tau) = \begin{pmatrix} 1 & -\tau & 0 & \ldots \\ 0 & 1 & -\tau & \ldots \\ 0 & 0 & 1 & \ldots \\ \ldots & \ldots & \ldots & \ldots \end{pmatrix}.$$

Lemma 1.21. *Let $\tau_1, \ldots, \tau_\ell$ be distinct points on $\mathbf{T}$ and let $\alpha_1, \ldots, \alpha_\ell$ be positive integers. Then*

$$\operatorname{Ker} T\left(\xi_{\tau_1}^{\alpha_1} \cdots \xi_{\tau_\ell}^{\alpha_\ell}\right) = \{0\} \text{ on } \ell^p \quad (1 \le p < \infty) \tag{1.29}$$

and

$$\operatorname{Ker} T\left(\xi_{\tau_1}^{\alpha_1}\dots\xi_{\tau_\ell}^{\alpha_\ell}\right) = \operatorname{lin}\{w_{\tau_1},\dots,w_{\tau_\ell}\} \text{ on } \ell^\infty, \tag{1.30}$$

where $(w_\tau)_n := 1/\tau^n$.

Proof. Put $\xi = \xi_{\tau_1}^{\alpha_1}\dots\xi_{\tau_\ell}^{\alpha_\ell}$ and write

$$\xi(t) = a_0 + a_1\frac{1}{t} + a_2\frac{1}{t^2} + \dots + a_N\frac{1}{t^N}.$$

The equation $T(\xi)x = 0$ is the difference equation

$$a_0x_n + a_1x_{n+1} + \dots + a_Nx_{n+N} = 0 \quad (n \ge 0),$$

which is satisfied if and only if

$$x_n = \sum_{k=0}^{\alpha_1-1}\gamma_k^{(1)}\frac{n^k}{\tau_1^n} + \dots + \sum_{k=0}^{\alpha_\ell-1}\gamma_k^{(\ell)}\frac{n^k}{\tau_\ell^n} \tag{1.31}$$

with complex numbers $\gamma_k^{(j)}$. The sequence given by (1.31) belongs to ℓ^p $(1 \le p < \infty)$ if and only if it is identically zero, which proves (1.29), and it is in ℓ^∞ if and only if it is of the form

$$x_n = \gamma_0^{(1)}\frac{1}{\tau_1^n} + \dots + \gamma_0^{(\ell)}\frac{1}{\tau_\ell^n},$$

which gives (1.30). □

Given $\lambda \in b(\mathbf{T})$, we denote the distinct zeros of $b - \lambda$ on $\mathbf{T}$ by $\tau_1,\dots,\tau_\ell$ and their multiplicities by $\alpha_1,\dots,\alpha_\ell$. We extract the zeros by "anti-analytic" linear factors, that is, we write $b - \lambda$ in the form

$$b(t) - \lambda = \prod_{j=1}^{\ell}\left(1 - \frac{\tau_j}{t}\right)^{\alpha_j} c(t), \tag{1.32}$$

where $c(t) \ne 0$ for $t \in \mathbf{T}$.

Proposition 1.22. *Let* $1 \le p < \infty$. *Suppose* $\lambda \in b(\mathbf{T})$ *and write* $b - \lambda$ *in the form* (1.32). *Then* λ *is an eigenvalue of the operator* $T(b)$ *on* ℓ^p *if and only if* $\operatorname{wind} c = -m < 0$, *in which case* $\operatorname{Ker}(T(b) - \lambda I)$ *is of the dimension* m *and all eigenvectors are exponentially decaying.*

Proof. By Proposition 1.4,

$$T(b - \lambda) = \prod_{j=1}^{\ell}[T(\xi_{\tau_j})]^{\alpha_j}T(c). \tag{1.33}$$

From (1.29) we see that Ker $T(b-\lambda) = \text{Ker}\, T(c)$, and Proposition 1.20 therefore gives the assertion. □

We now turn to the case $p = \infty$. A sequence $\{x_n\}_{n=0}^{\infty}$ is called *extended* if

$$\limsup_{n\to\infty} |x_n| > 0.$$

Proposition 1.23. *Let $\lambda \in b(\mathbf{T})$ and write $b-\lambda$ in the form (1.32). Then λ is an eigenvalue of $T(b)$ on ℓ^∞ if and only if* wind $c = -m < \ell$. *In that case the dimension of* Ker $(T(b)-\lambda I)$ *is $m+\ell$. There is a basis in* Ker $(T(b)-\lambda I)$ *whose elements enjoy the following properties:*

(a) *if $m > 0$, then m elements of the basis decay exponentially and ℓ elements have zeros in the first m places and are extended;*

(b) *if $m \le 0$, then all the $\ell - |m|$ elements of the basis are extended.*

Proof. Combining (1.30) and (1.33), we see that $T(b-\lambda)x = 0$ if and only if there are complex numbers γ_j such that

$$T(c)x = \sum_{j=1}^{\ell} \gamma_j w_{\tau_j}. \tag{1.34}$$

Let $c = c_-\chi_{-m}c_+$ be a Wiener-Hopf factorization of c. It can be readily checked that $T(c_-^{-1})w_\tau = c_-^{-1}(1/\tau)w_\tau$. Thus, setting $\delta_j = \gamma_j c_-^{-1}(1/\tau_j)$, we can rewrite (1.34) in the form

$$T(\chi_{-m})T(c_+)x = \sum_{j=1}^{\ell} \delta_j w_{\tau_j}. \tag{1.35}$$

If $m \ge 0$, then (1.35) holds if and only if

$$T(c_+)x \in \text{lin}\,\{e_0, \dots, e_{m-1}, T(\chi_m)w_{\tau_1}, \dots, T(\chi_m)w_{\tau_\ell}\},$$

which is equivalent to the requirement that x be in

$$\text{lin}\,\{T(c_+^{-1})e_0, \dots, T(c_+^{-1})e_{m-1}, T(\chi_m)T(c_+^{-1})w_{\tau_1}, \dots, T(\chi_m)T(c_+^{-1})w_{\tau_\ell}\}.$$

The sequences $T(c_+^{-1})e_j$ decay exponentially (Lemma 1.17), and since

$$[T(c_+^{-1})w_\tau]_n = \frac{1}{\tau^n}\sum_{k=0}^{n}(c_+^{-1})_k\tau^k,$$

it follows that the sequences $T(c_+^{-1})w_{\tau_j}$ are extended. This completes the proof in the case $m \ge 0$.

Let now $m < 0$. In that case (1.35) is satisfied if and only if

$$\sum_{j=1}^{\ell} \delta_j (1/\tau_j)^n = 0 \ \text{ for } n = 0, 1, \dots, |m|-1 \tag{1.36}$$

and

$$x = \sum_{j=1}^{\ell} \delta_j T(c_+^{-1}) T(\chi_{-|m|}) w_{\tau_j}. \tag{1.37}$$

Equations (1.36) are a Vandermonde system for the δ_j's. If $|m| \geq \ell$, then (1.36) has only the trivial solution. If $|m| < \ell$, we can choose $\delta_1, \dots, \delta_{\ell-|m|}$ arbitrarily. The numbers $\delta_{\ell-|m|+1}, \dots, \delta_\ell$ are then uniquely determined. This shows that the set of all x of the form (1.37) has the dimension $\ell - |m|$ and that all nonzero x in this set are extended. □

In geometric terms, the winding number of the function c in (1.32) can be determined as follows. Choose a number $\varrho > 1$ and consider the function b_ϱ defined by $b_\varrho(t) = b(\varrho t)$ ($t \in \mathbf{T}$). If ϱ is sufficiently close to 1, then $b_\varrho - \lambda$ has no zeros on $\mathbf{T}$ and hence wind (b_ϱ, λ) is well defined.

Proposition 1.24. *We have* wind $c = \lim_{\varrho \to 1+0}$ wind (b_ϱ, λ).

Proof. Let $c_\varrho(t) := c(\varrho t)$. From (1.32) we obtain $b_\varrho(t) - \lambda = \prod_{j=1}^{\ell} (1 - \frac{\tau_j}{\varrho t})^{\alpha_j} c_\varrho(t)$, whence wind $(b_\varrho, \lambda) =$ wind c_ϱ. Since wind c_ϱ converges to wind c as $\varrho \to 1$, we arrive at the assertion. □

Frequently, the following observation is very useful.

Proposition 1.25. *Let $\lambda \in b(\mathbf{T})$ and suppose Ω is a connected component of $\mathbf{C} \setminus b(\mathbf{T})$ whose boundary contains λ. If* wind $(b, z) = \varkappa$ *for $z \in \Omega$, then* wind $c \geq \varkappa$.

Proof. For $z \in \Omega$, we have

$$b(t) - z = b_s t^{-r} \prod_{j=1}^{r+\varkappa} (t - \delta_j(z)) \prod_{k=1}^{s-\varkappa} (t - \mu_k(z))$$

with $|\delta_j(z)| < 1$ and $|\mu_k(z)| > 1$. Now let $z \in \Omega$ approach $\lambda \in \partial\Omega$. The some of the $\delta_j(z)$, say $\delta_1(z), \dots, \delta_m(z)$, move onto the unit circle, while the remaining $\delta_j(z)$ stay in the open unit disk. Analogously, some of the $\mu_k(z)$, say $\mu_1(z), \dots, \mu_\ell(z)$, attain modulus 1, whereas the remaining $\mu_k(z)$ keep modulus greater than 1. We can write $b(t) - \lambda$ as

$$b_s t^{-r} \prod_{j=1}^{m} (t - \delta_j(\lambda)) \prod_{j=m+1}^{r+\varkappa} (t - \delta_j(\lambda)) \prod_{k=1}^{\ell} (t - \mu_k(\lambda)) \prod_{k=\ell+1}^{s-\varkappa} (t - \mu_k(\lambda)).$$

The zeros of $b_\varrho - \lambda$ are $\delta_j(\lambda)/\varrho$ and $\mu_k(\lambda)/\varrho$. If $\varrho > 1$, then $|\delta_j(\lambda)| < 1$ for all j. If, in addition, $\varrho > 1$ is sufficiently close to 1, then $|\mu_k(\lambda)/\varrho| > 1$ for $k = \ell + 1, \dots, s - \varkappa$ and $|\mu_k(\lambda)/\varrho| < 1$ for $k = 1, \dots, \ell$. The result is that

$$\lim_{\varrho \to 1+0} \text{wind}\,(b_\varrho - \lambda) = -r + (r + \varkappa) + \ell = \varkappa + \ell \geq \varkappa.$$

From Proposition 1.24 we so obtain that wind $c \geq \varkappa$. □

Corollary 1.26. *If λ lies on the boundary of a connected component Ω of $\mathbf{C} \setminus b(\mathbf{T})$ such that* wind $(b, z) \geq 0$ *for $z \in \Omega$, then λ cannot be an eigenvalue of $T(b)$ on ℓ^p $(1 \leq p < \infty)$.*

Proof. Immediate from Propositions 1.22 and 1.25. □

Corollary 1.27. *If b is a real-valued Laurent polynomial, then $T(b)$ as an operator on ℓ^p $(1 \leq p < \infty)$ has no eigenvalues.*

Proof. This is a straightforward consequence of Corollary 1.26. Here is an alternative proof. Let $\lambda \in b(\mathbf{T})$ and write $b(t) - \lambda = b_r t^{-r} \prod_{j=1}^{2r}(t - z_j)$. Since $b(t) - \lambda$ is real valued, passage to the complex conjugate gives $b(t) - \lambda = b_r t^{-r} \prod_{j=1}^{2r}(t - 1/\overline{z}_j)$. Thus, if $b - \lambda$ has $\ell \geq 1$ distinct zeros $\tau_1, \ldots, \tau_\ell$ on $\mathbf{T}$ and $n \geq 0$ distinct zeros $\mu_1, \ldots, \mu_n$ of modulus greater than 1, then

$$b(t) - \lambda = b_r t^{-r} \prod_{j=1}^{n}[(t - \mu_j)(t - 1/\overline{\mu}_j)]^{\beta_j} \prod_{j=1}^{\ell}(t - \tau_j)^{\alpha_j} = \prod_{j=1}^{\ell}\left(1 - \frac{\tau_j}{t}\right)^{\alpha_j} c(t)$$

with

$$c(t) = b_r t^{-r} t^{\alpha_1 + \cdots + \alpha_\ell} \prod_{j=1}^{n}[(t - \mu_j)(t - 1/\overline{\mu}_j)]^{\beta_j}.$$

Clearly, wind $c = -r + \alpha_1 - \cdots + \alpha_\ell + \beta_1 + \cdots + \beta_n$. Since $\alpha_1 + \cdots + \alpha_\ell + 2\beta_1 + \cdots + 2\beta_n = 2r$ and $\alpha_j \geq 1$ for all j, we get $\beta_1 + \cdots + \beta_n < r$ and thus wind $c = r - \beta_1 - \cdots - \beta_n > 0$. The assertion now follows from Proposition 1.22. □

Example 1.28. We remark that Corollary 1.27 is not true for $p = \infty$: the sequence $\{1, 0, -1, 0, 1, 0, -1, 0, \ldots\}$ is obviously in the kernel of the operator

$$T(\chi_{-1} + \chi_1) = \begin{pmatrix} 0 & 1 & 0 & 0 & \ldots \\ 1 & 0 & 1 & 0 & \ldots \\ 0 & 1 & 0 & 1 & \ldots \\ 0 & 0 & 1 & 0 & \ldots \\ \ldots & \ldots & \ldots & \ldots & \ldots \end{pmatrix}.$$

For this operator, things are as follows. The symbol is $b(t) = t^{-1} + t$ and hence sp $T(b) = b(\mathbf{T}) = [-2, 2]$. For $\lambda \in [-2, 2]$,

$$b(t) - \lambda = \left(1 - \frac{\tau_1(\lambda)}{t}\right)\left(1 - \frac{\tau_2(\lambda)}{t}\right) t,$$

where $\tau_1(\lambda), \tau_2(\lambda) \in \mathbf{T}$ are given by

$$\tau_{1,2}(\lambda) = \frac{\lambda}{2} \pm i\sqrt{1 - \frac{\lambda^2}{4}}.$$

Thus, if $\lambda \in (-2, 2)$, then Proposition 1.23 (with wind $c = 1$ and $\ell = 2$) implies that $T(b) - \lambda I$ has a one-dimensional kernel in ℓ^∞ whose nonzero elements are extended, and if $\lambda \in \{-2, 2\}$, then Proposition 1.23 (with wind $c = 1$ and $\ell = 1$) shows that the kernel of $T(b) - \lambda I$ in ℓ^∞ is trivial. □

Example 1.29. Let $b(t) = (1 + 1/t)^3$. The image $b(\mathbf{T})$ is the solid curve in the left picture of Figure 1.2; this curve is traced out in the clockwise direction. The curve intersects itself at the point -1. We see that $\mathbf{C} \setminus b(\mathbf{T})$ has two bounded connected components Ω_1 and Ω_2 with wind $(b, \lambda) = -1$ for $\lambda \in \Omega_1$ and wind $(b, \lambda) = -2$ for $\lambda \in \Omega_2$. Thus, $\operatorname{sp} T(b) = \Omega_1 \cup \Omega_2 \cup b(\mathbf{T})$.

By Proposition 1.20, the points in $\Omega_1 \cup \Omega_2$ are eigenvalues. Looking at Figure 1.2 and using Proposition 1.24, we see that wind $c = -1$ for the points on the two small open arcs of $b(\mathbf{T})$ that join 0 and -1. Thus, by virtue of Propositions 1.22 and 1.23, these points are also eigenvalues. The points of $b(\mathbf{T})$ which are boundary points of the unbounded connected component of $\mathbf{C} \setminus b(\mathbf{T})$, including the point -1, are not eigenvalues if $1 \le p < \infty$ (Corollary 1.26), but they are eigenvalues if $p = \infty$, because wind $c = 0$ (Propositions 1.23 and 1.24). Finally, for $\lambda = 0$ we have $\ell = 1$, and the right picture of Figure 1.2 reveals that wind $c = 0$. Consequently, $\lambda = 0$ is not an eigenvalue if $1 \le p < \infty$ (Proposition 1.22) and is an eigenvalue if $p = \infty$ (Proposition 1.23). □

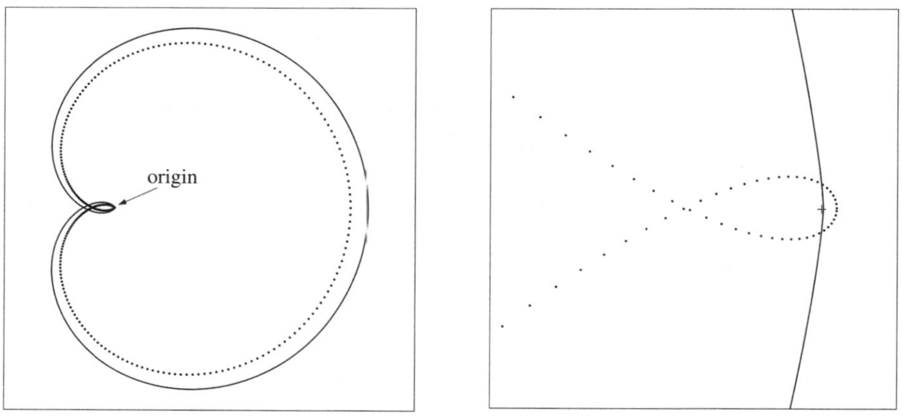

Figure 1.2. *The curves $b(\mathbf{T})$ (solid) and $b_\varrho(\mathbf{T})$ with $\varrho = 1.05$ (dotted). Both curves are traced out clockwise. The right picture is a close-up (with a magnification about 4300) of the left picture in a neighborhood of the origin, which is marked by $+$.*

Remark 1.30. Let b be a real-valued Laurent polynomial and suppose λ is a point in $b(\mathbf{T})$. We know from Corollary 1.27 that $\operatorname{Ker} T(b - \lambda) = \{0\}$ in ℓ^p $(1 \le p < \infty)$. This implies that if $1 < p < \infty$, then the range $\operatorname{Im} T(b - \lambda)$ is not closed but dense in ℓ^p. In other words, $T(b)$ has no residual spectrum on ℓ^p for $1 < p < \infty$. However, the polynomial $b = \chi_{-1} + \chi_1$ is an example of a symbol for which $\operatorname{Ker} T(b) \ne \{0\}$ in ℓ^∞ and thus $\operatorname{Im} T(b)$ is not dense in ℓ^1. Consequently, there are b's such that $T(b)$ has a residual spectrum on ℓ^1.

1.9 Selfadjoint Operators

We now consider Toeplitz operators on the space ℓ^2. Obviously, $T(b)$ is selfadjoint if and only if $\overline{b}_n = b_{-n}$ for all n, that is, if and only if b is real valued. Thus, let

$$b(e^{ix}) = \sum_{k=-s}^{s} b_k e^{ikx} = b_0 + \sum_{n=1}^{s}(a_n \cos nx + c_n \sin nx),$$

where b_0, a_n, c_n are real numbers.

The resolution of the identity. Let A be a bounded selfadjoint operator on the space ℓ^2. Then the operator $f(A)$ is well defined for every bounded Borel function f on $\mathbf{R}$. For $\lambda \in \mathbf{R}$, put $E(\lambda) = \chi_{(-\infty,\lambda]}(A)$, where $\chi_{(-\infty,\lambda]}$ is the characteristic function of $(-\infty, \lambda]$. The family $\{E(\lambda)\}_{\lambda\in\mathbf{R}}$ is called the *resolution of the identity* for A. Stone's formula states that

$$\frac{1}{2}(E(\lambda+0) + E(\lambda-0))x$$
$$= \lim_{\varepsilon\to 0+0} \frac{1}{2\pi i}\int_{-\infty}^{\lambda} \left((A-(\lambda+i\varepsilon)I)^{-1} - (A-(\lambda-i\varepsilon)I)^{-1}\right) x\, d\lambda \qquad (1.38)$$

for every $x \in \ell^2$. Let $\ell^2_{\text{pp}}, \ell^2_{\text{ac}}, \ell^2_{\text{sing}}$ denote the set of all $x \in \ell^2$ for which the measure $d_x(\lambda) := d(E(\lambda)x, x)$ is a pure point measure, is absolutely continuous with respect to Lebesgue measure, and is singular continuous with respect to Lebesgue measure, respectively. The sets $\ell^2_{\text{pp}}, \ell^2_{\text{ac}}, \ell^2_{\text{sing}}$ are closed subspaces of ℓ^2 whose orthogonal sum is all of ℓ^2. Moreover, each of the spaces $\ell^2_{\text{pp}}, \ell^2_{\text{ac}}, \ell^2_{\text{sing}}$ is an invariant subspace of A. The spectra of the restrictions $A|\ell^2_{\text{ac}}$ and $A|\ell^2_{\text{sing}}$ are referred to as the *absolutely continuous spectrum* and the *singular continuous spectrum* of A, respectively. The *point spectrum* of A is defined as the set of the eigenvalues of A (and *not* as the spectrum of the restriction $A|\ell^2_{\text{pp}}$). It is well known that the spectrum of A is the union of the absolutely continuous spectrum, the singular continuous spectrum, and the closure of the point spectrum.

The spectrum of $T(b)$ is the line segment $b(\mathbf{T}) =: [m, M]$. Corollary 1.27 tells us that the point spectrum of $T(b)$ is empty unless b is a constant. As the following theorem shows, the singular continuous spectrum is also empty.

Theorem 1.31 (Rosenblum). *The spectrum of a Toeplitz operator generated by a real-valued nonconstant Laurent polynomial is purely absolutely continuous.*

Proof. Let b be a real-valued nonconstant Laurent polynomial. We may without loss of generality assume that the highest coefficient b_s is 1. Fix $\lambda \in \mathbf{R}$ and $\varepsilon > 0$, and put $z = \lambda + i\varepsilon$. As in Section 1.4, we can write

$$b(t) - z = \prod\left(1 - \frac{\delta_j}{t}\right)\prod(t - \mu_j),$$

where $|\delta_j| = |\delta_j(z)| < 1$ and $|\mu_j| = |\mu_j(z)| > 1$. Passing to the complex conjugate, we get

$$b(t) - \overline{z} = \prod \left(1 - \frac{1}{\overline{\mu}_j t}\right) \prod \left(t - \frac{1}{\overline{\delta}_j}\right).$$

Proposition 1.19 now implies that, for $\alpha \in \mathbf{D}$,

$$(T^{-1}(b - z)w_\alpha, w_\alpha) = \frac{1}{f(z)} \frac{1}{1 - |\alpha|^2}, \quad (T^{-1}(b - \overline{z})w_\alpha, w_\alpha) = \frac{1}{\overline{f(z)}} \frac{1}{1 - |\alpha|^2},$$

where $f(z) := \prod(1 - \delta_j \alpha) \prod(\overline{\alpha} - \mu_j)$. We have

$$\left| \frac{1}{f(z)} - \frac{1}{\overline{f(z)}} \right| = 2 \left| \frac{\operatorname{Im} f(z)}{|f(z)|^2} \right| \le \frac{2}{|f(z)|} = \frac{2}{\prod |1 - \delta_j \alpha| \prod |\overline{\alpha} - \mu_j|} \le \frac{2}{(1 - |\alpha|)^{2s}},$$

because $|1 - \delta_j \alpha| \ge 1 - |\alpha|$ and $|\overline{\alpha} - \mu_j| \ge 1 - |\alpha|$ for all j. Since $E(\lambda - 0) = E(\lambda + 0)$, formula (1.38) gives

$$|(E(\lambda_2)w_\alpha, w_\alpha) - (E(\lambda_1)w_\alpha, w_\alpha)| \le \frac{1}{2\pi} \int_{\lambda_1}^{\lambda_2} \frac{2}{(1 - |\alpha|)^{2s}} \, d\lambda,$$

which shows that the function $\lambda \mapsto (E(\lambda)w_\alpha, w_\alpha)$ is absolutely continuous for each $\alpha \in \mathbf{D}$. It follows that $w_\alpha \in \ell^2_{\rm ac}$ for each $\alpha \in \mathbf{D}$, and as the linear hull of the set $\{w_\alpha\}_{\alpha \in \mathbf{D}}$ is dense in ℓ^2, we arrive at the conclusion that $\ell^2_{\rm ac} = \ell^2$, which is the assertion. □

The problem of diagonalizing selfadjoint bounded Toeplitz operators is solved. More or less explicit formulas can be found in [227], [228], [229], and [288]. We here confine ourselves to a few simple observations.

Chebyshev polynomials. We denote by $\{U_n\}_{n=0}^\infty$ the normalized Chebyshev polynomials of the second kind:

$$U_n(\cos\theta) = \sqrt{\frac{2}{\pi}} \, \frac{\sin(n+1)\theta}{\sin\theta}.$$

The polynomials $\{U_n\}_{n=0}^\infty$ constitute an orthonormal basis in the Hilbert space $L^2((-1, 1), \sqrt{1 - \lambda^2}) =: L^2(\sigma)$,

$$\int_{-1}^{1} U_j(\lambda) U_k(\lambda) \sqrt{1 - \lambda^2} \, d\lambda = \delta_{jk},$$

and they satisfy the identities

$$\lambda U_n(\lambda) = \frac{1}{2} U_{n+1}(\lambda) + \frac{1}{2} U_{n-1}(\lambda), \quad U_{-1}(\lambda) := 0. \tag{1.39}$$

For $\alpha \in \mathbf{T}$, we define $V_\alpha : \ell^2 \to L^2(\sigma)$ by

$$(V_\alpha x)(\lambda) = \sum_{n=0}^{\infty} x_n \alpha^n U_n(\lambda), \quad \lambda \in (-1, 1).$$

Clearly, V_α is unitary and $V_\alpha^{-1} : L^2(\sigma) \to \ell^2$ acts by the rule

$$(V_\alpha^{-1} f)_n = \frac{1}{\alpha^n} \int_{-1}^{1} f(\lambda) U_n(\lambda) \sqrt{1-\lambda^2}\, d\lambda, \quad n \geq 0.$$

We denote by $M_{f(\lambda)}$ the operator of multiplication by the function $f(\lambda)$ on $L^2(\sigma)$.

Proposition 1.32. *Let*

$$b(e^{ix}) = \overline{b}_1 e^{-ix} + b_0 + b_1 e^{ix} = b_0 + a_1 \cos x + c_1 \sin x$$

be a real valued trinomial. Put

$$\alpha = \sqrt{\frac{\overline{b}_1}{b_1}} = \sqrt{\frac{a_1 + ic_1}{a_1 - ic_1}}, \quad \beta = 2|b_1| = \frac{1}{2}\sqrt{a_1^2 + c_1^2}.$$

Then $T(b) = V_\alpha^{-1} M_{b_0 + \beta\lambda} V_\alpha$.

Proof. Using (1.39) and the orthonormality of the polynomials U_n we obtain

$$\begin{aligned}
(V_\alpha^{-1} M_\lambda V_\alpha x)_n &= \frac{1}{\alpha^n} \int_{-1}^{1} \lambda (V_\alpha x)(\lambda) U_n(\lambda) \sqrt{1-\lambda^2}\, d\lambda \\
&= \frac{1}{\alpha^n} \sum_{k=0}^{\infty} x_k \alpha^k \int_{-1}^{1} \lambda U_k(\lambda) U_n(\lambda) \sqrt{1-\lambda^2}\, d\lambda \\
&= \frac{1}{\alpha^n} \sum_{k=0}^{\infty} x_k \alpha^k \int_{-1}^{1} \left(\frac{1}{2} U_k(\lambda) U_{n+1}(\lambda) + \frac{1}{2} U_k(\lambda) U_{n-1}(\lambda) \right) \sqrt{1-\lambda^2}\, d\lambda \\
&= \frac{1}{\alpha^n} \left(x_{n+1} \frac{\alpha^{n+1}}{2} + x_{n-1} \frac{\alpha^{n-1}}{2} \right) = \frac{\alpha}{2} x_{n+1} + \frac{1}{2\alpha} x_{n-1},
\end{aligned}$$

where $x_{-1} := 0$. Equivalently,

$$V_\alpha^{-1} M_\lambda V_\alpha = T\left(\frac{\alpha}{2} \chi_{-1} + \frac{1}{2\alpha} \chi_1 \right).$$

This implies that $V_\alpha^{-1} M_{b_0 + 2|b_1|\lambda} V_\alpha$ is the Toeplitz operator with the symbol

$$b_0 + \sqrt{\frac{\overline{b}_1}{b_1}}\, |b_1| \chi_{-1} + \sqrt{\frac{b_1}{\overline{b}_1}}\, |b_1| \chi_1 = b_0 + \overline{b}_1 \chi_{-1} + b_1 \chi_1. \qquad \square$$

In particular,

$$T(\cos x) := T\left(\frac{1}{2} \chi_{-1} + \frac{1}{2} \chi_1 \right) = V_1^{-1} M_\lambda V_1, \tag{1.40}$$

$$T(\sin x) := T\left(\frac{i}{2} \chi_1 - \frac{i}{2} \chi_{-1} \right) = V_i^{-1} M_\lambda V_i. \tag{1.41}$$

Diagonalization of symmetric and skewsymmetric Toeplitz matrices. The polynomials

$$g(x) = b_0 + \sum_{n=1}^{s} a_n \cos nx \quad \text{and} \quad u(x) = \sum_{n=1}^{s} c_n \sin nx$$

generate symmetric ($A^\top = A$) and skewsymmetric ($A^\top = -A$) Toeplitz matrices, respectively. From identities (1.39) we infer that

$$\lambda^2 U_n(\lambda) = \frac{1}{4}U_{n+2}(\lambda) + \frac{1}{2}U_n(\lambda) + \frac{1}{4}U_{n-2}(\lambda),$$
$$\lambda^3 U_n(\lambda) = \frac{1}{8}U_{n+3}(\lambda) + \frac{3}{8}U_{n+1}(\lambda) + \frac{3}{8}U_{n-1}(\lambda) + \frac{1}{8}U_{n-3}(\lambda),$$

and so on, where $U_{-2}(\lambda) = U_{-3}(\lambda) = \cdots = 0$. Consequently, as in the proof of Proposition 1.32,

$$T\left(\frac{1}{2} + \frac{1}{4}\cos 2x\right) = T\left(\frac{1}{4}\chi_{-2} + \frac{1}{2}\chi_0 + \frac{1}{4}\chi_2\right) = V_1^{-1} M_{\lambda^2} V_1,$$
$$T\left(\frac{3}{8}\cos x + \frac{1}{8}\cos 3x\right) = T\left(\frac{1}{8}\chi_{-3} + \frac{3}{8}\chi_{-1} + \frac{3}{8}\chi_1 + \frac{1}{8}\chi_3\right) = V_1^{-1} M_{\lambda^3} V_1,$$

etc. This shows that we can find coefficients $\gamma_0, \gamma_1, \ldots, \gamma_s$ such that

$$T\left(b_0 + \sum_{n=1}^{s} a_n \cos nx\right) = V_1^{-1} M_{\gamma_0 + \gamma_1\lambda + \cdots + \gamma_s \lambda^s} V_1.$$

One can diagonalize the skewsymmetric matrices $T(u)$ analogously.

Resolution of the identity for Toeplitz operators. Let A be a bounded selfadjoint operator on ℓ^2 and suppose we have a unitary operator V such that VAV^{-1} is multiplication by λ on $L^2(\sigma) := L^2((-1,1), \sqrt{1-\lambda^2})$. Then the resolution of the identity for A can be computed from the formula $E(\lambda) = V M_{\chi_{(-\infty,\lambda]}} V^{-1}$. Clearly, we can think of $E(\lambda)$ as an infinite matrix $(E_{jk}(\lambda))_{j,k=0}^{\infty}$.

Proposition 1.33. *The resolution of the identity for* $T(\cos x) = T\left(\frac{1}{2}\chi_{-1} + \frac{1}{2}\chi_1\right)$ *is given by* $E(\lambda) = 0$ *for* $\lambda \in (-\infty, -1]$, $E(\lambda) = I$ *for* $\lambda \in [1, \infty)$, *and*

$$E_{jk}(\lambda) = \begin{cases} \dfrac{1}{\pi}\left(\dfrac{\sin(j+k+2)\theta}{j+k+2} - \dfrac{\sin(j-k)\theta}{j-k}\right) & \text{if } j \neq k \\ \dfrac{1}{\pi}\left(\dfrac{\sin(2j+2)\theta}{2j+2} - \theta + \pi\right) & \text{if } j = k \end{cases}$$

with $\theta = \arccos\lambda$ *for* $\lambda \in (-1, 1)$.

Proof. It suffices to consider λ in $(-1, 1)$. Let e_n be the nth element of the standard basis of ℓ^2. By virtue of (1.40),

$$\begin{aligned}
E_{jk}(\lambda) &= (E(\lambda)e_k, e_j) = (V_1^{-1} M_{\chi(-\infty,\lambda)} V_1 e_k, e_j) \\
&= (M_{\chi(-\infty,\lambda)} V_1 e_k, V_1 e_j) = (M_{\chi(-\infty,\lambda)} U_k, U_j) \\
&= \int_{-1}^{\lambda} U_k(\mu) U_j(\mu) \sqrt{1-\mu^2}\, d\mu \\
&= \frac{2}{\pi} \int_{\theta}^{\pi} \frac{\sin(k+1)\varphi}{\sin\varphi} \frac{\sin(j+1)\varphi}{\sin\varphi} \sin^2\varphi\, d\varphi \\
&= \frac{1}{\pi} \int_{\theta}^{\pi} [\cos(j-k)\varphi - \cos(j+k+2)\varphi]\, d\varphi,
\end{aligned}$$

which implies the assertion. □

From (1.40) we also deduce that if f is any continuous function on $[-1, 1]$, then the j, k entry of $f(T(\cos x))$ is

$$\begin{aligned}
[f(T(\cos x))]_{jk} &= (V_1^{-1} M_{f(\lambda)} V_1 e_k, e_j) = (M_{f(\lambda)} U_k, U_j) \\
&= \int_{-1}^{1} f(\lambda) U_k(\lambda) U_j(\lambda) \sqrt{1-\lambda^2}\, d\lambda \\
&= \frac{2}{\pi} \int_{\theta}^{\pi} f(\cos\theta) \frac{\sin(k+1)\theta}{\sin\theta} \frac{\sin(j+1)\theta}{\sin\theta} \sin^2\theta\, d\theta \\
&= \frac{1}{\pi} \int_{\theta}^{\pi} f(\cos\theta)[\cos(j-k)\theta - \cos(j+k+2)\theta]\, d\theta.
\end{aligned}$$

For example, the nonnegative square root of

$$T(2-2\cos x) = \begin{pmatrix} 2 & -1 & 0 & \dots \\ -1 & 2 & -1 & \dots \\ 0 & -1 & 2 & \dots \\ \dots & \dots & \dots & \dots \end{pmatrix}$$

has j, k entry

$$\begin{aligned}
&\frac{1}{\pi} \int_{\theta}^{\pi} \sqrt{2-2\cos\theta}\, [\cos(j-k)\theta - \cos(j+k+2)\theta]\, d\theta \\
&= \frac{2}{\pi} \int_{\theta}^{\pi} \left(\sin\frac{\theta}{2}\right) [\cos(j-k)\theta - \cos(j+k+2)\theta]\, d\theta \\
&= \frac{1}{\pi} \int_{-1}^{1} \left[\sin\left(j-k+\frac{1}{2}\right)\theta - \sin\left(j-k-\frac{1}{2}\right)\theta \right. \\
&\qquad \left. - \sin\left(j+k+2+\frac{1}{2}\right)\theta + \sin\left(j+k+2-\frac{1}{2}\right)\theta\right] d\theta \\
&= \frac{1}{\pi}\left(\frac{1}{j-k+\frac{1}{2}} - \frac{1}{j-k-\frac{1}{2}} + \frac{1}{j+k+2+\frac{1}{2}} - \frac{1}{j+k+2-\frac{1}{2}}\right),
\end{aligned}$$

and this equals

$$\frac{4}{\pi}\left(\frac{1}{4(j+k+2)^2+1} - \frac{1}{4(j-k)^2+1}\right).$$

Exercises

1. (a) Find a function $a \in L^\infty(\mathbf{T})$ whose Fourier coefficients a_n $(n \in \mathbf{Z})$ are just $a_n = 1/(n+1/2)$.

   (b) Show that the infinite Toeplitz matrix

$$\begin{pmatrix} 1 & -\frac{1}{2} & -\frac{1}{3} & \dots \\ \frac{1}{2} & 1 & -\frac{1}{2} & \dots \\ \frac{1}{3} & \frac{1}{2} & 1 & \dots \\ \dots & \dots & \dots & \dots \end{pmatrix}$$

   induces a bounded operator on ℓ^2 but not on ℓ^1.

   (c) Show that the infinite Toeplitz matrix

$$\begin{pmatrix} 1 & \frac{1}{2} & \frac{1}{3} & \dots \\ \frac{1}{2} & 1 & \frac{1}{2} & \dots \\ \frac{1}{3} & \frac{1}{2} & 1 & \dots \\ \dots & \dots & \dots & \dots \end{pmatrix}$$

   does not generate a bounded operator on ℓ^2.

   (d) Show that the infinite upper-triangular Toeplitz matrix

$$\begin{pmatrix} 1 & \frac{1}{2} & \frac{1}{3} & \dots \\ & 1 & \frac{1}{2} & \dots \\ & & 1 & \dots \\ & & & \dots \end{pmatrix}$$

   does not define a bounded operator on ℓ^2 but that the infinite Hankel matrix

$$\begin{pmatrix} 1 & \frac{1}{2} & \frac{1}{3} & \dots \\ \frac{1}{2} & \frac{1}{3} & \dots & \dots \\ \frac{1}{3} & \dots & \dots & \dots \\ \dots & \dots & \dots & \dots \end{pmatrix}$$

   is the matrix of a bounded operator on ℓ^2.

2. Prove that $H(ab) = H(a)\,T(\tilde{b}) + T(a)\,H(b)$ for all $a, b \in W$.

3. Find a Wiener-Hopf factorization of $6t - 41 + 31t^{-1} - 6t^{-2}$.

4. Let $b_1, \dots, b_m \in \mathcal{P}^+$ have no common zero on $\mathbf{T}$. Prove that there are $c_1, \dots, c_m \in W$ such that $T(c_1)T(b_1) + \dots + T(c_m)T(b_m) = I$. Can one choose the $c_1, \dots, c_m$ as rational functions without poles on $\mathbf{T}$?

5. Let $b(t) = 1 + 2t + \gamma t^3$. Show that there is no $\gamma \in \mathbf{C}$ for which $T(b)$ is invertible.

6. Let

$$b(t) = \det \begin{pmatrix} 2 & 1 & 0 & 0 & 0 \\ 1 & 2 & 1 & 0 & 0 \\ 0 & 1 & 2 & 1 & 0 \\ 0 & 0 & 1 & 2 & 1 \\ 1 & t & t^2 & t^3 & t^4 \end{pmatrix}.$$

Show that b has no zeros on $\mathbf{T}$ and that wind $b = 4$. Try to prove the analogue of this if the 5×5 determinant is replaced by an $n \times n$ determinant in the obvious way.

7. Let $b(t) = 4 + \sum_{j=-5}^{5} t^j$. Show that $T(b)$ is invertible.

8. Let b be a Laurent polynomial and $1 \le p \le \infty$. Show that $T(b) : \ell^p \to \ell^p$ has closed range if and only if either b is identically zero or b has no zeros on $\mathbf{T}$.

9. Let $b_n(t) = 1 + \frac{1}{2}(t + t^{-1}) + \frac{1}{3}(t^2 + t^{-2}) + \cdots + \frac{1}{n}(t^n + t^{-n})$. Show that

$$2 \log n + 0.1544 \le \|T(b_n)\|_4 \le 2 \log n + 0.1545$$

for all sufficiently large n.

10. Prove that $\|T(a) + K\|_p \ge \|T(a)\|_p$ for every $a \in W$ and every compact operator on ℓ^p $(1 \le p \le \infty)$. Deduce that the zero operator is the only compact Toeplitz operator.

11. Prove that $\|T^n(a)\|_2 = \|T(a^n)\|_2$ for every $a \in W$.

12. Show that there exist Laurent polynomials b such that $\|H(b)\|_2 < \|b\|_\infty$ and $\|H(b)\|_1 < \|b\|_W$.

13. For $b = \sum_j b_j \chi_j \in \mathcal{P}$, define

$$S_n b = \sum_{|j| \le n-1} b_j \chi_j, \quad \sigma_n b = \frac{1}{n}(S_1 b + \cdots + S_n b).$$

Prove that always $\|T(\sigma_n b)\|_2 \le \|T(b)\|_2$ but that there exist b and n such that $\|T(S_n b)\|_2 > 100\, \|T(b)\|_2$.

14. Show that if $a \in W$ and $T(a)$ is a unitary operator on ℓ^2, then a is a unimodular constant.

15. Let $B = (b_{jk})_{j,k=1}^{\infty}$ with $b_{23} = b_{32} = -1$ and $b_{jk} = 0$ otherwise. Show that the operator $T(\chi_{-1} + \chi_1) + B \in \mathcal{B}(\ell^p)$ $(1 \le p \le \infty)$ has eigenvalues in $(-2, 2)$.

16. Let $A = T(\chi_{-1} + \chi_1) + \operatorname{diag}(v_j)_{j=1}^{\infty}$. Show that if $v_j = o(1/j)$, then $A \in \mathcal{B}(\ell^2)$ has at most finitely many eigenvalues in each segment $[\alpha, \beta] \subset (-2, 2)$ and that if $v_j = o(1/j^{1+\varepsilon})$ with some $\varepsilon > 0$, then the only possible eigenvalues of $A \in \mathcal{B}(\ell^2)$ are -2 and 2.

17. Show that the positive square root of $T(2+2\cos x)$ is the Toeplitz-plus-Hankel matrix

$$\frac{4}{\pi}\left(\frac{(-1)^{j-k+1}}{(2j-2k-2)(2j-2k+1)}+\frac{(-1)^{j+k+2}}{(2j+2k+3)(2j+2k+5)}\right)_{j,k=0}^{\infty}.$$

Notes

In his 1911 paper [267], Otto Toeplitz considered doubly infinite matrices of the form $(a_{j-k})_{j=-\infty}^{\infty}$ and proved that the spectrum of the corresponding operator on $\ell^2(\mathbf{Z})$ is just the curve

$$\left\{\sum_{k=-\infty}^{\infty} a_k t^k : t \in \mathbf{T}\right\}.$$

The matrices $L(a) := (a_{j-k})_{j=-\infty}^{\infty}$ are nowadays called Laurent matrices. In a footnote of [267], Toeplitz established that the simply infinite matrix $(a_{j-k})_{j=0}^{\infty}$ induces a bounded operator on $\ell^2(\mathbf{Z}_+)$ if *and only if* the doubly infinite matrix $(a_{j-k})_{j=-\infty}^{\infty}$ generates a bounded operator on $\ell^2(\mathbf{Z})$. This is why the matrices $(a_{j-k})_{j=0}^{\infty}$ now bear his name.

The material of Sections 1.1 to 1.7 is standard. The books [71] and [130] may serve as introductions to the basic phenomena in connection with infinite Toeplitz matrices. A nice source is also [150]. In [25], infinite systems with a banded Toeplitz matrix $T(a)$ are treated with the tools of the theory of difference equations; in this book, we find formulas for the entries of the inverses in terms of the zeros of $a(z)$ ($z \in \mathbf{C}$) and solvability criteria in the spaces of sequences $x = \{x_n\}_{n=1}^{\infty}$ subject to the condition $x_n = O(\varrho^n)$. Advanced topics in the theory of infinite Toeplitz matrices (= Toeplitz operators) are treated in the monographs [70], [103], [195], [196]. The standard texts on Hankel matrices are [196], [201], [204], [213].

Full proofs of Theorems 1.5, 1.6, 1.7 can be found in [103] or [230], for example. Theorem 1.8 as it is stated is due to Mark Krein [184]. The method of Wiener-Hopf factorization was introduced by N. Wiener and E. Hopf in 1931. What we call Wiener-Hopf factorization has its origin in the work of Gakhov [123], although the basic idea (in the case of vanishing winding number) was already employed by Plemelj [205]. Mark Krein [184] was the first to understand the operator theoretic essence and the Banach algebraic background of Wiener-Hopf factorization and to present the method in a crystal-clear manner.

The results of Section 1.5 are also due to Krein [184]. However, it had been known a long time before that $T(a)$ is Fredholm of index $-\text{wind}\, a$ whenever a has no zeros on $\mathbf{T}$; this insight is more or less explicit in works by F. Noether, S. G. Mikhlin, N. I. Muskhelishvili, F. D. Gakhov, V. V. Ivanov, A. P. Calderón, F. Spitzer, H. Widom, A. Devinatz, G. Fichera, and certainly others. Moreover, in 1952, Israel Gohberg [128] had already proved that $T(a)$ is Fredholm if *and only if* a has no zeros on $\mathbf{T}$. From this result it is only a small step (from the present-day understanding of the matter) to the formula $\text{Ind}\, T(a) = -\text{wind}\, a$.

Section 1.8 is based on known results of [129], [130], [184].

Rosenblum's papers [226], [227], [228] are the classics on selfadjoint Toeplitz operators. The monograph [229] contains very readable material on the topic. In these works one can also find precise references to previous work on selfadjoint Toeplitz operators. For example, in [229] it is pointed out that the diagonalization (1.40) was carried out by Hilbert (1912) and Hellinger (1941). Proposition 1.19 is from [81] and [226] and Theorem 1.31 was established in [226]. The results around Proposition 1.32 are special cases of more general results in [227], [228]. Part of Rosenblum's theory was simplified and generalized by Vreugdenhil [288]. We took Proposition 1.33 and the example following after it from [288].

Exercises 5 and 6 are from [203]. Exercises 15 and 16 are results of the papers [182], [183]. Actually, these two papers are devoted to the following more general problem: If λ is not an eigenvalue for $T(b) \in \mathcal{B}(\ell^p)$, for which perturbations $B \in \mathcal{B}(\ell^p)$ is λ not an eigenvalue of $T(b)+B$? In [183] it is in particular proved that if $B = (b_{jk})_{j,k=1}^{\infty}$ is such that $b_{jk} = 0$ for $j > k$ and $(j^{1+\varepsilon} b_{jk})_{j,k=1}^{\infty}$ induces a bounded operator on ℓ^p $(1 \le p < \infty)$, then the interval $(-2, 2)$ contains no eigenvalues of $T(\chi_{-1} + \chi_1) + B$. As Exercise 15 shows, the requirement that B be upper-triangular is essential. A solution to Exercise 17 is in [288].

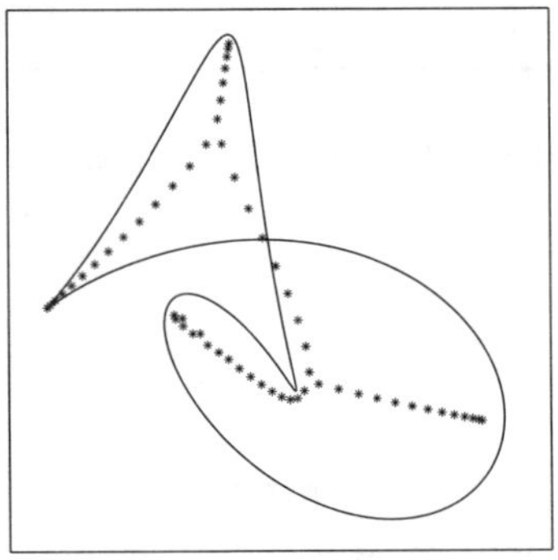

Chapter 2
Determinants

In this chapter, the main actors of this book enter the scene: finite Toeplitz matrices. For a in the Wiener algebra W and $n \in \{1, 2, 3, \dots\}$, we define the $n \times n$ Toeplitz matrix $T_n(a)$ as the principal $n \times n$ section of $T(a)$, that is, by

$$T_n(a) := (a_{j-k})_{j,k=1}^{n} = \begin{pmatrix} a_0 & a_{-1} & \dots & a_{-(n-1)} \\ a_1 & a_0 & \dots & a_{-(n-2)} \\ \vdots & \vdots & \ddots & \vdots \\ a_{n-1} & a_{n-2} & \dots & a_0 \end{pmatrix}. \tag{2.1}$$

If a finite Toeplitz matrix is a circulant matrix, then nearly every piece of information on its spectral properties is explicitly available. We also provide formulas for the eigenvalues and eigenvectors of tridiagonal Toeplitz matrices. Things are significantly more complicated for general Toeplitz matrices.

The focus of this chapter is on the determinants $D_n(a) := \det T_n(a)$. We establish several exact and asymptotic formulas for these determinants, including the Szegö-Widom limit theorem and the Geronimo-Case-Borodin-Okounkov formula. Clearly, nowadays nobody would determine the eigenvalues of $T_n(a)$ by computing the zeros of the polynomial $D_n(a - \lambda) = \det(T_n(a) - \lambda I) = \det T_n(a - \lambda)$. However, the results of Chapter 11 on the asymptotic distribution of the eigenvalues of $T_n(a)$ in the limit $n \to \infty$ are heavily based on consideration of determinants and, independently of eigenvalues, Toeplitz determinants are a hot topic in statistical physics.

2.1 Circulant Matrices

Circulant matrices are the "periodic cousins" of Toeplitz matrices. While Toeplitz matrices usually emerge in stationary problems with zero boundary conditions, circulant matrices arise in connection with periodic boundary conditions. From the viewpoint of spectral theory, circulant matrices are much simpler than (noncirculant) Toeplitz matrices.

Given $a_0, a_1, \dots, a_{n-1} \in \mathbf{C}$, we denote by $\operatorname{circ}(a_0, a_1, \dots, a_{n-1})$ the *circulant matrix* whose first column is $(a_0 \; a_1 \; \dots \; a_{n-1})^\top$,

$$\operatorname{circ}(a_0, a_1, \dots, a_{n-1}) = \begin{pmatrix} a_0 & a_{n-1} & a_{n-2} & \dots & a_1 \\ a_1 & a_0 & a_{n-1} & \dots & a_2 \\ a_2 & a_1 & a_0 & \dots & a_3 \\ \vdots & \vdots & \vdots & \ddots & \vdots \\ a_{n-1} & a_{n-2} & a_{n-3} & \dots & a_0 \end{pmatrix}.$$

Let $\omega_n = \exp(2\pi i/n)$ and put

$$F_n = \frac{1}{\sqrt{n}} \begin{pmatrix} 1 & 1 & 1 & \dots & 1 \\ 1 & \omega_n & \omega_n^2 & \dots & \omega_n^{n-1} \\ 1 & \omega_n^2 & \omega_n^4 & \dots & \omega_n^{2(n-1)} \\ \vdots & \vdots & \vdots & & \vdots \\ 1 & \omega_n^{n-1} & \omega_n^{2(n-1)} & \dots & \omega_n^{(n-1)(n-1)} \end{pmatrix}.$$

The matrix F_n is called the *Fourier matrix* of order n. Obviously, F_n is unitary. By a straightforward computation one can readily verify that

$$\operatorname{circ}(a_0, a_1, \dots, a_{n-1}) = F_n^* \operatorname{diag}(a(1), a(\omega_n), \dots, a(\omega_n^{n-1}))\, F_n, \tag{2.2}$$

where

$$a(z) := a_0 + a_1 z + \dots + a_{n-1} z^{n-1}. \tag{2.3}$$

Identity (2.3) tells us that the eigenvalues of $\operatorname{circ}(a_0, a_1, \dots, a_{n-1})$ are

$$a(1),\; a(\omega_n),\; a(\omega_n^2),\; \dots,\; a(\omega_n^{n-1})$$

and that

$$\frac{1}{\sqrt{n}} \left(1 \; \overline{\omega}_n^j \; \overline{\omega}_n^{2j} \; \dots \; \overline{\omega}_n^{(n-1)j} \right)^\top$$

is a (normalized) eigenvector to $a(\omega_n^j)$. Notice that the eigenvectors are extended, which means that their moduli do not show any kind of exponential decay.

Let $z_1, \dots, z_{n-1}$ be the zeros of the polynomial (2.3). For the determinant, we obtain

$$\begin{aligned} \det \operatorname{circ}(a_0, a_1, \dots, a_{n-1}) &= \prod_{j=0}^{n-1} a(\omega_n^j) \\ = \prod_{j=0}^{n-1} a_{n-1} \prod_{k=1}^{n-1} (\omega_n^j - z_k) &= \prod_{j=0}^{n-1} a_{n-1} (-1)^{n-1} \prod_{k=1}^{n-1} (z_k - \omega_n^j) \\ = a_{n-1}^n \prod_{k=1}^{n-1} \prod_{j=0}^{n-1} (z_k - \omega_n^j) &= a_{n-1}^n \prod_{k=1}^{n-1} (z_k^n - 1) \\ = a_{n-1}^n (-1)^{n-1} \prod_{k=1}^{n-1} (1 - z_k^n). \end{aligned} \tag{2.4}$$

Now let b be a Laurent polynomial of the form

$$b(t) = \sum_{j=-r}^{s} b_j t^j, \quad r \geq 1, \quad s \geq 1, \quad b_{-r} b_s \neq 0. \tag{2.5}$$

If $n \geq \max(r, s) + 1$, then the first and last columns of $T_n(b)$ are

$$(\, b_0 \; b_1 \; \dots \; b_s \; 0 \; \dots \; 0\,)^\top \quad \text{and} \quad (\, 0 \; \dots \; 0 \; b_{-r} \; \dots \; b_{-1} \; b_0\,)^\top,$$

respectively. For $n \geq r + s + 1$, we define the circulant matrix $C_n(b)$ as

$$C_n(b) = \operatorname{circ}(b_0, b_1, \dots, b_s, 0, \dots, 0, b_{-r}, \dots, b_{-1}).$$

Thus, $C_n(b)$ results from $T_n(b)$ by "periodization." For example, if

$$b(t) = b_{-2}t^{-2} + b_{-1}t^{-1} + b_0 + b_1 t + b_2 t^2 + b_3 t^3,$$

then

$$C_8(b) = \begin{pmatrix} b_0 & b_{-1} & b_{-2} & 0 & 0 & b_3 & b_2 & b_1 \\ b_1 & b_0 & b_{-1} & b_{-2} & 0 & 0 & b_3 & b_2 \\ b_2 & b_1 & b_0 & b_{-1} & b_{-2} & 0 & 0 & b_3 \\ b_3 & b_2 & b_1 & b_0 & b_{-1} & b_{-2} & 0 & 0 \\ 0 & b_3 & b_2 & b_1 & b_0 & b_{-1} & b_{-2} & 0 \\ 0 & 0 & b_3 & b_2 & b_1 & b_0 & b_{-1} & b_{-2} \\ b_{-2} & 0 & 0 & b_3 & b_2 & b_1 & b_0 & b_{-1} \\ b_{-1} & b_{-2} & 0 & 0 & b_3 & b_2 & b_1 & b_0 \end{pmatrix}.$$

Proposition 2.1. *If $n \geq r + s + 1$, then*

$$C_n(b) = F_n^* \operatorname{diag}(b(1), b(\omega_n), \dots, b(\omega_n^{n-1}))\, F_n, \tag{2.6}$$

the eigenvalues of $C_n(b)$ are

$$b(1), \; b(\omega_n), \; \dots, \; b(\omega_n^{n-1}),$$

and a (normalized) eigenvector for $b(\omega_n^j)$ is

$$\frac{1}{\sqrt{n}} \left(\, 1 \; \overline{\omega}_n^{\,j} \; \overline{\omega}_n^{\,2j} \; \dots \; \overline{\omega}_n^{\,(n-1)j}\,\right)^\top.$$

Proof. In this case the polynomial (2.3) is

$$a(z) = b_0 + b_1 z + \dots + b_s z^s + b_{-r} z^{n-r} + \dots + b_{-1} z^{n-1}.$$

Since

$$a(\omega_n^j) = b_0 + b_1 \omega_n^j + \dots + b_s \omega_n^{js} + b_{-r} \omega_n^{-jr} + \dots + b_{-1} \omega_n^{-j} = b(\omega_n^j),$$

the assertions follow from the corresponding result on $\operatorname{circ}(a_0, a_1, \dots, a_{n-1})$. □

Proposition 2.2. *If $n \geq r+s+1$, then the determinant of $C_n(b)$ is*

$$\det C_n(b) = b_s^n (-1)^{s(n-1)} \prod_{k=1}^{r+s} (1 - z_k^n),$$

where $z_1, \ldots, z_{r+s}$ are the zeros of the polynomial $z^r b(z)$.

Proof. Using Proposition 2.1 we get

$$\begin{aligned}
\det C_n(b) &= \prod_{j=0}^{n-1} b(\omega_n^j) = \prod_{j=0}^{n-1} b_s \omega_n^{-jr} \prod_{k=1}^{r+s} (\omega_n^j - z_k) \\
&= b_s \left(\prod_{j=0}^{n-1} \omega_n^{-j} \right)^r \prod_{j=0}^{n-1} \prod_{k=1}^{r+s} (\omega_n^j - z_k) \\
&= b_s^n (-1)^{(n-1)r} \prod_{j=0}^{n-1} (-1)^{r+s} \prod_{k=1}^{r+s} (z_k - \omega_n^j) \\
&= b_s^n (-1)^{(n-1)r} (-1)^{(r+s)n} \prod_{k=1}^{r+s} \prod_{j=0}^{n-1} (z_k - \omega_n^j) \\
&= b_s^n (-1)^{ns-r} \prod_{k=1}^{r+s} (z_k^n - 1) = b_s^n (-1)^{ns-r} (-1)^{r+s} \prod_{k=1}^{r+s} (1 - z_k^n) \\
&= b_s^n (-1)^{s(n-1)} \prod_{k=1}^{r+s} (1 - z_k^n). \qquad \square
\end{aligned}$$

Example 2.3. Let $b(t) = t + \alpha^2 t^{-1}$, where $\alpha \in (0, 1]$. Since $b(e^{ix}) = (1+\alpha^2)\cos x + i(1-\alpha^2)\sin x$, the values $b(t)$ trace out an ellipse in the counterclockwise direction as t moves along the unit circle in the counterclockwise direction; for $\alpha = 1$, the ellipse degenerates to the line segment $[-2, 2]$. The eigenvalues of $C_n(b)$ are quite regularly distributed on this ellipse, and the eigenvectors are all extended. The zeros of $zb(z) = z^2 + \alpha^2$ are $\pm i\alpha$, and hence for $n \geq 3$, the determinant $\det C_n(b)$ equals

$$(-1)^{n-1}(1 - (i\alpha)^n)(1 - (-i\alpha)^n) = (-1)^{n-1}\left(1 - 2\alpha^n \cos\frac{n\pi}{2} + \alpha^{2n}\right). \qquad \square$$

2.2 Tridiagonal Toeplitz Matrices

By a tridiagonal Toeplitz matrix we understand a matrix of the form

$$T(a) = \begin{pmatrix} a_0 & a_{-1} & 0 & 0 & \cdots \\ a_1 & a_0 & a_{-1} & 0 & \cdots \\ 0 & a_1 & a_0 & a_{-1} & \cdots \\ 0 & 0 & a_1 & a_0 & \cdots \\ \cdots & \cdots & \cdots & \cdots & \cdots \end{pmatrix}.$$

The symbol of this matrix is $a(t) = a_{-1}t^{-1} + a_0 + a_1 t$. Suppose $a_{-1} \neq 0$ and $a_1 \neq 0$. We fix any value $\alpha = \sqrt{a_{-1}/a_1}$ and define $\sqrt{a_1/a_{-1}} := 1/\alpha$ and $\sqrt{a_1 a_{-1}} := a_1\alpha$. Recall that $T_n(a)$ is the principal $n \times n$ block of $T(a)$.

Theorem 2.4. *The eigenvalues of $T_n(a)$ are*

$$\lambda_j = a_0 + 2\sqrt{a_1 a_{-1}} \cos \frac{\pi j}{n+1} \quad (j = 1, \dots, n), \tag{2.7}$$

and an eigenvector for λ_j is $x_j = (x_1^{(j)} \ \dots \ x_n^{(j)})^\top$ with

$$x_k^{(j)} = \left(\sqrt{\frac{a_1}{a_{-1}}}\right)^k \sin \frac{k\pi j}{n+1} \quad (k = 1, \dots, n). \tag{2.8}$$

Proof. Put $b(t) = t + \alpha^2 t^{-1}$. Thus,

$$T(b) = \begin{pmatrix} 0 & \alpha^2 & 0 & 0 & \dots \\ 1 & 0 & \alpha^2 & 0 & \dots \\ 0 & 1 & 0 & \alpha^2 & \dots \\ 0 & 0 & 1 & 0 & \dots \\ \dots & \dots & \dots & \dots & \dots \end{pmatrix}.$$

Since, obviously, $T_n(a) = a_0 + a_1 T_n(b)$, it suffices to prove that $T_n(b)$ has the eigenvalues

$$\mu_j = 2\alpha \cos \frac{\pi j}{n+1} \quad (j = 1, \dots, n)$$

and that $x_j = (x_1^{(j)} \ \dots \ x_n^{(j)})^\top$ with

$$x_k^{(j)} = \alpha^{-k} \sin \frac{k\pi j}{n+1} \quad (k = 1, \dots, n)$$

is an eigenvector for μ_j. This is equivalent to proving the equalities

$$\begin{aligned} &\alpha^2 x_2^{(j)} = \mu_j x_1^{(j)}, \\ &x_k^{(j)} + \alpha^2 x_{k+2}^{(j)} = \mu_j x_{k+1}^{(j)} \quad (k = 1, \dots, n-2), \\ &x_{n-1}^{(j)} = \mu_j x_n^{(j)}. \end{aligned} \tag{2.9}$$

But these equalities can easily be verified: For example, (2.9) amounts to

$$\alpha^{-k} \sin \frac{k\pi j}{n+1} + \alpha^2 \alpha^{-k-2} \sin \frac{(k+2)\pi j}{n+1} = 2\alpha\alpha^{-k-1} \cos \frac{\pi j}{n+1} \cos \frac{(k+1)\pi j}{n+1},$$

which follows from the identity

$$\sin \beta + \sin \gamma = 2 \cos \frac{\beta - \gamma}{2} \sin \frac{\beta + \gamma}{2}. \qquad \square$$

Example 2.5. Let b be as in Example 2.3. In contrast to the situation for $C_n(b)$, the eigenvalues of $T_n(b)$ are distributed along the interval $(-2\alpha, 2\alpha)$, which is the interval between the foci of the ellipse $b(\mathbf{T})$. Also notice that the eigenvectors are localized and exponentially decaying from the right for $\alpha \in (0, 1)$ (non-Hermitian case, $b(\mathbf{T})$ is a nondegenerate ellipse) and that they are extended for $\alpha = 1$ (Hermitian case, $b(\mathbf{T})$ degenerates to $[-2, 2]$). □

Theorem 2.6. *Let q_1, q_2 be the zeros of the polynomial $q^2 - a_0 q + a_1 a_{-1}$. Then*

$$D_n(a) = \frac{q_2^{n+1} - q_1^{n+1}}{q_2 - q_1} \quad \text{if} \quad q_1 \neq q_2, \tag{2.10}$$

$$D_n(a) = (n+1)q^n \quad \text{if} \quad q_1 = q_2 = q. \tag{2.11}$$

Proof. It suffices to prove (2.10), because (2.11) results from (2.10) by the limit passage $q_2 \to q_1$. We have

$$\begin{aligned} D_1(a) &= a_0 = q_1 + q_2, \\ D_2(a) &= a_0^2 - a_{-1}a_1 = (q_1+q_2)^2 - (q_1 q_2)^2 = q_1^2 + q_1 q_2 + q_2^2, \\ D_n(a) &= a_0 D_{n-1}(a) - a_{-1}a_1 D_{n-2}(a) \quad (n \geq 3). \end{aligned}$$

Let δ_n denote the right-hand side of (2.10). Since

$$\begin{aligned} &\delta_1 = q_1 + q_2, \quad \delta_2 = q_1^2 + q_1 q_2 + q_2^2, \\ &\delta_n - a_0 \delta_{n-1} + a_{-1} a_1 \delta_{n-2} = 0 \quad (n \geq 3), \end{aligned}$$

it follows that $D_n(a) = \delta_n$ for all $n \geq 1$. □

In particular, for the matrix

$$T(\chi_{-1} + \chi_1) = \begin{pmatrix} 0 & 1 & 0 & 0 & \dots \\ 1 & 0 & 1 & 0 & \dots \\ 0 & 1 & 0 & 1 & \dots \\ 0 & 0 & 1 & 0 & \dots \\ \dots & \dots & \dots & \dots & \dots \end{pmatrix}$$

we obtain $q_1 = i$, $q_2 = -i$ and thus Theorems 2.4 and 2.6 give

$$D_n(\chi_{-1} + \chi_1) = 2^n \prod_{j=1}^{n} \cos \frac{\pi j}{n+1} = \begin{cases} 0 & \text{if} \quad n \equiv 1, 3 \pmod 4, \\ -1 & \text{if} \quad n \equiv 2 \pmod 4, \\ 1 & \text{if} \quad n \equiv 0 \pmod 4. \end{cases}$$

2.3 The Baxter-Schmidt Formula

Let

$$a(z) = a_0 + a_1 z + a_2 z^2 + \cdots, \quad a_0 \neq 0,$$

be a function that is analytic and nonzero in some open neighborhood of the origin. In this neighborhood, we can consider the analytic function

$$c(z) := 1/a(z) = c_0 + c_1 z + c_2 z^2 + \cdots, \quad c_0 = a_0^{-1}.$$

We denote by $D_n(z^{-r}a(z))$ and $D_r(z^{-n}c(z))$ the determinants of the $n \times n$ and $r \times r$ principal submatrices of the Toeplitz matrices

$$\begin{pmatrix} a_r & a_{r-1} & \dots & a_0 & 0 & 0 & \dots \\ a_{r+1} & a_r & \dots & a_1 & a_0 & 0 & \dots \\ a_{r+2} & a_{r+1} & \dots & a_2 & a_1 & a_0 & \dots \\ \dots & \dots & \dots & \dots & \dots & \dots & \dots \end{pmatrix}$$

and

$$\begin{pmatrix} c_n & c_{n-1} & \dots & c_0 & 0 & 0 & \dots \\ c_{n+1} & c_n & \dots & c_1 & c_0 & 0 & \dots \\ c_{n+2} & c_{n+1} & \dots & c_2 & c_1 & c_0 & \dots \\ \dots & \dots & \dots & \dots & \dots & \dots & \dots \end{pmatrix},$$

respectively.

Theorem 2.7 (Baxter and Schmidt). *If* $n, r \geq 1$, *then*

$$a_0^{-r} D_n(z^{-r}a(z)) = (-1)^{rn} c_0^{-n} D_r(z^{-n}c(z)). \tag{2.12}$$

Proof. Since (2.12) is symmetric in n and r, we may without loss of generality assume that $n \geq r$. Put

$$A = \left(\begin{array}{ccc|ccc} a_r & \dots & a_1 & a_0 & \dots & 0 \\ a_{r+1} & \dots & a_2 & a_1 & \dots & 0 \\ \vdots & & \vdots & \vdots & & \vdots \\ a_{n-1} & \dots & a_{n-r} & a_{n-r-1} & \dots & a_0 \\ \hline a_n & \dots & a_{n-r+1} & a_{n-r} & \dots & a_1 \\ \vdots & & \vdots & \vdots & & \vdots \\ a_{n+r-1} & \dots & a_n & a_{n-1} & \dots & a_r \end{array}\right)$$

and

$$C = \left(\begin{array}{ccc|ccc} c_0 & \dots & 0 & 0 & \dots & 0 \\ c_1 & \dots & 0 & 0 & \dots & 0 \\ \vdots & & \vdots & \vdots & & \vdots \\ c_{r-1} & \dots & c_0 & 0 & \dots & 0 \\ \hline c_r & \dots & c_1 & 1 & \dots & 0 \\ \vdots & & \vdots & \vdots & & \vdots \\ c_{n-1} & \dots & c_{n-r} & 0 & \dots & 1 \end{array}\right).$$

Taking into account that $a(z)c(z) = 1$, we get

$$AC = \begin{pmatrix} 0 & D \\ -R & * \end{pmatrix},$$

where D is a lower-triangular $(n-r) \times (n-r)$ matrix whose diagonal elements are all equal to a_0 and where R is the $r \times r$ matrix

$$R = \begin{pmatrix} a_0 c_n & \dots & a_0 c_{n-r+1} \\ \vdots & & \vdots \\ a_{r-1} c_n + \dots + a_0 c_{n-r+1} & \dots & a_{r-1} c_{n-r+1} + \dots + a_0 c_n \end{pmatrix}$$

$$= \begin{pmatrix} a_0 & & & \\ a_1 & a_0 & & \\ \vdots & \vdots & \ddots & \\ a_{r-1} & a_{r-2} & \dots & a_0 \end{pmatrix} \begin{pmatrix} c_n & c_{n-1} & \dots & c_{n-r+1} \\ c_{n+1} & c_n & \dots & c_{n-r+2} \\ \vdots & \vdots & \ddots & \vdots \\ c_{n+r-1} & c_{n+r-2} & \dots & c_n \end{pmatrix}.$$

Clearly,

$$\det A = D_n(z^{-r} a(z)), \quad \det C = c_0^r,$$
$$\det AC = (-1)^{r(n-r)} a_0^{n-r} (-1)^r \det R,$$
$$\det R = a_0^r D_r(z^{-n} c(z)),$$

whence

$$c_0^r D_n(z^{-r} a(z)) = (-1)^{r(n-r)+r} a_0^n D_r(z^{-n} c(z)).$$

Since $r(n-r) + r \equiv rn \pmod 2$, $c_0 = a_0^{-1}$, $a_0 = c_0^{-1}$, we therefore arrive at the desired formula (2.12). □

2.4 Widom's Formula

Let now b be a Laurent polynomial,

$$b(t) = \sum_{j=-r}^{s} b_j t^j \quad (t \in \mathbf{T}).$$

We are interested in a formula for the determinant $D_n(b)$ of the Toeplitz band matrix $T_n(b)$ whose complexity is independent of n. If $r \le 0$ or $s \le 0$, then $T_n(b)$ is triangular and hence $D_n(b)$ is the nth power of the entry on the main diagonal. Thus, assume $r \ge 1$, $s \ge 1$, $b_{-r} \ne 0$, and $b_s \ne 0$. We can write

$$b(t) = b_s t^{-r} \prod_{j=1}^{r+s} (t - z_j) \quad (t \in \mathbf{T}), \tag{2.13}$$

where $z_1, \dots, z_{r+s}$ are the zeros of the polynomial

$$z^r b(z) = b_{-r} + b_{-r+1} z + \dots + b_s z^{r+s}. \tag{2.14}$$

Theorem 2.8 (Widom). *If the zeros $z_1, \dots, z_{r+s}$ are pairwise distinct then, for every $n \ge 1$,*

$$D_n(b) = \sum_M C_M w_M^n, \tag{2.15}$$

where the sum is over all $\binom{r+s}{s}$ *subsets* $M \subset \{1, 2, \ldots, r+s\}$ *of cardinality* $|M| = s$ *and, with* $\overline{M} := \{1, 2, \ldots, r+s\} \setminus M$,

$$w_M := (-1)^s b_s \prod_{j \in \overline{M}} z_j, \quad C_M := \prod_{j \in M} z_j^r \prod_{\substack{j \in M \\ k \in \overline{M}}} (z_j - z_k)^{-1}.$$

Proof. From (2.13) we see that

$$D_n(b) = b_s^n D_n(z^{-r} a(z)) \tag{2.16}$$

where

$$a(z) = (z - z_1) \cdots (z - z_{r+s}) = a_0 + a_1 z + \cdots + z^{r+s}.$$

Put

$$c(z) = 1/a(z) = c_0 + c_1 z + c_2 z^2 + \cdots .$$

The Baxter-Schmidt formula (2.12) gives

$$D_n(z^{-r} a(z)) = (-1)^{rn} a_0^r c_0^{-n} D_r(z^{-n} c(z)) = (-1)^{rn} a_0^{r+n} D_r(z^{-n} c(z)). \tag{2.17}$$

We decompose $c(z)$ into partial fractions:

$$c(z) = \frac{1}{a(z)} = \frac{1}{(z - z_1) \cdots (z - z_{r+s})} = \sum_i \frac{B_i}{z_i - z}$$

$$= \sum_i \frac{B_i}{z_i} \frac{1}{1 - z/z_i} = \sum_i \frac{B_i}{z_i} \sum_\ell \frac{z^\ell}{z_i^\ell} = \sum_\ell \left(\sum_i \frac{B_i}{z_i^{\ell+1}} \right) z^\ell,$$

which holds whenever $|z| < |z_i|$ for all i (notice that $b_{-r} \neq 0$ so that $|z_i| \neq 0$ for all indices i). Using the Cauchy-Binet formula and the formula for Vandermonde determinants, we obtain

$$D_n(z^{-n} c(z)) = \begin{vmatrix} c_n & \cdots & c_{n-r+1} \\ \vdots & \ddots & \vdots \\ c_{n+r-1} & \cdots & c_n \end{vmatrix}$$

$$= \begin{vmatrix} \sum_i B_i / z_i^{n+1} & \cdots & \sum_i B_i / z_i^{n-r+2} \\ \vdots & \ddots & \vdots \\ \sum_i B_i / z_i^{n+r} & \cdots & \sum_i B_i / z_i^{n+1} \end{vmatrix}$$

$$= \begin{vmatrix} B_1 & \cdots & B_{r+s} \\ \vdots & & \vdots \\ B_1 / z_1^{r-1} & \cdots & B_{r+s} / z_{r+s}^{r-1} \end{vmatrix} \begin{vmatrix} 1/z_1^{n+1} & \cdots & 1/z_1^{n-r+2} \\ \vdots & & \vdots \\ 1/z_{r+s}^{n+1} & \cdots & 1/z_{r+s}^{n-r+2} \end{vmatrix}$$

$$= \sum_{1 \le i_1 \le \cdots \le i_r \le r+s} \begin{vmatrix} B_{i_1} & \cdots & B_{i_r} \\ \vdots & & \vdots \\ B_{i_1} / z_{i_1}^{r-1} & \cdots & B_{i_r} / z_{i_r}^{r-1} \end{vmatrix} \begin{vmatrix} 1/z_{i_1}^{n+1} & \cdots & 1/z_{i_r}^{n-r+2} \\ \vdots & & \vdots \\ 1/z_{i_r}^{n+1} & \cdots & 1/z_{i_r}^{n-r+2} \end{vmatrix}$$

$$
\begin{aligned}
&= \sum_{i \in \overline{M}} \frac{B_{i_1} \cdots B_{i_r}}{(z_{i_1} \cdots z_{i_r})^{n+1}} \begin{vmatrix} 1 & \dots & 1 \\ \vdots & & \vdots \\ 1/z_{i_1}^{r-1} & \dots & 1/z_{i_r}^{r-1} \end{vmatrix} \begin{vmatrix} 1 & \dots & z_{i_1}^{r-1} \\ \vdots & & \vdots \\ 1 & \dots & z_{i_r}^{r-1} \end{vmatrix} \\
&= \sum_{i \in \overline{M}} \frac{B_{i_1} \cdots B_{i_r}}{(z_{i_1} \cdots z_{i_r})^{n+1}} \prod_{\beta > \alpha} \left(\frac{1}{z_{i_\beta}} - \frac{1}{z_{i_\alpha}} \right) \prod_{\beta > \alpha} (z_{i_\beta} - z_{i_\alpha}) \\
&= \sum_{i \in \overline{M}} \frac{B_{i_1} \cdots B_{i_r}}{(z_{i_1} \cdots z_{i_r})^{n+1}} \prod_{\beta \neq \alpha} (z_{i_\beta} - z_{i_\alpha}) \prod_{\beta} \frac{1}{z_{i_\beta}^{r-1}} \\
&= \sum_{\overline{M}} \left(\prod_{i \in \overline{M}} B_i \right) \left(\prod_{i \in \overline{M}} z_i^{-n-1} \right) \left(\prod_{\substack{i,k \in \overline{M} \\ k \neq i}} (z_i - z_k) \right) \prod_{i \in \overline{M}} z_i^{-r-1} \\
&= \sum_{\overline{M}} \left(\prod_{i \in \overline{M}} B_i \right) \left(\prod_{i \in \overline{M}} z_i^{-n-r} \right) \left(\prod_{\substack{i,k \in \overline{M} \\ k \neq i}} (z_i - z_k) \right). \qquad (2.18)
\end{aligned}
$$

The coefficients B_i of the partial fraction decomposition are

$$B_i = -\prod_{\ell \neq i} (z_i - z_\ell).$$

Hence

$$
\begin{aligned}
&\left(\prod_{i \in \overline{M}} B_i \right) \left(\prod_{\substack{i,k \in \overline{M} \\ k \neq i}} (z_i - z_k) \right) \\
&\quad = (-1)^r \prod_{\substack{i \in \overline{M} \\ j \in M}} (z_i - z_j)^{-1} \prod_{\substack{i,\ell \in \overline{M} \\ \ell \neq i}} (z_i - z_\ell)^{-1} \prod_{\substack{i,k \in \overline{M} \\ k \neq i}} (z_i - z_k) \\
&\quad = (-1)^{r+rs} \prod_{\substack{i \in \overline{M} \\ j \in M}} (z_i - z_j)^{-1}. \qquad (2.19)
\end{aligned}
$$

Combining (2.16), (2.17), (2.18), and (2.19), we arrive at the equality

$$D_n(b) = b_s^n (-1)^{rn} a_0^{r+n} \sum_{\overline{M}} \left(\prod_{i \in \overline{M}} z_i^{-n-r} \right) (-1)^{r-rs} \left(\prod_{\substack{i \in \overline{M} \\ j \in M}} (z_i - z_j)^{-1} \right).$$

Since $a_0 = (-1)^{r+s} z_1 \cdots z_{r+s}$, it follows that

$$D_n(b) = b_s^n (-1)^{rn+(r+s)(r+n)+r+rs} \sum_{\overline{M}} \left(\prod_{j\in M} z_i^{n+r} \right) \left(\prod_{\substack{i\in\overline{M} \\ j\in M}} (z_i - z_j)^{-1} \right).$$

As $rn + (r+s)(r+n) + r + rs \equiv 2rn + 2rs + r(r+1) + sn \equiv sn \pmod 2$ and

$$\sum_{\overline{M}} \left(\prod_{j\in M} z_i^{n+r} \right) \left(\prod_{\substack{i\in\overline{M} \\ j\in M}} (z_i - z_j)^{-1} \right) = \sum_{M} \left(\prod_{j\in M} z_i^{n+r} \right) \left(\prod_{\substack{k\in\overline{M} \\ j\in M}} (z_i - z_k)^{-1} \right),$$

we finally obtain (2.15). □

Example 2.9. Let $T(a)$ be the tridiagonal matrix considered in Section 2.2 and write $a(t) = t^{-1}(t - z_1)(t - z_2)$. Suppose $z_1 \neq z_2$. There are exactly two sets $M \subset \{1, 2\}$ of cardinality $|M| = 1$, namely $M = \{1\}$ and $M = \{2\}$. Thus, by Theorem 2.8,

$$w_{\{1\}} = (-1)a_1 z_1, \quad C_{\{1\}} = z_1(z_1 - z_2)^{-1},$$
$$w_{\{2\}} = (-1)a_1 z_2, \quad C_{\{2\}} = z_2(z_2 - z_1)^{-1},$$
$$D_n(a) = \frac{z_1}{z_1 - z_2}(-1)^n a_1^n z_1^n + \frac{z_2}{z_2 - z_1}(-1)^n a_1^n z_2^n = (-1)^n a_1^n \frac{z_1^{n+1} - z_2^{n+1}}{z_2 - z_1}.$$

We leave it as an exercise to verify that this is in accordance with the formula provided by Theorem 2.6. □

2.5 Trench's Formula

Let b be as in Section 2.4. We now establish a formula for $D_n(b)$ that is also applicable to the case where the polynomial (2.14) has multiple zeros.

We denote by $g_n(z)$ the row

$$g_n(z) = (\,1 \;\; z \;\; \ldots \;\; z^{r-1} \;\; z^{r+n} \;\; z^{r+n+1} \;\; \ldots \;\; z^{r+n+s-1}\,).$$

Let $\xi_1, \ldots, \xi_m$ be the distinct roots of the polynomial (2.14) and let $\alpha_1, \ldots, \alpha_m$ be their multiplicities. We define G_n as the determinant of the $(r+s) \times (r+s)$ matrix A_{r+s} whose first α_1 rows are $g_n(\xi_1), g_n'(\xi_1), \ldots, g_n^{(\alpha_1 - 1)}(\xi_1)$, whose next α_2 rows are $g_n(\xi_2), g_n'(\xi_2), \ldots, g_n^{(\alpha_2 - 1)}(\xi_2)$, and so on.

Theorem 2.10 (Trench). *We have $G_0 \neq 0$ and*

$$D_n(b) = (-1)^{ns} b_s^n \frac{G_n}{G_0} \quad \text{for every } n \geq 1. \tag{2.20}$$

Confluent Vandermonde determinants. Before turning to the proof of this theorem, we cite the formula for the so-called confluent Vandermonde determinants. Let $h_\beta(z)$ be the row $(z^{\beta} \quad z^{\beta+1} \quad \dots \quad z^{\beta+\alpha-1})$. Given m distinct number $\xi_1, \dots, \xi_m$ and natural numbers $\alpha_1, \dots, \alpha_m$ such that $\alpha_1 + \cdots + \alpha_m = \alpha$, we let

$$V(\xi_1(\beta_1, \alpha_1), \dots, \xi_m(\beta_m, \alpha_m))$$

denote the determinant of the $\alpha \times \alpha$ matrix whose rows are

$$h_{\beta_1}(\xi_1), h'_{\beta_1}(\xi_1), \dots, h_{\beta_1}^{(\alpha_1-1)}(\xi_1), \dots, h_{\beta_m}(\xi_m), h'_{\beta_m}(\xi_m), \dots, h_{\beta_m}^{(\alpha_m-1)}(\xi_m).$$

In the case $\alpha_1 = \cdots = \alpha_m = 1$, this is a pure Vandermonde determinant and hence

$$V(\xi_1(\beta_1, 1), \dots, \xi_\alpha(\beta_\alpha, 1)) = \prod_i \xi_i^{\beta_i} \prod_{j>i} (\xi_j - \xi_i).$$

Appropriate limit passages and l'Hospital's rule therefore yield

$$V(\xi_1(\beta_1, \alpha_1), \dots, \xi_m(\beta_m, \alpha_m)) = \prod_i \xi_i^{\alpha_i \beta_i} \prod_i G(\alpha_i + 1) \prod_{j>i} (\xi_j - \xi_i)^{\alpha_j \alpha_i}, \tag{2.21}$$

where $G(\alpha + 1) := (\alpha - 1)!(\alpha - 2)! \dots 1!0!$.

Proof of Theorem 2.10. By formula (2.21),

$$G_0 = V(\xi_1(0, \alpha_1), \dots, \xi_m(0, \alpha_m)) = \prod_i G(\alpha_i + 1) \prod_{j>i} (\xi_j - \xi_i)^{\alpha_j \alpha_i} \neq 0.$$

Suppose first that the zeros $z_1, \dots, z_{r+s}$ of the polynomial (2.14) are all simple. We show that in this case the right-hand side of (2.20) coincides with the right-hand side of (2.15). Let M range over all subsets $\{j_1, \dots, j_s\}$ of $\{1, 2, \dots, r+s\}$ whose cardinality is s and put $\{1, 2, \dots, r+s\} \setminus M =: \{k_1, \dots, k_r\}$. Using Laplace's expansion theorem, we obtain that G_n is

$$\sum_M (-1)^{1+\cdots+r+k_1+\cdots+k_r} \begin{vmatrix} 1 & z_{k_1} & \dots & z_{k_1}^{r-1} \\ \vdots & \vdots & & \vdots \\ 1 & z_{k_r} & \dots & z_{k_r}^{r-1} \end{vmatrix} \begin{vmatrix} z_{j_1}^{r+n} & \dots & z_{j_1}^{r+n+s-1} \\ \vdots & & \vdots \\ z_{j_s}^{r+n} & \dots & z_{j_s}^{r+n+s-1} \end{vmatrix}$$

$$= \sum_M (-1)^{1+\cdots+r+k_1+\cdots+k_r} (z_{j_1} \dots z_{j_s})^{r+n} \prod_{\alpha>\beta} (z_{k_\alpha} - z_{k_\beta}) \prod_{\gamma>\delta} (z_{j_\gamma} - z_{j_\delta}). \tag{2.22}$$

Permuting the rows $1, 2, \dots, r+s$ of G_0 to $k_1, \dots, k_r, j_1, \dots, j_s$ shows that G_0 equals

$$(-1)^{(k_1-1)+(k_2-2)+\cdots+(k_r-r)} \prod_{\alpha>\beta} (z_{k_\alpha} - z_{k_\beta}) \prod_{\gamma>\delta} (z_{j_\gamma} - z_{j_\delta}) \prod_{\gamma,\alpha} (z_{j_\gamma} - z_{k_\alpha}). \tag{2.23}$$

From (2.22) and (2.23) we get

$$\begin{aligned}\frac{G_n}{G_0} &= \sum_M (z_{j_1} \dots z_{j_s})^{r+n} \prod_{\gamma,\alpha} (z_{j_\gamma} - z_{k_\alpha})^{-1} \\ &= \sum_M \left(\prod_{j\in M} z_j \right)^n \prod_{j\in M} z_j^r \prod_{\substack{j\in M \\ k\in \overline{M}}} (z_j - z_k)^{-1} \\ &= \sum_M ((-1)^s b_s^{-1} w_M)^n C_M = (-1)^{sn} b_s^{-n} \sum_M C_M w_M^n,\end{aligned}$$

which completes the proof in the case of simple zeros. If there are multiple zeros among $z_1, \dots, z_{r+s}$, then the formula follows from what was just proved by the appropriate limit passages. □

2.6 Szegö's Strong Limit Theorem

Let $b(t) = \sum_{j=-r}^{s} b_j t^j$ be a Laurent polynomial and suppose

$$b(t) \neq 0 \text{ for } t \in \mathbf{T} \text{ and } \operatorname{wind} b = 0. \tag{2.24}$$

In that case there is a function $\log b \in C(\mathbf{T})$ such that $b = \exp(\log b)$. Clearly, $\log b$ is determined uniquely up to an additive constant in $2\pi i\mathbf{Z}$. Let $(\log b)_k$ be the kth Fourier coefficient of $\log b$ and put

$$G(b) = \exp(\log b)_0, \tag{2.25}$$

$$E(b) = \exp \sum_{k=1}^{\infty} k (\log b)_k (\log b)_{-k}. \tag{2.26}$$

Since $b \in C^\infty(\mathbf{T})$, the function $\log b$ also belongs to $C^\infty(\mathbf{T})$. This implies that $(\log b)_k = O(1/|k|^m)$ for every $m \geq 1$ and shows that the series in (2.26) converges absolutely. The constants $G(b)$ and $E(b)$ are obviously independent of the particular choice of $\log b$.

We know from Section 1.4 that if (2.24) holds, then

$$b(t) = b_s \prod_{i=1}^{k} \left(1 - \frac{\delta_i}{t}\right)^{\varrho_i} \prod_{j=1}^{\ell} (t - \mu_j)^{\sigma_j},$$

where $\delta_1, \dots, \delta_k, \mu_1, \dots, \mu_\ell$ are distinct, $|\delta_i| < 1$, $|\mu_j| > 1$, $\varrho_1 + \dots + \varrho_k = r$, and $\sigma_1 + \dots + \sigma_\ell = s$. On writing

$$\log b(t) = \log b_s + \sum_{i=1}^{k} \varrho_i \log\left(1 - \frac{\delta_i}{t}\right) + \sum_{j=1}^{\ell} \sigma_j \log(-\mu_j) + \sum_{j=1}^{\ell} \sigma_j \log\left(1 - \frac{t}{\mu_j}\right)$$

and using the formula $\log(1-z) = -z - z^2/2 - z^3/3 - \cdots$ $(|z| < 1)$, we get

$$(\log b)_0 = \log b_s + \sum_{j=1}^{\ell} \sigma_j \log(-\mu_j),$$

$$(\log b)_n = -\sum_{j=1}^{\ell} \frac{\sigma_j}{n\mu_j^n} \quad (n \geq 1),$$

$$(\log b)_{-n} = -\sum_{i=1}^{k} \frac{\varrho_i \delta_i^n}{n} \quad (n \geq 1).$$

Thus, the constants (2.25) and (2.26) are

$$G(b) = b_s(-1)^s \prod_{j=1}^{\ell} \mu_j^{\sigma_j}, \tag{2.27}$$

$$E(b) = \prod_{i=1}^{k}\prod_{j=1}^{\ell} \left(1 - \frac{\delta_i}{\mu_j}\right)^{-\varrho_i \sigma_j}. \tag{2.28}$$

Theorem 2.11 (Szegö's strong limit theorem). *If b is a Laurent polynomial satisfying* (2.24), *then*

$$D_n(b) = G(b)^n E(b)(1 + O(q^n)) \quad as \quad n \to \infty \tag{2.29}$$

with some constant $q \in (0, 1)$.

Proof. Formula (2.20) and Laplace expansion of the determinant G_n show that

$$D_n(b) = \frac{(-1)^{ns} b_s^b}{G_0} V(\delta_1(0, \varrho_1), \ldots, \delta_k(0, \varrho_k))$$
$$\times V(\mu_1(r+s, \sigma_1), \ldots, \mu_\ell(r+n, \sigma_\ell))\,(1 + O(q^n))$$

with

$$\max |\delta_i| \,/\, \min |\mu_j| < q < 1. \tag{2.30}$$

By formula (2.21),

$$V(\delta_1(0, \varrho_1), \ldots, \delta_k(0, \varrho_k))$$
$$= \prod_i G(\varrho_i + 1) \prod_{i_2 > i_1} (\delta_{i_2} - \delta_{i_1})^{\varrho_{i_2}\varrho_{i_1}},$$

$$V(\mu_1(r+s, \sigma_1), \ldots, \mu_\ell(r+n, \sigma_\ell))$$
$$= \prod_j \mu_j^{\sigma_i(r+n)} \prod_j G(\sigma_j + 1) \prod_{j_2 > j_1} (\mu_{j_2} - \mu_{j_1})^{\sigma_{j_2}\sigma_{j_1}},$$

$$\begin{aligned} G_0 &= V(\delta_1(0,\varrho_1),\ldots,\delta_k(0,\varrho_k),\mu_1(0,\sigma_1),\ldots,\mu_\ell(0,\sigma_\ell)) \\ &= \prod_i G(\varrho_i - 1)\prod_j G(\sigma_j + 1) \\ &\times \prod_{i_2>i_1}(\delta_{i_2}-\delta_{i_1})^{\varrho_{i_2}\varrho_{i_1}} \prod_{j_2>j_1}(\mu_{j_2}-\mu_{j_1})^{\sigma_{j_2}\sigma_{j_1}} \prod_j\prod_i(\mu_j-\delta_i)^{\sigma_j\varrho_i}, \end{aligned}$$

whence

$$\begin{aligned} &\frac{(-1)^{ns}b_s^b}{G_0} V(\delta_1(0,\varrho_1),\ldots,\delta_k(0,\varrho_k))\, V(\mu_1(r+s,\sigma_1),\ldots,\mu_\ell(r+n,\sigma_\ell)) \\ &= (-1)^{ns}b_s^n\left(\prod_j \mu_j^{\sigma_j}\right)^n \prod_j \mu_j^{\sigma_j r}\prod_j\prod_i(\mu_j-\delta_i)^{-\sigma_j\varrho_i}, \end{aligned}$$

and since

$$\begin{aligned} &\prod_j \mu_j^{\sigma_j r}\prod_j\prod_i(\mu_j-\delta_i)^{-\sigma_j\varrho_i} \\ &= \prod_j \mu_j^{\sigma_j r}\prod_j\left(u_j^{-\sigma_j r}\prod_i\left(1-\frac{\delta_i}{\mu_j}\right)^{-\sigma_j\varrho_i}\right) = \prod_j\prod_i\left(1-\frac{\delta_i}{\mu_j}\right)^{-\sigma_j\varrho_i}, \end{aligned}$$

formulas (2.27) and (2.28) imply the assertion. □

2.7 The Szegö-Widom Theorem

The road from Baxter-Schmidt through Widom and Trench led to a completely elementary but rather computational proof of Szegö's strong limit theorem. With a little bit of operator theory, this theorem can be proved in a straightforward and very nifty way.

Trace class. Let $(c_{jk})_{j,k=0}^\infty$ be an infinite matrix that defines a compact operator K on ℓ^2. The operator K is said to be a *Hilbert-Schmidt operator* if

$$\sum_{j,k=0}^\infty |c_{jk}|^2 < \infty,$$

and K is called a *trace class operator* if it is the product of two Hilbert-Schmidt operators. The Hilbert-Schmidt and trace class operators form two-sided ideals in the algebra of all bounded operators.

Operator determinants. Let K be a trace class operator and let $\{\lambda_j(K)\}_{j=1}^N$ (N finite or $N=\infty$) be the sequence of its eigenvalues, each eigenvalue repeated according to its algebraic multiplicity. Then

$$\sum_{j=1}^N |\lambda_j(K)| < \infty,$$

and the *determinant* $\det(I+K)$ is defined by

$$\det(I+K) = \prod_{j=1}^{N}(1+\lambda_j(K)).$$

The operator $I+K$ is invertible if and only if $\det(I+K) \neq 0$. If K and L are trace class operators, then

$$\det(I+K)(I+L) = \det(I+K)\det(I+L). \tag{2.31}$$

If C is an invertible operator, then

$$\det(I+CKC^{-1}) = \det C(I+K)C^{-1} = \det(I+K). \tag{2.32}$$

For $n \geq 1$, let $P_n : \ell^2 \to \ell^2$ be the projection defined by

$$P_n : \{x_0, x_1, x_2, \dots\} \mapsto \{x_0, x_1, \dots, x_{n-1}, 0, \dots\}.$$

We identify $P_n(I+K)P_n$ with the $n \times n$ matrix $(\delta_{jk}+c_{jk})_{j,k=1}^{n-1}$, where δ_{jk} is the Kronecker delta. Then, if K is of trace class,

$$\lim_{n\to\infty} \det P_n(I+K)P_n = \det(I+K). \tag{2.33}$$

Prologue. Recall that $H(a)$ is the Hankel operator given by the infinite matrix $(a_{j+k+1})_{j,k=0}^{\infty}$. Obviously,

$$H(a) \text{ is Hilbert-Schmidt} \iff \sum_{n=1}^{\infty} n|a_n|^2 < \infty. \tag{2.34}$$

Now let b be a Laurent polynomial satisfying (2.24). We then can write $b = b_- b_+$ with

$$b_-(t) = \prod_{i=1}^{r}\left(1-\frac{\delta_i}{t}\right), \quad b_+(t) = b_s\prod_{j=1}^{s}(t-\mu_j), \tag{2.35}$$

where $|\delta_i| < 1$ and $|\mu_j| > 1$. The matrix $T(b_-^{-1})$ is upper triangular with 1 on the main diagonal, while $T(b_+^{-1})$ is lower triangular with $(b_s\prod(-\mu_j))^{-1}$ on the main diagonal. This implies that

$$P_nT(b_-^{-1})P_n = T(b_-^{-1})P_n, \quad P_nT(b_+^{-1})P_n = P_nT(b_+^{-1}) \tag{2.36}$$

and

$$\det T_n(b_-^{-1}) = 1, \quad \det T_n(b_+^{-1}) = \left(b_s\prod_{j=1}^{s}(-\mu_j)\right)^{-n} = G(b)^{-n} \tag{2.37}$$

(recall (2.27)). By Proposition 1.3,

$$T(b)T(b^{-1}) = I - H(b)H(\widetilde{b^{-1}}) \tag{2.38}$$

and

$$\begin{aligned} T(b_+^{-1})T(b)T(b_-^{-1}) &= T(b_-)T(b_-^{-1}) - H(b_+^{-1})H(\widetilde{b})T(b_-^{-1}) \\ &= I - H(b_+^{-1})H(\widetilde{b})T(b_-^{-1}). \end{aligned}$$

We therefore deduce from (2.34) that both $T(b)T(b^{-1})$ and $T(b_+^{-1})T(b)T(b_-^{-1})$ are of the form I + trace class operator.

Theorem 2.12 (Szegö-Widom limit theorem). *If b is a Laurent polynomial without zeros on the unit circle and with winding number zero, then*

$$\lim_{n\to\infty} \frac{D_n(b)}{G(b)^n} = \det T(b)T(b^{-1}).$$

Proof. From (2.36) and (2.37) we see that

$$\begin{aligned} \det P_n T(b_+^{-1})T(b)T(b_-^{-1})P_n &= \det P_n T(b_+^{-1})P_n T(b)P_n T(b_-^{-1})P_n \\ &= \det T_n(b_+^{-1})T_n(b)T_n(b_-^{-1}) = D_n(b)/G(b)^n. \end{aligned} \tag{2.39}$$

On the other hand, since $T(b_+^{-1})T(b)T(b_-^{-1}) - I$ is of trace class, we infer from (2.33) that

$$\lim_{n\to\infty} \det P_n T(b_+^{-1})T(b)T(b_-^{-1})P_n = \det T(b_+^{-1})T(b)T(b_-^{-1}),$$

and (2.32) shows that

$$\det T(b_+^{-1})T(b)T(b_-^{-1}) = \det T(b)T(b_-^{-1})T(b_+^{-1}) = \det T(b)T(b^{-1}). \qquad \square$$

To compute the operator determinant $\det T(b)T(b^{-1})$ we need two auxiliary results. We abbreviate $\det T(b)T(b^{-1})$ to $E(b)$.

Lemma 2.13. *Let $f_\pm, g_\pm, h_\pm$ be Laurent polynomials in $W_\pm$ without zeros on* **T** *and with winding number zero. Then*

$$E(f_-g_+h_+) = E(f_-g_+)E(f_-h_+), \quad E(g_-h_-f_+) = E(g_-f_+)E(h_-f_+). \tag{2.40}$$

Proof. Using (2.31) and (2.32), we obtain

$$\begin{aligned} E(f_-g_+h_+) &= \det T(f_-g_+h_+)T(f_-^{-1}h_+^{-1}g_+^{-1}) \\ &= \det T(f_-)T(g_+)T(h_+)T(f_-^{-1})T(h_+^{-1})T(g_+^{-1}) \\ &= \det T(g_+^{-1})T(f_-)T(g_+)T(h_+)T(f_-^{-1})T(h_+^{-1}) \\ &= \det T(g_+^{-1})T(f_-)T(g_+)T(f_-^{-1}) \det T(f_-)T(h_+)T(f_-^{-1})T(h_+^{-1}) \\ &= \det T(f_-)T(g_+)T(f_-^{-1})T(g_+^{-1}) \det T(f_-)T(h_+)T(f_-^{-1})T(h_+^{-1}) \\ &= E(f_-g_+)E(f_-h_+). \end{aligned}$$

The proof of the second equality of (2.40) is analogous. $\square$

Lemma 2.13 implies that

$$E(b) = \prod_{i=1}^{r} \prod_{j=1}^{s} E\left[\left(1 - \frac{\delta_i}{t}\right)(t - \mu_j)\right], \tag{2.41}$$

and hence we are left with computing the factors on the right-hand side of (2.41).

Lemma 2.14. *If* $|\delta| < 1$ *and* $|\mu| > 1$, *then*

$$E\left[\left(1 - \frac{\delta}{t}\right)(t - \mu)\right] = \left(1 - \frac{\delta}{\mu}\right)^{-1}. \tag{2.42}$$

Proof. Put $c(t) = (1 - \delta/t)/(t - \mu)$. By virtue of Proposition 1.3,

$$E(c) = \det T(c)T(c^{-1}) = \det(I - H(c)H(\widetilde{c^{-1}})).$$

Clearly, $I - H(c)H(\widetilde{c^{-1}})$ equals

$$I - \begin{pmatrix} 1 & 0 & 0 & \dots \\ 0 & 0 & \dots & \\ 0 & \dots & & \\ \dots & & & \end{pmatrix} \begin{pmatrix} (c^{-1})_{-1} & (c^{-1})_{-2} & (c^{-1})_{-3} & \dots \\ (c^{-1})_{-2} & (c^{-1})_{-3} & \dots & \\ (c^{-1})_{-3} & \dots & & \\ \dots & & & \end{pmatrix}$$

$$= \begin{pmatrix} 1 - (c^{-1})_{-1} & * & * & \dots \\ 0 & 1 & * & \dots \\ 0 & 0 & 1 & \dots \\ \dots & \dots & \dots & \dots \end{pmatrix},$$

which shows that $E(c) = 1 - (c^{-1})_{-1}$. The -1st Fourier coefficient of c^{-1} is

$$(c^{-1})_{-1} = \left[\left(1 + \frac{\delta}{t} + \frac{\delta^2}{t^2} + \cdots\right)\left(-\frac{1}{\mu}\right)\left(1 + \frac{t}{\mu} + \frac{t^2}{\mu^2} + \cdots\right)\right]_0$$

$$= -\frac{1}{\mu}\left(\delta + \frac{\delta^2}{\mu} + \frac{\delta^3}{\mu^2} + \cdots\right) = -\frac{\delta}{\mu}\frac{1}{1 - \delta/\mu},$$

whence $1 - (c^{-1})_{-1} = 1/(1 - \delta/\mu$. □

Combining (2.41) and (2.42) we arrive at the formula

$$\det T(b)T(b^{-1}) = \prod_{i=1}^{r} \prod_{j=1}^{s} \left(1 - \frac{\delta_i}{\mu_j}\right)^{-1},$$

which is in accordance with (2.28).

2.8 Geronimo, Case, Borodin, Okounkov

Jacobi's theorem says that if A is an invertible $m \times m$ matrix, then the determinant of the upper-left $n \times n$ block of A^{-1} is equal to the determinant of the lower-right $(m-n) \times (m-n)$ block of A divided by $\det A$. Let $Q_n : \ell^2 \to \ell^2$ be the projection acting by the rule

$$Q_n : \{x_0, x_1, x_2, \dots\} \mapsto \{0, \dots, 0, x_n, x_{n+1}, \dots\}.$$

Thus, $Q_n = I - P_n$. If $A = I + K$ with a trace class operator K, then

$$\det P_n A^{-1} P_n = \frac{\det Q_n A Q_n}{\det A}. \tag{2.43}$$

Indeed, with A replaced by $P_m A P_m$, this is Jacobi's theorem, and for general A the formula follows from (2.33) after the limit passage $m \to \infty$.

Now let $b = b_- b_+$, where b_- and b_+ are given by (2.35). Put

$$u = b_- b_+^{-1}, \quad v = b_-^{-1} b_+.$$

Since $uv = 1$, we have $T(u)T(v) = I - H(u)H(\widetilde{v})$. Notice that $H(u)H(\widetilde{v})$ is in the trace class.

The remarkable formula contained in the following theorem was established by Geronimo and Case [127] in 1979 (for positive symbols b) and rediscovered by Borodin and Okounkov [31] in 2000.

Theorem 2.15. *Let b be a Laurent polynomial without zeros on the unit circle and with winding number zero. Then for all $n \geq 1$,*

$$\frac{D_n(b)}{G(b)^n} = \frac{\det Q_n T(u)T(v) Q_n}{\det T(u)T(v)} = \frac{\det(I - Q_n H(u)H(\widetilde{v}) Q_n)}{\det(I - H(u)H(\widetilde{v}))}. \tag{2.44}$$

Proof. By (2.39) and (2.43),

$$\frac{D_n(b)}{G(b)^n} = \det P_n T(b_+^{-1}) T(b) T(b_-^{-1}) P_n = \frac{\det Q_n T(b_-) T^{-1}(b) T(b_+) Q_n}{\det T(b_-) T^{-1}(b) T(b_+)},$$

and as

$$T(b_-)T(b)T(b_+) = T(b_-)T(b_-^{-1})T(b_-^{-1})T(b_+) = T(u)T(v) = I - H(u)H(\widetilde{v}),$$

we arrive at the assertion. □

Since $\det(I - Q_n H(u)H(\widetilde{v}) Q_n) \to 1$ as $n \to \infty$, formula (2.44) implies at once the Szegö-Widom limit theorem:

$$\begin{aligned}
\lim_{n\to\infty} \frac{D_n(b)}{G(b)^n} &= \frac{1}{\det T(u)T(v)} = \det T^{-1}(v) T^{-1}(u) \\
&= \det T(b_+^{-1}) T(b_-) T(b_+) T(b_-^{-1}) \\
&= \det T(b_-) T(b_+) T(b_-^{-1}) T(b_+^{-1}) = \det T(b) T(b^{-1}).
\end{aligned}$$

Exercises

1. Prove that

$$\det\begin{pmatrix} d_1 & x & \dots & x \\ x & d_2 & \dots & x \\ \vdots & \vdots & \ddots & \vdots \\ x & x & \dots & d_n \end{pmatrix}$$

$$= x(d_1 - x)\cdots(d_n - x)\left(\frac{1}{x} + \frac{1}{d_1 - x} + \cdots + \frac{1}{d_n - x}\right).$$

2. Prove that

$$\det\begin{pmatrix} x & a_1 & a_2 & \dots & a_n \\ a_1 & x & a_2 & \dots & a_n \\ a_1 & a_2 & x & \dots & a_n \\ \vdots & \vdots & \vdots & \ddots & \vdots \\ a_1 & a_2 & a_3 & \dots & x \end{pmatrix}$$

$$= (x + a_1 + a_2 + \cdots + a_n)(x - a_1)(x - a_2)\cdots(x - a_n).$$

3. Let $b(t) = x + t + t^{-1}$. Prove that

$$D_n(b) = x^n - \binom{n-1}{1}x^{n-2} + \binom{n-2}{2}x^{n-4} - + \cdots .$$

4. Let $\omega_n = e^{2\pi i/n}$. Prove that

$$\det\begin{pmatrix} 1 & 1 & 1 & \dots & 1 \\ 1 & \omega_n & \omega_n^2 & \dots & \omega_n^{n-1} \\ 1 & \omega_n^2 & \omega_n^4 & \dots & \omega_n^{2(n-1)} \\ \vdots & \vdots & \vdots & & \vdots \\ 1 & \omega_n^{n-1} & \omega_n^{2(n-1)} & \dots & \omega_n^{(n-1)(n-1)} \end{pmatrix} = n^{n/2}\, i^{-(n-1)(n+2)/2}.$$

5. Prove that

$$\det\begin{pmatrix} 1 & \frac{1}{2} & \frac{1}{3} & \cdots & \frac{1}{n} \\ \frac{1}{2} & \frac{1}{3} & \frac{1}{4} & \cdots & \frac{1}{n+1} \\ \vdots & \vdots & \vdots & & \vdots \\ \frac{1}{n} & \frac{1}{n+1} & \frac{1}{n+2} & \cdots & \frac{1}{2n-1} \end{pmatrix} = \frac{[1!\,2!\,\cdots\,(n-1)!]^3}{n!\,(n+1)!\,\cdots\,(2n-1)!}$$

and that, with $\mathbf{G}(m) := (m-2)!\cdots 2!\,1!\,0!$, this is the same as

$$\frac{(\mathbf{G}(n+1))^4}{\mathbf{G}(2n+1)}.$$

6. Let $a_1 = 1, a_2 = 3, a_3 = 240, a_4 = 1512000$. Which of the three numbers

$$1512030752000, \quad 1536288768000, \quad 1541291254000$$

is a_5?

7. Let $b(t) = 8t^2 - 54t + 101 - 54t^{-1} + 8t^{-2}$. Prove that $D_n(b) > 2^{6n-1}$ for all sufficiently large n.

8. Let $b \in \mathcal{P}$ satisfy (2.24) and let $b = b_- b_+$ be a Wiener-Hopf factorization.

   (a) Show that

$$T(b)T(b^{-1}) = e^{T(\log b_-)} e^{T(\log b_+)} e^{-T(\log b_-)} e^{-T(\log b_-)}.$$

   (b) Show that

$$\operatorname{tr}\,(T(\log b_-)T(\log b_+) - T(\log b_+)T(\log b_-))$$
$$= \operatorname{tr} H(\log b)H((\log b)\tilde{}\,) = \sum_{k=1}^{\infty} k(\log b)_k(\log b)_{-k}.$$

9. Let $b \in \mathcal{P}$ and suppose $T(b)$ is invertible. Show that $\det P_n T^{-1}(b) P_n = 1/G(b)^n$ and that, therefore, the Szegö-Widom limit theorem can also be written as

$$\det P_n T(b) P_n \det P_n T^{-1}(b) P_n \to \det T(b)T(b^{-1}).$$

10. Let $b \in \mathcal{P}_r$ and suppose that b has no zeros on $\mathbf{T}$ and winding number zero. Prove that

$$D_n(b^{-1}) = G(b^{-1})^n \det T(b^{-1})T(b)$$

for all $n \geq r$.

11. Let $|\alpha_i| < 1$ and $|\beta_j| < 1$. Prove that

$$D_n\left(\frac{1}{\prod(1-\alpha_j t)\prod(1-\beta_j t^{-1})}\right) = \prod_{i,j} \frac{1}{1-\alpha_i\beta_j}$$

for all $n \geq 1$.

12. Let $a \in W$ and suppose that $a(t) > 0$ for $t \in \mathbf{T}$. Prove that if there are nonzero constants $\widetilde{G}(a)$ and $\widetilde{E}(a)$ such that

$$D_n(a) = \widetilde{G}(a)^n \widetilde{E}(a) \quad \text{for all } n \geq r,$$

then $a^{-1} \in \mathcal{P}_r$.

13. (a) Show that if $H(a)$ is a trace class operator, then $|a_1| + |a_2| + |a_3| + \cdots < \infty$.

    (b) Show that if $|a_1| + 2\,|a_2| + 3\,|a_3| + \cdots < \infty$, then $H(a)$ is of trace class.

    (c) Prove that if $a \in C^2(\mathbf{T})$, then $H(a)$ is a trace class operator.

14. Let $b(t) = t^{-70}(t-\alpha)^{40}(t-\beta)^{60}$ with distinct points α and β on $\mathbf{T}$. Prove that

$$D_n(b) = \frac{\mathbf{G}(31)\mathbf{G}(21)\mathbf{G}(11)}{\mathbf{G}(60)} \frac{\alpha^{400}\beta^{600}}{(\alpha-\beta)^{1000}} (\alpha^{10}\beta^{20})^n n^{1100} \left(1 + O\left(\frac{1}{n}\right)\right).$$

15. Let

$$b(t) = \left(1 - \frac{\tau_1}{t}\right)^{\delta_1} \left(1 - \frac{t}{\tau_1}\right)^{\gamma_1} \left(1 - \frac{\tau_2}{t}\right)^{\delta_2} \left(1 - \frac{t}{\tau_2}\right)^{\gamma_2},$$

where $\tau_1 \in \mathbf{T}$, $\tau_2 \in \mathbf{T}$, $\tau_1 \neq \tau_2$, and $\delta_1, \gamma_1, \delta_2, \gamma_2$ are nonnegative integers. Find all quadruples $(\delta_1, \gamma_1, \delta_2, \gamma_2)$ such that $\delta_1 + \gamma_1 = 9$, $\delta_2 + \gamma_2 = 3$, and $D_n(b)/n^{\delta_1\gamma_1+\delta_2\gamma_2}$ converges to a finite and nonzero limit as $n \to \infty$.

16. Let b be a nonzero Laurent polynomial. Prove that $D_n(b)$ equals

$$\frac{1}{(2\pi)^n n!} \int_0^{2\pi} \cdots \int_0^{2\pi} \prod_{1 \le j < k \le n} \left|e^{i\theta_j} - e^{i\theta_k}\right|^2 \prod_{j=1}^n b(e^{i\theta_j})\, d\theta_1 \dots d\theta_n.$$

17. Let $b \in \mathcal{P}$ and suppose that $b(t) \geq 0$ for $t \in \mathbf{T}$. Prove that there exists an analytic polynomial $c_+ \in \mathcal{P}^+$ such that $b(t) = |c_+(t)|^2$ for $t \in \mathbf{T}$.

18. Let b be a nonzero Laurent polynomial and suppose $b(t) \geq 0$ for $t \in \mathbf{T}$. We denote by $L^2(b)$ the Hilbert space of all f on $\mathbf{T}$ for which

$$\|f\|_b^2 := \frac{1}{2\pi} \int_0^{2\pi} |f(e^{i\theta})|^2 b(e^{i\theta}) d\theta < \infty.$$

Let

$$\varphi_n(t) = t^n + \varphi_{n-1}^{(n)} t^{n-1} + \dots + \varphi_0^{(n)} \quad (n = 0, 1, 2, \dots)$$

be the system of polynomials in $\mathcal{P}^+$ determined by the orthogonality relations

$$\frac{1}{2\pi} \int_0^{2\pi} \varphi_j(e^{i\theta}) \overline{\varphi_k(e^{i\theta})} b(e^{i\theta}) d\theta = 0 \text{ for } j \neq k.$$

(a) Show that

$$T_n(b) \begin{pmatrix} \varphi_0^{(n)} \\ \vdots \\ \varphi_{n-1}^{(n)} \end{pmatrix} = - \begin{pmatrix} b_{-n} \\ \vdots \\ b_{-1} \end{pmatrix}.$$

(b) Prove that

$$D_n(b) = \|\varphi_0\|_b^2 \, \|\varphi_1\|_b^2 \cdots \|\varphi_{n-1}\|_b^2.$$

(c) Show that

$$\min \left\{ \|g\|_b^2 : g(t) = t^n + g_{n-1} t^{n-1} + \dots + g_0 \right\} = \frac{D_{n+1}(b)}{D_n(b)}$$

and that the minimum is attained at $g = \varphi_n$.

(d) Deduce that

$$\|\varphi_n\|_b^2 \to G(b) := \exp\left(\frac{1}{2\pi}\int_0^{2\pi} \log b(e^{i\theta})d\theta\right).$$

(e) Let $\chi_n(e^{i\theta}) := e^{in\theta}$. Deduce that the set

$$\operatorname{clos}_{L^2(b)} \bigcup_{n=0}^{\infty} \operatorname{lin}\{\chi_0, \chi_1, \ldots, \chi_n\}$$

does not coincide with $L^2(b)$.

Notes

The material of Sections 2.1 and 2.2 is well known. From [161] we learned that Theorem 2.4 goes back to D. E. Rutherford [239], [240]. In [117], [118], the result is referred to as Bloch's theorem. For real-valued symbols, Theorem 2.4 is essentially also in [145] and [176], and paper [243] shows that an arbitrary tridiagonal Toeplitz matrix is similar to a Hermitian tridiagonal Toeplitz matrix (see formula (11.1)).

The Baxter-Schmidt formula (Theorem 2.7) was established in [19]. The proof given here is based on a proof by Tismenetsky [266] and a remark of a referee of [266]. Theorem 2.8 is a result of Widom [290] in the reformulation of [243]. Our proof is from [67]. Theorem 2.10 was proved by Trench [278] for $n \geq r + s + 1$ and subsequently by Berg [24] for all $n \geq 1$. For more on Sections 2.3 to 2.5, the reader may also consult paper [32].

In 1915, Szegö [262] proved what is now called his "first limit theorem": If $b \in L^1$, $b \geq 0$, and $\log b \in L^1$, then $D_n(b)/D_{n-1}(b) \to G(b)$ as $n \to \infty$. At the turn of the 1950s, he was told by S. Kakutani that it would be desirable to have a second-order asymptotic formula for $D_n(b)$, because this would prove a formula by Lars Onsager for the spontaneous magnetization of the two-dimensional Ising model (we learned of this story from [13]). As a consequence of this, Szegö published his "strong limit theorem" in 1952 [263]: If $b > 0$ and the derivative b' satisfies a Hölder condition, then $D_n(b)/G(b)^n \to E(b)$ as $n \to \infty$. Baxter [18], Hirschman [164], and Devinatz [102] then became to understand that the positivity of b may be replaced by the "index zero condition" (2.24), and they gave different proofs of the formula $D_n(b) = G(b)^n E(b)(1 + o(1))$ for symbols b from classes much larger than the class of Laurent polynomials. The development culminated with Widom's paper [294]. Basor [13] writes: "The proofs of the various Szegö theorems were for the most part difficult, indirect, and worst of all gave no 'natural' indication why the terms in the expansion, especially the $E(b)$, occurred. Fortunately, this state of affairs was considerably altered in 1976 by Widom [294], whose elegant application of ideas from operator theory extended Szegö's theorem to the block case and gave easy proofs of the results." For more details on Szegö's limit theorem we refer the reader to the books [70] and [71]. Comprehensive studies of trace class operators and infinite determinants are in [133], [217], [254].

The story with formula (2.44) started in 1979, but let us begin 20 years later. In June 1999, during an MSRI workshop on random matrices, Alexander Its and Percy Deift raised the question of whether there is a general formula that expresses the determinant of the Toeplitz matrix $T_n(a)$ as the Fredholm determinant of an operator $I - K$, where K acts on $\ell^2\{n, n+1, \dots\}$. Borodin and Okounkov [31] then showed that such a formula indeed exists. The form in which we cited their formula in Theorem 2.15 is due to Harold Widom. The original proof by Borodin and Okounkov is based on representation theory and combinatorics, in particular on results by Okounkov on infinite wedge and random partitions and a theorem by I. M. Gessel expressing a Toeplitz determinant as a sum over partitions of products of Schur functions. Two other proofs were subsequently given by Basor and Widom [16]. The first of these proofs uses an identity for $\det T_{n-1}(a) / \det T_n(a)$ containing just $H(b)H(\tilde{c})$, which was established by Widom in 1973, and the second is a further development of the argument employed by Basor and J. W. Helton in 1980 to prove the Szegö-Widom limit theorem. The two proofs by Basor and Widom are operator-theoretic and very lucid. A third operator-theoretic proof was found [37]. It follows from the identity

$$T_n^{-1}(a) = T_n(a_+^{-1})(I - P_n T(c) Q_n X Q_n T(b) P_n) T_n(a_-^{-1}),$$

where $X = (I - Q_n H(b) H(\tilde{c}) Q_n)^{-1}$. This identity, which lifts formula (2.44) from the determinant level to the matrix level, was obtained in 1980 by B. Silbermann and one of the authors [63]. The proof given here is from [38]. It is a modification of the second proof of [16] and is the probably shortest proof of identity (2.44).

In July 2003, the people involved in formula (2.44) and its proof since 1999 received an email from Percy Deift. This email was as follows. "Recently Jeff Geronimo showed me a 1979 paper of his with Ken Case in which they wrote down the Borodin-Okounkov formula in the context of proving strong Szegö. The reference is [127]. See, in particular, formula VII.28 on page 308. It's quite remarkable that the formula was already known in 1979. The proof of the formula by Geronimo-Case is inverse-scattering theoretic and is the analogue of Dyson's second-derivative log det formula for the Schrödinger case." There is nothing we can add, except for our congratulations to Jeff Geronimo and Ken Case on their brilliant feat, for our gratification of the eventual recognition of their great success, and for expressing our regret that their names are missing in [16], [31], [37], [38].

Simon's book [255] contains a proof of (2.44) under the most general (and natural) smoothness conditions on b: It is only required that $\sum_{k=-\infty}^{\infty} |k|\,|b_k|^2 < \infty$. This proof is based on the proof given here and on additional technical arguments due to Rowan Killip and Percy Deift.

Exercises 1 to 5 are from [113] and [208]. The observation of Exercise 10 was probably already known to Szegö (see [145]). Exercise 11 is due to Baxter [17] (see also [79] and [174]). Solutions to Exercises 10 and 12 are in [63] and [67]. The problem of characterizing the symbols a for which $H(a)$ is of trace class had been open for a long time before it was completely solved by Peller [202], [203] (see also his recent capital monograph [204]). Simple solutions to Exercise 13 can be found in [67]. For Exercise 15 see [64]. The formula of Exercise 16 goes back to Szegö [262]. The result of Exercise 17 is due to Fejér and F. Riesz; see, e.g., [145]. Exercise 18 summarizes classical results by Szegö and we refer the reader to [108], [145], [255] for proofs and more material.

The link between Exercises 8(a) and 8(b) is the formula

$$\det e^A e^B e^{-A} e^{-B} = e^{\operatorname{tr}(AB-BA)}, \tag{2.45}$$

which is true whenever A and B are bounded on ℓ^2 and $AB - BA$ is a trace class operator on ℓ^2. Formula (2.45) appeared first in [158] and [206]. It was used by Widom in [294] to show that

$$\det T(b)T(b^{-1}) = \exp \sum_{k=1}^{\infty} k(\log b)_k (\log b)_{-k}$$

and thus to recover Szegö's original strong limit theorem from Theorem 2.12. The proofs of (2.45) given in [158] and [206] are difficult. Recently Torsten Ehrhardt [106] found a remarkably simple proof of the stronger identity

$$\det e^A e^B e^{-A-B} = e^{\operatorname{tr} \frac{1}{2}(AB-BA)}. \tag{2.46}$$

This identity is true for arbitrary $A, B \in \mathcal{B}(\ell^2)$ for which $AB - BA$ is of trace class. Note that (2.46) implies that

$$\det e^{A+B} e^{-A} e^{-B} = e^{\operatorname{tr} \frac{1}{2}(AB-BA)}, \tag{2.47}$$

and that multiplication of (2.46) and (2.47) yields (2.45).

Further results: Fisher-Hartwig symbols. The formulas established by Widom and Trench can be employed to compute the determinants $D_n(b)$ for an arbitrary Laurent polynomial b. However, in order to understand the asymptotic behavior of $D_n(b)$, additional work must be done (see, for example, Exercise 14). In the following we cite some asymptotic results for Laurent polynomials in the so-called Fisher-Hartwig class.

Let first

$$b(t) = \left(1 - \frac{\tau}{t}\right)^{\delta} \left(1 - \frac{t}{\tau}\right)^{\gamma} c(t), \quad t \in \mathbf{T},$$

where $\tau \in \mathbf{T}$, δ and γ are nonnegative integers, and c is a Laurent polynomial without zeros on $\mathbf{T}$ and with $\operatorname{wind} c = 0$. In the case where c is identically 1 (and thus absent), an exact formula for $D_n(b)$ was established in [68]:

$$\begin{aligned} &D_n\left[\left(1 - \frac{\tau}{t}\right)^{\delta} \left(1 - \frac{t}{\tau}\right)^{\gamma}\right] \\ &\quad = \frac{\mathbf{G}(1+\delta)\mathbf{G}(1+\gamma)}{\mathbf{G}(1+\gamma+\delta)} \frac{\mathbf{G}(n+1)\mathbf{G}(n+1+\gamma+\delta)}{\mathbf{G}(n+1+\delta)\mathbf{G}(n+1+\gamma)}, \end{aligned} \tag{2.48}$$

where $\mathbf{G}(1) = 1$ and $\mathbf{G}(m) = (m-2)! \cdots 2!\,1!\,0!$ for $m \geq 2$. (Two elementary proofs of this identity can be found in [73].) The right-hand side of (2.48) is

$$\frac{\mathbf{G}(1+\delta)\mathbf{G}(1+\gamma)}{\mathbf{G}(1+\gamma+\delta)} n^{\delta\gamma}(1 + o(1)) \quad \text{as } n \to \infty.$$

Now suppose that c is present. We put

$$G(c) = \exp(\log c)_0, \quad E(c) = \exp \sum_{k=1}^{\infty} k(\log c)_k (\log c)_{-k},$$

$$c_+(t) = \exp \sum_{k=1}^{\infty} (\log c)_k t^k, \quad c_-(t) = \exp \sum_{k=1}^{\infty} (\log c)_{-k} t^{-k}.$$

Clearly, $c = G(c)c_-c_+$. In [69], it is shown that

$$D_n\left[\left(1-\frac{\tau}{t}\right)^{\delta}\left(1-\frac{t}{\tau}\right)^{\gamma}c(t)\right] \Big/ D_n\left[\left(1-\frac{\tau}{t}\right)^{\delta}\left(1-\frac{t}{\tau}\right)^{\gamma}\right]$$
$$= \frac{G(c)^n E(c)}{c_-(\tau)^{\gamma} c_+(\tau)^{\delta}}\left(1+O\left(\frac{1}{n}\right)\right).$$

Things are essentially more complicated for symbols with more than one zero on the unit circle. Let now

$$b(t) = \prod_{j=1}^{N}\left(1-\frac{\tau_j}{t}\right)^{\delta_j}\left(1-\frac{t}{\tau_j}\right)^{\gamma_j} c(t), \quad t \in \mathbf{T},$$

where $\tau_1, \ldots, \tau_N$ are distinct points on $\mathbf{T}$, δ_j and γ_j are nonnegative integers, and c is as above. The asymptotics of $D_n(b)$ in this case was obtained in [64]. The result is as follows. Without loss of generality assume that

$$\gamma := \gamma_1 + \cdots + \gamma_N \le \delta_1 + \cdots + \delta_N =: \delta$$

(otherwise pass to the adjoint matrix). Put $z_j = \gamma_j + \delta_j$,

$$\mathcal{M} = \{(m_1, \ldots, m_N) : m_j \in \mathbf{Z},\ 0 \le m_j \le z_j,\ m_1 + \cdots + m_n = \gamma\},$$

$$Q = \max\left\{\sum_{j=1}^{N}(m_j z_j - m_j^2) : (m_1, \ldots, m_N) \in \mathcal{M}\right\},$$

$$\mathcal{M}^* = \left\{(m_1, \ldots, m_N) \in \mathcal{M} : \sum_{j=1}^{N}(m_j z_j - m_j^2) = Q\right\}.$$

Then

$$D_n(b) = G(c)^n E(c)\, n^Q\, (1 + O(1/n))$$
$$\times \left[\sum_{(m_1,\ldots,m_n)\in\mathcal{M}^*} A_{m_1,\ldots,m_N}\left(\frac{\tau_1^{m_1}\cdots\tau_N^{m_N}}{\tau_1^{\gamma_1}\cdots\tau_N^{\gamma_N}}\right)^n + O\left(\frac{1}{n}\right)\right], \tag{2.49}$$

where $G(c)$ and $E(c)$ are as above and

$$A_{m_1,\ldots,m_N} = \prod_{i\ne j}\left(1-\frac{\tau_i}{\tau_j}\right)^{(m_i - z_i)m_j}$$
$$\times \prod_{j=1}^{N}\frac{\mathbf{G}(z_j - m_j + 1)\mathbf{G}(m_j + 1)}{\mathbf{G}(z_j + 1)} \prod_{j=1}^{N}\frac{1}{c_-(\tau_j)^{m_j} c_+(\tau_j)^{z_j - m_j}}. \tag{2.50}$$

The constellation where $\delta_j = \gamma_j$ for all j is especially interesting. In this case the asymptotics of $D_n(b)$ was described earlier by Widom [293]. Thus, let

$$b(t) = \prod_{j=1}^{N}|t-\tau_j|^{2\gamma_j} c(t), \quad t \in \mathbf{T}.$$

If $(m_1, \ldots, m_N) \in \mathcal{M}$, then

$$\sum_{j=1}^{N}(m_j \cdot 2\gamma_j - m_j^2) \le \sum_{j=1}^{N}(m_j^2 + \gamma_j^2 - m_j^2) = \sum_{j=1}^{N}\gamma_j^2 = \sum_{j=1}^{N}(\gamma_j \cdot 2\gamma_j - \gamma_j^2)$$

and equality holds if and only if $m_j = \gamma_j$ for all j. Hence $Q = \gamma_1^2 + \cdots + \gamma_N^2$ and $\mathcal{M}^* = \{(\gamma_1, \ldots, \gamma_N)\}$. Formulas (2.49) and (2.50) therefore yield

$$\begin{aligned} D_n(b) &= G(c)^n E(c)\, n^{\gamma_1^2+\cdots+\gamma_N^2}\, A \left(\frac{\tau_1^{\gamma_1}\cdots\tau_N^{\gamma_N}}{\tau_1^{\gamma_1}\cdots\tau_N^{\gamma_N}}\right)^n (1 + O(1/n)) \\ &= G(c)^n E(c)\, A\, n^{\gamma_1^2+\cdots+\gamma_N^2}(1 + O(1/n)) \end{aligned} \tag{2.51}$$

with

$$A = \prod_{i \ne j}\left(1 - \frac{\tau_i}{\tau_j}\right)^{-\gamma_i\gamma_j} \prod_{j=1}^{N} \frac{\mathbf{G}(2\gamma_i - \gamma_i + 1)\mathbf{G}(\gamma_i + 1)}{\mathbf{G}(2\gamma_i + 1)} \prod_{j=1}^{N} \frac{1}{c_-(\tau_j)^{\gamma_j} c_+(\tau_j)^{\gamma_j}}.$$

Since

$$\begin{aligned} \prod_{i \ne j}\left(1 - \frac{\tau_i}{\tau_j}\right)^{-\gamma_i\gamma_j} &= \prod_{i>j}\left[\left(1 - \frac{\tau_i}{\tau_j}\right)^{-\gamma_i\gamma_j}\left(1 - \frac{\tau_j}{\tau_i}\right)^{-\gamma_j\gamma_i}\right] \\ = \prod_{i>j}\left|1 - \frac{\tau_i}{\tau_j}\right|^{-2\gamma_i\gamma_j} &= \prod_{i>j}|\tau_i - \tau_j|^{-2\gamma_i\gamma_j} = \prod_{i \ne j}|\tau_i - \tau_j|^{-\gamma_i\gamma_j}, \end{aligned}$$

we obtain

$$A = G(c)^\gamma \prod_{i \ne j}|\tau_i - \tau_j|^{-\gamma_i\gamma_j} \prod_{j=1}^{N}\frac{1}{c(\tau_j)^{\gamma_j}} \prod_{j=1}^{N}\frac{\mathbf{G}(\gamma_j + 1)^2}{\mathbf{G}(2\gamma_j + 1)}. \tag{2.52}$$

As already said, (2.51) and (2.52) are already in [293].

Chapter 3
Stability

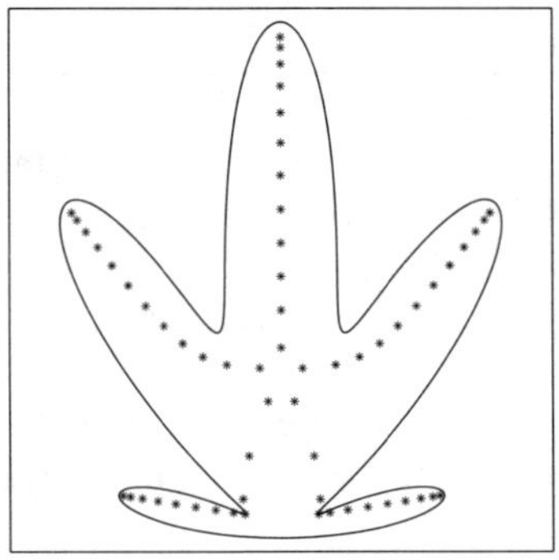

For $a \in W$ and $n \in \{1, 2, 3, \dots\}$, the $n \times n$ Toeplitz matrix $T_n(a)$ is defined by (2.1). We now consider the sequence $\{T_n(a)\}_{n=1}^{\infty}$ as an entity associated with $T(a)$. In this chapter, we study the following problem: Is there an $n_0 \geq 1$ and an $M \in (0, \infty)$ such that $T_n(a)$ is invertible for all $n \geq n_0$ and

$$\|T_n^{-1}(a)\|_p \leq M \text{ for all } n \geq n_0 \text{ ?} \tag{3.1}$$

This problem plays a key role in the theory of large Toeplitz matrices. Since

$$\|T_n^{-1}(a)\|_2 = \frac{1}{\sigma_{\min}(T_n(a))} \geq \text{rad } T_n^{-1}(a),$$

where $\sigma_{\min}(T_n(a))$ is the smallest singular value of $T_n(a)$ and rad $T_n^{-1}(a)$ stands for the spectral radius of $T_n^{-1}(a)$, question (3.1) for $p = 2$ is equivalent to asking whether $\sigma_{\min}(T_n(a))$ stays away from zero, and if the answer is yes, we can conclude that the eigenvalues of $T_n(a)$ are bounded away from zero as $n \to \infty$.

3.1 Strong and Weak Convergence

Throughout this section, X and Y are Banach spaces. We denote by $\mathcal{B}(X, Y)$ the Banach space of all bounded linear operators from X to Y, and we let $\mathcal{K}(X, Y)$ stand for the Banach space of all compact operators from X to Y. As usual, we put $\mathcal{B}(X, X) =: \mathcal{B}(X)$ and $\mathcal{K}(X, X) =: \mathcal{K}(X)$.

Let $\{A_n\}_{n=1}^{\infty}$ be a sequence of operators $A_n \in \mathcal{B}(X, Y)$. We say that A_n converges to an operator $A \in \mathcal{B}(X, Y)$

uniformly if $\|A_n - A\| \to 0$,
strongly if $\|A_n x - Ax\| \to 0$ for all $x \in X$,
weakly if $|(A_n x, y) - (Ax, y)| \to 0$ for all $x \in X$ and all $y \in Y^*$;

here Y^* is the dual space of Y, and for $z \in Y$ and $y \in Y^*$, we let (z, y) denote the value of the functional y at z.

Theorem 3.1 (Banach and Steinhaus). *Let $\{A_n\}_{n=1}^\infty$ be a sequence of operators $A_n \in \mathcal{B}(X, Y)$ such that $\{A_n x\}_{n=1}^\infty$ is a convergent sequence in Y for each $x \in X$. Then $\sup_{n\geq 1} \|A_n\| < \infty$, the operator A defined by $Ax := \lim_{n\to\infty} A_n x$ belongs to $\mathcal{B}(X, Y)$, and $\|A\| \leq \liminf_{n\to\infty} \|A_n\|$.*

This theorem, which is also known as the uniform boundedness principle, is proved in every text on functional analysis.

Proposition 3.2. *If $K \in \mathcal{K}(X, Y)$ and the operators $A_n \in \mathcal{B}(Y)$ converge strongly to $A \in \mathcal{B}(Y)$, then the operators $A_n K$ converge uniformly to AK.*

Proof. Let $B_1 := \{x \in X : \|x\| \leq 1\}$ and $S_1 := \{x \in X : \|x\| = 1\}$. Fix $\varepsilon > 0$. Since K maps B_1 to a set whose closure is a compact subset of Y, there exists some finite collection of elements $x_1, \ldots, x_N \in B_1$ such that for each $x \in S_1$ we can find an x_j satisfying $\|Kx - Kx_j\| < \varepsilon$. Clearly,

$$\begin{aligned}\|A_n Kx - AKx\| &\leq \|A_n\| \, \|Kx - Kx_j\| + \|A_n Kx_j - AKx_j\| + \|A\| \, \|Kx_j - Kx\| \\ &\leq \|A_n\| \, \varepsilon + \|A_n Kx_j - AKx_j\| + \|A\| \, \varepsilon.\end{aligned}$$

Theorem 3.1 implies that $\|A_n\| \leq M < \infty$ for all n, and as A_n converges strongly to A, the norms $\|A_n Kx_j - AKx_j\|$ are less than ε for all j provided n is sufficiently large. Thus, for n large enough,

$$\|A_n Kx - AKx\| \leq (M + 1 + \|A\|) \, \varepsilon = (M + 1 + \|A\|)\varepsilon\|x\|,$$

which shows that $\|A_n K - AK\| \to 0$. □

Proposition 3.3. *If $K \in \mathcal{K}(X, Y)$ and the operators $A_n \in \mathcal{B}(X)$ converge weakly to $A \in \mathcal{B}(X)$, then the operators KA_n converge strongly to KA.*

Proof. Fix an arbitrary $x \in X$. Since $(A_n x, y) \to (Ax, y)$ for all $x \in X$ and all $y \in Y^*$, we see that the operators T_n defined by $T_n : Y^* \to \mathbf{C}$, $y \mapsto (A_n x, y)$ converge strongly to the operator $T : y \mapsto (Ax, y)$. Thus, by Theorem 3.1, $\sup_{n\geq 1} \|A_n x\| = \sup_{n\geq 1} \|T_n\| < \infty$. This shows that the sequence $\{A_n x\}_{n=1}^\infty$ is bounded. Contrary to what we want, let us assume that $KA_n x$ does not converge to KAx. Then there exist an $\varepsilon > 0$ and a sequence $\{n_k\}$ such that

$$\|KA_{n_k} x - KAx\| \geq \varepsilon \text{ for all } k. \tag{3.2}$$

Since K is compact and $\{A_{n_k}\}_{k=1}^\infty$ is bounded, there exists a subsequence $\{n_{k_j}\}$ of $\{n_k\}$ such that $\{KA_{n_{k_j}} x\}$ has a limit z in Y. The weak convergence of $KA_{n_{k_j}}$ to KA implies that $z = KAx$, whence $\|KA_{n_{k_j}} x - KAx\| \to 0$ as $j \to \infty$. But this contradicts (3.2). □

Let ℓ^p $(1 \leq p \leq \infty)$ be the spaces introduced in Section 1.2. We denote by $\|\cdot\|_p$ both the norm in ℓ^p and $\mathcal{B}(\ell^p)$. Furthermore, we let c_0 denote the closed subspace of ℓ^∞ that is constituted by the sequences that converge to zero. The norm in c_0 is the $\|\cdot\|_\infty$ norm.

For $n = 1, 2, \ldots$, let P_n and Q_n be the projections on ℓ^p and c_0 defined by

$$P_n : \{x_0, x_1, x_2, \ldots\} \mapsto \{x_0, x_1, \ldots, x_{n-1}, 0, 0, \ldots\}, \tag{3.3}$$

$$Q_n : \{x_0, x_1, x_2, \ldots\} \mapsto \{\underbrace{0, \ldots, 0}_{n}, x_n, x_{n+1}, \ldots\}. \tag{3.4}$$

Note that $P_n + Q_n = I$. It is clear that the operators P_n do not converge uniformly. The operators P_n converge strongly to I on ℓ^p if $1 \le p < \infty$ and on c_0, but they do not converge strongly on ℓ^∞.

3.2 Stable Sequences

Throughout this section we assume that the underlying Banach space X is c_0 or ℓ^p ($1 \le p \le \infty$).

For $n = 1, 2, \ldots$, we denote by X_n the space $\mathbf{C}^n$ with the ℓ^p norm if $X = \ell^p$ and with the ℓ^∞ norm if $X = c_0$. On identifying

$$\{x_0, x_1, \ldots, x_{n-1}\} \quad \text{and} \quad \{x_0, x_1, \ldots, x_{n-1}, 0, 0, \ldots\},$$

we can identify X_n with the image of X under the projection P_n given by (3.3). If A is an $n \times n$ matrix, we think of A as an operator on (the column space) X_n and therefore $\|A\|$ is well defined. Notice that $\|A\| = \|P_n A P_n\|$, where $\|A\|$ is the norm of A as an element of $\mathcal{B}(X_n)$ and $\|P_n A P_n\|$ stands for the norm of the (well-defined) operator $P_n A P_n \in \mathcal{B}(X)$.

Let $\{A_n\} := \{A_n\}_{n=1}^\infty$ be a sequence of $n \times n$ matrices A_n. The sequence $\{A_n\}$ is said to be *stable on* X if

$$\limsup_{n\to\infty} \|A_n^{-1}\| < \infty,$$

where, by convention, $\|A_n^{-1}\| := \infty$ if A_n is not invertible. Equivalently, $\{A_n\}$ is stable on X if and only if there exist $n_0 \ge 1$ and $M \in (0, \infty)$ such that A_n is invertible for all $n \ge n_0$ and $\|A_n^{-1}\| \le M$ for all $n \ge n_0$.

Lemma 3.4. *Suppose $X = c_0$ or $X = \ell^p$ with $1 < p < \infty$. Let $\{A_n\}$ be a sequence of $n \times n$ matrices A_n and assume there is an operator $A \in \mathcal{B}(X)$ such that $A_n \to A$ and $A_n^* \to A^*$ strongly. If*

$$\liminf_{n\to\infty} \|A_n^{-1}\| < \infty \tag{3.5}$$

then A is invertible.

Proof. If (3.5) holds, there are $n_1 < n_2 < \cdots$ and $M \in (0, \infty)$ such that $\|A_{n_k}^{-1}\| \le M$. For every $x \in X$, $\|P_{n_k} x\| = \|A_{n_k}^{-1} A_{n_k} P_{n_k} x\| \le M \|A_{n_k} P_{n_k} x\|$. Since $P_{n_k} \to I$ and $A_{n_k} \to A$ strongly, it follows that $\|x\| \le M\|Ax\|$. Hence, A is injective and has a closed range.

The dual space $(\ell^p)^*$ may be identified with ℓ^q $(1/p + 1/q = 1)$ and the dual space $(c_0)^*$ with ℓ^1. Since $\|(A_n^*)^{-1}\| = \|A_n^{-1}\|$, we can repeat the argument of the preceding paragraph to obtain that $\|x\| \le M\|A^* x\|$ for all $x \in X^*$. This shows that A^* is injective.

Hence, A has dense range. Together with what was already proved, this implies that the operator A is invertible. □

Proposition 3.5. *Suppose X is c_0 or ℓ^p with $1 \le p < \infty$. Let $\{A_n\}$ be a sequence of $n \times n$ matrices A_n and assume that $A_n \to A$ strongly on X for some invertible operator $A \in \mathcal{B}(X)$. Then the following are equivalent:*

(i) *$\{A_n\}$ is stable;*

(ii) *A_n is invertible for all sufficiently large n and $A_n^{-1} \to A^{-1}$ strongly on X.*

Proof. To show (ii) $\Rightarrow$ (i), note that if $A_n^{-1} \to A^{-1}$ strongly, then $\limsup \|A_n^{-1}\|$ is finite by virtue of Theorem 3.1. So let us prove the implication (i) $\Rightarrow$ (ii). For every $y \in X$,

$$\|A_n^{-1}P_n y - A^{-1}y\| \le \|A_n^{-1}P_n y - P_n A^{-1}y\| + \|P_n A^{-1}y - A^{-1}y\|. \tag{3.6}$$

The second term on the right of (3.6) goes to zero, because $P_n \to I$ strongly. The first term on the right of (3.6) is

$$\|A_n^{-1}(P_n y - A_n P_n A^{-1}y)\| \le M\|P_n y - A_n P_n A^{-1}y\| = o(1),$$

since $A_n P_n A^{-1} \to AA^{-1} = I$ strongly. □

3.3 The Baxter-Gohberg-Feldman Theorem

In this section we show that if $a \in W$, then the sequence $\{T_n(a)\}$ is stable if and only if $T(a)$ is invertible.

Toeplitz operators on c_0. In the previous section, we have seen that in connection with stability problems the passage to adjoint operators is useful. The dual space of ℓ^1 may be identified with ℓ^∞, and as soon as the space ℓ^∞ enters the scene, difficulties arise, because the projections P_n do not converge strongly on ℓ^∞. To tackle questions located in the space ℓ^1, it is often advantageous to think of ℓ^1 as the dual space of c_0.

Let us denote by $\|\cdot\|_0$ both the norm in c_0 and $\mathcal{B}(c_0)$. It is easily seen that the infinite Toeplitz matrix $T(a)$ induces a bounded operator on c_0 if and only if $a \in W$, in which case $\|T(a)\|_0 = \|a\|_W$. As in the proofs of Propositions 1.1 and 1.2 one can show that if $a \in W$, then the Hankel operator $H(a)$ is compact on c_0. Finally, the reasoning of the proof of Theorem 1.9 gives that if $a \in W$, then $T(a)$ is Fredholm on c_0 if and only if a has no zeros on $\mathbf{T}$, in which case $\operatorname{Ind} T(a) = -\operatorname{wind} a$ and $T(a)$ is invertible if $\operatorname{wind} a = 0$.

Lemma 3.6. *Suppose X is a linear space, P and Q are complementary projections on X (that is, $P^2 = P$, $Q^2 = Q$, $P + Q = I$), and A is an invertible linear operator on X. Then the compression $PAP \,|\, \operatorname{Im} P$ of A to the range* $\operatorname{Im} P$ *of P is invertible if and only if the compression $QA^{-1}Q \,|\, \operatorname{Im} Q$ of A^{-1} to the range* $\operatorname{Im} Q$ *of Q is invertible. In that case*

$$(PAP)^{-1}P = PA^{-1}P - PA^{-1}Q(QA^{-1}Q)^{-1}QA^{-1}P.$$

Proof. We have

$$\begin{aligned}
&PAP(PA^{-1}P - PA^{-1}Q(QA^{-1}Q)^{-1}QA^{-1}P)\\
&= PA(I-Q)A^{-1}P - PA(I-Q)A^{-1}Q(QA^{-1}Q)^{-1}QA^{-1}P\\
&= P - PAQA^{-1}P - 0 + PAQA^{-1}P = P,
\end{aligned}$$

and similarly, $(PA^{-1}P - PA^{-1}Q(QA^{-1}Q)^{-1}QA^{-1}P)PAP = P$. □

Theorem 3.7 (Baxter, Gohberg, and Feldman). *Let X be one of the spaces c_0 or ℓ^p $(1 \le p \le \infty)$ and let $a \in W$. Then*

$$\limsup_{n\to\infty} \|T_n^{-1}(a)\| < \infty \quad \text{if } T(a) \text{ is invertible}, \tag{3.7}$$

$$\lim_{n\to\infty} \|T_n^{-1}(a)\| = \infty \quad \text{if } T(a) \text{ is not invertible}. \tag{3.8}$$

Thus, $\{T_n(a)\}_{n=1}^\infty$ is stable if and only if a has no zeros on $\mathbf{T}$ and $\operatorname{wind} a = 0$.

Proof. Suppose first that $X = c_0$ or $X = \ell^p$ with $1 < p < \infty$.

Assertion (3.8) is immediate from Lemma 3.4. So assume $T(a)$ is invertible. By Proposition 1.3, $T(a)T(a^{-1}) = I - H(a)H(\widetilde{a}^{-1})$, whence

$$T^{-1}(a) = T(a^{-1}) + T^{-1}(a)H(a)H(\widetilde{a}^{-1}) =: T(a^{-1}) + K.$$

The operator K is compact due to Proposition 1.2. Recall that $Q_n := I - P_n$. Since $Q_n \to 0$ strongly, we deduce from Proposition 3.2 that

$$\|Q_n K Q_n \,|\, \operatorname{Im} Q_n\| = \|Q_n K Q_n\| \le \|Q_n K\| = o(1),$$

which implies that

$$Q_n T^{-1}(a) Q_n \,|\, \operatorname{Im} Q_n = Q_n T(a^{-1}) Q_n \,|\, \operatorname{Im} Q_n + K_n$$

with $\|K_n\| \to 0$. The operator $Q_n T(a^{-1}) Q_n \,|\, \operatorname{Im} Q_n$ has the same matrix as $T(a^{-1})$, and as $T(a^{-1})$ is invertible together with $T(a)$ (Theorem 1.9), it follows that $Q_n T^{-1}(a) Q_n \,|\, \operatorname{Im} Q_n$ is invertible for all sufficiently large n and that for every $\varepsilon > 0$ there is an $n_0(\varepsilon)$ such that

$$\|(Q_n T^{-1}(a) Q_n)^{-1} Q_n\| < (1+\varepsilon)\|T^{-1}(a^{-1})\|$$

for all $n \ge n_0(\varepsilon)$. For these n we obtain from Lemma 3.6 that the matrices $T_n(a) = P_n T(a) P_n \,|\, \operatorname{Im} P_n$ are invertible and that

$$(P_n T(a) P_n)^{-1} P_n = P_n T^{-1}(a) P_n - P_n T^{-1}(a) Q_n (Q_n T^{-1}(a) Q_n)^{-1} Q_n T^{-1}(a) P_n.$$

This implies that

$$\|T_n^{-1}(a)\| \le \|T^{-1}(a)\| + (1+\varepsilon)\|T^{-1}(a)\|\,\|T^{-1}(a^{-1})\|\,\|T^{-1}(a)\|$$

and hence gives (3.7).

Let now X be the space ℓ^1. From what was already proved, we know that $\limsup \|T_n^{-1}(\overline{a})\|_0 < \infty$ if $T(\overline{a})$ is invertible on c_0 and that $\lim \|T_n^{-1}(\overline{a})\|_0 = \infty$ if $T(\overline{a})$ is not invertible on c_0. As $\|T_n^{-1}(\overline{a})\|_0 = \|T_n^{-1}(a)\|_1$ and as $T(\overline{a})$ is invertible on c_0 if and only if $T(a)$ is invertible on ℓ^1, we arrive at (3.7) and (3.8) for $X = \ell^1$. Finally, the preceding argument can be employed to reduce the case $X = \ell^\infty$ to the (by now settled) case $X = \ell^1$. □

Corollary 3.8. *Let X be c_0 or ℓ^p with $1 \le p < \infty$ and $a \in W$. If $T(a)$ is invertible, then the operators $T_n^{-1}(a)P_n$ converge strongly to $T^{-1}(a)$ on X.*

Proof. The proof is immediate from Proposition 3.5 and Theorem 3.7. □

Corollary 3.8 tells us that if $T(a)$ is invertible, we can approximately solve the infinite system $T(a)x = y$ by replacing it with the finite system $T_n(a)x^{(n)} = P_n y$, $x^{(n)} \in \operatorname{Im} P_n$. This is called the *finite section method*. In this connection it is of interest to know something about the first n_0 such that $T_n(a)$ is invertible for all $n \ge n_0$. In the case of banded matrices, the following result reveals that this n_0 is in general not an astronomic number.

Proposition 3.9. *Let b be a Laurent polynomial and suppose b has no zeros on the unit circle* $\mathbf{T}$ *and* $\operatorname{wind} b = 0$. *Choose a number α satisfying* (1.23). *Then there is a constant $C(b, \alpha)$, depending only on b and α, such that $T_n(b)$ is invertible whenever $C(b, \alpha)\, e^{-2\alpha n} < 1$.*

Proof. By Lemma 3.6, $T_n(b) = P_n T(b) P_n \,|\, \operatorname{Im} P_n$ is invertible if and only if the operator $A_n := Q_n T^{-1}(b) Q_n \,|\, \operatorname{Im} Q_n$ is invertible. We have

$$T^{-1}(b) = T(b_+^{-1})T(b_-^{-1}) = T(b^{-1}) - K, \quad K := H(b_+^{-1})H(\widetilde{b_-^{-1}}).$$

Since the operator $Q_n T(b^{-1}) Q_n \,|\, \operatorname{Im} Q_n$ has the same matrix as $T(b^{-1})$, the operator A_n is certainly invertible if $\|Q_n K Q_n\|\, \|T^{-1}(b^{-1})\| < 1$. But from Lemma 1.17 we infer that

$$\begin{aligned} \|Q_n K Q_n\| &\le \|Q_n H(b_+^{-1})\|\, \|H(\widetilde{b_-^{-1}}) Q_n\| \\ &\le \left(\sum_{j=0}^{\infty} (j+1)\left|(b_+^{-1})_{n+j}\right| \right) \left(\sum_{j=0}^{\infty} (j+1)\left|(b_-^{-1})_{-n-j}\right| \right) = O(e^{-2\alpha n}). \end{aligned}$$ □

3.4 Silbermann Theory

In this section we present Silbermann's approach to the stability problem. This approach yields another proof to Theorem 3.7 and, moreover, allows us to extend this theorem to Toeplitz matrices with certain perturbations.

We begin with the analogue of Proposition 1.3 for finite Toeplitz matrices. For $n = 1, 2, 3, \ldots$, we define the operators W_n on ℓ^p by

$$W_n : \{x_0, x_1, x_2, \ldots\} \mapsto \{x_{n-1}, \ldots, x_1, x_0, 0, 0, \ldots\}. \tag{3.9}$$

Obviously, $W_n^2 = P_n$. It is also easy to verify that

$$W_n T_n(a) W_n = T_n(\widetilde{a}), \tag{3.10}$$

where, as usual, $\widetilde{a}(t) := a(1/t)$ $(t \in \mathbf{T})$.

Proposition 3.10 (Widom). *If* $a, b \in W$ *then*

$$T_n(ab) = T_n(a)T_n(b) + P_n H(a)H(\widetilde{b})P_n + W_n H(\widetilde{a})H(b)W_n.$$

Proof. From Proposition 1.3 we obtain

$$T_n(ab) = P_n T(ab)P_n = P_n T(a)T(b)P_n + P_n H(a)H(\widetilde{b})P_n,$$

and since

$$\begin{aligned} P_n T(a)T(b)P_n &= P_n T(a)P_n T(b)P_n + P_n T(a)Q_n T(b)P_n \\ &= T_n(a)T_n(b) + P_n T(a)Q_n T(b)P_n, \end{aligned}$$

it suffices to check that

$$P_n T(a)Q_n T(b)P_n = W_n H(\widetilde{a})H(b)W_n. \tag{3.11}$$

But it is easily seen that the j, k entry $(1 \le j, k \le n)$ of each side of (3.11) is

$$a_{-n+j-1}b_{n-k+1} + a_{-n+j-2}b_{n-k+2} + \cdots . \qquad \square$$

We now suppose that X is c_0 or ℓ^p with $1 < p < \infty$. Let $\mathbf{F}$ denote the set of all sequences $\{A_n\} := \{A_n\}_{n=1}^{\infty}$ of $n \times n$ matrices A_n such that $\sup_{n \ge 1} \|A_n\| < \infty$. It is easily seen that $\mathbf{F}$ is a Banach algebra with the algebraic operations

$$\{A_n\} + \{B_n\} := \{A_n + B_n\}, \quad \alpha\{A_n\} := \{\alpha A_n\}, \quad \{A_n\}\{B_n\} := \{A_n B_n\}$$

and the norm $\|\{A_n\}\| := \sup_{n \ge 1} \|A_n\|$. We define $\mathbf{S}$ as the subset of $\mathbf{F}$ that is constituted by the sequences $\{A_n\}$ for which there exist two operators $A \in \mathcal{B}(X)$ and $\widetilde{A} \in \mathcal{B}(X)$ such that

$$A_n \to A, \quad A_n^* \to A^*, \quad W_n A_n W_n \to \widetilde{A}, \quad W_n A_n^* W_n \to \widetilde{A}^*$$

strongly. It is not difficult to check that $\mathbf{S}$ is a closed subalgebra of $\mathbf{F}$, and hence $\mathbf{S}$ itself is a Banach algebra. Finally, let $\mathbf{J}$ be the set of all sequences $\{A_n\} \in \mathbf{F}$ that are of the form

$$A_n = P_n K P_n + W_n L W_n + C_n \tag{3.12}$$

with compact operators K and L and with $\|C_n\| \to 0$ as $n \to \infty$.

Lemma 3.11 (Silbermann). $\mathbf{J}$ *is a closed two-sided ideal of* $\mathbf{S}$.

Proof. Let A_n be of the form (3.12). Since $W_n \to 0$ weakly, we deduce from Proposition 3.3 that $A_n \to K$ strongly. Similarly, $A_n^* \to K^*$, $W_n A_n W_n \to L$, $W_n A_n^* W_n \to L^*$ strongly. Thus, $\mathbf{J}$ is a subset of $\mathbf{S}$. It is clear that $\mathbf{J}$ is a linear space.

To prove that $\mathbf{J}$ is closed, notice first that, by Theorem 3.1,

$$\|K\| \le \liminf_{n\to\infty} \|A_n\|, \quad \|L\| \le \liminf_{n\to\infty} \|W_n A_n W_n\|.$$

Consequently, if $\{A_n^{(j)}\}_{j=1}^{\infty}$ is a Cauchy sequence in $\mathbf{J}$, then the sequences $\{K^{(j)}\}_{j=1}^{\infty}$ and $\{L^{(j)}\}_{j=1}^{\infty}$ are Cauchy sequences in $\mathcal{K}(X)$. This implies that there exist operators K and L in $\mathcal{K}(X)$ such that $K^{(j)} \to K$ and $L^{(j)} \to L$ uniformly. Now it follows easily that $\|\{A_n^{(j)}\} - \{A_n\}\| \to 0$ as $j \to \infty$ for some $\{A_n\} \in \mathbf{S}$. This proves that $\mathbf{J}$ is closed.

Finally, let $\{B_n\} \in \mathbf{S}$ and $\{A_n\} \in \mathbf{J}$. Then

$$\begin{aligned} B_n A_n &= B_n P_n K P_n + B_n W_n L W_n + B_n C_n \\ &= P_n B_n K P_n + W_n (W_n B_n W_n) P_n L W_n + B_n C_n \\ &= P_n B K P_n + W_n \widetilde{B} L W_n + C_n' \end{aligned}$$

with $\|C_n'\| \to 0$, since $B_n \to BK$ and $W_n B_n W_n P_n L \to \widetilde{B} L$ uniformly due to Proposition 3.2. Consideration of adjoints shows that $\{A_n B_n\}$ is also in $\mathbf{J}$. □

Theorem 3.12 (Silbermann). *Let X be c_0 or ℓ^p with $1 < p < \infty$. A sequence $\{A_n\} \in \mathbf{S}$ is stable if and only if A and $\widetilde{A}$ are invertible in $\mathcal{B}(X)$ and the coset $\{A_n\} + \mathbf{J}$ is invertible in the quotient algebra $\mathbf{S}/\mathbf{J}$. Moroever, if $\{R_n\} + \mathbf{J}$ is the inverse of $\{A_n\} + \mathbf{J}$ in $\mathbf{S}/\mathbf{J}$, then*

$$B_n := R_n + P_n (A^{-1} - R) P_n + W_n (\widetilde{A}^{-1} - \widetilde{R}) W_n$$

satisfies $A_n B_n = P_n + C_n'$ and $B_n A_n = P_n + C_n''$ with $\|C_n'\| \to 0$ and $\|C_n''\| \to 0$.

Proof. Suppose the sequence $\{A_n\}$ is stable. Then A is invertible by Lemma 3.4. Since $\|(W_n A_n W_n)^{-1}\| = \|A_n^{-1}\|$, the sequence $\{W_n A_n W_n\}$ is also stable, and hence, again by Lemma 3.4, $\widetilde{A}$ is invertible. The stability of $\{A_n\}$ implies the existence of a sequence $\{B_n\} \in \mathbf{F}$ such that $B_n A_n = P_n + C_n$ with $\|C_n\| \to 0$ (simply take $B_n = A_n^{-1}$ for all sufficiently large n). Using Proposition 3.5, it is not difficult to see that $\{B_n\}$ belongs to $\mathbf{S}$. Thus, $\{B_n\} + \mathbf{J}$ is a left inverse of $\{A_n\} + \mathbf{J}$. Analogously one can show the $\{A_n\} + \mathbf{J}$ is invertible from the right.

Conversely, suppose now that A and $\widetilde{A}$ are invertible and that $\{R_n\} + \mathbf{J}$ is the inverse of $\{A_n\} + \mathbf{J}$. Then

$$A_n R_n = P_n + P_n K P_n + W_n L W_n + C_n,$$

where K and L are compact and $\|C_n\| \to 0$. Passage to the strong limit $n \to \infty$ gives $AR = I + K$ and $\widetilde{A}\widetilde{R} = I + L$. Thus, $S := A^{-1} - R = -A^{-1}K$ and $T := \widetilde{A}^{-1} - \widetilde{R} = -\widetilde{A}L$ are also compact. Put

$$B_n = R_n + P_n S P_n + W_n T W_n.$$

Then $\{B_n\} \in \mathbf{S}$ and, by Proposition 3.2,

$$\begin{aligned} A_n B_n &= P_n + P_n K P_n + W_n L W_n + C_n + A_n P_n S P_n + A_n W_n T W_n \\ &= P_n + P_n (K + A_n P_n S) P_n + W_n (L + W_n A_n W_n T) W_n + C_n \\ &= P_n + P_n (K + AS) P_n + W_n (L + \widetilde{A} T) W_n + C_n' \\ &= P_n + C_n' \end{aligned}$$

with $\|C_n'\| \to 0$. Consideration of adjoints yields that $B_nA_n = P_n + C_n''$ with $\|C_n''\| \to 0$. □

An infinite matrix is said to have *finite support* if at most finitely many of its entries are nonzero. Clearly, infinite matrices with finite support generate compact operators. Recall that $\widetilde{a}$ is defined by $\widetilde{a}(t) := a(1/t)$ $(t \in \mathbf{T})$.

Theorem 3.13 (Silbermann). *Let X be c_0 or ℓ^p with $1 \le p \le \infty$ and let*

$$A_n = T_n(a) + P_nKP_n + W_nLW_n,$$

where $a \in W$ and K and L are matrices with finite support. Put

$$A = T(a) + K, \quad \widetilde{A} = T(\widetilde{a}) + L.$$

Then

$$\limsup_{n\to\infty} \|A_n^{-1}\| < \infty \text{ if } A \text{ and } \widetilde{A} \text{ are invertible,} \tag{3.13}$$

$$\lim_{n\to\infty} \|A_n^{-1}\| = \infty \text{ if } A \text{ or } \widetilde{A} \text{ is not invertible.} \tag{3.14}$$

Consequently, $\{A_n\}_{n=1}^\infty$ is stable if and only if both A and $\widetilde{A}$ are invertible.

Proof. We first consider the case where $X = c_0$ or $X = \ell^p$ with $1 < p < \infty$.

If $\liminf \|A_n^{-1}\| < \infty$, A is invertible by virtue of Lemma 3.4. Since $\|(W_nA_nW_n)^{-1}\| = \|A_n^{-1}\|$, we can apply Lemma 3.4 to the sequence $\{W_nA_nW_n\}$ to obtain that $\widetilde{A}$ is invertible whenever $\liminf \|A_n^{-1}\| < \infty$. This completes the proof of (3.14).

Let now A and $\widetilde{A}$ be invertible. It is clear that $\{A_n\} \in \mathbf{S}$. Thus, (3.13) will follow from Theorem 3.12 once we have shown that $\{A_n\} + \mathbf{J}$ is invertible in $\mathbf{S}/\mathbf{J}$. As $\{A_n\} + \mathbf{J} = \{T_n(a)\} + \mathbf{J}$, we are left to prove that $\{T_n(a)\} + \mathbf{J}$ is invertible. But if $A = T(a) + K$ is invertible, then $T(a)$ is Fredholm and $a \in GW$. By Proposition 3.10,

$$T_n(a^{-1})T_n(a) = P_n - P_nH(a^{-1})H(\widetilde{a})P_n - W_nH(\widetilde{a}^{-1})H(a)W_n,$$

and since all occurring Hankel operators are compact, it follows that

$$\{P_nH(a^{-1})H(\widetilde{a})P_n + W_nH(\widetilde{a}^{-1})H(a)W_n\} \in \mathbf{J}.$$

Thus, $\{T_n(a^{-1})\}+\mathbf{J}$ is a left inverse of $\{T_n(a)\}+\mathbf{J}$. Similarly one can show that $\{T_n(a^{-1})\}+\mathbf{J}$ is a right inverse of $\{T_n(a)\} + \mathbf{J}$. This proves that $\{T_n(a)\} + \mathbf{J}$ is invertible, as desired.

To dispose of the case where $X = \ell^1$, we proceed as in the proof of Theorem 3.7. Define $B_n := T_n(\overline{a}) + P_nK^*P_n + W_nL^*W_n$, where K^* and L^* are the usual Hermitian adjoints of the matrices K and L. By what was already proved, $\limsup \|B_n^{-1}\|_0 < \infty$ if $B = T(\overline{a}) + K^*$ and $\widetilde{B} = T(\widetilde{\overline{a}}) + L^*$ are invertible on c_0, while $\liminf \|B_n^{-1}\|_0 = \infty$ if B or $\widetilde{B}$ is not invertible on c_0. But $\|B_n^{-1}\|_0 = \|(B_n^{-1})^*\|_1 = \|A_n^{-1}\|_1$, and B and $\widetilde{B}$ are invertible on c_0 if and only if $B^* = A$ and $\widetilde{B}^* = \widetilde{A}$ are invertible on ℓ^1. This gives (3.13) and (3.14) for $X = \ell^1$. The ℓ^∞ case can be reduced to the ℓ^1 case by the same reasoning. □

Corollary 3.14. *Let X be c_0 or ℓ^p with $1 \le p < \infty$ and let $A_n = T_n(a)+P_nKP_n+W_nLW_n$ where $a \in W$ and K and L are finitely supported. Then the operators $A_n^{-1}P_n$ converge strongly to $(T(a) + K)^{-1}$ on X if and only if both $T(a) + K$ and $T(\widetilde{a}) + L$ are invertible.*

Proof. This follows from Proposition 3.5 and Theorem 3.13. □

3.5 Asymptotic Inverses

The sequence $\{B_n\}$ delivered by Theorem 3.12 is an *asymptotic inverse* for A_n: we have $A_n^{-1} = B_n + C_n$ with $\|C_n\| \to 0$ (notice that if $\|P_n - A_n B_n\| \to 0$ and $\{A_n\}$ is stable, then $\|A_n^{-1} - B_n\| \le \|A_n^{-1}\| \, \|P_n - A_n B_n\| \to 0$). In the special case $A_n = T_n(a)$ we have $A = T(a)$ and $\widetilde{A} = T(\widetilde{a})$, and when proving Theorem 3.13, we saw that $R_n = T_n(a^{-1})$ does the desired job, whence $R = T(a^{-1})$ and $\widetilde{R} = T(\widetilde{a}^{-1})$. Thus, on defining

$$K(a) := T^{-1}(a) - T(a^{-1}), \quad K(\widetilde{a}) := T^{-1}(\widetilde{a}) - T(\widetilde{a}^{-1}), \tag{3.15}$$

for $a \in GW$ with wind $a = 0$, we obtain

$$T_n^{-1}(a) = T_n(a^{-1}) + P_n K(a) P_n + W_n K(\widetilde{a}) W_n + C_n \tag{3.16}$$

with $\|C_n\| \to 0$ on c_0 and ℓ^p for $1 \le p < \infty$ (note that the proof of the convergence $\|C_n'\| \to 0$ in Theorem 3.12 also works for ℓ^1).

Here are some **alternative expressions** for the operators (3.15). From Proposition 1.3 we infer that $T(a)T(a^{-1}) = I - H(a)H(\widetilde{a}^{-1})$, whence $T^{-1}(a) - T(a^{-1}) = T^{-1}(a)H(a)H(\widetilde{a}^{-1})$. This shows that

$$K(a) = T^{-1}(a)H(a)H(\widetilde{a}^{-1}), \quad K(\widetilde{a}) = T^{-1}(\widetilde{a})H(\widetilde{a})H(a^{-1}). \tag{3.17}$$

Combining (3.17) and Proposition 1.2 we see that $K(a)$ and $K(\widetilde{a})$ are compact. Analogously, starting with $T(a^{-1})T(a) = I - H(a^{-1})H(\widetilde{a})$, we get $T^{-1}(a) - T(a^{-1}) = H(a^{-1})H(\widetilde{a})T^{-1}(a)$ and thus

$$K(a) = H(a^{-1})H(\widetilde{a})T^{-1}(a), \quad K(\widetilde{a}) = H(\widetilde{a}^{-1})H(a)T^{-1}(\widetilde{a}). \tag{3.18}$$

Given a Wiener-Hopf factorization $a = a_- a_+$, we can also write

$$K(a) = T(a_+^{-1})T(a_-^{-1}) - T(a_+^{-1}a_-^{-1}) = -H(a_+^{-1})H(\widetilde{a}_-^{-1}), \tag{3.19}$$

$$K(\widetilde{a}) = T(\widetilde{a}_-^{-1})T(\widetilde{a}_+^{-1}) - T(\widetilde{a}_-^{-1}\widetilde{a}_+^{-1}) = -H(\widetilde{a}_-^{-1})H(a_+^{-1}). \tag{3.20}$$

Finally, from (3.19) and Exercise 2 of Chapter 1 we obtain

$$\begin{aligned} K(a) &= -H(a_+^{-1})H(\widetilde{a}_-^{-1}) = -H(a^{-1}a_-)H(\widetilde{a}_+\widetilde{a}^{-1}) \\ &= -H(a^{-1})T(\widetilde{a}_-)T(\widetilde{a}_+)H(\widetilde{a}^{-1}) = -H(a^{-1})T^{-1}(\widetilde{a})H(\widetilde{a}^{-1}) \end{aligned} \tag{3.21}$$

and analogously,

$$K(\widetilde{a}) = -H(\widetilde{a}^{-1})T^{-1}(a)H(a^{-1}). \tag{3.22}$$

In the case of Toeplitz band matrices, we can show that the norm of the matrices C_n in (3.16) decays exponentially.

Theorem 3.15. *Let b be a Laurent polynomial without zeros on $\mathbf{T}$ and with winding number zero. Choose an α satisfying* (1.23) *and let $1 \le p \le \infty$. Then* (3.16) *holds with a replaced by b and with $\|C_n\|_p = O(e^{-\alpha n})$.*

Proof. From Proposition 3.10 and formula (3.10) we see that $T_n^{-1}(b)$ equals

$$T_n(b^{-1}) + T_n^{-1}(b)P_nH(b)H(\widetilde{b}^{-1})P_n + W_nT_n^{-1}(\widetilde{b})P_nH(\widetilde{b})H(b^{-1})W_n. \tag{3.23}$$

We have

$$\begin{aligned} T_n^{-1}(b)P_nH(b)H(\widetilde{b}^{-1})P_n &= T_n^{-1}(b)P_n(I - T(b)T(b^{-1}))P_n \\ &= T_n^{-1}(b)P_n(T(b)T^{-1}(b) - T(b)T(b^{-1}))P_n = T_n^{-1}(b)P_nT(b)K(b)P_n \\ &= T_n^{-1}(b)P_nT(b)P_nK(b)P_n + T_n^{-1}(b)P_nT(b)Q_nK(b)P_n \\ &= P_nK(b)P_n + T_n^{-1}(b)P_nT(b)Q_nK(b)P_n. \end{aligned} \tag{3.24}$$

Let $b = b_-b_+$ be the Wiener-Hopf factorization (1.19). By virtue of (3.19), $K(b) = -H(b_+^{-1})H(\widetilde{b}_-^{-1})$. Using Lemma 3.5 we obtain as in the proof of Proposition 3.9 that

$$\|Q_nK(b)\|_p \le M_1 \sum_{j=0}^{\infty}(j+1)\left|(b_+^{-1})_{n+j}\right| \le M_1 \sum_{j=0}^{\infty} M_2 e^{-\alpha(n+j)} = O(e^{-\alpha n}).$$

Since $T_n^{-1}(b) = O(1)$ due to Theorem 3.7, we arrive at the conclusion that (3.24), that is, the second term of (3.23), is $P_nK(b)P_n + O(e^{-\alpha n})$. Analogously, one can show that the third term of (3.23) is $W_nK(\widetilde{b})W_n + O(e^{-\alpha n})$. □

Corollary 3.16. *Let b be a Laurent polynomial without zeros on* **T** *and suppose the winding number is zero. Let α satisfy* (1.23) *and let $1 \le p \le \infty$. Then for each natural number k,*

$$\|T_n^{-1}(b)P_k - T^{-1}(b)P_k\|_p = O(e^{-\alpha n}) \quad \text{as } n \to \infty.$$

Proof. For $j \ge 0$, let $e_j \in \ell^p$ be the sequence defined by $(e_j)_k = 1$ for $k = j$ and $(e_j)_k = 0$ for $k \ne j$. Clearly, it suffices to prove that

$$\|T_n^{-1}(b)e_j - T^{-1}(b)e_j\|_p = O(e^{-\alpha n})$$

for each fixed j. Let $n \ge j$. By (3.16),

$$\begin{aligned} &T_n^{-1}(b)e_j - T^{-1}(b)e_j \\ &= T_n(b^{-1})e_j + P_n(T^{-1}(b) - T(b^{-1}))P_ne_j - T^{-1}(b)e_j + W_nK(\widetilde{b})W_ne_j + C_ne_j \\ &= -Q_nT^{-1}(b)e_j + W_nK(\widetilde{b})W_ne_j + C_ne_j. \end{aligned} \tag{3.25}$$

Let $b(t) = \sum_{j=-r}^{r} b_jt^j$. Then $H(b) = H(b)P_r$. Since $W_ne_j = e_{n-j}$, we therefore obtain from (3.18) that $\|W_nK(\widetilde{b})W_ne_j\|_p \le \|H(\widetilde{b}^{-1})H(b)\|_p\,\|P_rT^{-1}(\widetilde{b})e_{n-r}\|_p$. As $\|P_rT^{-1}(\widetilde{b})e_{n-r}\|_p$ is the ℓ^p norm of the first r components of the $(n-r)$th column of the matrix $T^{-1}(\widetilde{b})$, we deduce from Lemma 1.17 and Proposition 1.18 that the norm $\|P_rT^{-1}(\widetilde{b})e_{n-r}\|_p$ is $O(e^{-\alpha n})$. Also by Lemma 1.17 and Proposition 1.18, $\|Q_nT^{-1}(b)\|_p = O(e^{-\alpha n})$. Finally, because $\|C_ne_j\|_p = O(e^{-\alpha n})$ by virtue of Theorem 3.15, all three terms in (3.25) have norm $O(e^{-\alpha n})$. □

Corollary 3.17. *Under the hypotheses of Corollary* 3.16,

$$\|P_kT_n^{-1}(b)P_n - P_kT^{-1}(b)\|_p = O(e^{-\alpha n}) \quad \text{as } n \to \infty.$$

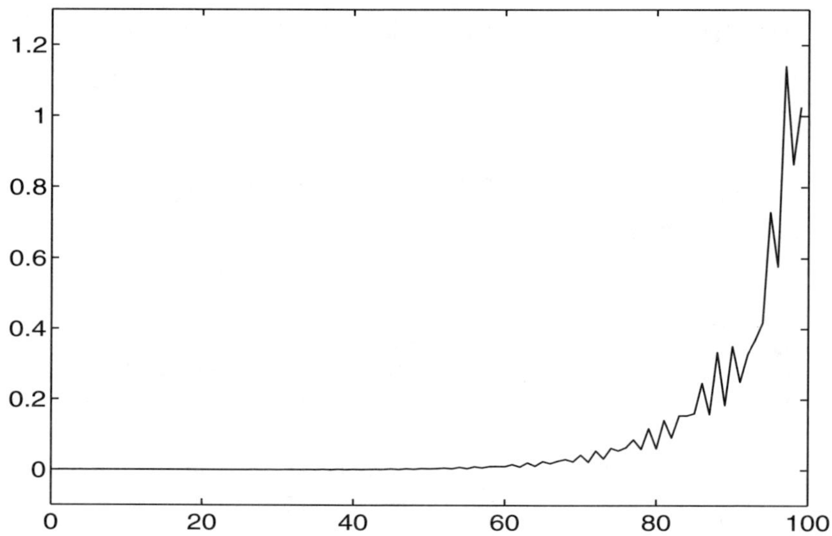

Figure 3.1. *The componentwise error* $|x_j - x_j^{(99)}|$ *against* $j = 0, \dots, 98$.

Proof. Put $E_n = T_n^{-1}(\bar{b})P_k - T^{-1}(\bar{b})P_k$. Corollary 3.16 says that if we consider E_n on c_0 or ℓ^p $(1 \le p \le \infty)$, then $\|E_n\| = O(e^{-\alpha n})$. This implies that $\|E_n^*\| = O(e^{-\alpha n})$ for the adjoint operator $E_n^* = P_k T_n^{-1}(b)P_n - P_k T^{-1}(b)$. As $\ell^1 = c_0^*$ and $\ell^p = (\ell^q)^*$ with $1/p + 1/q = 1$ for $1 < p \le \infty$, we arrive at the assertion. □

Example 3.18. The symbol $b(t) = -2 - 3it - 2t^2 - 3t^3 + t^{-1} - 2it^{-2} + 3t^{-3}$ generates an invertible Toeplitz operator $T(b)$. (To see this, plot $b(\mathbf{T})$.) Choose $x \in \ell^2$ by $x_j = 1$ for $0 \le j \le 99$ and $x_j = 0$ for $j \ge 100$ and put $y = T(b)x$. Obviously, only the first 103 components of y are nonzero. The equation $T_{99}(b)x^{(99)} = P_{99}y$ has a unique solution, and Figure 3.1 shows the componentwise error $|x_j - x_j^{(99)}|$ versus $j = 0, \dots, 98$. It is clearly seen that $P_{50}x^{(99)}$ is a good approximation to $P_{50}x$, but that $x^{(99)}$ and $P_{99}x$ differ heavily in their last components. For example, the last two components of $x^{(99)}$ are $0.5082 - 0.7108i$ and $2.0239 - 0.0622i$ (while those of $P_{99}x$ are 1 and 1). Things change dramatically when passing from 99 to 100: we have $x^{(100)} = P_{100}x$ exactly and MATLAB reports $\|x^{(100)} - P_{100}x\|_\infty < 2 \cdot 10^{-15}$. Figure 3.2 shows the norms $\|x^{(n)} - P_n x\|_\infty$ and $\|x^{(n)} - P_n x\|_2$ versus n. □

Example 3.19. Let $b(t) = 5 + t - t^2 + t^3 - t^{-2}$. The operator $T(b)$ is invertible and hence we can solve the equation $T(b)x = y$ for each $y \in \ell^2$ approximately by passing to the truncated systems $T_n(b)x^{(n)} = P_n y$. We choose 200 vectors x_j of length 100 randomly from the unit sphere $\mathbf{S}^{99}$ of the Euclidean space $\mathbf{R}^{100}$ with the uniform distribution, extend x_j by zeros to a sequence in ℓ^2, and put $y_j = T(b)x_j$. Then we solve the 200 systems $T_n(b)x_j^{(n)} = P_n y_j$ for $n = 30$, $n = 60$, and $n = 90$. The logarithmic error $\log_{10}\|P_n x_j - x_j^{(n)}\|_2$ of the computational results mildly fluctuates around -2 for $n = 30$, $n = 60$, $n = 90$, that is, there is no significant improvement when increasing n. However, the improvement in a

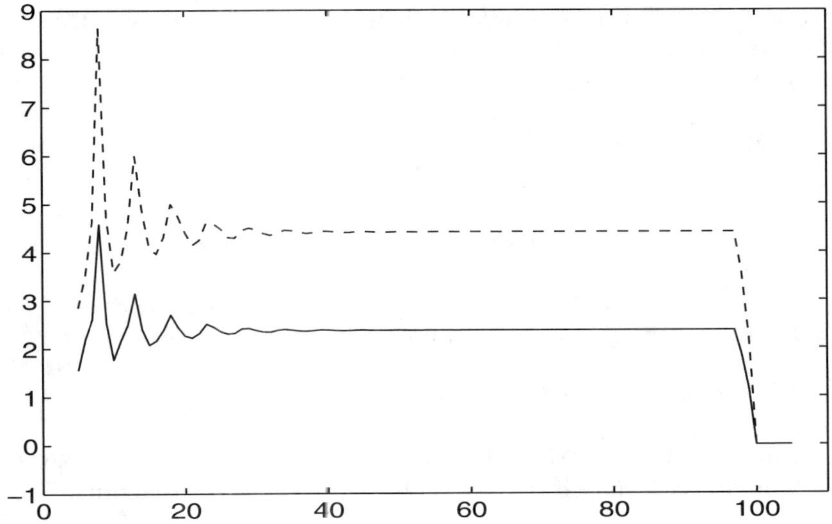

Figure 3.2. *The norms* $\|x^{(n)} - P_n x\|_\infty$ *(solid) and* $\|x^{(n)} - P_n x\|_2$ *(dashed) versus* n *for* $5 \le n \le 105$.

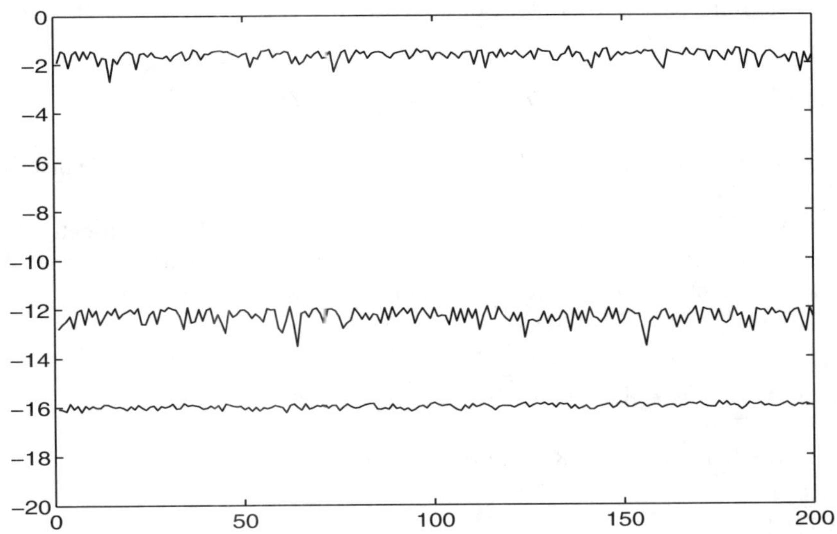

Figure 3.3. *Let* $T(b)x_j = y_j$ *with* 200 *random vectors* x_j *of length* 100. *The picture shows* $\log_{10} \|P_{30}x_j - P_{30}x_j^{(n)}\|_2$ *versus* $j = 1, \ldots, 200$ *for the* 200 *systems* $T_n(b)x_j^{(n)} = P_n y_j$ *with* $n = 30$ *(upper curve),* $n = 60$ *(middle curve), and* $n = 90$ *(lower curve).*

fixed set of components is drastic. Figure 3.3 shows that $\log_{10} \|P_{30}x_j - P_{30}x_j^{(n)}\|_2$ fluctuates around the values -2, -12, and -16 for $n = 30$, $n = 60$, and $n = 90$, respectively. □

It is well known that in many instances Toeplitz matrices generated by the inverses of Laurent polynomials have much better properties than those generated by Laurent polynomials. The following result illustrates this phenomenon.

Theorem 3.20. *Let $b(t) = \sum_{j=-r}^{s} b_j t^j$ be a Laurent polynomial without zeros on the unit circle* $\mathbf{T}$ *and with winding number zero. Put $a = b^{-1}$. Then*

$$T_n^{-1}(a) = T_n(b) + P_n K(a) P_n + W_n K(\widetilde{a}) W_n \tag{3.26}$$

for all $n \geq n_0 :=$ $\max(r, s)$. Moreover, all entries outside the $n_0 \times n_0$ upper-left blocks of $K(a)$ and $K(\widetilde{a})$ are zero. If $b = b_- b_+$ is a Wiener-Hopf factorization of b, then

$$K(a) = -H(b_+)H(\widetilde{b}_-), \quad K(\widetilde{a}) = -H(\widetilde{b}_-)H(b_+).$$

Proof. Let $b = b_- b_+$ be the Wiener-Hopf factorization (1.19). By Lemma 3.6, $T_n(a) = P_n T(a) P_n \mid \operatorname{Im} P_n$ is invertible if and only if $Q_n T^{-1}(a) Q_n \mid \operatorname{Im} Q_n$ is invertible. We have

$$Q_n T^{-1}(a) Q_n = Q_n T(b_+) T(b_-) Q_n = Q_n T(b) Q_n - Q_n H(b_+) H(\widetilde{b}_-) Q_n,$$

and from (3.19) we know that $H(b_+)H(b_-) = -K(a)$. Since b_+ and b_- are analytic polynomials of degree s and r in t and t^{-1}, respectively, we see that $K(a)$ has nonzero entries in the principal $n_0 \times n_0$ block only. This implies that $Q_n K(a) = 0$ for $n \geq n_0$, whence

$$Q_n T^{-1}(a) Q_n = Q_n T(b) Q_n$$

for $n \geq n_0$. As $Q_n T(b) Q_n \mid \operatorname{Im} Q_n$ has the same matrix as $T(b)$, we conclude that $T_n(a)$ is invertible for all $n \geq n_0$.

From (3.20) we see that $K(\widetilde{a}) = -H(\widetilde{b}_-)H(b_+)$, and all nonzero entries of this matrix are clearly concentrated in the principal $n_0 \times n_0$ block. Finally, as in the proof of Theorem 3.15 we obtain that $T_n^{-1}(a)$ is equal to

$$\begin{aligned} &T_n(b^{-1}) + P_n K(a) P_n + T_n^{-1}(a) P_n T(a) Q_n K(a) P_n \\ &+ W_n K(\widetilde{a}) W_n + W_n T_n^{-1}(\widetilde{a}) P_n T(\widetilde{a}) Q_n K(\widetilde{a}) W_n, \end{aligned}$$

and since $Q_n K(a) = 0$ and $Q_n K(\widetilde{a}) = 0$ for $n \geq n_0$, we arrive at (3.26). □

Exercises

1. Let $b \in \mathcal{P}$ and suppose b has no zeros on $\mathbf{T}$ and winding number zero. Let

$$T_n(b) \begin{pmatrix} x_0^{(n)} \\ x_1^{(n)} \\ \vdots \\ x_{n-1}^{(n)} \end{pmatrix} = \begin{pmatrix} 1 \\ 0 \\ \vdots \\ 0 \end{pmatrix}, \quad T(b) \begin{pmatrix} x_0 \\ x_1 \\ \vdots \end{pmatrix} = \begin{pmatrix} 1 \\ 0 \\ \vdots \end{pmatrix}.$$

Show that

$$x_0^{(n)} = \frac{D_{n-1}(b)}{D_n(b)}, \quad x_0 = \frac{1}{G(b)},$$

deduce that $D_n(b)/D_{n-1}(b)$ converges to $G(b)$ exponentially fast, and use this insight to conclude that $D_n(b)/G(b)^n$ converges to a finite and nonzero limit.

2. Show that Corollary 3.8 is not true for $p = \infty$.

3. Let $a \in W$ and $Q_N x = 0$. Put $y = T(a)x$. Show that if $n \geq N$ and $T_n(a)$ is invertible, then the solution $x^{(n)}$ of $T_n(a)x^{(n)} = P_n y$ satisfies the equality $P_N x^{(n)} = P_N x$.

4. Explain the straight horizontal parts and the sudden descents of the two curves of Figure 3.2.

5. In this exercise, ℓ^1 is the real Banach space of all real sequences in ℓ^1 over the scalar field $\mathbf{R}$. Let $y \in \ell^1$ and let $b_+(t) = \sum_{j=0}^{r} b_j t^j$ be an analytic polynomial with real coefficients. Suppose that b_+ has no zeros on $\mathbf{T}$. Put

$$d = \inf_{x \in \ell^1} \|y - T(b_+)x\|_1.$$

(a) Show that there exists an $x_0 \in \ell^1$ such that $\|y - T(b_+)x_0\|_1 = d$.

(b) Show that

$$d_n := \min_{x \in \ell^1} \|P_n y - P_n T(b_+) P_{n-r} x\|_1$$

converges to d as $n \to \infty$.

(c) Let κ be the number of zeros (counted with multiplicities) of b_+ in the open unit disk. Show that

$$d_n^0 := \min_{x \in \ell^1} \|P_n y - P_n T(b_+) P_{n-\kappa} x\|_1$$

converges to d as $n \to \infty$.

6. (a) Let $\{B_n\}$ be a sequence of $n \times n$ matrices, let B be an invertible operator on ℓ^2, and let K be a trace class operator on ℓ^2. Suppose $B_n^{-1} P_n \to B^{-1}$ strongly. Prove that

$$\lim_{n \to \infty} \frac{\det (B_n + P_n K P_n)}{\det B_n} = \det (I + B^{-1} K).$$

(b) Let $a, b \in \mathcal{P}$ and suppose b has no zeros on $\mathbf{T}$ and winding number zero. Let $T_n(b)$ and $H_n(a)$ denote the $n \times n$ truncations of the Toeplitz matrix $T(b)$ and the Hankel matrix $H(a)$, respectively. Prove that $T(b^{-1})T(b) + T(b^{-1})H(a)$ is of the form identity plus trace class operator and show that

$$\lim_{n \to \infty} \frac{\det (T_n(b) + H_n(a))}{G(b)^n} = \det (T(b^{-1})T(b) + T(b^{-1})H(a)).$$

(c) Let a, b be as in (b) and let $b = b_- b_+$ be a Wiener-Hopf factorization. Put $c = b_-^{-1}\widetilde{b}_+^{-1}a$. Prove that

$$\det\left(T(b^{-1})T(b) + T(b^{-1})H(a)\right) = \det T(b^{-1})T(b)\,\det\left(I + H(c)\right).$$

7. Let $a(z, w) = \sum a_{mn} z^m w^n$ be a Laurent polynomial of two variables z and w. The quarter-plane Toeplitz operator $T^{(2)}(a)$ is defined on $\ell^2(\mathbf{N} \times \mathbf{N})$ by

$$(T^{(2)}(a)x)_{ij} = \sum_{k,\ell=1}^{\infty} a_{i-k,j-\ell}\, x_{k\ell}, \quad i, j \geq 1.$$

Suppose $a(z, w) = b(zw^{-1})$ with some Laurent polynomial b of a single variable.

(a) Prove that $T^{(2)}(a)$ is Fredholm if and only if $\{T_n(b)\}_{n=1}^{\infty}$ is stable.

(b) Prove that $T^{(2)}(a)$ is invertible if and only if $\{T_n(b)\}_{n=1}^{\infty}$ is stable and $D_n(b) \neq 0$ for all $n \geq 1$.

8. A Wiener-Hopf integral operator is an operator of the form

$$(Af)(x) = \int_0^{\infty} k(x - t) f(t)\, dt, \quad x > 0.$$

Suppose $k \in L^1(\mathbf{R})$ and $k(x) = 0$ for $|x| > M$. Let $\{e_n\}_{n=0}^{\infty}$ be the orthonormal basis in $L^2(0, \infty)$ constituted by the Laguerre functions

$$e_n(x) = \frac{e^{x/2}}{n!} \frac{d^n}{dx^n}(x^n e^{-x}).$$

To solve the equation $f + Af = g$ approximately, one can look for an approximate solution in the form

$$f^{(n)} = \gamma_1^{(n)} e_1 + \cdots + \gamma_n^{(n)} e_n$$

and determine the coefficients $\gamma_1^{(n)}, \ldots, \gamma_n^{(n)}$ from the n linear equations

$$\gamma_j^{(n)} + (Af^{(n)}, e_j) = (g, e_j) \quad \text{for} \quad j = 1, \ldots, n.$$

(This is called a Galerkin method.)

(a) Prove that A is bounded on $L^2(0, \infty)$ and that the matrix representation of A in the basis $\{e_n\}_{n=0}^{\infty}$ is the Toeplitz matrix $T(a) = (a_{j-k})_{j,k=1}^{\infty}$ with

$$a_n = \frac{1}{2\pi} \int_{-\infty}^{\infty} \hat{k}(\xi) \left(\frac{\xi + i/2}{\xi - i/2}\right)^n \frac{d\xi}{\xi^2 + 1/4},$$

where $\hat{k}(\xi) := \int_{-\infty}^{\infty} k(x) e^{i\xi x} dx$.

(b) Prove that $a \in W$.

(c) Suppose $I + A$ is invertible on $L^2(0, \infty)$. Let $g \in L^2(0, \infty)$ and let $f \in L^2(0, \infty)$ be the solution of the equation $f + Af = g$. Show that if the approximate solutions $f^{(n)}$ are computed as described above, then $f^{(n)}$ converges in $L^2(0, \infty)$ to f.

(d) Prove that $I + A$ is invertible on $L^2(0, \infty)$ if and only if $1 + \hat{k}(\xi) \neq 0$ for all $\xi \in \mathbf{R}$ and the winding number of $1 + \hat{k}$ about the origin is zero.

9. The Cauchy singular integral operator S is defined by

$$(Sf)(t) = \frac{1}{\pi i} \text{ v.p.} \int_{\mathbf{T}} \frac{f(\tau)}{\tau - t} \, d\tau, \quad t \in \mathbf{T}.$$

(a) Show that $S\chi_n = \chi_n$ for $n \geq 0$ and $S\chi_n = -\chi_n$ for $n < 0$.

(b) Let $a, b \in \mathcal{P}$. Compute the matrix representation $R = (r_{jk})_{j,k=-\infty}^{\infty}$ of the so-called singular integral operator $aI + bS$ in the orthonormal basis $\{(1/(2\pi)\, \chi_n\}_{n=-\infty}^{\infty}$ of $L^2(\mathbf{T})$.

(c) Put $R_n = (r_{jk})_{j,k=-n}^{n}$. Prove that $\{R_n\}$ is stable on $\ell^2(\mathbf{Z})$ if and only if $a + b$ and $a - b$ have no zeros on $\mathbf{T}$ and winding number zero.

(d) Use (b) and (c) to establish an approximation method for the singular integral equation $af + bSf = g$ on $L^2(\mathbf{T})$.

10. Let $L^2(\mathbf{D})$ denote the Hilbert space of all functions f on $\mathbf{D}$ that satisfy

$$\|f\|^2 := \frac{1}{\pi} \int_{\mathbf{D}} |f(z)|^2 dA(z) := \frac{1}{\pi} \int_{\mathbf{D}} |f(re^{i\theta})|^2 \, r \, dr \, d\theta < \infty.$$

The Bergman space $A^2(\mathbf{D})$ is the set of all functions in $L^2(\mathbf{D})$ that are analytic in $\mathbf{D}$. It is well known that $A^2(\mathbf{D})$ is a closed subspace of $L^2(\mathbf{D})$ and that the orthogonal projection of $L^2(\mathbf{D})$ onto $A^2(\mathbf{D})$ acts by the rule

$$(Pf)(z) = \frac{1}{\pi} \int_{\mathbf{D}} \frac{f(w)}{(1 - z\overline{w})^2} \, dA(w), \quad z \in \mathbf{D}.$$

The Bergman space Toeplitz operator $T^B(a)$ induced by a function $a \in W$ is defined by

$$T^B(a) : A^2(\mathbf{D}) \to A^2(\mathbf{D}), \quad f \mapsto P(\hat{a} f),$$

where $\hat{a}$ is the harmonic extension of a into $\mathbf{D}$,

$$\hat{a}(re^{i\theta}) = \sum_{j=-\infty}^{\infty} a_j r^{|j|} e^{ij\theta}.$$

An orthonormal basis in $A^2(\mathbf{D})$ is built by the functions $\{e_n\}_{n=1}^{\infty}$ defined by $e_n(z) = \sqrt{n}\, z^{n-1}$.

(a) Show that the matrix representation of $T^B(a)$ in the basis $\{e_n\}_{n=1}^{\infty}$ is $(b_{jk})_{j,k=1}^{\infty}$ with

$$b_{jk} = \frac{2\sqrt{jk}}{|j-k| + j + k} a_{j-k},$$

where $a_n = \frac{1}{2\pi} \int_0^{2\pi} a(e^{i\theta}) e^{-in\theta} d\theta$.

(b) Show that the matrix $(b_{jk} - a_{j-k})_{j,k=1}^{\infty}$ induces a compact operator on the space ℓ^2.

(c) Use (a) and (b) to establish an approximation method for the equation $T^B(a) f = g$.

Notes

The material of Section 3.1 can be found in every standard text on functional analysis. Proposition 3.5 is usually referred to as Polski's theorem [207]. For a more thorough discussion of the issues of Section 3.2 we recommend the books [71], [130], and [149].

Lemma 3.6 and its various disguises were discovered repeatedly and successfully employed in several contexts by many people. We learned of it from Anatoli Kozak in the second half of the 1970s. Theorem 3.7 was established by Baxter [18] and Reich [218] for $X = \ell^1$ and by Gohberg and Feldman for $X = c_0$ and $X = \ell^p$. There exist numerous different proofs of this basic theorem. The proof given here is from [67]. In [25], Theorem 3.7 is proved for banded Toeplitz matrices with methods from the theory of difference equations.

The approach presented in Section 3.4 originated from a lucky constellation that emerged in the late 1970s. Silbermann was then embarking on Szegö's limit theorem and read Widom's paper [294] on this occasion. That paper contained in particular Proposition 3.10. At the same time, Silbermann and one of the authors were able to extend Theorem 3.7 to Toeplitz operators whose symbols are piecewise continuous with a finite number of jumps [65], and Silbermann was anxious to further extend the result to symbols with countably many jumps, that is, for general piecewise continuous symbols. In this connection, he understood that Widom's formula is the perfect tool to carry out so-called localization over a central subalgebra provided everything is appropriately adjusted. This led him to consider the Banach algebra **S** and its ideal **J**, and in the groundbreaking paper [253], which contains all the results of Section 3.4, he solved the basic problems concerning the finite section method for Toeplitz operators that had been open at that time. Moreover, this paper laid the foundation for a new level of application of Banach algebra techniques to numerical analysis and thus to an approach that has led to plenty of impressive results during the last 25 years.

Except for Theorem 3.15 and its two corollaries, the results of Section 3.5 are in principle already in [253] and [294]. We have no explicit reference for Theorem 3.15 and Corollaries 3.16 and 3.17, but these results are well known to specialists. Strohmer's paper [260] is a very readable account of several aspects of the inversion and of the inverses of positive definite (infinite and finite) Toeplitz matrices.

For Exercises 1, 8, and 9 we refer to [130]. Exercise 7 is based on [104], and Exercise 10 is from [33].

Exercise 5 is a result of [168]. We remark that the prevailing operators in discrete-time linear time-invariant systems are lower-triangular Toeplitz operators $T(b_+)$. If the inputs and outputs are in ℓ^2 (that is, of finite energy), then $T(b_+)$ must be considered on ℓ^2 and its norm is $\|T(b_+)\|_2 = \|b_+\|_\infty$. This leads to H^∞ control. However, if the inputs and outputs are from ℓ^∞ (persistent perturbations bounded in magnitude), then the relevant norm is $\|T(b_+)\|_\infty = \|b_+\|_W$, that is, the ℓ^1 norm of the coefficients of b_+. This is the origin of what is called ℓ^1 control and provides a good example for the necessity of studying Toeplitz operators not only on ℓ^2.

For the observation made in Exercise 6, see [70, Proposition 10.25]. Recently Basor and Ehrhardt [14] studied in detail the case $a = b$ and showed that

$$\det(I + T^{-1}(b)H(b)) = E(b)F(b),$$

where $E(b)$ is defined by (2.26) and, with the Wiener-Hopf factorization $b = G(b)b_-b_+$,

$$F(b) = \left(\frac{b_+(1)}{b_+(-1)}\right)^{1/2} \exp\left(-\frac{1}{2}\sum_{k=1}^{\infty} k(\log b)_k^2\right)$$
$$= \exp\left(\sum_{k=1}^{\infty}(\log b)_{2k-1} - \frac{1}{2}\sum_{k=1}^{\infty} k(\log b)_k^2\right).$$

Further results: effective solution of Toeplitz systems. Theorem 3.7 and its refinements, such as the formulas of Section 3.5, are the tools we need to study asymptotic spectral properties of large Toeplitz matrices. A big business deals with exact formulas for $T_n^{-1}(a)$ and with fast ($O(n^2)$ operations) or superfast ($O(n(\log n)^\delta)$ operations) algorithms for the solution of finite Toeplitz systems $T_n(a)x = y$. As this is not the topic of the present book, we confine ourselves to a few modest remarks. Trench [276], [277] established recursion formulas that allowed him to obtain all entries of $T_n^{-1}(a)$ from the entries of the first and last columns of $T_n^{-1}(a)$. This result was independently found by Gohberg and Sementsul [134], who, moreover, formulated the result in beautiful matrix language. The Gohberg-Sementsul formula is as follows: if $T_n(a)$ is invertible, if $(x_1 \ \dots \ x_n)^\top$ and $(y_1 \ \dots \ y_n)^\top$ are the first and last columns of $T_n^{-1}(a)$, respectively, and if $x_1 \neq 0$, then

$$T_n^{-1}(a) = \frac{1}{x_1}\begin{pmatrix} x_1 & & & \\ x_2 & x_1 & & \\ \vdots & \vdots & \ddots & \\ x_n & x_{n-1} & \dots & x_1 \end{pmatrix}\begin{pmatrix} y_n & y_{n-1} & \dots & y_1 \\ & y_n & \dots & y_2 \\ & & \ddots & \vdots \\ & & & y_n \end{pmatrix}$$
$$- \frac{1}{x_1}\begin{pmatrix} 0 & & & & \\ y_1 & 0 & & & \\ y_2 & y_1 & 0 & & \\ \vdots & \vdots & \ddots & & \\ y_{n-1} & y_{n-2} & \dots & y_1 & 0 \end{pmatrix}\begin{pmatrix} 0 & x_n & x_{n-1} & \dots & x_2 \\ & 0 & x_n & \dots & x_3 \\ & & & \ddots & \vdots \\ & & & & x_n \\ & & & & 0 \end{pmatrix}.$$

This representation of $T_n^{-1}(a)$ via triangular matrices can be used to design effective algorithms for inverting $T_n(a)$ or for solving systems of the form $T_n(a)x = y$. There exist many other formulas of the above type, in particular, formulas that do not need any additional assumptions such as $x_1 \neq 0$. We refer the reader to Heinig and Rost's books [155], [157] and the references therein for more on this subject.

Formulas for $T_n^{-1}(a)$ in which the triangular matrices of the Gohberg-Sementsul formula are replaced by circulant matrices were detected by Lerer and Tismenetsky [188] and by Ammar and Gader [3]. These lead to different effective algorithms for finite Toeplitz systems. Finally, formulas involving only diagonal matrices and a few discrete Fourier transforms as well as the associated algorithms for systems with Toeplitz matrices can be found in [156] and the references listed there.

Further results: approximating inverses of Toeplitz matrices by circulants. Given a Laurent polynomial b, define the circulant matrix $C_n(b)$ as in Section 2.1. Suppose b has no zeros on $\mathbf{T}$ and wind $b = 0$. Then $T_n(b)$ and $C_n(b)$ are invertible for all sufficiently large n.

Theorem 3.7 implies that the first column of $T_n^{-1}(b)$ converges to the first column of $T^{-1}(b)$. If $b = b_- b_+$ is a Wiener-Hopf factorization subject to the normalization $(b_-)_0 = 1$, then the first column of $T^{-1}(b)$ is

$$T^{-1}(b)e_0 = T(b_+^{-1})T(b_-^{-1})e_0 = T(b_+^{-1})e_0 = (\ (b_+^{-1})_0\ \ (b_+^{-1})_1\ \ (b_+^{-1})_2\ \ \ldots\ )^\top.$$

On the other hand, using (2.6) it easy to see that the first column of $C_n^{-1}(b)$ converges to

$$\frac{1}{2\pi}(\ (b^{-1})_0\ \ (b^{-1})_1\ \ (b^{-1})_2\ \ \ldots\ )^\top.$$

Consequently, in general one cannot approximate the first and last columns of $T_n^{-1}(b)$ by the corresponding columns of $C_n^{-1}(b)$. However, things are different for the central columns. Let $t_j^{(n)}$ and $c_j^{(n)}$ denote the jth column of $T_n^{-1}(b)$ and $C_n^{-1}(b)$, respectively. Fix a natural number K. Then $\|t_{j_n}^{(n)} - c_{j_n}^{(n)}\|_2 \to 0$ exponentially fast as $n \to \infty$ whenever $|j_n - n/2| \le K$ for all n. Thus, the middle columns of the inverses of banded Toeplitz matrices are approximated by the corresponding columns of appropriate circulant matrices exponentially well. This was first observed in [260] and [261] for positive symbols b and was proved in [60] for symbols with no zeros on $\mathbf{T}$ and with winding number zero. Figure 3.4 illustrates the phenomenon. We took the same symbol b as in Example 3.18 and computed $\|t_j^{(n)} - c_j^{(n)}\|_2$ $(1 \le j \le n)$ for $n = 50$, $n = 100$, and $n = 150$.

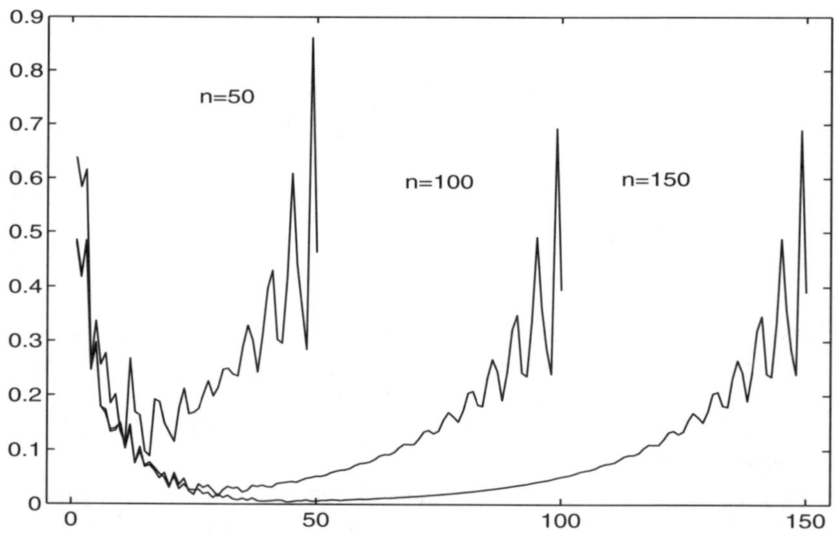

Figure 3.4. *The norms* $\|t_j^{(n)} - c_j^{(n)}\|_2$ *versus* $1 \le j \le n$ *for three choices of* n.

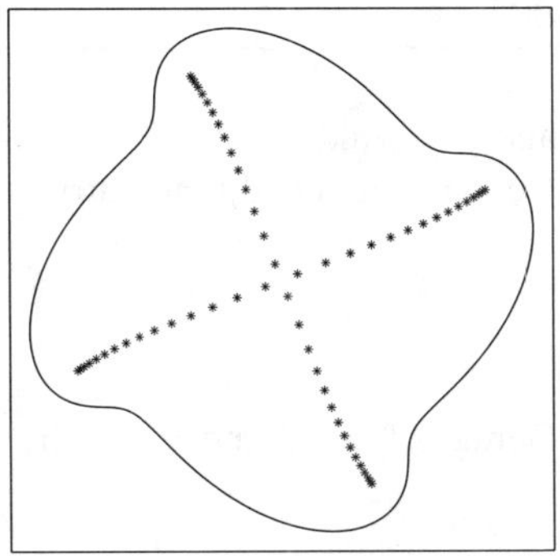

Chapter 4

Instability

We know from the Baxter-Gohberg-Feldman theorem that $\|T_n^{-1}(b)\|_p \to \infty$ if $T(b)$ is a noninvertible Toeplitz band matrix. This chapter is devoted to estimates of the growth of $\|T_n^{-1}(b)\|_p$ as $n \to \infty$. We can slightly restate the problem. Namely, let $\lambda \in \mathbf{C}$ and consider $\|T_n^{-1}(b-\lambda)\|_p$. If $\lambda \notin \operatorname{sp} T(b)$, then the theorem tells us that $\|T_n^{-1}(b-\lambda)\|_p$ remains bounded. In this chapter we consider the case where $\lambda \in \operatorname{sp} T(b)$. We first embark on the case where $\lambda \in \operatorname{sp} T(b) \setminus \operatorname{sp}_{\mathrm{ess}} T(b)$, and after this we pass to numbers λ in $\operatorname{sp}_{\mathrm{ess}} T(b)$.

4.1 Outside the Essential Spectrum

Let b be a Laurent polynomial which does not vanish on $\mathbf{T}$ and write b in the form (1.11). As in Section 1.7, we define $\delta \in (0,1)$ and $\mu \in (1,\infty)$ by

$$\delta = \max(|\delta_1|, \dots, |\delta_J|), \quad \mu = \min(|\mu_1|, \dots, |\mu_K|).$$

Finally, let ℓ_n^p stand for $\mathbf{C}^n$ with the ℓ^p norm.

Theorem 4.1. *Let b be a Laurent polynomial and suppose* $\operatorname{wind} b \neq 0$. *Let further $1 \le p \le \infty$. Then for every*

$$\alpha < \min\left(\log\frac{1}{\delta}, \log\mu\right)$$

there is a constant C_α depending only on α (and b, p) such that

$$\|T_n^{-1}(b)\|_p \ge C_\alpha e^{\alpha n} \text{ for all } n \ge 1. \tag{4.1}$$

Proof. Put $\chi_k(t) = t^k$ $(t \in \mathbf{T})$. We have $b = \chi_\kappa c_- c_+$ with

$$c_-(t) = b_s \prod_{j=1}^{J}\left(1 - \frac{\delta_j}{t}\right), \quad c_+(t) = \prod_{k=1}^{K}(t - \mu_k),$$

and $\kappa = \text{wind}\, b \in \mathbf{Z} \setminus \{0\}$. We assume without loss of generality that $\kappa < 0$, since otherwise we may pass to adjoint matrices. By Lemma 1.17,

$$c_+^{-1}(t) = \sum_{\ell=0}^{\infty} d_\ell t^\ell \quad \text{with} \quad d_\ell = O(e^{-\alpha \ell}).$$

Define $x^{(n)} \in \ell_n^p$ and $x \in \ell^p$ by $x^{(n)} = (d_0, \dots, d_{n-1})$ and $x = \{d_0, d_1, \dots\}$. Clearly,

$$T_n(b)x^{(n)} = P_n T(c_-)T(\chi_\kappa)T(c_+)x^{(n)}. \tag{4.2}$$

Since $\kappa < 0$, we get $P_n(T(c_-)T(\chi_\kappa)e_0) = 0$, and as $e_0 = T(c_+)x$, it follows that

$$P_n T(c_-)T(\chi_\kappa)T(c_+)x = 0. \tag{4.3}$$

From (4.2) and (4.3) we obtain that

$$T_n(b)x^{(n)} = P_n T(c_-)T(\chi_\kappa)T(c_+)(x^{(n)} - x) = P_n T(b)(x^{(n)} - x),$$

and since $\|x^{(n)} - x\|_p = O(e^{-\alpha n})$, we arrive at the estimate

$$\|T_n(b)x^{(n)}\|_p \le \|T(b)\|_p \|x^{(n)} - x\|_p \le D_\alpha e^{-\alpha n}.$$

Taking into account that $\|x^{(n)}\|_p \to \|x\|_p > 0$ as $n \to \infty$, we see that, for all sufficiently large n,

$$\|T_n^{-1}(b)\|_p \ge \frac{\|x^{(n)}\|_p}{\|T_n(b)x^{(n)}\|_p} \ge \frac{(1/2)\|x\|_p}{D_\alpha e^{-\alpha n}} = B_\alpha e^{\alpha n}. \qquad \square$$

Figure 4.1 shows the norms $\|T_n^{-1}(b - \lambda)\|_2$ $(5 \le n \le 80)$ for $b(t) = t^{-2} + 0.75 \cdot t^{-1} + 0.65 \cdot t$ and $\lambda = -0.5, 0.82, 0.83 + 0.7i$ (top pictures and left picture in the middle) and for $b(t) = t^{-2} - 2t^{-1} + 1.25 \cdot t^3$ and $\lambda = -3.405, 1.48, 0.995 + 3i$ (right middle picture and bottom pictures). The curve $b(\mathbf{T})$ and the point λ are indicated in the lower right corners of the pictures.

4.2 Exponential Growth Is Generic

Theorem 4.1 provides us with a general lower bound for the norms $\|T_n^{-1}(b)\|_p$ in case b has no zeros but nonvanishing winding number. In this section we consider the problem of finding upper bounds. If $b(t) = t$, then $T_n(b)$ is not invertible and hence $\|T_n^{-1}(b)\|_p = \infty$ for all $n \ge 1$. This indicates that a universal upper bound for the growth of $\|T_n^{-1}(b)\|_p$ will hardly exist.

Hadamard's inequality says that if a matrix $X \in \mathbf{C}^{n \times n}$ has the columns $x_1, \dots, x_n$, then

$$|\det X| \le \|x_1\|_2 \cdots \|x_n\|_2,$$

where $\|x_j\|_2$ is the ℓ^2 norm of x_j. Note that if $b(t) = \sum_{j=-r}^{s} b_j t^j$ $(t \in \mathbf{T})$, then

$$\sum_{j=-r}^{s} |b_j|^2 = \frac{1}{2\pi} \int_0^{2\pi} |b(e^{i\theta})|^2 d\theta = \|b\|_2^2. \tag{4.4}$$

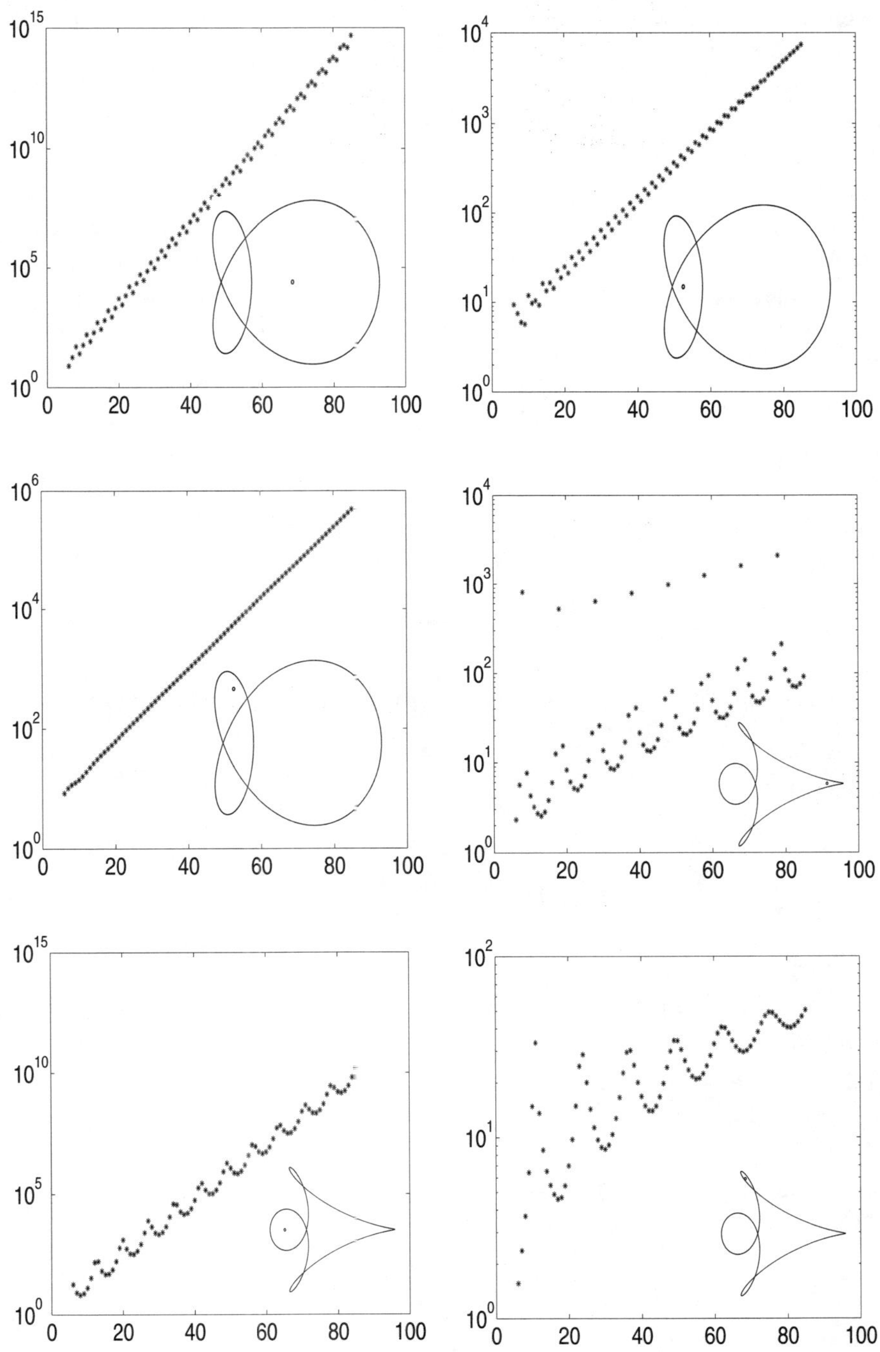

Figure 4.1. *Norms $\|T_n^{-1}(b-\lambda)\|_2$ for two symbols b and three λ's.*

Theorem 4.2. *Let b be a Laurent polynomial and suppose that the determinant $D_n(b) :=$ $\det T_n(b)$ is nonzero. Then for every $1 \le p \le \infty$,*

$$\|T_n^{-1}(b)\|_p \le \frac{1}{|D_n(b)|}\, n^{1/q}\, \|b\|_2^{n-1}, \tag{4.5}$$

where $1/p + 1/q = 1$ and $n^{1/\infty} := 1$.

Proof. We have $T_n^{-1}(b) = (1/D_n(b))A_n(b)$, where $A_n(b) = (a_{jk}^{(n)}(b))_{j,k=1}^n$ and the number $a_{jk}^{(n)}(b)$ is $(-1)^{j+k}$ times the determinant of the matrix arising from $T_n(b)$ by deleting the kth row and the jth column. Simple application of Hölder's inequality shows that

$$\|A_n(b)\|_p \le \max_{1\le j\le n} \left(|a_{j1}^{(n)}(b)|^q + \cdots + |a_{jn}^{(n)}(b)|^q\right)^{1/q}. \tag{4.6}$$

By Hadamard's inequality,

$$|a_{jk}^{(n)}(b)| \le \underbrace{\left(\sum_{j=-r}^{s} |b_j|^2\right)^{1/2} \cdots \left(\sum_{j=-r}^{s} |b_j|^2\right)^{1/2}}_{n-1},$$

and (4.4) therefore gives $|a_{jk}^{(n)}(b)| \le \|b\|_2^{n-1}$. Combining this estimate and (4.6) we arrive at (4.5). □

Example 4.3. Let

$$T(b) = \begin{pmatrix} 0 & -4 & 0 & \dots \\ 1 & 0 & -4 & \dots \\ 0 & 1 & 0 & \dots \\ \dots & \dots & \dots & \dots \end{pmatrix}.$$

Then $b(t) = t - 4t^{-1} = t^{-1}(t-2)(t+2)$, which shows that $\operatorname{wind} b = -1$. From Example 2.9 we get

$$|D_n(b)| = \frac{|2^{n+1} - (-2)^{n+1}|}{|2-(-2)|} = \begin{cases} 2^n & \text{if } n \text{ is even,} \\ 0 & \text{if } n \text{ is odd.} \end{cases}$$

By virtue of (4.4), $\|b\|_2 = \sqrt{17}$. Hence, (4.5) implies that $\|T_n^{-1}(b)\|_2$ is at most

$$\frac{1}{\sqrt{17}}\, n^{1/2} \left(\frac{\sqrt{17}}{2}\right)^n = \frac{1}{\sqrt{17}}\, n^{1/2}\, e^{n \log(\sqrt{17}/2)} < 0.25\, n\, e^{0.723\, n} < 0.25\, e^{0.73\, n}$$

for all sufficiently large even n. On the other hand, (4.1) with $\alpha = \log 2 - 0.001$ shows that $\|T_n^{-1}(b)\|_2 \ge Ce^{n(\log 2 - 0.001)} > Ce^{0.69\, n}$ with some constant $C > 0$ for all n. In summary, there are $C_1, C_2 \in (0,\infty)$ such that $C_1 e^{0.69\, n} \le \|T_n^{-1}(b)\|_2 \le C_2\, e^{0.73\, n}$ for all even n, and we have $\|T_n^{-1}(b)\|_2 = \infty$ for all odd n. □

Let $\mathcal{D}$ be the set of all Laurent polynomials of the form

$$b(t) = b_s t^{-r}(t - z_1) \cdots (t - z_{r+s}) \quad (t \in \mathbf{T}) \tag{4.7}$$

with $b_s \neq 0$ and $0 < |z_1| < |z_2| < \cdots < |z_{r+s}|$.

Corollary 4.4. *Let $1 \leq p \leq \infty$. If $b \in \mathcal{D}$, then there are constants $\gamma \in (0, \infty)$ and $D_\gamma \in (0, \infty)$ depending only on b and p such that*

$$\|T_n^{-1}(b)\|_p \leq D_\gamma e^{\gamma n} \text{ for all } n \geq 1. \tag{4.8}$$

Proof. From Theorem 2.8 we deduce that

$$|D_n(b)| = A\,|b_s|^n |z_{r+1} \cdots z_{r+s}|^n (1 + O(q^n)) \tag{4.9}$$

with some $A \neq 0$ and some $q \in (0, 1)$. Combining (4.9) and (4.5), we get (4.8). □

Corollary 4.5. *Let $1 \leq p \leq \infty$ and let $\mathcal{E}$ be the set of all Laurent polynomials that have no zeros on $\mathbf{T}$ and whose winding number is nonzero.*

(a) *$\mathcal{E} \cap \mathcal{D}$ is a dense and open subset of the set $\mathcal{E}$ (with the uniform metric).*

(b) *If $b \in \mathcal{E} \cap \mathcal{D}$, then there are constants $C_1, C_2 \in (0, \infty)$ and $\gamma_1, \gamma_2 \in (0, \infty)$ depending only on b and p such that*

$$C_1 e^{\gamma_1 n} \leq \|T_n^{-1}(b)\|_p \leq C_2 e^{\gamma_2 n} \text{ for all } n \geq 1. \tag{4.10}$$

Proof. (a) It is clear that $\mathcal{E} \cap \mathcal{D}$ is an open subset of $\mathcal{E}$. Every Laurent polynomial b can be written in the form (4.7). Put

$$b_{\varepsilon_1,\ldots,\varepsilon_{r+s}}(t) = b_s t^{-r}(t - z_1 - \varepsilon_1) \cdots (t - z_{r+s} - \varepsilon_{r+s}).$$

Clearly, given any $\varepsilon > 0$, we can find $\varepsilon_1, \ldots, \varepsilon_{r+s} \in \mathbf{C}$ such that the moduli

$$|z_1 + \varepsilon_1|, \ldots, |z_{r+s} + \varepsilon_{r+s}|$$

are pairwise distinct and such that $\|b - b_{\varepsilon_1,\ldots,\varepsilon_{r+s}}\|_\infty < \varepsilon$. If $b \in \mathcal{E}$, then $b_{\varepsilon_1,\ldots,\varepsilon_{r+s}}$ can obviously also be chosen in $\mathcal{E}$. This proves that $\mathcal{E} \cap \mathcal{D}$ is dense in $\mathcal{E}$.

(b) The lower estimate of inequality (4.10) holds for every $b \in \mathcal{E}$ by virtue of Theorem 4.1, and the upper estimate of (4.10) is satisfied for every $b \in \mathcal{D}$ due to Corollary 4.4. □

In view of the preceding corollary, we may say that exponential growth of the norms $\|T_n^{-1}(b)\|_p$ is generic for Toeplitz band matrices which generate Fredholm operators of nonzero index.

4.3 Arbitrarily Fast Growth

In the previous section we observed that if $T(b)$ is Fredholm of nonzero index, then there may be infinitely many n such that $\|T_n^{-1}(b)\|_p = \infty$. Example 4.3 gives rise to the following question: Is there a kind of function $\varphi : \mathbf{N} \to \mathbf{N}$ (for example, $\varphi(n) = e^{\gamma n}$) such that either

$\|T_n^{-1}(b)\|_p = \infty$ or $\|T_n^{-1}(b)\|_p \le C\varphi(n)$ with some constant $C \in (0, \infty)$ independent of n? The purpose of the present section is to show that the answer to this question is no.

Pick $\alpha \in (0, 1)$ and put

$$b(t) = t + \alpha^2 t^{-1} = t^{-1}(t + i\alpha)(t - i\alpha) \quad (t \in \mathbf{T}). \tag{4.11}$$

Since $b(e^{i\theta}) = (1 + \alpha^2)\cos\theta + i(1 - \alpha^2)\sin\theta$, we see that $b(\mathbf{T})$ is an ellipse with the foci -2α and 2α. If $\lambda \in (-2\alpha, 2\alpha)$, then $b - \lambda$ has no zeros on $\mathbf{T}$ and wind $(b - \lambda)$ equals 1.

Theorem 4.6. *Let $\varphi : \mathbf{N} \to \mathbf{N}$ be any monotonically increasing function, for example, $\varphi(n) = \exp(n^n)$, and let $1 \le p \le \infty$. Then, with b given by* (4.11), *there exists a number $\lambda \in (-2\alpha, 2\alpha)$ such that $\|T_n^{-1}(b)\|_p < \infty$ for all $n \ge 1$ and $\|T_{n_k}^{-1}(b)\|_p > n_k\,\varphi(n_k)$ for infinitely many $n_k \in \mathbf{N}$.*

Proof. Every $\lambda \in (-2\alpha, 2\alpha)$ can be written in the form $\lambda = 2\alpha\cos y$ with $y \in (0, \pi)$. By Theorem 2.4, the eigenvalues of $T_n(b - \lambda)$ are

$$\lambda_j^{(n)} := 2\alpha\cos\frac{\pi j}{n+1} - 2\alpha\cos y, \quad j \in \{1, \dots, n\}. \tag{4.12}$$

Notice that

$$|\lambda_j^{(n)}| = 4\alpha\left|\sin\frac{1}{2}\left(\frac{\pi j}{n+1} - y\right)\sin\frac{1}{2}\left(\frac{\pi j}{n+1} + y\right)\right| \le 2\alpha\pi\left|\frac{y}{\pi} - \frac{j}{n+1}\right|. \tag{4.13}$$

Let $\lambda(n)$ be the minimum of $|\lambda_1^{(n)}|, \dots, |\lambda_n^{(n)}|$. Since

$$\|T_n^{-1}(b - \lambda)\|_p \ge \text{rad}\,\left(T_n^{-1}(b - \lambda)\right) = 1/\lambda(n),$$

where rad $(\cdot)$ denotes the spectral radius, it suffices to prove that there is a y such that $0 < \lambda(n)$ for all $n \ge 1$ and $\lambda(n_k) < 1/(n_k\varphi(n_k))$ for infinitely many n_k.

Pick any natural number $N_1 \ge 1$ and choose natural numbers $N_1 < N_2 < N_3 < \cdots$ successively by requiring that

$$2\alpha\pi\left(10^{N_1+\cdots+N_k} - 1\right)\varphi\left(10^{N_1+\cdots+N_k} - 1\right) < 10^{N_{k+1}} \quad (k \ge 1). \tag{4.14}$$

Put $n_k := 10^{N_1+\cdots+N_k} - 1$ and $y/\pi := 10^{-N_1} + 10^{-N_1-N_2} + 10^{-N_1-N_2-N_3} + \cdots$. Obviously, y/π is irrational. This implies that none of the eigenvalues (4.12) is zero, and hence $\lambda(n) > 0$ for all $n \ge 1$. As

$$0 < 10^{-N_1} + 10^{-N_1-N_2} + \cdots + 10^{-N_1-\cdots-N_k} < 1,$$

it follows that $10^{-N_1} + 10^{-N_1-N_2} + \cdots + 10^{-N_1-\cdots-N_k} = j_k\,10^{-N_1-\cdots-N_k}$ with a natural number j_k satisfying $1 \le j_k \le 10^{N_1+\cdots+N_k} - 1 = n_k$. We have

$$\begin{aligned}\frac{y}{\pi} - \frac{j_k}{n_k+1} &= \left(10^{-N_1} + 10^{-N_1-N_2} + \cdots\right) \\ &\quad - \left(10^{-N_1} + 10^{-N_1-N_2} + \cdots + 10^{-N_1-\cdots-N_k}\right) \\ &= 10^{-N_1-\cdots-N_{k+1}} + 10^{-N_1-\cdots-N_{k+2}} + \cdots,\end{aligned}$$

which shows that

$$0 < \frac{y}{\pi} - \frac{j_k}{n_k + 1} < 2 \cdot 10^{-N_1 - \cdots - N_{k+1}} < 10^{-N_{k+1}} < \frac{1}{\alpha\pi\, n_k \varphi(n_k)},$$

the last inequality resulting from (4.14). Thus, by (4.13),

$$\lambda(n) \le \left|\lambda_{j_k}^{(n)}\right| < \alpha\pi / (\alpha\pi\, n_k \varphi(n_k)) = 1/(n_k \varphi(n_k)). \qquad \square$$

In Figure 4.2 we see plots of $\|T_n^{-1}(b-\lambda)\|_2$ for $b(t) = t + 0.49\, t^{-1}$ and six different choices of λ ($\lambda = 1.4$ and $\lambda = 0.01$ in the top, $\lambda = 0$ and $\lambda = 0.7$ in the middle, and $\lambda = 0.06$ and $\lambda = 1.294$ in the bottom). If $\|T_n^{-1}(b-\lambda)\|_2$ exceeded 10^{15}, we plotted the value 10^{17}.

4.4 Sequences Versus Polynomials

This section serves as the preparation of the remaining sections of this chapter, which are devoted to the growth of $\|T_n^{-1}(b-\lambda)\|_2$ in case λ belongs to $\mathrm{sp}_{\mathrm{ess}} T(b) = b(\mathbf{T})$.

In Section 1.6, we identified ℓ^2 with the Hardy space H^2. In what follows, it will be convenient to identify ℓ_n^2 (= the linear space $\mathbf{C}^n$ with the ℓ^2 norm) with the space $\mathcal{P}_n^+$ of all analytic polynomials of degree at most $n-1$. Thus, with $x = (x_0, x_1, \ldots, x_{n-1}) \in \ell_n^2$ we associate the polynomial

$$f(e^{i\theta}) = x_0 + x_1 e^{i\theta} + \cdots + x_{n-1} e^{i(n-1)\theta}. \tag{4.15}$$

The scalar product and the norm in $\mathcal{P}_n^+$ are those of $L^2 = L^2(\mathbf{T})$:

$$(f, g) = \int_0^{2\pi} f(e^{i\theta})\overline{g(e^{i\theta})}\,\frac{d\theta}{2\pi}, \quad \|f\|_2^2 = \int_0^{2\pi} |f(e^{i\theta})|^2\,\frac{d\theta}{2\pi}.$$

Lemma 4.7. *Let $a \in W$. If $x, y \in \ell_n^2$ and f, g are the corresponding polynomials in $\mathcal{P}_n^+$, then*

$$(T_n(a)x, y) = \frac{1}{2\pi}\int_0^{2\pi} a(e^{i\theta}) f(e^{i\theta})\overline{g(e^{i\theta})}\,d\theta. \tag{4.16}$$

In particular,

$$(T_n(a)x, x) = \frac{1}{2\pi}\int_0^{2\pi} a(e^{i\theta})|f(e^{i\theta})|^2\,d\theta, \tag{4.17}$$

$$\|x\|_2 = \|f\|_2. \tag{4.18}$$

Proof. We have

$$\int_0^{2\pi} a(e^{i\theta}) f(e^{i\theta})\overline{g(e^{i\theta})}\,d\theta = \int_0^{2\pi} \left(\sum_\ell a_\ell e^{i\ell\theta}\right)\left(\sum_k x_k e^{ik\theta}\right)\left(\sum_j \overline{y}_j e^{-ij\theta}\right) d\theta$$

$$= \sum_{\ell,j,k} a_\ell x_k \overline{y}_j \int_0^{2\pi} e^{i(\ell+k-j)}\,d\theta = 2\pi \sum_{j,k} a_{j-k} x_k \overline{y}_j = 2\pi (T_n(a)x, y). \qquad \square$$

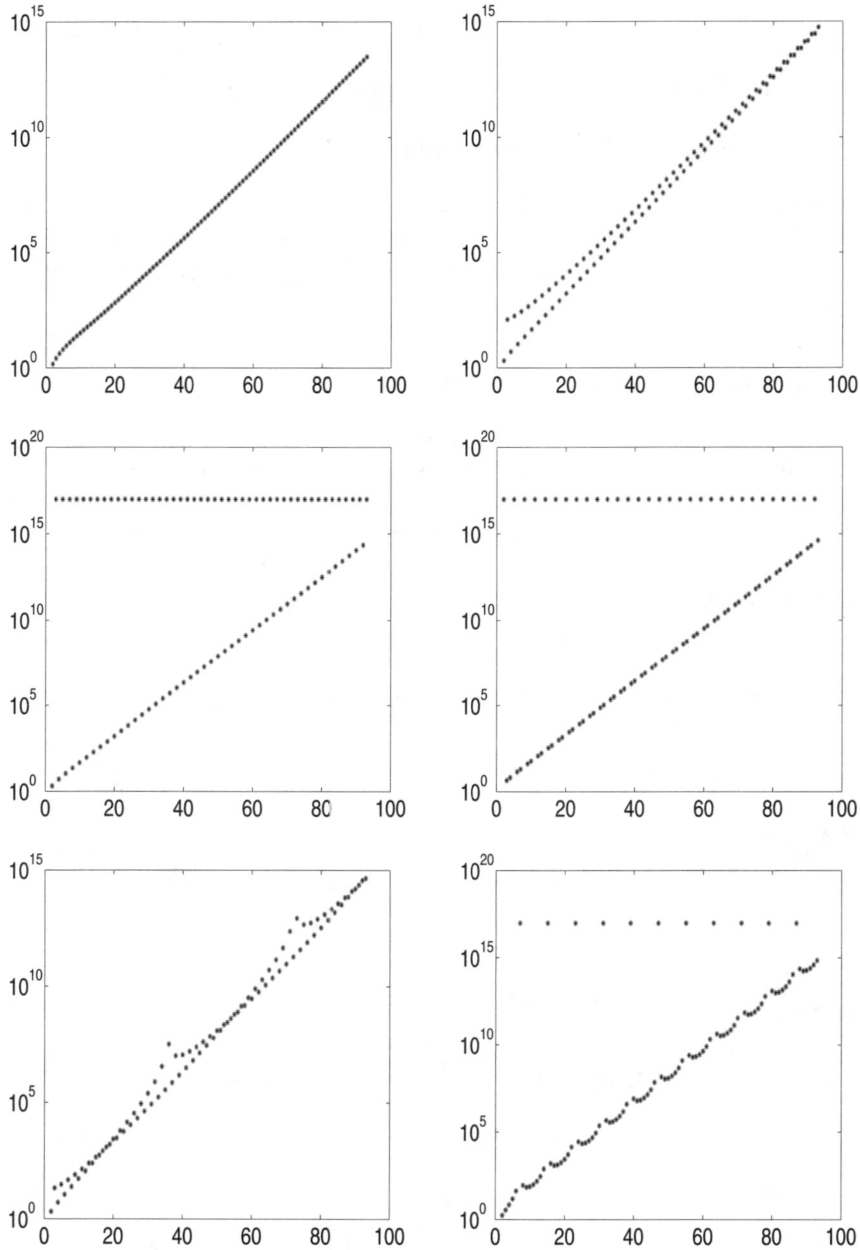

Figure 4.2. *The norms* $\|T_n^{-1}(b-\lambda)\|_2$ *against* n *for* $b(t) = t + 0.49\,t^{-1}$ *and six different choices of* λ.

Lemma 4.8. *If $a \in W$ and if $x \in \ell^2_n$ and $f \in \mathcal{P}^+_n$ are related by* (4.15), *then*

$$\|T_n(a)x\|_2 \le \|af\|_2. \tag{4.19}$$

Proof. The jth Fourier coefficient of af is $\sum_{k=0}^{n-1} a_{j-k}x_k$. As the map (1.17) is an isometry, we therefore get

$$\|af\|_2^2 = \sum_{j=-\infty}^{\infty} \left| \sum_{k=0}^{n-1} a_{j-k}x_k \right|^2 \ge \sum_{j=0}^{n-1} \left| \sum_{k=0}^{n-1} a_{j-k}x_k \right|^2 = \|T_n(a)x\|_2^2. \qquad \square$$

Lemma 4.9. *If $a \in W$ and $f \in \mathcal{P}^+_n$, then $\|T_n^{-1}(a)\|_2 \ge \|f\|_2/\|af\|_2$.*

Proof. By the definition of the norm,

$$\|T_n^{-1}(a)\|_2 = \sup_{x\ne 0} \frac{\|T_n^{-1}(a)x\|_2}{\|x\|_2} = \sup_{y\ne 0} \frac{\|y\|_2}{\|T_n(a)y\|_2}.$$

Hence, if $y \in \ell^2_n$ corresponds to $f \in \mathcal{P}^+_n$, then $\|T_n^{-1}(a)\|_2 \ge \|y\|_2/\|T_n(a)y\|_2$. From (4.18) and (4.19) we finally get $\|y\|_2/\|T_n(a)y\|_2 \ge \|f\|_2/\|af\|_2$. $\square$

A class of good test polynomials. For $j, m \in \mathbf{N}$, consider the Laurent polynomial

$$p_m^j(e^{i\theta}) = \left(1 + e^{i\theta} + \cdots + e^{im\theta}\right)^j. \tag{4.20}$$

Obviously,

$$p_m^j(e^{i\theta}) = \left(\frac{1 - e^{i(m+1)\theta}}{1 - e^{i\theta}}\right)^j = e^{imj\theta/2} \left(\frac{\sin \frac{m+1}{2}\theta}{\sin \frac{\theta}{2}}\right)^j. \tag{4.21}$$

From (4.20) we see that $p_m^j \in \mathcal{P}^+_{mj+1}$. Both (4.20) and (4.21) immediately show that

$$\|p_m^j\|_\infty = (m+1)^j.$$

From (4.18) we deduce that

$$\|p_m^1\|_2^2 = 1^2 + \cdots + 1^2 = m + 1. \tag{4.22}$$

The following lemma gives good estimates for $\|p_m^j\|_2$ in the case $j \ge 2$.

Lemma 4.10. *For each $j \ge 1$, there is a constant $D_j \in (0, \infty)$ such that*

$$(1/D_j)\, m^{2j-1} \le \|p_m^j\|_2^2 \le D_j\, m^{2j-1}.$$

Proof. For $j = 1$, this follows from (4.22). So let $j \ge 2$. Then

$$\begin{aligned} 2\pi\, \|p_m^j\|_2^2 &= \int_{-\pi}^{\pi} \left(\frac{\sin((m+1)\theta/2)}{\theta/2}\right)^{2j} \left(\frac{\theta/2}{\sin(\theta/2)}\right)^{2j} d\theta \\ &= \int_{-\pi}^{\pi} \left(\frac{\sin((m+1)\theta/2)}{\theta/2}\right)^{2j} \left(1 + O(\theta^2)\right) d\theta, \end{aligned}$$

and the substitution $x = (m+1)\theta/2$ gives

$$\int_{-\pi}^{\pi} \left(\frac{\sin((m+1)\theta/2)}{\theta/2}\right)^{2j} O(\theta^2)d\theta$$
$$= O\left(8(m+1)^{2j-3}\int_{-\pi(m+1)/2}^{\pi(m+1)/2}\left(\frac{\sin x}{x}\right)^{2j} x^2 dx\right) = O(m^{2j-3})$$

and

$$\int_{-\pi}^{\pi} \left(\frac{\sin((m+1)\theta/2)}{\theta/2}\right)^{2j} d\theta = 2(m+1)^{2j-1}\int_{-\pi(m+1)/2}^{\pi(m+1)/2}\left(\frac{\sin x}{x}\right)^{2j} dx$$
$$= 2(m+1)^{2j-1}\left(\int_{-\infty}^{\infty}\left(\frac{\sin x}{x}\right)^{2j} dx + o(1)\right).$$

This shows that

$$\lim_{m\to\infty} m^{-(2j-1)}\|p_m^j\|_2^2 = \frac{1}{\pi}\int_{-\infty}^{\infty}\left(\frac{\sin x}{x}\right)^{2j} dx. \qquad \square$$

4.5 Symbols with Zeros: Lower Estimates

Let b be a Laurent polynomial. We now consider $\|T_n^{-1}(b-\lambda)\|_2$ for $\lambda \in \mathrm{sp}_{\mathrm{ess}} T(b) = b(\mathbf{T})$. Equivalently, we study the behavior of $\|T_n^{-1}(b)\|_2$ in the case where b has zeros on $\mathbf{T}$.

So suppose $b(t_0) = 0$ for some $t_0 = e^{i\theta_0} \in \mathbf{T}$. Also assume that b does not vanish identically. The function b is analytic in $\mathbf{C} \setminus \{0\}$, and there is a smallest natural number $\alpha \geq 1$ such that the αth derivative $b^{(\alpha)}(t_0)$ is nonzero. We then have

$$b(z) = \frac{b^{(\alpha)}(t_0)}{\alpha!}(z-t_0)^\alpha + O\left((z-t_0)^{\alpha+1}\right)$$

as $z \in \mathbf{C}$ approaches t_0. We call α the *order of the zero* t_0. Clearly, if b has a zero of order α at t_0, then

$$|b(t)| \leq K|t-t_0|^\alpha \quad \text{for all } t \in \mathbf{T}$$

and hence

$$|b(e^{i\theta})| \leq K|\theta-\theta_0|^\alpha \quad \text{for all } \theta \in \mathbf{R}$$

with some constant $K \in (0,\infty)$.

Theorem 4.11. *Let b be a Laurent polynomial and suppose that b has a zero of order α at $t_0 \in \mathbf{T}$. Then there is a constant $C \in (0,\infty)$ independent of n such that $\|T_n^{-1}(b)\|_2 \geq C n^\alpha$ for all $n \geq 1$.*

Proof. Put $c(t) = b(t/t_0)$ $(t \in \mathbf{T})$. Then

$$T_n(c) = \mathrm{diag}\,(1, t_0^{-1}, \dots, t_0^{-(n-1)})\, T_n(b)\, \mathrm{diag}\,(1, t_0, \dots, t_0^{n-1}),$$

whence $\|T_n^{-1}(b)\|_2 = \|T_n^{-1}(c)\|_2$. Thus, we may assume without loss of generality that $t_0 = 1$. Then $|b(e^{i\theta})| \le K\,|\theta|^\alpha$ for all $\theta \in \mathbf{R}$. Let $n \ge \alpha + 2$. There is a unique $m \in \mathbf{N}$ such that $n = m(\alpha+1) + k$ with $k \in \{1, \dots, \alpha+1\}$. We show that

$$\|bp_m^{\alpha+1}\|_2^2 \,/\, \|p_m^{\alpha+1}\|_2^2 \le D\,m^{-2\alpha} \tag{4.23}$$

for some constant $D \in (0, \infty)$ independent of m. Since $p_m^{\alpha+1}$ belongs to $\mathcal{P}_n^+$ and $m \ge (n-\alpha-1)/(\alpha+1)$, the assertion then follows from Lemma 4.9.

We have

$$2\pi\,\|bp_m^{\alpha+1}\|_2^2 = \int_{-\pi}^{\pi} |b(e^{i\theta})|^2\,|p_m^{\alpha+1}(e^{i\theta})|^2 d\theta.$$

Since $\|p_m^{\alpha+1}\|_\infty = (m+1)^{\alpha+1}$, it follows that

$$\int_{-1/m}^{1/m} |b(e^{i\theta})|^2\,|p_m^{\alpha+1}(e^{i\theta})|^2 d\theta \le K^2 \frac{1}{m^{2\alpha}}(m+1)^{2\alpha+2} \int_{-1/m}^{1/m} d\theta \le M_1 m$$

with some $M_1 < \infty$ independent of m. As

$$|p_m^{\alpha+1}(e^{i\theta})|^2 < \left(\sin\frac{\theta}{2}\right)^{-(2\alpha+2)} < \left(\frac{\pi}{\theta}\right)^{2\alpha+2}$$

for $0 < |\theta| < \pi$, we get

$$\int_{1/m<|\theta|<\pi} |b(e^{i\theta})|^2\,|p_m^{\alpha+1}(e^{i\theta})|^2 d\theta < 2K^2 \int_{1/m}^{\pi} \theta^{2\alpha}(\pi/\theta)^{2\alpha+2} d\theta$$
$$= 2K^2\pi^{2\alpha+2} \int_{1/m}^{\pi} \theta^{-2} d\theta < 2K^2\pi^{2\alpha+2}\frac{1}{m^2} \int_{1/m}^{\pi} d\theta \le M_2 < \infty.$$

Thus,

$$\|bp_m^{\alpha+1}\|_2^2 \le M_1 m + M_2 \le (M_1 + M_2) m. \tag{4.24}$$

From Lemma 4.10 we know that

$$\|p_m^{\alpha+1}\|_2^2 \ge (1/D_{\alpha+1})\, m^{2\alpha+1}. \tag{4.25}$$

Combining (4.24) and (4.25) we arrive at (4.23). □

Corollary 4.12. *Let b be a Laurent polynomial and assume the zeros of b on $\mathbf{T}$ are $t_1, \dots, t_k$ with the orders $\alpha_1, \dots, \alpha_k$. Then*

$$\|T_n^{-1}(b)\|_2 \ge C\, n^{\max(\alpha_1, \dots, \alpha_k)} \quad \textit{for all } n \ge 1,$$

where $C \in (0, \infty)$ is a constant independent of n. □

4.6 Symbols with Zeros: Upper Estimates

In the case where b has zeros on $\mathbf{T}$, the problem of obtaining upper estimates for $\|T_n^{-1}(b)\|_2$ is more difficult than the problem of establishing lower estimates. In this section we establish a sharp upper estimate for $\|T_n^{-1}(b)\|_2$ under the assumption that b is of the form

$$b(t) = \left(1 - \frac{1}{t}\right)^{\gamma} (1-t)^{\delta} c(t) \quad (t \in \mathbf{T}),$$

where $\gamma, \delta \in \mathbf{Z}_+$ and c is a Laurent polynomial without zeros on $\mathbf{T}$ and with winding number zero. The main result of this section, Theorem 4.17, will be used in the following section to obtain upper estimates for $\|T_n^{-1}(b)\|_2$ in the case where b is an arbitrary Laurent polynomial with a single zero on $\mathbf{T}$.

For $\beta \in \mathbf{Z}_+$, we define the Laurent polynomials ξ_β and η_β by

$$\xi_\beta(t) = \left(1 - \frac{1}{t}\right)^{\beta} = \sum_{j=0}^{\beta} (-1)^j \binom{\beta}{j} t^{-j},$$

$$\eta_\beta(t) = (1-t)^{\beta} = \sum_{j=0}^{\beta} (-1)^j \binom{\beta}{j} t^{j}.$$

The (triangular) Toeplitz matrices $T(\xi_\beta)$ and $T(\eta_\beta)$ induce bounded operators on the space ℓ^2. The numbers $(-1)^j\binom{\beta}{j}$ make sense for arbitrary $\beta \in \mathbf{Z}$, and hence we can define $T(\xi_\beta)$ and $T(\eta_\beta)$ for every $\beta \in \mathbf{Z}$ as follows: $T(\xi_\beta)$ is the upper-triangular Toeplitz matrix whose first row is

$$\left(\binom{\beta}{0}, -\binom{\beta}{1}, \binom{\beta}{2}, -\binom{\beta}{3}, \dots\right)$$

and $T(\eta_\beta)$ is the lower-triangular Toeplitz matrix with the first column

$$\left(\binom{\beta}{0}, -\binom{\beta}{1}, \binom{\beta}{2}, -\binom{\beta}{3}, \dots\right)^{\top}.$$

Clearly, $T(\xi_\beta)$ and $T(\eta_\beta)$ do not generate bounded operators on ℓ^2 if $\beta < 0$. However, these matrices induce bounded operators between certain pairs of spaces with weight. For $\beta \in \mathbf{Z}$, let $\ell^2(\beta)$ be the Hilbert space of all sequences $x = \{x_n\}_{n=1}^{\infty}$ for which

$$\|x\|_{2,\beta}^2 := \sum_{n=1}^{\infty} |x_n|^2 n^{2\beta} < \infty.$$

Theorem 4.13 (Pomp). *If* $\beta \in \mathbf{Z}_+$, *then*

$$T(\xi_{-\beta}) \in \mathcal{B}(\ell^2(\beta), \ell^2) \quad \textit{and} \quad T(\eta_{-\beta}) \in \mathcal{B}(\ell^2, \ell^2(-\beta)).$$

Proof. We prove that $T(\eta_{-1}) \in \mathcal{B}(\ell^2(-\mu), \ell^2(-\mu-1))$ for every $\mu \in \mathbf{Z}_+$. Since $T(\eta_{-\beta}) = T(\eta_{-1}) \cdots T(\eta_{-1})$ (β factors), this implies that $T(\eta_{-\beta}) \in \mathcal{B}(\ell^2, \ell^2(\beta))$, and then passage to adjoints gives $T(\xi_{-\beta}) \in \mathcal{B}(\ell^2(\beta), \ell^2)$.

For $x = \{x_n\}_{n=1}^{\infty} \in \ell^2(-\mu)$,

$$T(\eta_{-1})x = \begin{pmatrix} 1 & & & \\ 1 & 1 & & \\ 1 & 1 & 1 & \\ \dots & \dots & \dots & \dots \end{pmatrix} \begin{pmatrix} x_1 \\ x_2 \\ x_3 \\ \dots \end{pmatrix} = \begin{pmatrix} x_1 \\ x_1 + x_2 \\ x_1 + x_2 + x_3 \\ \dots \end{pmatrix},$$

whence

$$\begin{aligned} \|T(\eta_{-1})x\|_{2,-\mu-1}^2 &= |x_1|^2 + |x_1 + x_2|^2 \frac{1}{2^{2\mu+2}} + |x_1 + x_2 + x_3|^2 \frac{1}{3^{2\mu+2}} + \cdots \\ &\le |x_1|^2 + \frac{1}{2^2}\left(\frac{|x_1|}{2^\mu} + \frac{|x_2|}{2^\mu}\right)^2 + \frac{1}{3^2}\left(\frac{|x_1|}{3^\mu} + \frac{|x_2|}{3^\mu} + \frac{|x_3|}{3^\mu}\right)^2 + \cdots \\ &\le |x_1|^2 + \frac{1}{2^2}\left(|x_1| + \frac{|x_2|}{2^\mu}\right)^2 + \frac{1}{3^2}\left(|x_1| + \frac{|x_2|}{2^\mu} + \frac{|x_3|}{3^\mu}\right)^2 + \cdots. \end{aligned} \tag{4.26}$$

Using Hardy's inequality, which says that

$$a_1^2 + \left(\frac{a_1 + a_2}{2}\right)^2 + \left(\frac{a_1 + a_2 + a_3}{3}\right)^2 + \cdots \le 4\left(a_1^2 + a_2^2 + a_3^2 + \cdots\right)$$

for every sequence $\{a_1, a_2, a_3, \dots\}$ of nonnegative numbers, we obtain that (4.26) does not exceed

$$4\left(|x_1|^2 + \frac{|x_2|^2}{2^{2\mu}} + \frac{|x_3|^2}{3^{2\mu}} + \cdots\right) = 4\|x\|_{2,-\mu}^2.$$

This proves that the operator $T(\eta_{-1}) : \ell^2(-\mu) \to \ell^2(-\mu - 1)$ is bounded (and of norm at most 2). □

For $\beta \in \mathbf{Z}_+$, we put

$$M_\beta = \operatorname{diag}\left(\mu_0^{(\beta)}, \mu_1^{(\beta)}, \mu_2^{(\beta)}, \dots\right) \quad \text{with} \quad \mu_n^{(\beta)} := \binom{\beta + n}{n}.$$

By a theorem of Weierstrass,

$$\prod_{k=1}^{\infty}\left(1 + \frac{\beta}{k}\right)e^{-\beta/k} = \frac{e^{-C_E\beta}}{\beta!}, \tag{4.27}$$

where $C_E = 0.577\dots$ is Euler's constant. Since

$$\frac{1}{1} + \frac{1}{2} + \cdots + \frac{1}{n} = C_E + \log n + o(1),$$

(4.27) implies that there is a constant $D_\beta \in (1, \infty)$ such that

$$(1/D_\beta)\, n^\beta \le |\mu_n^{(\beta)}| \le D_\beta\, n^\beta \quad \text{for all } n \ge 1. \tag{4.28}$$

Theorem 4.14 (Duduchava and Roch). *If* $\gamma, \delta \in \mathbf{Z}_+$, *then*

$$T(\eta_\gamma)M_{\gamma+\delta}T(\xi_\delta) = \Gamma_{\gamma,\delta}M_\delta T(\xi_\delta\eta_\gamma)M_\gamma, \tag{4.29}$$

where $\Gamma_{\gamma,\delta} = \gamma!\delta!/(\gamma+\delta)!$.

This is a set of nontrivial identities for binomial coefficients, and we invite the reader to try his or her hands. A full proof can be found in [70, Theorem 6.20]). □

Corollary 4.15. *If* $\gamma, \delta \in \mathbf{Z}_+$, *then* $T_n(\xi_\delta\eta_\gamma)$ *is invertible for all* $n \geq 1$ *and there is a constant* $C = C_{\gamma,\delta} \in (0,\infty)$ *such that*

$$\|T_n^{-1}(\xi_\delta\eta_\gamma)\|_2 \leq C\,n^{\gamma+\delta} \text{ for all } n \geq 1.$$

Proof. Let $\ell_n^2(\beta)$ be the space $\mathbf{C}^n$ with the $\ell^2(\beta)$ norm and let $M_{\beta,n}$ denote the operator $P_nM_\beta P_n|\mathrm{Im}\,P_n$. On multiplying (4.29) from the left and the right by P_n, we get

$$T_n(\eta_\gamma)M_{\gamma+\delta,n}T_n(\xi_\delta) = \Gamma_{\gamma,\delta}M_{\delta,n}T_n(\xi_\delta\eta_\gamma)M_{\gamma,n}. \tag{4.30}$$

As $\det T_n(\xi_\delta) = \det T_n(\eta_\gamma) = 1$, it follows that $\det T_n(\xi_\delta\eta_\gamma) \neq 0$. Furthermore, from (4.30) we obtain that

$$T_n^{-1}(\xi_\delta\eta_\gamma) = \Gamma_{\gamma\,\delta}M_{\gamma,n}T_n(\xi_{-\delta})M_{\gamma+\delta,n}^{-1}T_n(\eta_{-\gamma})M_{\delta,n}. \tag{4.31}$$

Consequently,

$$\begin{aligned}
&\|T_n^{-1}(\xi_\delta\eta_\gamma)\|_2 = \|T_n^{-1}(\xi_\delta\eta_\gamma)\|_{\mathcal{B}(\ell_n^2,\ell_n^2)}\\
&\leq |\Gamma_{\gamma,\delta}|\;\|M_{\gamma,n}\|_{\mathcal{B}(\ell_n^2,\ell_n^2)}\;\|T_n(\xi_{-\delta})\|_{\mathcal{B}(\ell_n^2(\delta),\ell_n^2)}\\
&\quad\times\|M_{\gamma+\delta,n}^{-1}\|_{\mathcal{B}(\ell_n^2(-\gamma),\ell_n^2(\delta))}\;\|T_n(\eta_{-\gamma})\|_{\mathcal{B}(\ell_n^2,\ell_n^2(-\gamma))}\;\|M_{\delta,n}\|_{\mathcal{B}(\ell_n^2,\ell_n^2)}.
\end{aligned}$$

Theorem 4.13 implies that

$$\begin{aligned}
\|T_n(\xi_{-\delta})\|_{\mathcal{B}(\ell_n^2(\delta),\ell_n^2)} &= \|P_nT(\xi_{-\delta})P_n\|_{\mathcal{B}(\ell_n^2(\delta),\ell_n^2)} = O(1),\\
\|T_n(\eta_{-\gamma})\|_{\mathcal{B}(\ell_n^2,\ell_n^2(-\gamma))} &= \|P_nT(\eta_{-\gamma})\|_{\mathcal{B}(\ell_n^2,\ell_n^2(-\gamma))} = O(1),
\end{aligned}$$

and (4.28) shows that

$$\begin{aligned}
&\|M_{\gamma,n}\|_{\mathcal{B}(\ell_n^2,\ell_n^2)} = O(n^\gamma), \quad \|M_{\delta,n}\|_{\mathcal{B}(\ell_n^2,\ell_n^2)} = O(n^\delta),\\
&\|M_{\gamma+\delta,n}^{-1}\|_{\mathcal{B}(\ell_n^2(-\gamma),\ell_n^2(\delta))} = O(1).
\end{aligned}$$

Thus, $\|T_n^{-1}(\xi_\delta\eta_\gamma)\|_2 = O(n^{\gamma+\delta})$. □

We now extend Corollary 4.15 to symbols of the form $\xi_\delta\eta_\gamma c$. This will be done on the basis of the following perturbation theorem.

Theorem 4.16. *Suppose*

(a) X, Y, Z, U *are Banach spaces and* P_n $(n = 1, 2, 3, \ldots)$ *are projections that are defined and bounded on each of the spaces* X, Y, Z, U;

(b) *$Z \subset Y$ and $X \subset U$, the embeddings being continuous;*

(c) *$A \in \mathcal{B}(X, Y)$ is invertible;*

(d) *the operators $A_n := P_n A P_n \in \mathcal{B}(P_n X, P_n Y)$ are invertible for all sufficiently large n and, for each $z \in Z$, $\|A_n^{-1} P_n z - A^{-1} z\|_U \to 0$ as $n \to \infty$;*

(e) *$K \in \mathcal{K}(U, Z)$;*

(f) *$A + K \in \mathcal{B}(X, Y)$ is invertible.*

Then

(g) *the operators $P_n(A+K)P_n \in \mathcal{B}(P_n X, P_n Y)$ are invertible for all n large enough;*

(h) *$\|(P_n(A+K)P_n)^{-1} P_n z - (A+K)^{-1} z\|_U \to 0$ as $n \to \infty$ for each $z \in Z$.*

Proof. The operator $I + A^{-1}K$ is an invertible element of $\mathcal{B}(X)$: the inverse is $(A+K)^{-1}A$. We claim that $I + A^{-1}K$ is also an invertible element of $\mathcal{B}(U)$. Since $A^{-1}K : U \to U$ is compact, it follows that $I + A^{-1}K$ is Fredholm of index zero on U. Thus, we must show that $I + A^{-1}K$ has a trivial kernel in U. Let $(I + A^{-1}K)u = 0$ for some $u \in U$. Then

$$u = -A^{-1}Ku \implies u \in X \implies Au = -Ku \implies (A+K)u = 0,$$

and this gives $u = 0$, as desired.

From (d) and (e) we conclude that $I + A_n^{-1} P_n K$ converges to $I + A^{-1}K$ uniformly on U (also recall Proposition 3.2). Therefore, by what was proved in the preceding paragraph, $I + A_n^{-1} P_n K$ is an invertible element of $\mathcal{B}(U)$ for all sufficiently large n, say $n \geq n_0$. Let $B_n \in \mathcal{B}(U)$ denote the inverse operator:

$$B_n + B_n A_n^{-1} P_n K = I, \quad B_n + A_n^{-1} P_n K B_n = I. \tag{4.32}$$

The second inequality of (4.32) implies that $B_n \in \mathcal{B}(X)$. It also shows that $P_n B_n P_n = B_n P_n$. Thus, for $y \in Y$,

$$\begin{aligned} P_n(A+K)P_n B_n A_n^{-1} P_n y &= A_n B_n A_n^{-1} P_n y + P_n K B_n A_n^{-1} P_n y \\ &= A_n B_n A_n^{-1} P_n y + (A_n - A_n B_n) A_n^{-1} P_n y = A_n A_n^{-1} P_n y = P_n y \end{aligned}$$

and, for $x \in X$,

$$\begin{aligned} B_n A_n^{-1} P_n P_n(A+K)P_n x &= B_n A_n^{-1} P_n A_n x + B_n A_n^{-1} P_n K P_n x \\ &= B_n P_n x + (I - B_n) P_n x = P_n x. \end{aligned}$$

It results that $P_n(A+K)P_n \in \mathcal{B}(P_n X, P_n Y)$ is invertible for all $n \geq n_0$, which is assertion (g), and that $B_n A_n^{-1} P_n = (I + A_n^{-1} P_n K)^{-1} A_n^{-1} P_n \in \mathcal{B}(P_n Y, P_n X)$ is the inverse operator.

Now let $z \in Z$ and $n \geq n_0$. Since $(I + A^{-1}K)^{-1} A^{-1} \in \mathcal{B}(Y, X)$ is the inverse of $A + K \in \mathcal{B}(X, Y)$, we get

$$\begin{aligned} &\|(P_n(A+K)P_n)^{-1} P_n z - (A+K)^{-1} P_n z\|_U \\ &\leq \|((I + A_n^{-1} P_n K)^{-1} - (I + A^{-1}K)^{-1}) A_n^{-1} P_n z\|_U \\ &\quad + \|(I + A^{-1}K)^{-1}(A_n^{-1} P_n z - A^{-1} z)\|_U \\ &\leq \|(I + A_n^{-1} P_n K)^{-1} - (I + A^{-1}K)^{-1}\|_{\mathcal{B}(U)} \|A_n^{-1} P_n\|_{\mathcal{B}(Z,U)} \|z\|_Z \\ &\quad + \|(I + A^{-1}K)^{-1}\|_{\mathcal{B}(U)} \|A_n^{-1} P_n z - A^{-1} z\|_U. \end{aligned} \tag{4.33}$$

We already observed that $\|A_n^{-1}P_nK - A^{-1}K\|_{\mathcal{B}(U)} \to 0$ as $n \to \infty$. Consequently,

$$\begin{aligned}&\|(I + A_n^{-1}P_nK)^{-1} - (I + A^{-1}K)^{-1}\|_{\mathcal{B}(U)}\\ &\le 2\,\|(I + A^{-1}K)^{-1}\|_{\mathcal{B}(U)}\,\|A_n^{-1}P_nK - A^{-1}K\|_{\mathcal{B}(U)}\end{aligned}$$

for all sufficiently large n. Now assertion (h) is immediate from (4.32). □

Theorem 4.17. *Let $\gamma, \delta \in \mathbf{Z}_+$ and let c be a Laurent polynomial without zeros on the unit circle $\mathbf{T}$ and with winding number zero. Put $b = \xi_\delta\eta_\gamma c$. Then $T_n(b)$ is invertible for all sufficiently large n and there exists a constant $C = C_{\gamma,\delta,c} \in (0, \infty)$ such that*

$$\|T_n^{-1}(b)\|_2 \le C\,n^{\gamma+\delta}$$

for all n large enough.

Proof. Let Y be the image of $T(\xi_\delta) : \ell^2 \to \ell^2$. It is readily verified that $T(\xi_1)$ and hence also $T(\xi_\delta) = (T(\xi_1))^\delta$ is injective on ℓ^2. Thus, on defining a norm in Y by $\|y\|_Y := \|T(\xi_{-\delta})y\|_2$ we make Y become a Banach space and $T(\xi_\delta) : \ell^2 \to Y$ become an isometric isomorphism. Let further X be the linear space of all sequences $x = \{x_n\}_{n=1}^\infty$ of complex numbers for which $T(\eta_\gamma)x \in \ell^2$. Since $T(\eta_\gamma)x = 0$ can only occur if $x = 0$, through $\|x\|_X := \|T(\eta_\gamma)x\|_2$ a norm is given in X and $T(\eta_\gamma)$ is an isometric isomorphism of X onto ℓ^2. Finally, put $Z = \ell^2(\delta)$ and $U = \ell^2(-\gamma)$. Theorem 4.13 is equivalent to the statement that Z is continuously embedded in Y and that X is continuously embedded in U. At this point we have assumption (b) of Theorem 4.16.

The usual projection P_n onto the first n coordinates is obviously bounded on Z and U, and it can be checked straightforwardly that P_n is also bounded on X and Y (notice that we do not require that $\sup_n \|P_n\| < \infty$). Thus, assumption (a) of Theorem 4.16 is also met.

Let $c = c_-c_+$ be a Wiener-Hopf factorization and put

$$A = T(c_+)T(\xi_\delta\eta_\gamma)T(c_-), \quad A + K = T(c_-)T(\xi_\delta\eta_\gamma)T(c_+).$$

Note that $A + K = T(b)$.

We claim that the operators $T(c_\pm)$ are bounded and invertible on X and Y. Clearly,

$$T(c_\pm) \in \mathcal{B}(X) \iff T(\eta_\gamma)T(c_\pm)T(\eta_{-\gamma}) \in \mathcal{B}(\ell^2).$$

Since $T(\eta_\gamma)T(c_+)T(\eta_{-\gamma}) = T(c_+)$, it is obvious that $T(c_+) \in \mathcal{B}(X)$. The argument of the proof of Proposition 1.3 gives

$$T(\eta_\gamma)T(c_-)T(\eta_{-\gamma}) = T(c_-) - H(\eta_\gamma)H(\widetilde{c}_-)T(\eta_{-\gamma}).$$

The operator $T(\eta_{-\gamma}) : \ell^2 \to \ell^2(-\gamma)$ is bounded due to Theorem 4.13. As $H(\eta_\gamma)$ and $H(\widetilde{c}_-)$ have only finitely many nonzero entries, we see that the linear operator $H(\eta_\gamma)H(\widetilde{c}_-) : \ell^2(-\gamma) \to \ell^2$ is also bounded. Consequently, $T(c_-) \in \mathcal{B}(X)$. Analogously, $T(c_\pm^{-1}) \in \mathcal{B}(X)$. This shows that the operators $T(c_\pm)$ are invertible on X. The boundedness and invertibility of $T(c_\pm)$ on Y can be proved similarly.

The result of the previous paragraph in conjunction with the equality $T(\xi_\delta \eta_\gamma) = T(\xi_\delta)T(\eta_\gamma)$ reveals that A and $A + K$ are bounded and invertible as operators from X to Y. Thus, we have assumptions (c) and (f) of Theorem 4.16. By Proposition 1.3,

$$\begin{aligned} K &= T(c_+ \xi_\delta \eta_\gamma c_-) - T(c_+)T(\xi_\delta \eta_\gamma)T(c_-) \\ &= T(c_+)H(\xi_\delta \eta_\gamma)H(\widetilde{c}_-) + H(c_+)H(\widetilde{\xi}_\delta \widetilde{\eta}_\gamma \widetilde{c}_+), \end{aligned}$$

and since the Hankel operators have at most a finite number of nonzero entries, we conclude that assumption (e) of Theorem 4.16 is also satisfied.

We are left with the verification of assumption (f) of Theorem 4.16. We have

$$A_n = P_n T(c_+)T(\xi_\delta \eta_\gamma)T(c_-)P_n = T_n(c_+)T_n(\xi_\delta \eta_\gamma)T_n(c_-).$$

From Corollary 4.15 we therefore deduce that A_n is invertible for all sufficiently large n and that

$$A_n^{-1} = T_n(c_-^{-1})T_n^{-1}(\xi_\delta \eta_\gamma)T_n(c_+^{-1}) = T(c_-^{-1})P_n T_n^{-1}(\xi_\delta \eta_\gamma)P_n T(c_+^{-1}).$$

Consequently, since $A^{-1} = T(c_-^{-1})T(\eta_{-\gamma})T(\xi_{-\delta})T(c_+^{-1})$, assumption (f) will follow as soon as we have shown that

$$\|T_n^{-1}(\xi_\delta \eta_\gamma)P_n z - T(\eta_{-\gamma})T(\xi_{-\delta})z\|_{2,-\gamma} \to 0 \text{ as } n \to \infty$$

for every $z \in \ell^2(\delta)$. Formulas (4.31) and (4.29) give

$$\begin{aligned} T_n^{-1}(\xi_\delta \eta_\gamma) &= \Gamma_{\gamma,\delta} M_\gamma P_n T(\xi_{-\delta})P_n M_{\gamma+\delta}^{-1} P_n T(\eta_{-\gamma})P_n M_\delta \\ &= \Gamma_{\gamma,\delta} M_\gamma T(\xi_{-\delta})M_{\gamma+\delta}^{-1} P_n T(\eta_{-\gamma})M_\delta \end{aligned}$$

and

$$T(\eta_{-\gamma})T(\xi_{-\delta}) = \Gamma_{\gamma,\delta} M_\gamma T(\xi_{-\delta})M_{\gamma+\delta}^{-1} T(\eta_{-\gamma})M_\delta.$$

Thus, with $\mu := |\Gamma_{\gamma,\delta}|$ and constants $C_j \in (0, \infty)$,

$$\begin{aligned} &\|T_n^{-1}(\xi_\delta \eta_\gamma)P_n z - T(\eta_{-\gamma})T(\xi_{-\delta})z\|_{2,-\gamma} \\ &= \mu \, \|M_\gamma T(\xi_{-\delta})M_{\gamma+\delta}^{-1} Q_n T(\eta_{-\gamma})M_\delta z\|_{2,-\gamma} \\ &\le \mu \, C_1 \, \|T(\xi_{-\delta})M_{\gamma+\delta}^{-1} Q_n T(\eta_{-\gamma})M_\delta z\|_2 \quad \text{(by (4.28))} \\ &\le \mu \, C_2 \, \|M_{\gamma+\delta}^{-1} Q_n T(\eta_{-\gamma})M_\delta z\|_{2,\delta} \quad \text{(by Theorem 4.13)} \\ &\le \mu \, C_3 \, \|Q_n T(\eta_{-\gamma})M_\delta z\|_{2,-\gamma} \quad \text{(by (4.28)).} \end{aligned}$$

Since $T(\eta_{-\gamma})M_\delta z \in \ell^2(-\gamma)$ by virtue of Theorem 4.13 and (4.28) and since Q_n converges strongly to zero on $\ell^2(-\gamma)$, it follows that $\|Q_n T(\eta_{-\gamma})M_\delta z\|_{2,-\gamma} \to 0$, as desired.

Theorem 4.16 now gives that $\|(P_n(A + K)P_n)^{-1}P_n z - (A + K)^{-1}z\|_{2,-\gamma} \to 0$ as $n \to \infty$ for every $z \in \ell^2(\delta)$. Hence, by Theorem 3.1,

$$\sup_n \|T_n^{-1}(b)\|_{\mathcal{B}(\ell_n^2(\delta), \ell_n^2(-\gamma))} = \sup_n \|(P_n(A + K)P_n)^{-1}\|_{\mathcal{B}(\ell_n^2(\delta), \ell_n^2(-\gamma))} < \infty.$$

Because

$$\|H\|_2 = \|H\|_{\mathcal{B}(\ell^2_n,\ell^2_n)} \le n^{\gamma+\delta}\,\|H\|_{\mathcal{B}(\ell^2_n(\delta),\ell^2_n(-\gamma))}$$

for every $n \times n$ matrix H, we finally get $\|T_n^{-1}(b)\|_2 = O(n^{\gamma+\delta})$. $\square$

Given two sequences $\{\alpha_n\}_{n=1}^{\infty}$ and $\{\beta_n\}_{n=1}^{\infty}$ of positive numbers, we write $\alpha_n \simeq \beta_n$ if there exists a constant $C \in (1,\infty)$ such that $(1/C)\beta_n \le \alpha_n \le C\beta_n$ for all sufficiently large n. Combining Theorems 4.11 and 4.17 we arrive at the conclusion that if $b = \xi_\delta \eta_\gamma c$ with a Laurent polynomial c that has no zeros on $\mathbf{T}$ and has winding number zero, then $\|T_n^{-1}(b)\|_2 \simeq n^{\gamma+\delta}$.

4.7 Inside the Essential Spectrum

Throughout this section we assume that b is a Laurent polynomial. We study the behavior of $\|T_n^{-1}(b-\lambda)\|_2$ in the case where $\lambda \in \mathrm{sp}_{\mathrm{ess}} T(b) = b(\mathbf{T})$.

Clearly, $\lambda \in b(\mathbf{T})$ if and only if $b-\lambda$ has zeros on $\mathbf{T}$. We denote by $S(b)$ the points λ for which $b-\lambda$ has at least two distinct zeros on $\mathbf{T}$. The points in $S(b)$ are met at least twice by $b(t)$ as t traces out the unit circle $\mathbf{T}$. If $\lambda \in b(\mathbf{T}) \setminus S(b)$, then $b(\mathbf{T})$ is an (analytic) arc in a sufficiently small neighborhood of λ.

Theorem 4.18. *Let* $\lambda \in b(\mathbf{T}) \setminus S(b)$ *and*

$$b(t) - \lambda = (t-t_0)^\beta t^k c(t), \quad t \in \mathbf{T}, \tag{4.34}$$

where $\beta \in \mathbf{N}$, $k \in \mathbf{Z}$, $c(t) \neq 0$ *for* $t \in \mathbf{T}$, *and* wind $(c,0) = 0$. *Then*

$$\|T_n^{-1}(b-\lambda)\|_2 \simeq n^\beta \ \text{ if } \ -\beta \le k \le 0, \tag{4.35}$$

and there are constants $C \in (0,\infty)$ *and* $\alpha \in (0,\infty)$ *such that*

$$\|T_n^{-1}(b-\lambda)\|_2 \ge C\,e^{\alpha n} \ \text{ if } \ k < -\beta \ \text{ or } \ k > 0. \tag{4.36}$$

Proof. Theorem 4.11 shows that the inequality $\|T_n^{-1}(b-\lambda)\|_2 \ge C\,n^\beta$ is true in either case.

Suppose first that $-\beta \le k \le 0$. We then can write

$$b(t) - \lambda = \left(1 - \frac{t}{t_0}\right)^{\beta-|k|}\left(1 - \frac{t_0}{t}\right)^{|k|}(-t_0)^{\beta-|k|}c(t)$$

with $\beta - |k| \ge 0$ and $|k| \ge 0$. Consequently, by Theorem 4.17,

$$\|T_n^{-1}(b-\lambda)\|_2 = O(n^{\beta-|k|+|k|}) = O(n^\beta).$$

This in conjunction with Theorem 4.11 completes the proof of (4.35).

Now suppose $k < -\beta$ or $k > 0$. Since

$$\widetilde{b}(t) - \lambda := b(1/t) - \lambda = (1-t_0 t)^\beta t^{-\beta-k} c(1/t), \quad t \in \mathbf{T},$$

and $\|T_n^{-1}(b-\lambda)\|_2 = \|T_n^{-1}(\widetilde{b}-\lambda)\|_2$, it suffices to consider the case $k < -\beta$. Thus, let $k = -\beta - m$ with $m > 0$. Let $c = c_- c_+$ be the Wiener-Hopf factorization introduced in the proof of Theorem 4.1 and define x and $x^{(n)}$ as in that proof. We have

$$b(t) - \lambda = (t - t_0)^\beta t^{-\beta-m} c_-(t)c_+(t) = \left(1 - \frac{t_0}{t}\right)^\beta t^{-m} c_-(t)c_+(t),$$

which can be written in the form $b(t) - \lambda = \xi_\beta(t)\chi_{-m}(t)c_-(t)c_+(t)$. As $T(c_+)x = e_0$ and $m > 0$, it results that $P_n T(b-\lambda)x = P_n T(\xi_\beta c_-)T(\chi_{-m})T(c_+)x = 0$. Hence, again as in the proof of Theorem 4.1,

$$\begin{aligned} \|T_n(b-\lambda)x^{(n)}\|_2 &= \|P_n T(b-\lambda)x^{(n)}\|_2 = \|P_n T(b-\lambda)(x^{(n)} - x)\|_2 \\ &\le \|T(b-\lambda)\|_2 \|x^{(n)} - x\|_2 \le D e^{-\alpha n} \end{aligned}$$

with certain constants $D, \alpha \in (0, \infty)$, which yields (4.36). □

Example 4.19. Consider the symbol $b(t) = (t-1)^2 t^k (2.001 + t + 0.49t^{-1})$. Figure 4.3 shows what happens in the five cases $k = -3, -2, -1, 0, 1$. In each picture we see the norm $\|T_n^{-1}(b)\|_2$ against n. We also plotted the shape of the curve $b(\mathbf{T})$ in the lower-right corner; the origin is marked by a big dot. As predicted by Theorem 4.18, the norms increase at least exponentially for $k = -3$ and $k = 1$, while the growth of the norms is polynomial for $-2 \le k \le 0$. In the picture in the bottom, we replaced values greater than 10^{15} by the value 10^{17}. □

Our next objective is to translate Theorem 4.18 into geometrical language. We label each connected component of $\mathbf{C} \setminus b(\mathbf{T})$ by the winding number of the oriented curve $b(\mathbf{T})$ about the points of the component. Let $\lambda \in b(\mathbf{T}) \setminus S(b)$. Then there is an open neighborhood $U_\lambda \subset \mathbf{C}$ of λ such that $U_\lambda \cap b(\mathbf{T}) =: \gamma_\lambda$ is an oriented analytic arc. Clearly, λ belongs to the boundaries of exactly two components Ω_λ^+ and Ω_λ^- of $\mathbf{C} \setminus b(\mathbf{T})$. We let Ω_λ^+ stand for the component on the left of γ_λ, and, accordingly, Ω_λ^- is the component on the right of γ_λ.

Lemma 4.20. *Let $\lambda \in b(\mathbf{T}) \setminus S(b)$. If m is the winding number of Ω_λ^+, then Ω_λ^- has the winding number $m - 1$.*

Proof. Let n be the winding number of Ω_λ^-. Fix a sufficiently small disk U_λ centered at λ, pick a point $\mu \in U_\lambda \cap \Omega_\lambda^+$, and replace $b(\mathbf{T})$ by the continuous curve δ_λ that coincides with $b(\mathbf{T})$ outside U_λ and with $\partial U_\lambda \cap \Omega_\lambda^+$ otherwise. It is obvious that wind $(\delta_\lambda, \mu) = m - 1$. On the other hand, since μ and Ω_λ^- are contained in the same connected component of $\mathbf{C} \setminus \delta_\lambda$, we have wind $(\delta_\lambda, \mu) = n$. Consequently, $n = m - 1$. □

For $r \in (0, 1)$, put $b_r(t) = b(rt)$ $(t \in \mathbf{T})$.

Lemma 4.21. *Let $\lambda \in b(\mathbf{T}) \setminus S(b)$ and let β be the order of the zero of $b - \lambda$ on $\mathbf{T}$. Suppose $r \in (0, 1)$ is sufficiently close to 1 and a point moves along the curve $b_r(\mathbf{T})$, following the orientation of this curve. Then, in a small neighborhood $U_\lambda \subset \mathbf{C}$ of λ, this point is first in $\Omega_\lambda^+ \cap U_\lambda$, then it encircles λ exactly $[(\beta - 1)/2]$ times in the clockwise direction, after which it is again in $\Omega_\lambda^+ \cap U_\lambda$.*

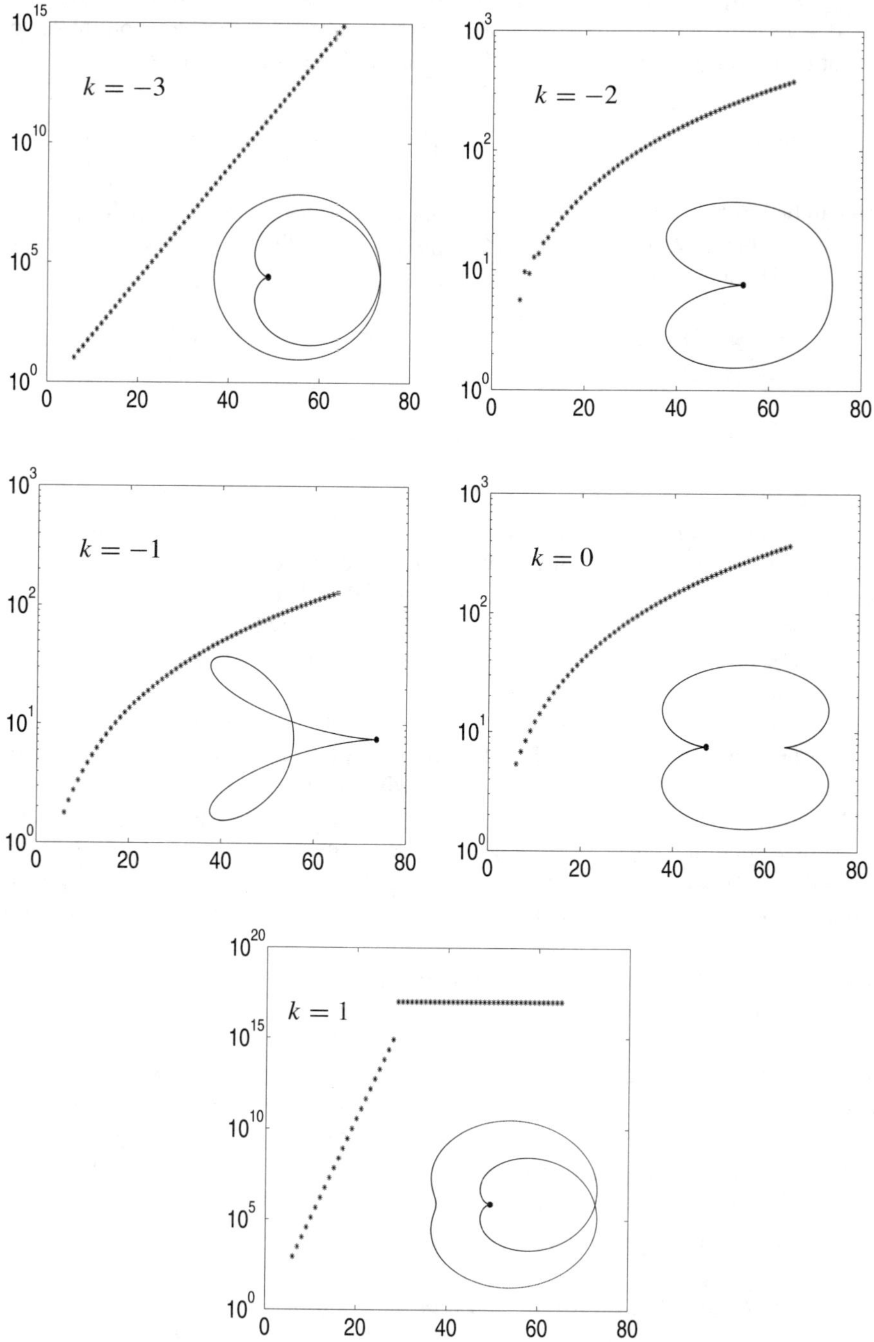

Figure 4.3. *Norms* $\|T_n^{-1}(b)\|_2$ *for several symbols b with zeros.*

Proof. This follows from the fact that the Riemann surface of $b(z) - \lambda$ at $z = t_0$ is locally homeomorphic to the Riemann surface of z^β at the origin and that $r\mathbf{T}$ lies "on the left" of the circle $\mathbf{T}$. □

Theorem 4.22. *Let $\lambda \in b(\mathrm{T}) \setminus S(b)$ Suppose the order of the zero of $b - \lambda$ is β and the winding number of Ω_λ^+ is m. Then*

$$\|T_n^{-1}(b-\lambda)\|_2 \simeq n^\beta \text{ if } -\left[\frac{\beta}{2}\right] \le m \le \left[\frac{\beta+1}{2}\right], \tag{4.37}$$

and there are constants $C \in (0, \infty)$ and $\alpha \in (0, \infty)$ such that

$$\|T_n^{-1}(b-\lambda)\,|_2 \ge C\, e^{\alpha n} \text{ if } m < -\left[\frac{\beta}{2}\right] \text{ or } m > \left[\frac{\beta+1}{2}\right]. \tag{4.38}$$

Proof. Write $b - \lambda$ in the form (4.34). Then $b(rt) - \lambda = (rt - t_0)^\beta r^k t^k c(rt)$ for $t \in \mathbf{T}$. If $r \in (0, 1)$ is sufficiently close to 1, then $\lambda \notin b_r(\mathbf{T})$ and

$$\begin{aligned} &\text{wind}\,(b(rt), \lambda) = \text{wind}\,(b(rt) - \lambda, 0) \\ &= \text{wind}\,((rt - t_0)^\beta, 0) + k + \text{wind}\,(c(rt), 0) = 0 + k + 0 = k. \end{aligned}$$

Evidently, $\|b_r - b\|_\infty \to 0$ as $r \to 1 - 0$. This in conjunction with Lemma 4.21 shows that if $\mu \in \Omega_\lambda^-$ is sufficiently close to λ, then

$$\text{wind}\,(b(rt), \lambda) = \text{wind}\,(b(rt), \mu) = \text{wind}\,(b(t), \mu) - \left[\frac{\beta-1}{2}\right].$$

Consequently, $k = \text{wind}\,(b, \mu) - [(\beta - 1)/2]$. By Lemma 4.20, $\text{wind}\,(b, \mu) = m - 1$. It results that $k = m - 1 - [(\beta - 1)/2]$, and since

$$-\beta \le m - 1 - \left[\frac{\beta-1}{2}\right] \le 0 \iff -\left[\frac{\beta}{2}\right] \le m \le \left[\frac{\beta+1}{2}\right],$$

(4.37) and (4.38) follow from (4.35) and (4.36). □

Here are two interesting special cases. Recall that, by Corollary 1.12,

$$\text{sp}\ T(b) = b(\mathbf{T}) \cup \left\{\lambda \in \mathbf{C} \setminus b(\mathbf{T}) : \ \text{wind}\ (b, \lambda) \ne 0\right\}.$$

Corollary 4.23. *If $\lambda \in b(\mathbf{T}) \setminus S(b)$ is located on $\partial\,\text{sp}\ T(b)$, the boundary of the spectrum of $T(b)$, and β is the order of the zero of $b - \lambda$, then*

$$\|T_n^{-1}(b-\lambda)\|_2 \simeq n^\beta.$$

Proof. By assumption, Ω_λ^+ or Ω_λ^- has the winding number zero. From Lemma 4.20 we deduce that the winding number m of Ω_λ^+ is 0 or 1. Since $[(\beta+1)/2)] \ge 1$ and $-[\beta/2] \le 0$ for every $\beta \in \mathbf{N}$, the assertion follows from (4.37). □

Corollary 4.24. *If λ belongs to $b(\mathbf{T}) \setminus S(b)$ and the order of the zero of $b - \lambda$ is 1, then $\|T_n^{-1}(b-\lambda)\|_2 \simeq n$ for $\lambda \in \partial\,\text{sp}\ T(b)$, while $\|T_n^{-1}(b-\lambda)\|_2$ increases at least exponentially in case $\lambda \notin \partial\,\text{sp}\ T(b)$.*

Proof. The first part of the assertion is immediate from Corollary 4.23. So suppose $\lambda \in b(\mathbf{T}) \setminus (S(b) \cup \partial \operatorname{sp} T(b))$ and let m be the winding number of Ω_λ^+. As neither Ω_λ^+ nor Ω_λ^- have the winding number zero, we infer from Lemma 4.20 that $m \neq 0$ and $m - 1 \neq 0$. Consequently, $m > 1$ or $m < 0$, and Theorem 4.22 with $\beta = 1$ completes the proof. □

Example 4.25. Let $b(t) = (t+1)^3$. We have $b(\mathbf{T}) = A \cup \{-1\} \cup B \cup \{0\}$, where A and B are as in Figure 4.4.

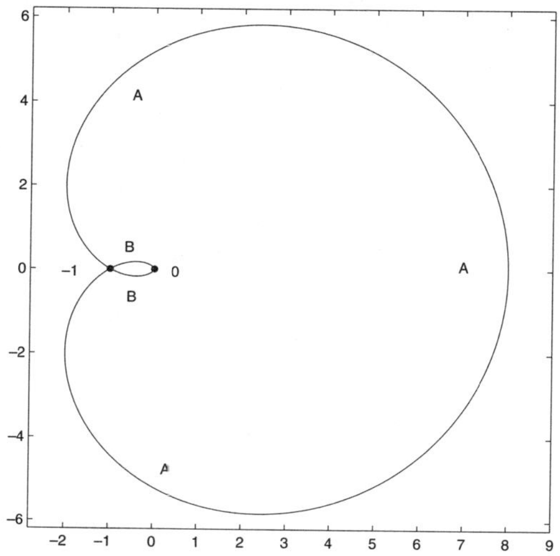

Figure 4.4. *The curve $b(\mathbf{T})$ for $b(t) = (t+1)^3$.*

Theorem 4.22 implies that

$$\|T_n^{-1}(b-\lambda)\|_2 \simeq \begin{cases} n & \text{for} \quad \lambda \in A, \\ n^3 & \text{for} \quad \lambda = 0, \end{cases}$$

and that $\|T_n^{-1}(b-\lambda)\|_2$ increases at least exponentially for $\lambda \in B$. We are left with $\lambda = -1$. In that case

$$b(t) - \lambda = (t+1)^3 + 1 = (t^2+t+1)(t+2) = (t-\omega)(t-\omega^2)(t+2),$$

where $\omega = e^{2\pi i/3}$. Since the two zeros of $b - \lambda$ on $\mathbf{T}$ are of order 1, Corollary 4.12 implies that $\|T_n^{-1}(b-\lambda)\|_2 \geq C\,n$ with some constant C. The results we have established so far cannot be used to obtain an upper estimate for $\|T_n^{-1}(b-\lambda)\|_2$. However, in the case at hand we can proceed as follows. Since

$$T_n(b(t) - \lambda) = T_n(t^2+t+1)T_n(t+2)$$

and $\sup_n \|T_n^{-1}(t+2)\|_2 < \infty$ by Theorem 3.7, it remains to estimate the norms $\|T_n^{-1}(t^2 + t + 1)\|_2$ from above. It can be checked straightforwardly that $T_n^{-1}(t^2+t+1)$ is the lower-triangular Toeplitz matrix whose first column is

$$(\,1,\ -1,\ 0,\ 1,\ -1,\ 0,\ 1,\ -1,\ 0,\ \ldots)^\top.$$

By computing the Frobenius norm of that matrix or by writing it as

$$T_n(1) + T_n(-t) + T_n(t^2) + T_n(-t^4) + T_n(t^6) + T_n(-t^7) + \cdots,$$

we see that the spectral norm is $O(n)$. In summary, $\|T_n(b-\lambda)\|_2 \simeq n$ for $\lambda = -1$. □

Example 4.26. Things are very complicated for points $\lambda \in S(b)$. Let, for example, $b(t) - \lambda = t^k(t-1)(t+1)$ ($t \in \mathbf{T}$). Corollary 4.12 gives $\|T_n^{-1}(b-\lambda)\|_2 \ge Cn$ for all $n \ge 1$. If $k \le -3$ or $k \ge 1$, then $T_n(b-\lambda)$ is triangular with zeros on the main diagonal and hence $\|T_n^{-1}(b-\lambda)\|_2 = \infty$. In the case $k = 0$, the inverse of $T_n(b-\lambda)$ is the lower-triangular Toeplitz matrix with the first column

$$(-1,\ 0,\ -1,\ 0,\ -1,\ 0,\ \ldots)^{\top},$$

and as in Example 4.25 we obtain that the spectral norm of this matrix is $O(n)$. Thus, $\|T_n^{-1}(b-\lambda)\|_2 \simeq n$ for $k = 0$. Passage to adjoints yields the same result for $k = -2$.

Finally, let $k = -1$. Then $T_n(b-\lambda)$ is a skew-symmetric tridiagonal Toeplitz matrix. Thus, $\|T_n^{-1}(b-\lambda)\|_2 = \infty$ if n is odd. For even n we may employ the fact that $T_n(b-\lambda)$ is normal and hence $\|T_n^{-1}(b-\lambda)\|_2 = 1/\mu_n$ where μ_n is the minimum of the moduli of the eigenvalues of $T_n(b-\lambda)$. From Theorem 2.4 we infer that the eigenvalues of $T_{2m}(b-\lambda)$ are $2i\cos \pi j/(2m+1)$ ($j = 1, \ldots, 2m$), which shows that

$$\mu_{2m} = 2\cos\frac{\pi m}{2m+1} = 2\sin\frac{\pi}{4m+2} \sim \frac{2\pi}{4m+2} \sim \frac{\pi}{2m} = \frac{\pi}{n}.$$

Consequently, $\|T_n^{-1}(b-\lambda)\|_2 \sim n/\pi$ for even n, where here and throughout the book $\alpha_n \sim \beta_n$ means that $\alpha_n/\beta_n \to 1$. □

4.8 Semi-Definite Matrices

A matrix $A \in \mathbf{C}^{n\times n}$ is said to be *positive semi-definite* if $\operatorname{Re}(Ax, x) \ge 0$ for all x in $\mathbf{C}^n$ and is called *positive definite* if there is an $\varepsilon > 0$ such that $\operatorname{Re}(Ax, x) \ge \varepsilon\|x\|_2^2$ for all $x \in \mathbf{C}^n$. From (4.17) we infer that if $a \in W$ and $\operatorname{Re} a(t) \ge 0$ for all $t \in \mathbf{T}$, then $T_n(a)$ is positive semi-definite, and that if $a \in W$ and $\operatorname{Re} a(t) \ge \varepsilon > 0$ for all $t \in \mathbf{T}$, then $T_n(a)$ is positive definite.

In this section we establish upper estimates for $\|T_n^{-1}(b)\|_2$ in terms of the zeros of the real part $\operatorname{Re} b$ of b provided b is positive semi-definite, that is, in the case where $\operatorname{Re} b(t) \ge 0$ for all $t \in \mathbf{T}$. The estimates we will obtain are very coarse in general, but they have some advantages. First, they imply that if b is a Laurent polynomial and $\lambda \in \partial \operatorname{conv} b(\mathbf{T})$ then $\|T_n^{-1}(b-\lambda)\|_2$ grows at most polynomially; note that λ is allowed to belong to $S(b)$. Second, our estimates for $\|T_n^{-1}(b)\|_2$ are sharp in case b is real valued and nonnegative.

For $a \in W$, let $\mathcal{R}(a) = a(\mathbf{T})$ be the range of a, let $\operatorname{conv}\mathcal{R}(a)$ stand for the convex hull of $\mathcal{R}(a)$, let $\partial \operatorname{conv}\mathcal{R}(a)$ denote the boundary of $\operatorname{conv}\mathcal{R}(a)$, and put $\operatorname{dist}(0, \operatorname{conv}\mathcal{R}(a)) := \min\{|z| : z \in \operatorname{conv}\mathcal{R}(a)\}$.

Proposition 4.27. *Suppose $a \in W$ does not vanish identically and $\mathcal{R}(a)$ is not a line segment containing the origin in its interior. If*

$$0 \notin \operatorname{conv}\mathcal{R}(a) \quad \text{or} \quad 0 \in \partial \operatorname{conv}\mathcal{R}(a), \tag{4.39}$$

then $T_n(a)$ is invertible for all $n \geq 1$.

Proof. Assume $T_n(a)$ is not invertible. Then $T_n(a)x = 0$ for some $x \in \mathbf{C}^n \setminus \{0\}$, and (4.17) implies that there exists a polynomial $f \in \mathcal{P}_n^+ \setminus \{0\}$ such that $\int a|f|^2 = 0$. By (4.39), we can find a number $\gamma \in \mathbf{T}$ such that $\operatorname{Re}(\gamma a) \geq 0$. As $\int \operatorname{Re}(\gamma a)|f|^2 = 0$ and $|f|^2 > 0$ almost everywhere, it follows that $\operatorname{Re}(\gamma a) = 0$ throughout $\mathbf{T}$. Consequently, $\mathcal{R}(\gamma a) = i[m, M]$ with real numbers $m < M$. Since $\int \operatorname{Im}(\gamma a)|f|^2 = 0$ and $|f|^2 > 0$ almost everywhere, we deduce that $m < 0$ and $M > 0$. However, this case was excluded. □

Corollary 4.28. *Let $a \in W$. If $\mathcal{R}(a)$ is a singleton, then $\operatorname{sp} T_n(a) = \mathcal{R}(a)$. If $\mathcal{R}(a)$ is a line segment $[z_1, z_2]$ with $z_1 \neq z_2$, then*

$$\operatorname{sp} T_n(a) \subset [z_1, z_2] \setminus \{z_1, z_2\}. \tag{4.40}$$

Finally, if $\mathcal{R}(a)$ is neither a singleton nor a line segment, then

$$\operatorname{sp} T_n(a) \subset \operatorname{conv} \mathcal{R}(a) \setminus \partial \operatorname{conv} \mathcal{R}(a). \tag{4.41}$$

Proof. The case where $\mathcal{R}(a)$ is a singleton is trivial. The inclusion (4.41) is immediate from Proposition 4.27. We are so left with the case where $\mathcal{R}(a)$ is a proper line segment. From Proposition 4.27 we deduce that $\operatorname{sp} T_n(a) \subset [z_1, z_2]$. Assume z_1 is in $\operatorname{sp} T_n(a)$. Then, as in the proof of Proposition 4.27, $\int (a - z_1)|f|^2 = 0$ for some nonzero $f \in \mathcal{P}_n^+$. Since $\mathcal{R}(\gamma(a - z_1)) \subset [0, \infty)$ for some $\gamma \in \mathbf{T}$, we therefore see that $a(t) = z_1$ for all $t \in \mathbf{T}$, which means that $\mathcal{R}(a)$ is a singleton. Thus, $z_1 \notin \operatorname{sp} T_n(a)$. Analogously one can show that $z_2 \notin \operatorname{sp} T_n(a)$. □

Theorem 4.29 (Brown and Halmos). *Let $a \in W$ and suppose*

$$d := \operatorname{dist}(0, \operatorname{conv} \mathcal{R}(a)) > 0. \tag{4.42}$$

Then $T(a)$ is invertible on ℓ^2 with

$$\|T^{-1}(a)\|_2 \leq 1/d \tag{4.43}$$

and $T_n(a)$ is invertible for all $n \geq 1$ with

$$\|T_n^{-1}(a)\|_2 \leq 1/d. \tag{4.44}$$

Proof. There is a $\gamma \in \mathbf{T}$ such that the set $\gamma \operatorname{conv} \mathcal{R}(a)$ is contained in the half-plane $\{z \in \mathbf{C} : \operatorname{Re} z \geq d\}$. Fix any $\varepsilon \in (0, d)$. If r is sufficiently large, then the disk $\{z \in \mathbf{C} : |z - (d + r - \varepsilon)| < r\}$ certainly contains the set $\gamma \operatorname{conv} \mathcal{R}(a)$. Thus $|\gamma\, a(t) - d - r + \varepsilon| < r$ for all $t \in \mathbf{T}$ and therefore

$$\left| \frac{\gamma\, a(t)}{d + r - \varepsilon} - 1 \right| < \frac{r}{d + r - \varepsilon}$$

for all $t \in \mathbf{T}$. Since

$$\frac{\gamma}{d + r - \varepsilon} T(a) = I + T\left(\frac{\gamma\, a}{d + r - \varepsilon} - 1\right)$$

and the norm of the Toeplitz operator on the right is less than $r/(d+r-\varepsilon) < 1$, it follows from expansion into the Neumann series that $T(a)$ is invertible and that

$$\frac{d+r-\varepsilon}{|\gamma|}\|T^{-1}(a)\|_2 < \frac{1}{1-\frac{r}{d+r-\varepsilon}} = \frac{d+r-\varepsilon}{d-\varepsilon},$$

whence $\|T^{-1}(a)\|_2 < 1/(d-\varepsilon)$. As $\varepsilon \in (0,d)$ can be chosen as small as desired, we arrive at estimate (4.43). Clearly, the argument remains true with $T(a)$ replaced by $T_n(a)$ and so also gives (4.44). □

Example 4.30. Consider the $n \times n$ Toeplitz matrix

$$T_n(b) = \begin{pmatrix} 2 & -1 & 0 & \dots & 0 \\ -1 & 2 & -1 & \dots & 0 \\ 0 & -1 & 2 & \dots & 0 \\ \vdots & \vdots & \vdots & \ddots & \vdots \\ 0 & 0 & 0 & \dots & 2 \end{pmatrix}.$$

The symbol is $b(e^{i\theta}) = -e^{-i\theta} + 2 - e^{i\theta} = 2(1-\cos\theta)$. Obviously, $\mathcal{R}(b) = [0,4]$ and hence $d = \operatorname{dist}(0, \operatorname{conv}\mathcal{R}(b)) = 0$. Thus, Theorem 4.29 is not applicable. However, let us replace b by $b + ig_n$, where

$$g_n(e^{i\theta}) = \cos n\theta = (e^{in\theta} + e^{-in\theta})/2.$$

As the Fourier coefficients $(g_n)_k$ of g_n are zero for $|k| \le n-1$, we have

$$T_n(b) = T_n(b + ig_n). \tag{4.45}$$

Clearly, $d_n := \operatorname{dist}(0, \operatorname{conv}\mathcal{R}(b+ig_n)) > 0$, and we can therefore apply Theorem 4.29 to the matrices (4.45).

Our aim is to estimate d_n from below. The graph of $b + ig_n$ in $\mathbf{C} = \mathbf{R}^2$ is given by

$$(2 - 2\cos\theta, \cos n\theta), \quad \theta \in (-\pi, \pi]. \tag{4.46}$$

Put

$$\varepsilon_n = \frac{4}{3}\left(1 - \cos\frac{\pi}{3n}\right) = \frac{8}{3}\sin^2\frac{\pi}{6n}. \tag{4.47}$$

The graph of

$$\left(2 - 2\cos\theta, \frac{1}{2} - \frac{1}{\varepsilon_n}(2 - 2\cos\theta)\right), \quad \theta \in \mathbf{R} \tag{4.48}$$

is the straight line $y = 1/2 - (1/\varepsilon_n)x$. We show that the range of $b + ig_n$ lies above this line. By (4.46) and (4.48) this is equivalent to showing that

$$\frac{1}{\varepsilon_n}(2 - 2\cos\theta) + \cos n\theta \ge \frac{1}{2} \tag{4.49}$$

for $\theta \in (-\pi, \pi]$. If $|n\theta| < \pi/3$, then $\cos n\theta > 1/2$ and hence (4.49) is true. If $|n\theta| \geq \pi/3$, then $\cos\theta \leq \cos(\pi/(3n))$, whence, by (4.47),

$$\frac{1}{\varepsilon_n}(2 - 2\cos\theta) + \cos n\theta \geq \frac{1}{\varepsilon_n}\left(2 - 2\cos\frac{\pi}{3n}\right) - 1 = \frac{1}{2},$$

which gives (4.49) again.

Thus, $d_n \geq D_n$ where D_n is the distance of the origin to the straight line $y = 1/2 - (1/\varepsilon_n)x$. Obviously,

$$D_n = \frac{\varepsilon_n}{4}\,\frac{1}{\sqrt{1/4 + \varepsilon_n^2/4}}.$$

Since $\varepsilon_n \to 0$ as $n \to \infty$, we have $D_n > \varepsilon_n/(2\sqrt{2})$ for all sufficiently large n. Taking into account (4.44), (4.45), (4.47) we therefore obtain

$$\|T_n^{-1}(b)\|_2 \leq \frac{1}{d_n} \leq \frac{1}{D_n} < \frac{2\sqrt{2}}{\varepsilon_n} = \frac{3\sqrt{2}}{4}\,\frac{1}{\sin^2(\pi/(6n))} < \frac{3\sqrt{2}}{4}\left(\frac{\pi}{2}\,\frac{6n}{\pi}\right)^2 < 10\,n^2. \qquad \square$$

To extend the trick employed in this example to symbols with more than one zero, we need a further result.

Lemma 4.31 (Dirichlet). *Let $\beta_1, \ldots, \beta_N$ be real numbers and $\mu > 0$. Then there exists a number $q \in \mathbf{N}$ such that $1 \leq q \leq ([1/\mu] + 1)^N$ and $q\beta_j \in \mathbf{Z} + (-\mu, \mu)$ for all $j \in \{1, \ldots, N\}$.*

Proof. For $x \in \mathbf{R}$, denote by $\{x\}$ the fractional part of x. Thus, $x = [x] + \{x\}$ with $[x] \in \mathbf{Z}$ and $\{x\} \in [0, 1)$. Put $K = [1/\mu] + 1$ and divide the cube $[0, 1)^N$ into K^N congruent cubes of the form

$$[i_1/K, (i_1 + 1)/K) \times \cdots \times [i_N/K, (i_N + 1)/K). \tag{4.50}$$

The $K^N + 1$ points $(\{\ell\beta_1\}, \ldots, \{\ell\beta_N\})$, $\ell = 0, 1, \ldots, K^N$ all belong to $[0, 1)^N$ and therefore two of them must be located in the same cube (4.50). Consequently, there are ℓ_1, ℓ_2 such that $0 \leq \ell_1 < \ell_2 \leq K^N$ and $|\{\ell_2\beta_j\} - \{\ell_1\beta_j\}| < 1/K$ for all j. Put $q = \ell_2 - \ell_1$ and $m_j = [\ell_2\beta_j] - [\ell_1\beta_j]$. Then

$$|q\beta_j - m_j| = |\ell_2\beta_j - [\ell_2\beta_j] - (\ell_1\beta_j - [\ell_1\beta_j])| = |\{\ell_2\beta_j\} - \{\ell_1\beta_j\}| < 1/K < \mu. \qquad \square$$

Theorem 4.32. *Let b be a Laurent polynomial and suppose $0 \in b(\mathbf{T})$. Assume that $\operatorname{Re} b \geq 0$ on $\mathbf{T}$ and that $\operatorname{Re} b$ is not identically zero. Then $\operatorname{Re} b$ has a finite number of zeros on $\mathbf{T}$ and the orders of these zeros are all even. If 2α is the maximal order of the zeros of $\operatorname{Re} b$ on $\mathbf{T}$, then*

$$\|T_n^{-1}(b)\|_2 \leq D\,n^{2\alpha} \quad \text{for all } n \geq 1$$

with some constant $D \in (0, \infty)$ independent of n.

Proof. Put $u(\theta) := \operatorname{Re} b(e^{i\theta})$. Clearly, u is also a Laurent polynomial. By assumption, u does not vanish identically, $u(\theta) \ge 0$ for all θ, and u has $N \ge 1$ zeros $\theta_1, \dots, \theta_N \in (-\pi, \pi]$ of even orders $2\alpha_1, \dots, 2\alpha_N$. Since $T_n(b)$ is invertible for all $n \ge 1$ due to Proposition 4.27, it suffices to prove the estimate $\|T_n^{-1}(b)\|_2 \le D\, n^{2\alpha}$ for all n large enough.

Using Lemma 4.31 with $\mu = 1/12$ and $\beta_j = n\theta_j/(2\pi)$, we get an integer q_n such that

$$1 \le q_n \le 13^N, \quad nq_n\theta_j \in 2\pi\mathbf{Z} + (-\pi/6, \pi/6). \tag{4.51}$$

We have

$$\cos(nq_n\theta) = \cos(nq_n\theta_j)\cos(nq_n(\theta - \theta_j)) - \sin(nq_n\theta_j)\sin(nq_n(\theta - \theta_j)),$$

and (4.51) shows that

$$\cos(nq_n\theta_j) > \cos\frac{\pi}{6} = \frac{\sqrt{3}}{2}, \quad \sin(nq_n\theta_j) < \sin\frac{\pi}{6} = \frac{1}{2}.$$

If $|nq_n(\theta - \theta_j)| < \pi/6$, then

$$\cos(nq_n(\theta - \theta_j)) > \cos\frac{\pi}{6} = \frac{\sqrt{3}}{2}, \quad \sin(nq_n(\theta - \theta_j)) < \sin\frac{\pi}{6} = \frac{1}{2}.$$

Hence,

$$\cos(nq_n\theta) > \frac{\sqrt{3}}{2}\frac{\sqrt{3}}{2} - \frac{1}{2}\frac{1}{2} = \frac{1}{2} \quad \text{for } |\theta - \theta_j| < \frac{\pi}{6nq_n}. \tag{4.52}$$

Choose $\delta > 0$ so that the sets $(\theta_j - \delta, \theta_j + \delta)$ are pairwise disjoint and put

$$\frac{1}{\omega_j(n)} := \min\left\{u(\theta) : \frac{1}{n} \le |\theta - \theta_j| < \delta\right\}. \tag{4.53}$$

Since θ_j is a zero of the order $2\alpha_j$, there is a constant $C_j \in (1, \infty)$ such that

$$(1/C_j)\, n^{2\alpha_j} \le \omega_j(n) \le C_j\, n^{2\alpha_j}. \tag{4.54}$$

Let $v(\theta) := \operatorname{Im} b(e^{i\theta})$ and put

$$\frac{1}{\varepsilon_{n,j}} := 3\,(\|v\|_\infty + 1)\,\omega_j\left(\frac{6nq_n}{\pi}\right), \quad M := 2\,(\|v\|_\infty + 1). \tag{4.55}$$

Consider the function

$$a_n(e^{i\theta}) := b(e^{i\theta}) + i\,M\,\cos(nq_n\theta).$$

Since $q_n \ge 1$, we have $T_n(b) = T_n(a_n)$. Now let n be so large that $\pi/(6nq_n) < \delta$. We claim that the range $\mathcal{R}(a_n|(\theta_j - \delta, \theta_j + \delta))$ lies above the straight line $y = 1 - x/\varepsilon_{n,j}$. As $a_n(e^{i\theta}) = u(\theta) + i(v(\theta) + M\cos(nq_n\theta))$, this is equivalent to claiming that

$$v(\theta) + M\cos(nq_n\theta) > 1 - u(\theta)/\varepsilon_{n,j}$$

for all $\theta \in (\theta_j - \delta, \theta_j + \delta)$. We prove that actually

$$u(\theta)/\varepsilon_{n,j} + M\cos(nq_n\theta) > 1 + \|v\|_\infty \tag{4.56}$$

for all $\theta \in (\theta_j - \delta, \theta_j + \delta)$.

If $|\theta - \theta_j| < \pi/(6nq_n)$, then (4.52), (4.54), and the nonnegativity of $u(\theta)$ give

$$u(\theta)/\varepsilon_{n,j} + M\cos(nq_n\theta) > \frac{m}{2} = \|v\|_\infty + 1.$$

So let $\pi/(6nq_n) \le |\theta - \theta_j| < \delta$. Then, by (4.53) and (4.55),

$$\begin{aligned} u(\theta)/\varepsilon_{n,j} + M\cos(nq_n\theta) &\ge \frac{1}{\varepsilon_{n,j}\,\omega_j(6nq_n/\pi)} - M \\ &= 3\,(\|v\|_\infty + 1) - 2\,(\|v\|_\infty + 1) = \|v\|_\infty + 1. \end{aligned}$$

This completes the proof of (4.56).

Thus, the range of the restriction of a_n to $\bigcup_j(\theta_j - \delta, \theta_j + \delta)$ lies above the line

$$y = 1 - x/\varepsilon_n, \quad \varepsilon_n := \min_j \varepsilon_{n,j} \tag{4.57}$$

(here we also took into account that $\operatorname{Re} a_n \ge 0$). The number η given by

$$\eta = \inf\left\{u(\theta) : \theta \in (-\pi, \pi] \Big\backslash \bigcup_{j=1}^{N}(\theta_j - \delta, \theta_j + \delta)\right\}$$

is positive. If $\theta \in (-\pi, \pi] \setminus \bigcup_j(\theta_j - \delta, \theta_j + \delta)$, then $a_n(e^{i\theta})$ is located on the right of the vertical line $x = \eta$. Since $1/\varepsilon_n \to \infty$ as $n \to \infty$, it follows that $\mathcal{R}(a_n)$ is contained in the half-plane above the line (4.57) for all sufficiently large n.

The distance of the origin to the line (4.57) is $D_n := \varepsilon_n/\sqrt{1+\varepsilon_n^2}$. Thus, $D_n > \varepsilon_n/2$ if only n is large enough. From Theorem 4.29 and (4.55), (4.57) we now obtain that

$$\|T_n^{-1}(b)\|_2 = \|T_n^{-1}(a_n)\|_2 \le \frac{1}{D_n} < \frac{2}{\varepsilon_n} = 6\,(\|v\|_\infty + 1)\,\max_j \omega_j\left(\frac{6nq_n}{\pi}\right),$$

whence, by (4.51) and (4.54), $\|T_n^{-1}(b)\|_2 = O(n^{\max(2\alpha_1,\dots,2\alpha_N)})$. □

Example 4.33. In general, there is a gap between Theorems 4.11 and 4.32. Let $b(t) = t$ and $\lambda = e^{i\theta_0} \in \mathbf{T}$. Since $b - \lambda$ has a zero of the order 1 on $\mathbf{T}$, Theorem 4.11 gives

$$\|T_n^{-1}(b-\lambda)\|_2 \ge C\,n.$$

We have

$$\operatorname{Re}\,(-\lambda^{-1}(b-\lambda)) = 1 - \operatorname{Re}\,(e^{-i\theta_0}e^{i\theta}) = 1 - \cos(\theta - \theta_0) \ge 0,$$

and $1 - \cos(\theta - \theta_0)$ has a zero of the order 2. Thus, by Theorem 4.32,

$$\|T_n^{-1}(b-\lambda)\|_2 = \|T_n^{-1}(-\lambda^{-1}(b-\lambda))\|_2 \le D\,n^2.$$

In fact

$$\|T_n^{-1}(b - \lambda)\|_2 \le n,$$

because

$$T_n^{-1}(b - \lambda) = -\frac{1}{\lambda} \begin{pmatrix} 1 & 0 & \dots & 0 \\ 1/\lambda & 1 & \dots & 0 \\ 1/\lambda^2 & 1/\lambda & \dots & 0 \\ \vdots & \vdots & \ddots & \vdots \\ 1/\lambda^{n-1} & 1/\lambda^{n-2} & \dots & 1 \end{pmatrix},$$

and on writing this matrix as

$$-\frac{1}{\lambda} T_n(1) - \frac{1}{\lambda^2} T_n(t) - \dots - \frac{1}{\lambda^{n-1}} T_n(t^{n-1}),$$

we see that its spectral norm is at most n. □

In certain cases combination of Theorems 4.11 and 4.32 yields a sharp result. Here is the most important of these cases.

Corollary 4.34. *Let b be a real-valued Laurent polynomial and suppose b is not constant. Then $\mathcal{R}(b) = [m, M]$ with $m < M$. If $\lambda \in \{m, M\}$ and the maximal order of the zeros of $b - \lambda$ on $\mathbf{T}$ is 2α, then $\|T_n^{-1}(b - \lambda)\|_2 \simeq n^{2\alpha}$.*

Proof. The proof is immediate from Theorems 4.11 and 4.32. □

Exercises

1. Try to prove Theorem 4.14.

2. (a) Find a $b \in \mathcal{P}$ such that $b(\mathbf{T}) = [m, M]$ with $m < 0$, $M > 0$, and $D_n(b) \ne 0$ for all $n \ge 1$.

   (b) Find a $b \in \mathcal{P}$ such that $b(\mathbf{T}) = [m, M]$ with $m < 0$ and $M > 0$ and such that there are infinitely many n with $D_n(b) = 0$ and infinitely many n with $D_n(b) \ne 0$.

3. Show that if $b \in \mathcal{P}$ does not vanish identically and $b(\mathbf{T}) \subset \mathbf{R}$, then $D_n(b) \ne 0$ for infinitely many n.

4. Let $a \in W$ and denote by $S_n a$ and $\sigma_n a$ the nth partial sum of the Fourier series of a and the nth Fejér mean, respectively.

   (a) Show that if $\operatorname{Re} a \ge \varepsilon > 0$ on $\mathbf{T}$, then

   $$\|T^{-1}(S_n a)\|_2 \to \|T^{-1}(a)\|_2, \quad \|T^{-1}(\sigma_n a)\|_2 \to \|T^{-1}(a)\|_2.$$

   (b) Show that if $a(e^{i\theta}) = \theta^2/4$ $(\theta \in (-\pi, \pi])$, then $a \in W$ and $T(S_n a)$ is not invertible whenever n is odd.

(c) Show that if $\operatorname{Re} a \geq 0$ on $\mathbf{T}$ and $\operatorname{Re} a$ has exactly N zeros of the orders $2\alpha_1, \ldots, 2\alpha_N$ on $\mathbf{T}$, then

$$\|T^{-1}(\sigma_n a)\|_2 = O\left(n^{2\max(\alpha_1,\ldots,\alpha_N)}\right).$$

5. Let $a \in W$ and suppose a vanishes identically on some subarc of $\mathbf{T}$. Show that there exist positive constants C and α such that

$$\|T_n^{-1}(a)\|_2 \geq Ce^{\alpha n}$$

for all $n \geq 1$.

6. Let $a \in W$ and suppose $\operatorname{Re} a \geq 0$ on $\mathbf{T}$. Assume that $\|T_n^{-1}(a)\|_2 = O(n^\alpha)$ for some $\alpha \geq 0$. We denote by ℓ^2_α the Hilbert space of all complex sequences $\{x_n\}_{n=1}^\infty$ for which $\sum n^{2\alpha}|x_n|^2 < \infty$. Prove that if y is an element of ℓ^2 such that the equation $T(a)x = y$ has a solution $x \in \ell^2_\alpha$, then the solution $x^{(n)}$ of $T_n(a)x^{(n)} = P_n y$ converges to x in the norm of ℓ^2.

7. We denote by ℓ^2_∞ the countably normed space of all complex sequences $x = \{x_n\}_{n=1}^\infty$ satisfying

$$\|x\|_{2,k}^2 := \sum_{n=1}^\infty n^{2k}|x_n|^2 < \infty$$

for all $k \in \mathbf{N}$. Let $b \in \mathcal{P}$.

(a) Show that $T(b)$ is bounded on ℓ^2_∞.

(b) Let $\tau_1, \ldots, \tau_m$ be the zeros of b on $\mathbf{T}$ (repeated according to their multiplicities). Then

$$b(t) = \prod_{j=1}^m \left(1 - \frac{\tau_j}{t}\right) c(t),$$

where $c(t) \neq 0$ for $t \in \mathbf{T}$. Show that $T(b)$ is invertible on ℓ^2_∞ if and only if $\operatorname{wind} c = 0$.

(c) Let K be a compact operator on ℓ^2_∞ and suppose that $T(b) + K$ is invertible on ℓ^2_∞. Show that for each $y \in \ell^2_\infty$ the equations $(T_n(a) + P_n K P_n)x^{(n)} = P_n y$ have a unique solution whenever n is large enough and that $x^{(n)}$ converges in ℓ^2_∞ to the solution $x \in \ell^2_\infty$ of the equation $(T(a) + K)x = y$.

8. Show that $T_n^{-1}(2 - t - t^{-1})$ equals

$$\begin{pmatrix} 1 & 1 & \ldots & \ldots & 1 \\ 1 & 2 & \ldots & \ldots & 2 \\ \vdots & \vdots & \ddots & & \vdots \\ \vdots & \vdots & & n-1 & n-1 \\ 1 & 2 & \ldots & n-1 & n \end{pmatrix} - \frac{1}{n+1}\begin{pmatrix} 1 \\ 2 \\ \vdots \\ n \end{pmatrix}\begin{pmatrix} 1 & 2 & \ldots & n \end{pmatrix}.$$

Notes

Estimates for the growth of $\|T_n^{-1}(b)\|_p$ in the case where b has zeros or nonzero winding number have been of great interest for a long time. For example, questions on random walks or on Toeplitz determinants with so-called Fisher-Hartwig symbols necessitate such estimates. As $\|T_n(b)\|_2\,\|T_n^{-1}(b)\|_2$ is just the (spectral) condition number of $T_n(b)$, the need for such estimates is of course also currently emerging in numerical analysis.

Theorem 4.1 as it is stated was established in our paper [47] (by different methods). However, at least for $p = 2$ and in a few other contexts, the theorem was known before. For example, the formulas of [25] for the entries of $T_n^{-1}(a)$ show that these entries grow exponentially for symbols with nonzero winding number. The abstract of Reichel and Trefethen's paper [219] begins as follows: "The eigenvalues of a nonhermitian Toeplitz matrix A are usually highly sensitive to perturbations, having condition numbers that increase exponentially with the dimension N. An equivalent statement is that the resolvent $(zI - A)^{-1}$ of a Toeplitz matrix may be much larger in norm than the eigenvalues alone would suggest - exponentially large as a function of N, even when z is far away from the spectrum."

The results of Sections 4.1 and 4.3 are also from [47].

The change between sequences and polynomials has been successfully employed since the earliest studies of Toeplitz matrices. In particular results like Lemmas 4.7 to 4.9 are already in [145]. Lemma 4.10 is taken from [46].

In the case of Hermitian matrices, a good deal of the results of Sections 4.5 and 4.8 are known. In particular, Corollary 4.34 goes back to Kac, Murdock, Szegö, Parter, and Widom for symbols with a single zero and to Serra Capizzano for symbols with several zeros. We will discuss this problem in detail in Chapter 10. In the form stated here, the results of Sections 4.5 and 4.8 were established in our paper [46].

Theorem 4.13 appeared in Andreas Pomp's preprints [209], [210] for the first time. People working on Toeplitz operators and matrices with "singular" symbols have come to appreciate Duduchava and Roch's formula (4.29) as a kind of magic wand. For $\gamma + \delta = 0$, the formula was established by Duduchava [105]. When studying Toeplitz determinants with Fisher-Hartwig symbols, Silbermann and one of the authors realized the need for the formula for $\mathrm{Re}\,(\gamma + \delta) > -1$. Steffen Roch was able to extend the formula to this significantly more general case. It was published in [68] for the first time. The original proofs by Duduchava and Roch were very complicated. A simpler proof is in [68] and [70]: This proof is based on expanding a hypergeometric function into a power series in two different ways and on subsequently comparing the coefficients of equal powers.

Corollary 4.15 and Theorems 4.16 and 4.17 are from [69], and the results of Section 4.7 are all taken from [47].

We thank André Eppler for Figures 4.1 to 4.3.

Further results: entries of the inverse. Let $c \in \mathcal{P}$ and suppose c is positive, $c(\mathbf{T}) \subset (0, \infty)$. For each natural number α, the matrices $T_n(|t-1|^{2\alpha}c(t))$ are positive definite banded Toeplitz matrices. The inverses $T_n^{-1}(|t-1|^{2\alpha}c(t))$ have been of considerable interest since papers by Spitzer and Stone [257] and Kesten [178], [179]. It has been well known for a long time, at least since [90], that for $1 \le k \le \ell \le n$ the k, ℓ entry of $T_n^{-1}(|t-1|^2)$ is

$$[T_n^{-1}(|t-1|^2)]_{k,\ell} = k\ell\left(\frac{1}{\ell} - \frac{1}{n+1}\right). \tag{4.58}$$

In the case $\alpha = 2$, we have for $1 \le k \le \ell \le n$ the formula

$$\frac{1}{k(k+1)\ell(\ell+1)}[T_n^{-1}(|t-1|^4)]_{k,\ell}$$
$$= \frac{1}{\ell+2} - \frac{1}{n+3} - \frac{k+\ell-1}{2}\left(\frac{1}{(\ell+1)(\ell+2)} - \frac{1}{(n+2)(n+3)}\right)$$
$$+ \frac{(k-1)(\ell-1)}{3}\left(\frac{1}{\ell(\ell+1)(\ell+2)} - \frac{1}{(n+1)(n+2)(n+3)}\right) \tag{4.59}$$

(see, e.g., [2]). Since the matrices $T_n^{-1}(|t-1|^{2\alpha})$ are Hermitian, the right-hand sides of (4.58) and (4.59) also give the entries $[T_n^{-1}(|t-1|^{2\alpha})]_{\ell,k}$ for $1 \le k \le \ell \le n$. Now take $k = [nx]$ and $\ell = [ny]$ with fixed $1 \le x \le y \le 1$. We here (and only here) denote by $[nx]$ and $[ny]$ the smallest integer in $\{1, \ldots, n\}$ that is greater than or equal to nx and ny, respectively. We so arrive at the asymptotic formulas

$$\frac{1}{n}[T_n^{-1}(|t-1|^2)]_{[nx],[ny]} = x(1-y) + o(1),$$
$$\frac{1}{n^3}[T_n^{-1}(|t-1|^4)]_{[nx],[ny]} = \frac{1}{6}x^2(1-y)^2(3y-2xy-x) + o(1).$$

Note that all these formulas concern the case where the function c is identically 1.

The presence of the factor $c(t)$ complicates things significantly. Recently Rambour and Seghier [215], [216] proved that if $x, y \in [0, 1]$, then

$$[T_n^{-1}(|t-1|^{2\alpha}c(t))]_{[nx],[ny]} = \frac{1}{c(1)}G_\alpha(x,y)n^{2\alpha-1} + o(n^{2\alpha-1}) \text{ as } n \to \infty, \tag{4.60}$$

uniformly with respect to x and y in $[0, 1]$ (see also [214] for the case $\alpha = 1$). The constant $G_\alpha(x, y)$ is independent of c. Thus, it is only the value of c at the zero of $|t-1|^{2\alpha}$ that enters the principal term of the right-hand side of (4.60). The constant $G_\alpha(x, y)$ satisfies

$$G_\alpha(x,y) = G_\alpha(y,x) = G_\alpha(1-x, 1-y) = G_\alpha(1-y, 1-x) \tag{4.61}$$

and hence it suffices to find $G_\alpha(x, y)$ for $0 \le x \le y \le 1$ or even only for $0 \le x \le 1$ and $y \ge \max(x, 1-x)$. Rambour and Seghier showed that, for these x and y,

$$G_1(x,y) = x(1-y),$$
$$G_2(x,y) = \frac{1}{6}x^2(1-y)^2(3y-x-2xy),$$

and they established a formula that allows at least in principle the computation $G_\alpha(x, y)$ for all α. In [40], the Duduchava-Roch formula (4.29) was used to prove (4.60) for $c = 1$ and to find the constant $G_\alpha(x, y)$ in "closed form." The result is as follows: If $0 \le x \le 1$ and $y \ge \max(x, 1-x)$, then

$$G_\alpha(x,y) = \frac{x^\alpha y^\alpha}{[(\alpha-1)!]^2}\int_y^1 \frac{(t-x)^{\alpha-1}(t-y)^{\alpha-1}}{t^{2\alpha}}\,dt. \tag{4.62}$$

In particular,

$$G_\alpha(x, x) = \frac{x^{2\alpha-1}(1-x)^{2\alpha-1}}{(2\alpha-1)[(\alpha-1)!]^2}. \tag{4.63}$$

Combining (4.60) and (4.63) we get the trace formula

$$\operatorname{tr} T_n^{-1}(|t-1|^{2\alpha}c(t)) = \frac{1}{c(1)} n^{2\alpha} \frac{(2\alpha-1)!(2\alpha-2)!}{(4\alpha-1)![(\alpha-1)!]^2} + o(n^{2\alpha}).$$

The sum of the entries of the inverse is also known to be of interest (see, e.g., [1], [242], [287]). In [40], it is shown that (4.60), (4.61), (4.62) yield the formula

$$\sum_{j,k=1}^{n} [T_n^{-1}(|t-1|^{2\alpha}c(t))]_{j,k} = \frac{1}{c(1)} \frac{1}{2\alpha+1} \left[\frac{\alpha!}{(2\alpha)!}\right]^2 n^{2\alpha+1} + o(n^{2\alpha+1}).$$

We will say more on $G_\alpha(x, y)$ and the accompanying story in Chapter 10.

Chapter 5

Norms

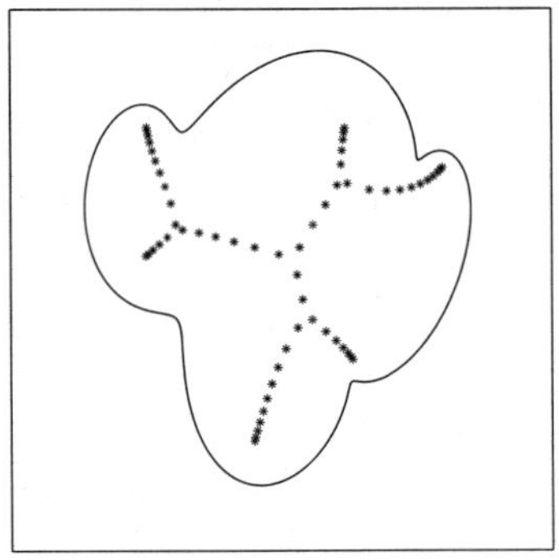

The norms of pure Toeplitz and of Toeplitz-like matrices approach a limit as the matrix dimension goes to infinity. In this chapter we identify this limit and give estimates for the speed of convergence.

5.1 A Universal Estimate

Let b be a Laurent polynomial. We write b in the form

$$b(t) = \sum_{j=-s}^{s} b_j t^j \quad (t \in \mathbf{T}), \tag{5.1}$$

assuming that at least one of the coefficients b_s and b_{-s} is nonzero. It is clear that

$$\|T_n(b)\|_1 = \|T_n(b)\|_\infty = \sum_{j=-s}^{s} |b_j| = \|T(b)\|_1 = \|T(b)\|_\infty$$

whenever $n \ge 2s+1$. So let $1 < p < \infty$. Obviously, $\|T_n(b)\|_p \le \|T(b)\|_p$. Since $T_n(b)$ converges strongly to $T(b)$, we deduce from Theorem 3.1 that

$$\|T(b)\|_p \le \liminf_{n\to\infty} \|T_n(b)\|_p.$$

Thus,

$$\lim_{n\to\infty} \|T_n(b)\|_p = \|T(b)\|_p. \tag{5.2}$$

The purpose of this section is to show that always

$$\|T_n(b)\|_p = \|T(b)\|_p + O(1/n).$$

We begin with a few elementary lemmas. As usual, $[\alpha]$ is the integral part of α.

Lemma 5.1. *Let n and s be natural numbers.*

(a) *The identity* $n = [4n/3] - [\frac{[4n/3]}{4}]$ *holds.*

(b) *If* $n \geq 12$ *then* $[4n/3] \geq 5n/4$.

(c) *If* $n \geq 8 + 12s$ *then* $[\frac{1}{s}[\frac{[4n/3]}{4}]] \geq \frac{n}{4s}$.

Proof. (a) For $n = 3k + \ell$ with $k \in \mathbf{N}$ and $\ell \in \{0, 1, 2\}$, the asserted identity is equivalent to the obvious identity $n = 4k + \ell - k$. (b) This is immediate from the inequality $[4n/3] \geq 4n/3 - 1$. (c) Let $n = 3k + \ell$ with $k \in \mathbf{N}$ and $\ell \in \{0, 1, 2\}$. Then

$$\left[\frac{1}{s}\left[\frac{[4n/3]}{4}\right]\right] = \left[\frac{1}{s}\left(k + \left[\frac{1}{4}\left[\frac{4}{3}\ell\right]\right]\right)\right] = \left[\frac{k}{s}\right] \geq \frac{k}{s} - 1$$
$$\geq \frac{1}{s}\,\frac{n-2}{3} - 1 = \frac{n - 2 - 3s}{3s} = \frac{n}{4s} + \frac{n - 8 - 12s}{12s}. \qquad \square$$

Lemma 5.2. *If* $1 < p < \infty$ *and* $0 \leq x \leq 1/2$, *then* $(1 - x)^{1/p} > 1 - 2x/p$.

Proof. A look at the graphs of the functions $y = (1 - x)^{1/p}$ and $y = 1 - 2x/p$ shows that it suffices to prove the asserted inequality for $x = 1/2$. Thus, we must prove that $1/p + 1/2^{1/p} > 1$. Consider the function $f(t) = t + 2^{-t}$. Since $f(0) = 1$ and $f'(t) = 1 - (\log 2)2^{-t} > 1 - \log 2 > 0$ for $t > 0$, it follows that $f(t) > 1$ for $t > 0$. Taking $t = 1/p$ we get $1/p + 1/2^{1/p} > 1$, as desired. $\square$

Lemma 5.3. *If* $1 < p < \infty$ *and* $1/p + 1/q = 1$, *then*

$$\|T_n(b)\|_p = \|T_n(\widetilde{b})\|_p = \|T_n(b)\|_q = \|T_n(\widetilde{b})\|_q.$$

Proof. From (3.10) we obtain

$$\|T_n(\widetilde{b})\|_p = \|W_n T_n(b) W_n\|_p \leq \|T_n(b)\|_p = \|W_n T_n(\widetilde{b}) W_n\|_p \leq \|T_n(\widetilde{b})\|_p,$$

whence $\|T_n(b)\|_p = \|T_n(\widetilde{b})\|_p$. As $T_n(\widetilde{b})$ is the transpose matrix of $T_n(b)$, we see that $\|T_n(\widetilde{b})\|_p = \|T_n(b)\|_q$. $\square$

Lemma 5.4. *Let* $1 < p < \infty$ *and let* s, ℓ, n *be natural numbers satisfying* $n \geq 3$ *and* $1 \leq s \leq \ell \leq n/3$. *If* $x \in \ell_n^p$ *is a unit vector,*

$$\|x\|_p^p = \sum_{j=0}^{n-1} |x_j|^p = 1,$$

then there is a natural number m *such that* $\ell + s \leq m \leq 3\ell - s$ *and*

$$\sum_{j=m-s}^{m+s-1} |x_j|^p \leq \left[\frac{\ell}{s}\right]^{-1}.$$

Proof. Put $d = [\ell/s]$. Since $\ell + 2ds \le 3\ell$, we have

$$1 \ge \sum_{j=\ell}^{3\ell-1} |x_j|^p \ge (|x_\ell|^p + \cdots + |x_{\ell+2s-1}|^p) \\ + (|x_{\ell+2s}|^p + \cdots + |x_{\ell+4s-1}|^p) + \cdots \\ + (|x_{\ell+2(d-1)s}|^p + \cdots + |x_{\ell+2ds-1}|^p).$$

As there are d terms on the right-hand side, at least one of them does not exceed $1/d$. Hence, there exist a $k_0 \in \{0, \dots, d-1\}$ such that

$$|x_{\ell+2k_0 s}|^2 + \cdots + |x_{\ell+2k_0 s+2s-1}|^2 \le \frac{1}{d}.$$

The assertion now follows with $m = \ell + 2k_0 s + s$. □

Theorem 5.5. *Let b be given by* (5.1) *and let $1 < p < \infty$. If $n \ge 40\,s$, then*

$$\|T(b)\|_p \left(1 - \frac{40\,s}{pn}\right) \le \|T_n(b)\|_p \le \|T(b)\|_p. \tag{5.3}$$

Proof. Put $M_0 = \|T(b)\|_p$ and $\varepsilon_n = M_0 - \|T_n(b)\|_p$. We already know that ε_n converges monotonically to zero. Thus, we are left with showing that

$$\varepsilon_n \le \frac{40\,s}{pn} M_0. \tag{5.4}$$

The assertion is trivial for $s = 0$. So let $s \ge 1$ and $n \ge 40\,s$. Choose a vector $x = (x_0, \dots, x_{n-1}) \in \ell_n^p$ so that

$$\|x\|_p = 1 \text{ and } \|T_n(b)x\|_p = M_0 - \varepsilon_n. \tag{5.5}$$

Set $\ell = [n/4]$. By Lemma 5.4, there exists a natural number m for which

$$\ell + s \le m \le 3\ell - s \text{ and } \sum_{j=m-s}^{m+s-1} |x_j|^p < \left[\frac{\ell}{s}\right]^{-1}. \tag{5.6}$$

We have

$$\|T_n(b)x\|_p^p = \sum_{j=0}^{m-1} |(T_n(b)x)_j|^p + \sum_{j=m}^{n-1} |(T_n(b)x)_j|^p. \tag{5.7}$$

Since $b_{j-k} = 0$ for $j \le m-1$ and $k \ge m+s$, we get

$$\sum_{j=0}^{m-1} |(T_n(b)x)_j|^p = \sum_{j=0}^{m-1} \left| \sum_{k=0}^{m+s-1} b_{j-k}x_k \right|^p \\ \le \sum_{j=0}^{m+s-1} \left| \sum_{k=0}^{m+s-1} b_{j-k}x_k \right|^p = \left\| T_{m+s}(b)(x_k)_{k=0}^{m+s-1} \right\|_p^p \\ \le (M_0 - \varepsilon_{m+s})^p \sum_{k=0}^{m+s-1} |x_k|^p. \tag{5.8}$$

Analogously, starting with the observation that $b_{j-k} = 0$ if $j \geq m$ and $k \leq m - s - 1$, we obtain

$$\begin{aligned}
\sum_{j=m}^{n-1} |(T_n(b)x)_j|^p &= \sum_{j=m}^{n-1} \left| \sum_{k=m-s}^{n-1} b_{j-k}x_k \right|^p = \sum_{j=m}^{n-1} \left| \sum_{r=0}^{n-m+s-1} b_{j-n+1+r}x_{n-1-r} \right|^p \\
&= \sum_{i=0}^{n-m-1} \left| \sum_{r=0}^{n-m+s-1} b_{n-1-i-n+1+r}x_{n-1-r} \right|^p = \sum_{i=0}^{n-m-1} \left| \sum_{r=0}^{n-m+s-1} b_{r-i}x_{n-1-r} \right|^p \\
&\leq \sum_{i=0}^{n-m+s-1} \left| \sum_{r=0}^{n-m+s-1} b_{r-i}x_{n-1-r} \right|^p = \left\| T_{n-m+s}(\widetilde{b})(x_{n-1-r})_{r=0}^{n-m+s-1} \right\|_p^p \\
&\leq (M_0 - \varepsilon_{n-m+s})^p \sum_{r=0}^{n-m+s-1} |x_{n-1-r}|^p \quad \text{(by Lemma 5.3)} \\
&= (M_0 - \varepsilon_{n-m+s})^p \sum_{k=m-s}^{n-1} |x_k|^p. \qquad (5.9)
\end{aligned}$$

Combining (5.7), (5.8), (5.9), we arrive at the inequality

$$\|T_n(b)x\|_p^p \leq (M_0 - \varepsilon_{m+s})^p \sum_{k=0}^{m+s-1} |x_k|^p + (M_0 - \varepsilon_{n-m+s})^p \sum_{k=m-s}^{n-1} |x_k|^p.$$

As $m + s \leq n - \ell$ and $n - m + s \leq n - \ell$, and as ε_k is monotonically decreasing, it follows that

$$\|T_n(b)x\|_p^p \leq (M_0 - \varepsilon_{n-\ell})^p \left(\sum_{k=0}^{m+s-1} |x_k|^p + \sum_{k=m-s}^{n-1} |x_k|^p \right).$$

The equality $\|x\|_p = 1$ and inequality (5.6) imply that

$$\sum_{k=0}^{m+s-1} |x_k|^p + \sum_{k=m-s}^{n-1} |x_k|^p \leq 1 + [\ell/s]^{-1}, \qquad (5.10)$$

whence

$$\|T_n(b)x\|_p^p \leq (M_0 - \varepsilon_{n-\ell})^p(1 + [\ell/s]) \leq (M_0 - \varepsilon_{n-\ell})^p + M_0^p[\ell/s]^{-1}.$$

Thus, by (5.5), $(M_0 - \varepsilon_n)^p \leq (M_0 - \varepsilon_{n-\ell})^p + M_0[\ell/s]^{-1}$ or, equivalently,

$$\left(1 - \frac{\varepsilon_n}{M_0}\right)^p \leq \left(1 - \frac{\varepsilon_{n-\ell}}{M_0}\right)^p + \left[\frac{\ell}{s}\right]^{-1}. \qquad (5.11)$$

Recall that $\ell = [n/4]$ and put $n_1 = [4n/3]$. In (5.11), we can replace n by n_1. Taking into consideration Lemma 5.1(a),(c) we get

$$\left(1 - \frac{\varepsilon_{n_1}}{M_0}\right)^p \leq \left(1 - \frac{\varepsilon_n}{M_0}\right)^p + \frac{4s}{n}; \qquad (5.12)$$

notice that $n \geq 40\,s > 8 + 12\,s$. Now substitute the n in (5.12) consecutively by

$$n_1 = \left[\frac{4n}{3}\right], \quad n_2 = \left[\frac{4n_1}{3}\right], \quad \ldots, \quad n_j = \left[\frac{4n_{j-1}}{3}\right],$$

add the corresponding inequalities and pass to the limit $j \to \infty$. What results is that

$$1 - \left(1 - \frac{\varepsilon_n}{M_0}\right)^p \leq \frac{4s}{n} + \sum_{j=1}^{\infty} \frac{4s}{n_j}.$$

By Lemma 5.1(b),

$$\frac{n}{n_1} \leq \frac{4}{5}, \quad \frac{n_1}{n_2} \leq \frac{4}{5}, \quad \frac{n_2}{n_3} \leq \frac{4}{5}, \quad \ldots .$$

Consequently,

$$1 - \left(1 - \frac{\varepsilon_n}{M_0}\right)^p \leq \frac{4s}{n} + \frac{4s}{n}\sum_{j=1}^{\infty}\left(\frac{4}{5}\right)^j = \frac{20\,s}{n}.$$

Since $n \geq 40\,s$, we infer from Lemma 5.2 that

$$\varepsilon_n \leq \left(1 - \left(1 - \frac{20\,s}{n}\right)^{1/p}\right) M_0 \leq \frac{40\,s}{pn}\, M_0,$$

which is (5.4). $\square$

Notice that, by Lemma 5.3, estimate (5.3) can be improved to

$$\|T(b)\|_p \left(1 - \frac{40\,s}{n\,\max(p,q)}\right) \leq \|T_n(b)\|_p \leq \|T(b)\|_p.$$

5.2 Spectral Norm of Toeplitz Matrices

Theorem 5.5 results from the techniques we employed to prove it, and this theorem can probably be sharpened. In this section we establish a significant improvement of Theorem 5.5 in the case $p = 2$.

If $b(t)$ takes a constant value b_0 for all $t \in \mathbf{T}$, then $\|T_n(b)\|_2 = |b_0| = \|T(b)\|_2$ for all $n \geq 1$. The following proposition describes the Laurent polynomials with constant modulus.

Proposition 5.6. *If b is a Laurent polynomial and $|b|$ is constant on $\mathbf{T}$, then $b(t) = \gamma t^m$ $(t \in \mathbf{T})$ with constants $\gamma \in \mathbf{C}$ and $m \in \mathbf{Z}$.*

Proof. The assertion is trivial if b vanishes identically. So suppose b is not identically zero and let b_m be the first nonzero coefficient of b. Then

$$t^{-m}b(t) = b_m + b_{m+1}t + \cdots + b_{m+k}t^k \quad (t \in \mathbf{T})$$

with $b_{m+k} \neq 0$. In the case $k = 0$ we are done. So assume $k \geq 1$. The function

$$f(z) := (b_m + b_{m+1}z + \cdots + b_{m+k}z^k)(\overline{b}_m + \overline{b}_{m+1}z^{-1} + \cdots + \overline{b}_{m+k}z^{-k})$$

is analytic in $\mathbf{C} \setminus \{0\}$ and takes the constant value of $|b|^2$ on $\mathbf{T}$. It follows that f is constant throughout $\mathbf{C} \setminus \{0\}$, which is impossible because $f(z) = b_{m+k}\overline{b}_m z^k + O(z^{k-1})$ as $z \to \infty$ and $b_{m+k}\overline{b}_m \neq 0$. □

Proposition 5.6 implies that if $|b|$ is constant on $\mathbf{T}$, then $\|T_n(b)\|_2 = |\gamma| = \|T(b)\|_2$ for all $n \geq m+1$, where $b(t) = \gamma t^m$.

We now turn to symbols whose modulus is not constant. We begin with semi-definite Hermitian matrices.

Lemma 5.7. *Let b be a nonconstant and nonnegative Laurent polynomial and let 2γ be the maximal order of the zeros of $\|b\|_\infty - b(t)$ on $\mathbf{T}$. Then there exist constants $0 < d_1 < d_2 < \infty$ independent of n such that*

$$\|b\|_\infty \left(1 - \frac{d_2}{n^{2\gamma}}\right) \leq \|T_n(b)\|_2 \leq \|b\|_\infty \left(1 - \frac{d_1}{n^{2\gamma}}\right) \quad \textit{for all } n \geq 1.$$

Proof. Let $M = \|b\|_\infty$. By Corollary 4.34, there are constants $0 < D_1 < D_2 < \infty$ such that

$$D_1 n^{2\gamma} \leq \|T_n^{-1}(M-b)\|_2 \leq D_2 n^{2\gamma}. \tag{5.13}$$

Given a positive definite Hermitian matrix A, we denote by $\lambda_{\min}(A)$ and $\lambda_{\max}(A)$ the minimal and maximal eigenvalue of A, respectively. Recall that rad $(\cdot)$ stands for the spectral radius. We have

$$\begin{aligned}\|T_n^{-1}(M-b)\|_2 &= \text{rad}\,(T_n^{-1}(M-b)) = 1/\lambda_{\min}(MI - T_n(b)) \\ &= 1/(M - \lambda_{\max}(T_n(b))) = 1/(M - \|T_n(b)\|_2).\end{aligned}$$

Inserting this into (5.13) we get

$$M\left(1 - \frac{1}{MD_2}\frac{1}{n^{2\gamma}}\right) \leq \|T_n(b)\|_2 \leq M\left(1 - \frac{1}{MD_1}\frac{1}{n^{2\gamma}}\right). \qquad \square$$

As the following result shows, Lemma 5.7 is almost literally true in the general case, too. Recall that $\|T(b)\|_2 = \|b\|_\infty$.

Theorem 5.8. *Let b be a Laurent polynomial, suppose $|b|$ is not constant on $\mathbf{T}$, and let 2γ be the maximal order of the zeros of $\|b\|_\infty - |b(t)|$ on $\mathbf{T}$. Then there are constants $0 < d_1 < d_2 < \infty$ which do not depend on n such that*

$$\|b\|_\infty \left(1 - \frac{d_2}{n^{2\gamma}}\right) \leq \|T_n(b)\|_2 \leq \|b\|_\infty \left(1 - \frac{d_1}{n^{2\gamma}}\right) \quad \textit{for all } n \geq 1. \tag{5.14}$$

Proof. Let $b(t) = \sum_{j=-s}^{s} b_j t^j$ $(t \in \mathbf{T})$ with $s \geq 1$. By Proposition 3.10,

$$T_n(b)T_n(\overline{b}) = T_n(|b|^2) - P_n K P_n - W_n L W_n,$$

where

$$K = H(b)H(\widetilde{\overline{b}}) = \begin{pmatrix} b_1 & b_2 & \dots \\ b_2 & \dots & \\ \dots & & \end{pmatrix} \begin{pmatrix} \overline{b}_1 & \overline{b}_2 & \dots \\ \overline{b}_2 & \dots & \\ \dots & & \end{pmatrix}$$

and

$$L = H(\widetilde{b})H(\overline{b}) = \begin{pmatrix} b_{-1} & b_{-2} & \dots \\ b_{-2} & \dots & \\ \dots & & \end{pmatrix} \begin{pmatrix} \overline{b}_{-1} & \overline{b}_{-2} & \dots \\ \overline{b}_{-2} & \dots & \\ \dots & & \end{pmatrix}.$$

Since $(Kx, x) \geq 0$ and $(Lx, x) \geq 0$ for all $x \in \ell^2$, it follows that

$$(T_n(b)T_n(\overline{b})x, x) \leq (T_n(|b|^2)x, x) \quad \text{for all} \quad x \in \ell^2.$$

As $T_n(b)T_n(\overline{b})$ and $T_n(|b|^2)$ are Hermitian, we therefore obtain

$$\|T_n(b)\|_2^2 = \|T_n(b)T_n(\overline{b})\|_2 \leq \|T_n(|b|^2)\|_2. \tag{5.15}$$

Lemma 5.7 implies that

$$\begin{aligned} \|T_n(|b|^2)\|_2 &\leq \|\, |b|^2 \,\|_\infty(1 - d/n^{2\gamma}) = \|b\|_\infty^2(1 - d/n^{2\gamma}) \\ &\leq \|b\|_\infty^2(1 - d/(2n^{2\gamma}))^2. \end{aligned} \tag{5.16}$$

Combining (5.15) and (5.16), we get the upper estimate in (5.14) with $d_1 = d/2$.

We are left to prove the lower estimate in (5.14). Put $M = \|b\|_\infty^2$. Since

$$T_n(b(t/t_0)) = \operatorname{diag}\,(1, t_0^{-1}, \dots, t_0^{-n+1})\, T_n(b)\, \operatorname{diag}\,(1, t_0, \dots, t_0^{n+1}),$$

we can without loss of generality assume that $M = |b(1)|^2$. By assumption, there is a constant $C \in (0, \infty)$ such that

$$M - |b(e^{i\theta})|^2 \leq C\,|\theta|^{2\gamma} \quad \text{for all} \quad \theta \in [-\pi, \pi]. \tag{5.17}$$

For $x \in \ell_n^2$,

$$\begin{aligned} &\|T_n(b)\|_2^2\|x\|_2^2 = \|T_n(b)T_n(\overline{b})\|_2\|x\|_2^2 \geq |(T_n(b)T_n(\overline{b})x, x)| \\ &= |(T_n(|b|^2)x, x) - (P_nKP_nx, x) - (W_nLW_nx, x)| \\ &\geq M\|x\|_2^2 - |((M - T_n(|b|^2))x, x)| - |(P_nKP_nx, x)| - |(W_nLW_nx, x)|. \end{aligned} \tag{5.18}$$

We now identify ℓ_n^2 and $\mathcal{P}_n^-$ as in Section 4.4. Put $j = \gamma + 1$, let m be the natural number given by $mj < n \leq (m+1)j$, define $f \in \mathcal{P}_n^+$ by

$$f(e^{i\theta}) = (1 + e^{i\theta} + \dots + e^{im\theta})^j = e^{imj\theta/2}\left(\frac{\sin\frac{m+1}{2}\theta}{\sin\frac{\theta}{2}}\right)^j \tag{5.19}$$

(recall (4.20) and (4.21)), and let $x \in \ell_n^2$ be the sequence of the Fourier coefficients of f. From Lemma 4.10 we know that

$$\|f\|_2 \geq C_1(m+1)^{j-1/2} \text{ with some } C_1 > 0. \tag{5.20}$$

We estimate the terms on the right of (5.18) separately. Since $b_j = 0$ for $|j| > s$, we have

$$|(P_n K P_n x, x)| = |(P_s K P_s x, x)| = |(K P_s x, P_s x)| \leq \|K\|_2 \|P_s x\|_2^2,$$

and from (5.19) we infer that $P_s x$ is independent of n provided n (and thus m) is large enough. Consequently, by (5.20) and Lemma 4.7,

$$\frac{|(P_n K P_n x, x)|}{\|x\|_2^2} \leq \frac{C_2}{(m+1)^{2j-1}} \leq \frac{C_2 j^{2j-1}}{n^{2j-1}} = \frac{C_2(\gamma+1)^{2\gamma+1}}{n^{2\gamma+1}} \tag{5.21}$$

with some $C_2 \in (0, \infty)$ independent of n. Analogously,

$$\frac{|(W_n L W_n x, x)|}{\|x\|_2^2} \leq \frac{C_3(\gamma+1)^{2\gamma+1}}{n^{2\gamma+1}}, \tag{5.22}$$

where $C_3 \in (0, \infty)$ does not depend on n. From (5.17), (5.19), and Lemma 4.15 we obtain

$$\begin{aligned}
&2\pi\, |((M - T_n(|b|^2))x, x)| = 2\pi\, |(T_n(M - |b|^2)x, x)| \\
&= \int_{-\pi}^{\pi} (M - |b(e^{i\theta})|^2) \left|\frac{\sin((m+1)\theta/2)}{\sin(\theta/2)}\right|^{2j} d\theta \\
&\leq \int_{|\theta|<1/(m+1)} C|\theta|^{2\gamma} \left(\frac{(m+1)|\theta|}{2}\right)^{2j} \left(\frac{\pi}{2}\frac{2}{|\theta|}\right)^{2j} d\theta \\
&\quad + \int_{|\theta|>1/(m+1)} C|\theta|^{2\gamma} \left(\frac{\pi}{2}\frac{2}{|\theta|}\right)^{2j} d\theta \\
&\leq C_3 \frac{(m+1)^{2j}}{(m+1)^{2\gamma+1}} + C_4 \frac{1}{(m+1)^{2j-2\gamma}} = C_3(m+1) + C_4 \frac{1}{(m+1)^2} \leq C_5 n
\end{aligned}$$

with constants $C_3, C_4, C_5 \in (0, \infty)$ independent of n. Taking into account (5.20) we therefore get

$$\frac{|((M - T_n(|b|^2))x, x)|}{\|x\|_2^2} \leq C_6 \frac{n}{n^{2j-1}} = C_6 \frac{1}{n^{2\gamma}}, \tag{5.23}$$

where $C_6 \in (0, \infty)$ is independent of n. Putting (5.18), (5.21), (5.22), (5.23) together we arrive at the estimate $\|T_n(b)\|_2^2 \geq M(1 - C_7/n^{2\gamma})$, where $C_7 \in (0, \infty)$ is independent of n, whence $\|T_n(b)\|_2 \geq \|b\|_\infty(1 - C_7/n^{2\gamma})$ as soon as $n^{2\gamma} > C_7$. □

An example is considered in Figure 5.1. The function $\|b\|_\infty - |b(t)|$ has two zeros, one of the order 4 and of the order 2. Hence, by Theorem 5.8, $\|b\|_\infty - \|T_n(b)\|_2$ decays as c_n/n^4 with a bounded sequence $\{c_n\}$. Figure 5.1 shows precisely this decay with c_n stabilizing very quickly at a constant value.

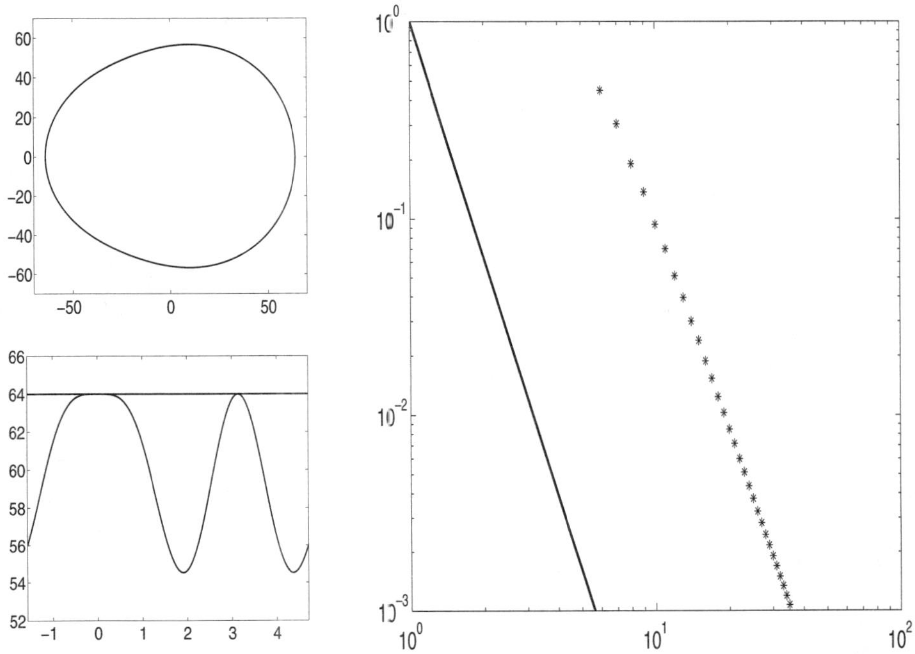

Figure 5.1. *The symbol is* $b(t) = (64 - |t-1|^4|t+1|^2)t$. *The set* $b(\mathbf{T})$ *is shown in the upper-left picture. The lower-left picture shows* $|b(e^{i\theta})|$ *for* $-\pi/2 \le \theta \le 3\pi/2$ *with the maximum line* $\|b\|_\infty = 64$. *In the right picture, the asterisks mark* $\|b\|_\infty - \|T_n(b)\|_2$ *versus* n. *As we have logarithmic scales, the slope* -4 *of the asterisks corresponds to a decay of* $\|b\|_\infty - \|T_n(b)\|_2$ *as* constant$/n^4$. *The straight line in the picture is simply a line with the slope* -4.

Corollary 5.9. *If* b *is a Laurent polynomial, then there is a constant* $d \in (0, \infty)$ *depending only on* b *such that*

$$\|b\|_\infty \left(1 - \frac{d}{n^2}\right) \le \|T_n(b)\|_2 \le \|b\|_\infty \quad \textit{for all } n \ge 1.$$

Proof. By Proposition 5.6, this is trivial if $|b|$ is constant on $\mathbf{T}$. So assume that $|b|$ is not constant on $\mathbf{T}$. Since the function $\|b\|_\infty - |b(t)|$ does always have a zero of order at least 2, the lower estimate follows from Theorem 5.8. The upper estimate is obvious. □

5.3 Fejér Means

In this section, we slightly change our view at Toeplitz matrices. Namely, given a sequence $b = \{b_j\}_{j=-\infty}^{\infty}$ of complex numbers, we denote by $T_n(b) := (b_{j-k})_{j,k=0}^{n-1}$ the $n \times n$ Toeplitz matrix generated by this *sequence*. We consider the Laurent polynomials

$$(S_n b)(e^{i\varphi}) = \sum_{|j|\le n-1} b_j e^{ij\varphi},$$

$$(\sigma_n b)(e^{i\varphi}) = \sum_{|j|\le n-1} b_j \left(1 - \frac{|j|}{n}\right) e^{ij\varphi}.$$

Clearly, if $\{b_j\}_{j=-\infty}^{\infty}$ is the sequence of the Fourier coefficients of an L^1 function a, then $S_n b$ is just the nth partial sum $S_n a$ of the Fourier series of a and $\sigma_n b$ is the nth Fejér mean $\sigma_n a$ of the Fourier series of a.

Since $T_n(b) = T_n(S_n b)$, it is clear that $\|T_n(b)\|_2 \le \|S_n b\|_\infty$ for all $n \ge 1$ and all sequences b. Is there a universal finite constant C such that $\|T_n(b)\|_2 \le C\|\sigma_n b\|_\infty$ for all $n \ge 1$ and b? The answer is negative: If the sequence b has only one nonzero term, say $b_{n-1} = 1$ $(n \ge 1)$, then $T_n(b)$ is the matrix with 1 in the lower-left corner and zeros elsewhere, so that $\|T_n(b)\|_2 = 1$, although $(\sigma_n b)(e^{i\varphi}) = e^{i(n-1)\varphi}/n$, which implies that $\|\sigma_n b\|_\infty = 1/n$.

What about estimates of $\|T_n(b)\|_2$ from below? Is there a universal constant $c > 0$ such that $c\|S_n b\|_\infty \le \|T_n(b)\|_2$ for all $n \ge 1$ and all b? The answer is again negative. Indeed, assume the answer is in the affirmative. Let a be a continuous function on the complex unit circle $\mathbf{T}$ and let $b = \{b_j\}$ be the sequence of the Fourier coefficients of a. Since $\|T_n(b)\|_2 \le \|a\|_\infty$ for all $n \ge 1$, it follows that $\|S_n a\|_\infty \le (1/c)\|a\|_\infty$ for all continuous functions a. This implies that $S_n a$ converges uniformly to a for every continuous function a, which is well known not to be true since du Bois-Reymond's 1876 paper [29] (also see [304, Theorem VIII.1.2]). On the other hand, we know from Fejér [119] that the means $\sigma_n a$ converge uniformly to a for every continuous function a. The following theorem provides us with a simple lower estimate for the spectral norm of a Toeplitz matrix through the Fejér mean.

Theorem 5.10. *The inequality $\|T_n(b)\|_2 \ge \|\sigma_n b\|_\infty$ holds for every $n \ge 1$ and every sequence b.*

Proof. Fix $t = e^{i\theta} \in \mathbf{T}$ and let $x_t \in \mathbf{C}^n$ be the vector $x_t = \frac{1}{\sqrt{n}}(1, \bar{t}, \ldots, \bar{t}^{\,n-1})$. Then $\|x_t\| = 1$ and hence $\|T_n(b)\|_2 \ge |(T_n(b)x_t, x_t)|$. Since

$$(T_n(b)x_t, x_t) = \frac{1}{n}\sum_{j,k=0}^{n-1} b_{j-k} e^{-ik\theta} e^{ij\theta}$$

$$= \frac{1}{n}\sum_{j,k=0}^{n-1} b_{j-k} e^{i(j-k)\theta} = \frac{1}{n}\sum_{j=-(n-1)}^{n-1} (n - |j|) b_j e^{ij\theta} = (\sigma_n b)(e^{i\theta}),$$

we get $\|T_n(b)\|_2 \ge |(\sigma_n b)(e^{i\theta})| = |(\sigma_n b)(t)|$. As $t \in \mathbf{T}$ can be chosen arbitrarily, it follows that $\|T_n(b)\|_2 \ge \|\sigma_n b\|_\infty$. □

5.4 Toeplitz-Like Matrices

In this section we consider sequences $\{B_n\}_{n=1}^{\infty}$ of $n \times n$ matrices B_n of the form

$$B_n = T_n(b) + P_n K P_n + W_n L W_n, \tag{5.24}$$

where $b(t) = \sum_{j=-s}^{s} b_j t^j$ $(t \in \mathbf{T})$ and where K and L have only a finite number of nonzero entries, which means that

$$P_{n_0} K P_{n_0} = K, \quad P_{n_0} L P_{n_0} = L \tag{5.25}$$

for some $n_0 \in \mathbf{N}$. Thus, the matrix B_n differs from $T_n(b)$ by the $n_0 \times n_0$ block K in the upper left and the $n_0 \times n_0$ block $W_n L W_n$ in the lower right corners. We refer to such matrices as *Toeplitz-like matrices*.

From Section 3.5 we know that the investigation of the inverses $T_n^{-1}(a)$ of Toeplitz matrices leads to matrices having the structure (5.24). Here is another context in which matrices of the form (5.24) emerge.

Proposition 5.11. *Let b_{jk} be a finite collection of Laurent polynomials. Then there exist matrices K and L satisfying (5.25) for some $n_0 \in \mathbf{N}$ such that*

$$\sum_j \prod_k T_n(b_{jk}) = T_n\left(\sum_j \prod_k b_{jk}\right) + P_n K P_n + W_n L W_n$$

for all sufficiently large n.

Proof. It suffices to prove that

$$T_n(b_1) \dots T_n(b_N) = T_n(b_1 \dots b_N) + P_n K_N P_n + W_n L_N W_n \tag{5.26}$$

for all sufficiently large n. Clearly, (5.26) holds with $K_N = L_N = 0$ if $N = 1$. So assume (5.26) is true with $P_{n_0} K_N P_{n_0} = K_N$ and $P_{n_0} L_N P_{n_0} = L_N$ for some $N \geq 1$. We have

$$\begin{aligned} &T_n(b_1) \dots T_n(b_N) T_n(b_{N+1}) - T_n(b_1 \dots b_N) T_n(b_{N+1}) \\ &= P_n K_N P_n T_n(b_{N+1}) + W_n L_N W_n T_n(b_{N+1}) \\ &= P_n K_N T(b_{N+1}) P_n - P_n K_N Q_n T(b_{N+1}) P_n \\ &\quad - W_n L_N T(\widetilde{b}_{N+1}) W_n - W_n L_N Q_n T(\widetilde{b}_{N+1}) W_n. \end{aligned}$$

Since $K_N Q_n = 0$ and $L_N Q_n = 0$ for $n \geq n_0$, we obtain from Proposition 3.10 that (5.26) is also true with N replaced by $N + 1$. $\square$

From (1.16) we infer that

$$\|B_n\|_1 = \max(\|T(b) + K\|_1, \|T(\widetilde{b}) + L\|_1), \tag{5.27}$$

$$\|B_n\|_\infty = \max(\|T(b) + K\|_\infty, \|T(\widetilde{b}) + L\|_\infty) \tag{5.28}$$

whenever $n \geq \max(2s + 1, 2n_0 + 1)$. Throughout what follows in this chapter we therefore assume that $1 < p < \infty$. Our objective is to show that

$$\lim_{n \to \infty} \|B_n\|_p = \max(\|T(b) + K\|_p, \|T(\widetilde{b}) + L\|_p).$$

Since $W_n \to 0$ weakly, we deduce from Proposition 3.3 that

$$W_n L W_n \to 0 \text{ strongly}, \qquad W_n K W_n \to 0 \text{ strongly}.$$

Consequently,

$$B_n \to T(b) + K \text{ strongly}, \qquad W_n B_n W_n \to T(\widetilde{b}) + L \text{ strongly}.$$

Because $\|W_n B_n W_n\|_p = \|B_n\|_p$, we deduce from Theorem 3.1 that

$$\max\left(\|T(b) + K\|_p, \|T(\widetilde{b}) + L\|_p\right) \le \liminf_{n\to\infty} \|B_n\|_p. \tag{5.29}$$

We define

$$M_p := \max\left(\|T(b) + K\|_p, \|T(\widetilde{b}) + L\|_p\right), \quad M_p^0 := \|T(b)\|_p. \tag{5.30}$$

Lemma 5.12. *We have $\|T(b) + K\|_p \ge \|T(b)\|_p$ and $\|T(\widetilde{b}) + L\|_p \ge \|T(b)\|_p$.*

Proof. For $n \in \mathbf{Z}$, let $\chi_n(t) = t^n$ $(t \in \mathbf{T})$. From Proposition 1.4 we obtain the equality $T(\chi_{-n})T(b)T(\chi_n) = T(b)$. On the other hand, from (5.25) we see that $KT(\chi_n) = 0$ for $n \ge n_0$. Hence, for $n \ge n_0$, $\|T(b)\|_p = \|T(\chi_{-n})(T(b)+K)T(\chi_n)\|_p$, and since $\|T(\chi_{\pm n})\|_p = 1$, it follows that $\|T(b)\|_p \le \|T(b)+K\|_p$. Analogously we get $\|T(\widetilde{b})\|_p \le \|T(\widetilde{b})+L\|_p$. As $\|T(b)\|_p = \lim_{n\to\infty} \|T_n(b)\|_p$ and $\|T(\widetilde{b})\|_p = \lim_{n\to\infty} \|T_n(\widetilde{b})\|_p$, we deduce from Lemma 5.3 that $\|T(\widetilde{b})\|_p = \|T(b)\|_p$. This gives the inequality $\|T(\widetilde{b}) + L\|_p \ge \|T(b)\|_p$. □

If $s = 0$, then

$$B_n = b_0 I + P_n K P_n + W_n L W_n,$$

and it is clear that

$$\|B_n\|_p = \max(\|b_0 I + K\|_p, |b_0|, \|b_0 I + L\|_p)$$

for $n \ge 2n_0 + 1$. As $\|b_0 I + K\|_p \ge |b_0|$ and $\|b_0 I + L\|_p \ge |b_0|$ by Lemma 5.12, it results that

$$\|B_n\|_p = \max(\|b_0 I + K\|_p, \|b_0 I + L\|_p) = M_p$$

for all $n \ge 2n_0 + 1$. Thus, in the following we will always assume that $s \ge 1$. The desired result is Corollary 5.14. The next theorem gives the difficult part of that corollary and, in addition, an upper estimate of $\|B_n\|_p$.

Theorem 5.13. *If $n \ge \max(8s + 8, 4n_0)$, then*

$$\|B_n\|_p \le M_p \left(1 + \frac{8s}{pn}\right).$$

Proof. Put $\ell = [n/4]$ and let $x = (x_j)_{j=0}^{n-1} \in \ell_n^p$ be any vector such that $\|x\|_p = 1$. By Lemma 5.4, there is an $m \in \mathbf{N}$ such that

$$\ell + s \le m \le 3\ell - s, \quad \sum_{j=m-s}^{m+s-1} |x_j|^p \le \left[\frac{\ell}{s}\right]^{-1}. \tag{5.31}$$

We now proceed as in the proof of Theorem 5.5. First, we have

$$\|B_n x\|_p^p = \sum_{j=0}^{m-1} |(B_n x)_j|^p + \sum_{j=m}^{n-1} |(B_n x)_j|^p. \tag{5.32}$$

It is clear that $\ell \le n/4$. By assumption, $n \ge 4n_0$, whence $\ell = [n/4] \ge n_0$. Thus,

$$n \ge 4\ell = \ell + 3\ell \ge n_0 + 3\ell > n_0 + 3\ell - s \ge n_0 + m. \tag{5.33}$$

If $j \le m-1$, then (5.33) implies that

$$n - 1 - j \ge n - m > n_0. \tag{5.34}$$

Let $K = (K_{ij})$ and $L = (L_{ij})$. Since

$$(W_n L W_n x)_j = (L W_n x)_{n-1-j} = \sum_{i=0}^{n_0-1} L_{n-1-j,i} x_{n-1-i},$$

we deduce from (5.25) and (5.34) that $(W_n L W_n x)_j = 0$ whenever $j \le m-1$. Consequently,

$$\begin{aligned}
\sum_{j=0}^{m-1} |(B_n x)_j|^p &= \sum_{j=0}^{m-1} \left| \sum_{i=0}^{m+s-1} b_{j-i} x_i + \sum_{i=0}^{n_0-1} K_{ji} x_i \right|^p \\
&= \|P_m (T(b) - K)(x_i)_{i=0}^{m+s-1}\|_p^p \\
&\le \|T(b) + K\|_p^p \sum_{i=0}^{m+s-1} |x_i|^p \le M_p^p \sum_{i=0}^{m+s-1} |x_i|^p.
\end{aligned} \tag{5.35}$$

Similarly, if $j \ge m$ then $(P_n K P_n x)_j = 0$, whence

$$\begin{aligned}
\sum_{j=m}^{n-1} |(B_n x)_j|^p &= \sum_{j=m}^{n-1} \left| \sum_{i=m-s}^{n-1} b_{j-i} x_i + \sum_{i=0}^{n_0-1} L_{n-1-j,i} x_{n-1-i} \right|^p \\
&= \sum_{j=m}^{n-1} \left| \sum_{r=0}^{n-m+s-1} b_{j-n+1+r} x_{n-1-r} + \sum_{r=0}^{n_0-1} L_{n-1-j,r} x_{n-1-r} \right|^p \\
&= \sum_{k=0}^{n-m-1} \left| \sum_{r=0}^{n-m+s-1} b_{n-1-k-n+1+r} x_{n-1-r} + \sum_{r=0}^{n_0-1} L_{n-1-n+1+k,r} x_{n-1-r} \right|^p \\
&= \sum_{k=0}^{n-m-1} \left| \sum_{r=0}^{n-m+s-1} b_{r-k} x_{n-1-r} + \sum_{r=0}^{n_0-1} L_{k,r} x_{n-1-r} \right|^p \\
&= \|P_{n-m}(T(\widetilde{b}) + L)(x_{n-1-r})_{r=0}^{n-m+s-1}\|_p^p \\
&\le \|T(\widetilde{b}) + L\|_p^p \sum_{r=0}^{n-m+s-1} |x_{n-1-r}|^p \le M_p^p \sum_{j=m-s}^{n-1} |x_j|^p.
\end{aligned} \tag{5.36}$$

Combining (5.32), (5.35), and (5.36) we get

$$\|B_n x\|_p^p \le M_p^p \left(\sum_{j=0}^{m+s-1} |x_j|^p + \sum_{j=m-s}^{n-1} |x_j|^p \right),$$

which, by (5.10), implies that $\|B_n x\|_p^p \le M_p^p(1 + [\ell/s]^{-1})$. As $[\ell/s] \ge n/(8s)$ for $n \ge 8 + 8s$, it results that

$$\|B_n\|_p \le M_p \left(1 + \frac{8s}{n}\right)^{1/p} \le M_p \left(1 + \frac{8s}{pn}\right). \qquad \square$$

Corollary 5.14. *We have*

$$\lim_{n\to\infty} \|B_n\|_p = M_p.$$

Proof. The proof is immediate from (5.29), (5.30), and Theorem 5.13. $\square$

5.5 Exponentially Fast Convergence Is Generic

We now turn to the problem of estimating the speed with which $\|B_n\|_p$ converges to M_p. The solution will be as follows: generically $\|B_n\|_p$ converges to M_p exponentially fast, but in some exceptional cases it may happen that $|\,\|B_n\|_p - M_p|$ decays only as $1/n^2$. Precise results are in this and the next sections.

Let B_n be as in the previous section and put $m_0 = \max(s, n_0)$ and $B = T(b) + K$. We know from Lemma 5.12 that always $\|B\|_p \ge \|T(b)\|_p$.

The following lemma shows the existence of a rapidly decaying sequence at which B attains its norm.

Lemma 5.15. *If $\|B\|_p > \|T(b)\|_p$, then there exists an $x_0 = \{x_j^{(0)}\}_{j=0}^{\infty} \in \ell^p$ such that*

$$\|x_0\|_p = 1, \quad \|Bx_0\|_p = \|B\|_p, \tag{5.37}$$

and

$$\left(\sum_{j=m}^{\infty} |x_j^{(0)}|^p \right)^{1/p} \le \left(\frac{\|T(b)\|_p}{\|B\|_p} \right)^{(m-(m_0+s))/(2s)} \tag{5.38}$$

for every $m \ge m_0 + s$.

Proof. Put $M_0 = \|T(b)\|_p$ and $M = \|B\|_p$. Pick any natural number $m \ge m_0$. There are $x_k = \{x_j^{(k)}\}_{j=0}^{\infty} \in \ell^p$ such that

$$\|x_k\|_p = 1, \quad \|Bx_k\|_p = M^p - \delta_k, \tag{5.39}$$

where δ_k goes monotonically to zero as $k \to \infty$. As in the proof of Theorem 5.13 we see that

$$Bx_k = P_m(T(b) + K)(x_j^{(k)})_{j=0}^{m+s-1} + Q_m T(b)(x_j^{(k)})_{j=m-s}^{\infty},$$

whence

$$M^p - \delta_k \le M^p \sum_{j=0}^{m+s-1} |x_j^{(k)}|^p + M_0^p \sum_{j=m-s}^{\infty} |x_j^{(k)}|^p. \tag{5.40}$$

On defining

$$O_m^{(k)} = \sum_{j=m}^{\infty} |x_j^{(k)}|^p, \quad \gamma_k = \delta_k / M^p, \quad q = (M_0/M)^p,$$

we can rewrite (5.40) in the form $1 - \gamma_k \le (1 - O_{m+s}^{(k)}) + q O_{m-s}^{(k)}$. Thus,

$$O_{m+s}^{(k)} \le q O_{m-s}^{(k)} + \gamma_k \text{ for all } m \ge m_0. \tag{5.41}$$

Since $O_m^{(k)} \le 1$, we obtain from (5.41) that $O_{m_0+s+v} \le q + \gamma_k$ for all v in the set $\{0, 1, \dots, 2s-1\}$. This and (5.41) give $O_{m_0+3s+v}^{(k)} \le q O_{m_0+s+v}^{(k)} + \gamma_k \le q^2 + (q+1)\gamma_k$ for $v \in \{0, 1, \dots, 2s-1\}$. Repeating this argument we arrive at the inequalities

$$O_{m_0+(2j+1)s+v} \le q^{j+1} + (q^j + \dots + 1)\gamma_k \le q^{j+1} + \frac{\gamma_k}{1-q} \tag{5.42}$$

for $v \in \{0, 1, \dots, 2s-1\}$. Let $r := q^{1/(2s)}$. Then (5.42) can be written as

$$O_{m_0+s+2js+v}^{(k)} \le q^{j+1} + \frac{\gamma_k}{1-q} = r^{2js+2s} + \frac{\gamma_k}{1-q},$$

and it results that

$$O_{m_0+s+\ell}^{(k)} \le r^\ell + \frac{\gamma_k}{1-q} \quad \text{for } \ell \ge 0. \tag{5.43}$$

We now show that $\{x_k\}_{k=1}^\infty$ can be taken to be a Cauchy sequence in ℓ^p. Given $\varepsilon > 0$, there is a natural number N such that $r^{(N-(m_0+s))/p} < \varepsilon/4$. On passing to a subsequence if necessary, we can assume that $\{P_N x_k\}_{k=1}^\infty$ is a Cauchy sequence in ℓ_N^p. Hence, there is a natural number R such that

$$\sum_{j=0}^{N-1} |x_j^{(k_1)} - x_j^{(k_2)}|^p < \left(\frac{\varepsilon}{4}\right)^p, \quad \frac{\gamma_{k_1} + \gamma_{k_2}}{1-q} < \frac{\varepsilon}{4}$$

whenever $k_1 \ge R$ and $k_2 \ge R$. From (5.43) we now obtain that $\|x_{k_1} - x_{k_2}\|_p$ does not exceed

$$\begin{aligned}
&\left(\sum_{j=0}^{N-1} |x_j^{(k_1)} - x_j^{(k_2)}|^p\right)^{1/p} + \left(\sum_{j=N}^{\infty} |x_j^{(k_1)}|^p\right)^{1/p} + \left(\sum_{j=N}^{\infty} |x_j^{(k_2)}|^p\right)^{1/p} \\
&\le \varepsilon/4 + \left(O_N^{(k_1)}\right)^{1/p} + \left(O_N^{(k_2)}\right)^{1/p} \\
&\le \varepsilon/4 + r^{(N-(m_0+s))/p} + r^{(N-(m_0+s))/p} + \frac{\gamma_{k_1} + \gamma_{k_2}}{1-q} \\
&< \varepsilon/4 + \varepsilon/4 + \varepsilon/4 + \varepsilon/4 = \varepsilon.
\end{aligned}$$

This shows that $\{x_k\}$ may be assumed to be a Cauchy sequence.

Let $x_0 \in \ell^p$ be the limit of x_k as $k \to \infty$. Passing to the limit $k \to \infty$ in (5.39) we arrive at (5.37). Repeating the above reasoning with x_k replaced by x_0 we get the analogue of (5.43) for x_0, that is,

$$\sum_{j=m_0+s+\ell}^{\infty} |x_j^{(0)}|^p \le r^\ell \quad \text{for } \ell \ge 0. \tag{5.44}$$

If $m \ge m_0 + s$, then $m = m_0 + s + \ell$ with $\ell \ge 0$, and (5.44) reads

$$\sum_{j=m}^{\infty} |x_j^{(0)}|^p \le r^{m-(m_0+s)} = \left(\frac{M_0}{M}\right)^{p(m-(m_0+s))/(2s)}. \qquad \square$$

The following theorem exhibits a case in which $\|B_n\|_p$ converges to M_p exponentially fast. Recall that $m_0 := \max(s, n_0)$.

Theorem 5.16. *Suppose $M_p^0 < M_p$. If $n \ge 4m_0 + 4s + 2$ then*

$$M_p(1 - c_p\varrho_p^n) \le \|B_n\|_p \le M_p(1 + c_p\varrho_p^n), \tag{5.45}$$

where

$$c_p = 2s\left(\frac{M_p}{M_p^0}\right)^{(2m_0+4s+1)/(4s)}, \quad \varrho_p = \left(\frac{M_p^0}{M_p}\right)^{1/(4s)}.$$

Proof. Put $M_0 = M_p^0$ and $M = M_p$. By Corollary 5.14, $\|B_n\|_p \to M$. Hence there are $x_n = (x_j^{(n)})_{j=0}^{n-1} \in \ell_n^p$ such that

$$\|x_n\|_p = 1, \quad \|B_n x_n\|_p = \|B_n\|_p, \quad \|B_n x_n\|_p^p = M_p^p + \varepsilon_n, \tag{5.46}$$

where $\varepsilon_n \in \mathbf{R}$ and $\varepsilon_n \to 0$ as $n \to \infty$. Again set $\ell = [n/4]$ and note that $\ell > m_0$ by assumption. For $\ell < m < 2\ell$, we have

$$\begin{aligned}
\|B_n\|_p^p &= \sum_{j=0}^{m-1}\left|\sum_{k=0}^{m+s-1} b_{j-k}x_k^{(n)} + \sum_{k=0}^{n_0-1} K_{jk}x_k^{(n)}\right|^p + \sum_{j=m}^{n-m-1}\left|\sum_{k=m-s}^{n-m+s-1} b_{j-k}x_k^{(n)}\right|^p \\
&\quad + \sum_{j=n-m}^{n-1}\left|\sum_{k=n-m+s}^{n-1} b_{j-k}x_k^{(n)} + \sum_{k=0}^{n_0-1}(W_nLW_n)_{jk}x_k^{(n)}\right|^p \\
&\le \|(T(b)+K)(x_k^{(n)})_{k=0}^{m+s-1}\|_p^p + \|T(b)(x_k^{(n)})_{k=m-s}^{n-m+s-1}\|_p^p \\
&\quad + \|(T_n(b) + W_nLW_n)(x_k^{(n)})_{k=n-m+s}^{n-1}\|_p^p.
\end{aligned}$$

Since $T_n(b) = W_nT_n(\widetilde{b})W_n$ and $\|W_n\|_p = 1$, it follows from (5.46) that

$$M^p + \varepsilon_n \le M^p \sum_{k=0}^{m+s-1} |x_k^{(n)}|^p + M_0^p \sum_{k=m-s}^{n-m+s-1} |x_k^{(n)}|^p + M^p \sum_{k=n-m+s}^{n-1} |x_k^{(n)}|^p.$$

With

$$O_m^{(n)} := \sum_{j=m}^{n-m-1} |x_j^{(n)}|^p,$$

we therefore get $M^p + \varepsilon_n \le M^p(1 - O_{m+s}^{(n)}) + M_0^p O_{m-s}^{(n)}$, whence $O_{m+s}^{(n)} \le q\, O_{m-s}^{(n)} - \gamma_n$ with $q := (M_0/M)^p$ and $\gamma_n := \varepsilon_n/M^p$. This inequality is of the form (5.41). Consequently, in analogy to (5.42) and (5.43) we have

$$O_{m_0+s+j}^{(n)} \le r^j - \frac{1-r^j}{1-q}\gamma_n, \quad r := q^{1/(2s)} \tag{5.47}$$

as long as $O_{m_0+s+j}^{(n)}$ is well defined, i.e., for all j satisfying $m_0+s+j \le n-(m_0+s+j)-1$ or, equivalently, for

$$j \le \frac{n-1}{2} - m_0 - s = \frac{n-2m_0-2s-1}{2}. \tag{5.48}$$

As the left-hand side of (5.47) is nonnegative provided (5.48) holds, we get

$$\begin{aligned}
\gamma_n &\le (1-q)\frac{r^{(n-2m_0-2s-1)/2}}{1-r^{(n-2m_0-2s-1)/2}} \le \frac{1-q}{1-r} r^{(n-2m_0-2s-1)/2} \\
&= \frac{1-\left((M_0/M)^{p/(2s)}\right)^{2s}}{1-(M_0/M)^{p/(2s)}} \left(\frac{M_0}{M}\right)^{\frac{p}{2s}\frac{n-2m_0-2s-1}{2}} \\
&\le 2s\left(\frac{M_0}{M}\right)^{\frac{p}{4s}(n-2m_0-2s-1)} \le 2s\left(\frac{M_0}{M}\right)^{\frac{1}{4s}(n-2m_0-2s-1)}.
\end{aligned} \tag{5.49}$$

Put $\eta_n := \|B_n\|_p - M$. By (5.46), $\eta_n = (M^p + \varepsilon_n)^{1/p} - M$. If $\varepsilon_n \ge 0$, then

$$\eta_n = M\left(1 + \frac{\varepsilon_n}{M^p}\right)^{1/p} - M \le M\left(1 + \frac{\varepsilon_n}{pM^p}\right) - M = \frac{\varepsilon_n}{pM^{p-1}},$$

and if $\varepsilon_n < 0$, then $\eta_n < 0$. Since $\gamma_n = \varepsilon_n/M^p$, we see from (5.49) that in either case

$$\eta_n \le \frac{M}{p} 2s \left(\frac{M}{M_0}\right)^{(2m_0+2s+1)/(4s)} \left(\frac{M_0}{M}\right)^{n/(4s)} \le M c_p \varrho_p^n.$$

This is the upper estimate in (5.45).

To prove the lower estimate in (5.45) we make use of Lemma 5.15. For the sake of definiteness, suppose $M = \|T(b) + K\|_p$ (since $\|B_n\|_p = \|W_n B_n W_n\|_p$, the case $M = \|T(\widetilde{b}) + L\|_p$ can be reduced to the case we consider). Let x_0 satisfy (5.37) and (5.38). Put $v = [n/2]$ ($> m_0$) and $y_n = P_v x_0$. Then $\|B_n\|_p$ equals

$$\begin{aligned}
&\|P_n(T(b)+K)y_n\|_p = \|(T(b)+K)y_n - Q_n(T(b)+K)y_n\|_p \\
&= \|(T(b)+K)(x_j^{(0)})_{j=0}^{\infty} - (T(b)+K)(x_j^{(0)})_{j=v}^{\infty} - Q_n(T(b)+K)(x_j^{(0)})_{j=v-s}^{v-1}\|_p
\end{aligned}$$

and this is clearly greater than or equal to

$$\|(T(b)+K)x_0\|_p - \|(T(b)+K)(x_j^{(0)})_{j=v}^{\infty}\|_p - \|Q_n(T(b)+K)(x_j^{(0)})_{j=v-s}^{v-1}\|_p$$

$$\geq M - \|(x_j^{(0)})_{j=v}^{\infty}\|_p - \|(x_j^{(0)})_{j=v-s}^{v-1}\|_p \geq M - 2M\left(\sum_{j=v-s}^{\infty}|x_j^{(0)}|^p\right)^{1/p}$$

$$\geq M - 2M\left(\frac{M_0}{M}\right)^{(v-s-(m_0+s))/(2s)} \geq M - 2M\left(\frac{M_0}{M}\right)^{((n-1)/2-s-(m_0+s))/(2s)},$$

which equals

$$M - 2M\left(\frac{M}{M_0}\right)^{(2m_0+4s+1)/(4s)}\left(\frac{M_0}{M}\right)^{(n/(4s))} \geq M - Mc_p\varrho_p^n.$$

Since $\|y_n\|_p \leq 1$, we finally obtain

$$\|B_n\|_p \geq \|B_n y_n\|_p/\|y_n\|_p \geq \|B_n y_n\|_p \geq M - Mc_p\varrho_p^n. \qquad \square$$

The crucial assumption of Theorem 5.16 is the strict inequality $M_p^0 < M_p$. In the case $p = 2$, the following argument can be employed to see that this inequality is generically true. Let $\mathcal{P}$ be the set of all Laurent polynomials and let $\mathcal{X}$ denote the set of all infinite matrices with only finitely many nonzero entries. We equip $\mathcal{P} \times \mathcal{X} \times \mathcal{X}$ with the norm $\|(b, K, L)\| := \max(\|T(b)\|_2, \|K\|_2, \|L\|_2)$. Recall that $\|T(b)\|_2 = \|b\|_\infty$ (Theorem 1.15).

Proposition 5.17. *The set*

$$\{(b, K, L) \in \mathcal{P} \times \mathcal{X} \times \mathcal{X} : \max(\|T(b)+K\|_2, \|T(\widetilde{b})+L\|_2) > \|T(b)\|_2\} \tag{5.50}$$

is an open and dense subset of $\mathcal{P} \times \mathcal{X} \times \mathcal{X}$.

Proof. It is clear that (5.50) is an open subset of $\mathcal{P} \times \mathcal{X} \times \mathcal{X}$. To show that it is dense in $\mathcal{P} \times \mathcal{X} \times \mathcal{X}$, it suffices to prove that the set of all $(b, K) \in \mathcal{P} \times \mathcal{X}$ satisfying $\|T(b)+K\|_2 > \|T(b)\|_2$ is dense in $\mathcal{P} \times \mathcal{X}$. Pick $(b, K) \in \mathcal{P} \times \mathcal{X}$ and suppose

$$\|T(b)+K\|_2 = \|T(b)\|_2 = \|b\|_\infty.$$

Given any $\varepsilon > 0$, we can find a $\lambda \in \mathbf{C}$ such that $|\lambda| < \varepsilon$ and $\|b+\lambda\|_\infty > \|b\|_\infty$. Then

$$\|T(b)+K+\lambda I\|_2 = \|T(b-\lambda)+K\|_2 \geq \|T(b+\lambda)\|_2 = \|b+\lambda\|_\infty > \|b\|_\infty,$$

and since $T(b)+K+\lambda P_n$ converges strongly to $T(b)+K+\lambda I$, it follows that

$$\liminf_{n\to\infty} \|T(b)+K+\lambda P_n\|_2 \geq \|T(b)+K+\lambda I\|_2 > \|T(b)\|_2.$$

Consequently, letting $K_0 := K + \lambda P_n$, we see that $\|T(b)+K_0\|_2 > \|T(b)\|_2$ for all sufficiently large n, while

$$\|(T(b)+K) - (T(b)+K_0)\|_2 \leq |\lambda|\,\|P_n\|_2 = |\lambda| < \varepsilon. \qquad \square$$

Now let B_n be given by (5.24) and let $p = 2$. Proposition 5.17 tells us that the strict inequality $M_2^0 < M_2$ represents the generic case, whereas the equality $M_2^0 = M_2$ is the exceptional case. Thus, Theorem 5.16 says that generically $\|B_n\|_2 - M_2$ decays to zero with exponential speed.

5.6 Slow Convergence

In concrete situations, we nevertheless often encounter the exceptional case where $M_p^0 = M_p$. For instance, we are definitely in this case whenever $B_n = T_n(b)$. Theorem 5.8 reveals that in the exceptional case it is possible that

$$M_2 - \|B_n\|_2 = M_2^0 - \|B_n\|_2 \geq d/n^2$$

with some $d > 0$ independent of n, which means that we do not have exponentially fast convergence.

The following theorem certainly does not give sharp estimates, but it provides us with a universal estimate with good constants.

Theorem 5.18. *If $s \geq 1$ and $n \geq \max(4n_0, 81\,s)$, then*

$$M_p^0\left(1 - \frac{81}{p}\frac{s}{n}\right) \leq \|B_n\|_p \leq M_p\left(1 + \frac{8}{p}\frac{s}{n}\right).$$

Proof. The upper estimate is immediate from Theorem 5.13. So we are left to prove the lower estimate.

By Theorem 5.5, there is an $x \in \ell^p_{[n/2]}$ such that $\|x\|_p = 1$ and

$$\|T_{[n/2]}(b)x\|_p \geq M_p^0\left(1 - \frac{40}{p}\frac{s}{[n/2]}\right).$$

Let $\ell = [n/4]$ and define the unit vector $y \in \ell^p_n$ by

$$y = (\underbrace{0, \ldots, 0}_{\ell}, x_0, x_1, \ldots, x_{[n/2]-1}, \underbrace{0, \ldots, 0}_{\ell'}).$$

By assumption, $\ell \geq n_0$ and $\ell' \geq n_0$. This implies that $B_n y = T_n(b)y$, whence

$$\|B_n y\|_p = \|T_n(b)y\|_p \geq \left(\sum_{j=\ell+1}^{\ell+[n/2]} |(T_n(b)y)_j|^p\right)^{1/p}$$

$$= \left(\sum_{j=\ell+1}^{\ell+[n/2]} \left|\sum_{k=\ell+1}^{\ell+[n/2]} b_{j-k}x_k\right|^p\right)^{1/p} = \|T_{[n/2]}(b)x\|_p$$

$$\geq M_p^0\left(1 - \frac{40}{p}\frac{s}{[n/2]}\right) \geq M_p^0\left(1 - \frac{40}{p}\frac{2s}{n-1}\right) \geq M_p^0\left(1 - \frac{81}{p}\frac{s}{n}\right),$$

the last estimate resulting from the assumption that $n \geq 81s \geq 81$. □

Corollary 5.19. *Suppose $M_p^0 = M_p$. If $s \geq 1$ and $n \geq \max(4n_0, 81\,s)$, then*

$$M_p\left(1 - \frac{81}{p}\frac{s}{n}\right) \leq \|B_n\|_p \leq M_p\left(1 + \frac{8}{p}\frac{s}{n}\right).$$

Proof. The proof is immediate from Theorem 5.18. □

5.7 Summary

Let us summarize the essence of this chapter. We are given matrices

$$B_n = T_n(b) + P_n K P_n + W_n L W_n,$$

where b is a Laurent polynomial and K and L have only finitely many nonzero entries. We put

$$M_p := \max\left(\|T(b) + K\|_p, \|T(\widetilde{b}) + L\|_p\right), \quad M_p^0 := \|T(b)\|_p.$$

The inequality $M_p \geq M_p^0$ is always true, and we know that at least for $p = 2$ the strict inequality $M_p > M_p^0$ represents the generic case.

If $1 < p < \infty$ and $M_p > M_p^0$, then there is a $\gamma > 0$ such that

$$\left| \|B_n\|_p - M_p \right| = O(e^{-\gamma n}), \tag{5.51}$$

while if $1 < p < \infty$ and $M_p = M_p^0$, then

$$\left| \|B_n\|_p - M_p \right| = O\left(\frac{1}{n}\right). \tag{5.52}$$

If $p = 1$ or $p = \infty$, then

$$\left| \|B_n\|_p - M_p \right| = 0 \tag{5.53}$$

for all sufficiently large n.

Now suppose that $B_n = T_n(b)$ is a pure Toeplitz band matrix. In this case, $M_p = M_p^0 = \|T(b)\|_p$. If $1 < p < \infty$, then

$$\left| \|T_n(b)\|_p - \|T(b)\|_p \right| = O\left(\frac{1}{n}\right). \tag{5.54}$$

In the case $p = 2$ this can be improved to

$$\left| \|T_n(b)\|_2 - \|T(b)\|_2 \right| = O\left(\frac{1}{n^2}\right). \tag{5.55}$$

The convergence in (5.55) may be faster, of the form $O(1/n^{2\gamma})$ with some natural number γ, but it is never exponentially fast unless $|b|$ is constant on $\mathbf{T}$ (which happens if and only if b has at most one nonzero coefficient). Finally, if $p = 1$ or $p = \infty$, then

$$\left| \|T_n(b)\|_p - \|T(b)\|_p \right| = 0 \tag{5.56}$$

for all n large enough.

Exercises

1. Let $A \in \mathcal{B}(\ell^2)$ be selfadjoint and positive definite. Put $A_n = P_n A P_n$. Prove that $\{A_n\}_{n=1}^{\infty}$ is a stable sequence and that

$$\|A_n\|_2 \le \|A\|_2 \text{ for all } n \ge 1, \quad \lim_{n\to\infty} \|A_n\|_2 = \|A\|_2,$$
$$\|A_n^{-1}\|_2 \le \|A^{-1}\|_2 \text{ for all } n \ge 1, \quad \lim_{n\to\infty} \|A_n^{-1}\|_2 = \|A^{-1}\|_2.$$

2. Let $b \in \mathcal{P}$ and let $P_{\text{circ}}(T_n(b))$ be the circulant matrix whose first column is

$$\operatorname{col}\left(\frac{(n-j)b_j + jb_{-(n-j)}}{n}\right)_{j=0}^{n-1} \quad (\text{with } b_{-n} := 0).$$

Prove that $P_{\text{circ}}(T_n(b))$ is the best approximation of $T_n(b)$ by a circulant matrix in the Frobenius norm.

3. Let $a, b \in \mathcal{P}$. Prove that

$$\|T_n(a)T_n(b) - T_n(b)T_n(a)\|_p = \|T(a)T(b) - T(b)T(a)\|_p$$

for all sufficiently large n.

4. Let α and β be positive integers and put

$$A_n = \begin{pmatrix} 0 & 1^\alpha & 2^\alpha & \dots & (n-1)^\alpha \\ 1^\beta & 0 & 1^\alpha & \dots & (n-2)^\alpha \\ 2^\beta & 1^\beta & 0 & \dots & (n-3)^\alpha \\ \vdots & \vdots & \vdots & \ddots & \vdots \\ (n-1)^\beta & (n-2)^\beta & (n-3)^\beta & \dots & 0 \end{pmatrix}.$$

Prove that $\|A_n\|_F^2/(n\|A_n\|_2^2) \to 0$ as $n \to \infty$, where $\|\cdot\|_F$ is the Frobenius norm.

5. Consider the sequence $\{2^{1^2}, 2^{2^2}, 2^{3^2}, 2^{4^2}, 2^{5^2}, \dots\}$. Define $a_k = k$ if k belongs to this sequence and $a_k = 0$ otherwise. Let A_n be the lower-triangular Toeplitz matrix

$$A_n = \begin{pmatrix} a_0 & & & \\ a_1 & a_0 & & \\ \vdots & \vdots & \ddots & \\ a_{n-1} & a_{n-2} & \dots & a_0 \end{pmatrix}.$$

Prove that for each number μ in the segment $[0, 1]$ there exists a sequence $\{n_j\}_{j=1}^{\infty}$ such that $\|A_{n_j}\|_F^2/(n_j\|A_{n_j}\|_2^2) \to \mu$ as $n_j \to \infty$.

6. Let a be a complex number with $|a| \ge 4$. Let $b(t) = 2 + t + t^{-1}$ and put

$$B_n = T_n(b) + \operatorname{diag}(a, 0, \dots, 0, a), \quad B = T(b) + \operatorname{diag}(a, 0, \dots).$$

Show that for $n \geq 10$,

$$\Big| \|B_n\|_2 - \|B\|_2 \Big| < \frac{1}{5}(|a|+2)^{\frac{11}{4}} \left(\frac{4}{\sqrt{|a|^2+1}} \right)^{\frac{n}{4}}.$$

7. Let

$$B = T(b) + K = \begin{pmatrix} 3 & 84 & 0 & \dots \\ -1 & 3 & -1 & \dots \\ 0 & -1 & 3 & \dots \\ \dots & \dots & \dots & \dots \end{pmatrix}$$

and $B_n = T_n(b) + P_n K P_n$.

(a) Show that $M_2^0 = 5$ and $80 \leq M_2 \leq 90$.

(b) Use Theorem 5.16 to show that for $n \geq 14$,

$$\Big| \|B_n\|_2 - \|B\|_2 \Big| < \frac{120150}{2^n}.$$

(c) MATLAB gives $\|B_{32}\|_2 = 84.1117$. Deduce that

$$\|B\|_2 = 84.1117 \pm 0.0001.$$

8. Show that if $\varphi \in L^\infty$ and

$$T_2(\varphi) = \begin{pmatrix} 0 & 1 \\ 1 & 0 \end{pmatrix},$$

then $\|\varphi\|_\infty \geq \sqrt{2}$.

Notes

Section 5.1 is from [58]. The results of Section 5.2 are well known in the case where b is real valued and, hence, $T_n(b)$ is Hermitian: See Kac, Murdock, and Szegö [176], Grenander and Szegö [145], Widom [290], Parter [198], and Serra Capizzano [247], [248]. In the form stated here, Theorem 5.8 was established in [58].

Theorem 5.10 probably first appears in our paper [53]. Our original proof of Theorem 5.10 used the Fejér kernel and gave only the inequality

$$\|T_n(b)\|_2 \geq (1/\sqrt{3\pi})\, \|\sigma_n b\|_\infty,$$

which, however, was sufficient for the purpose of our paper [53]. Then Stefano Serra Capizzano communicated an alternative proof to us which, moreover, showed that the constant $1/\sqrt{3\pi}$ can be improved to the (optimal) value 1. Subsequently we were able to modify (and even to simplify) our original proof so that it yielded the constant 1 as well, but this was no longer the point. Serra Capizzano's proof was based on arguments that are standard in the preconditioning literature (see, e.g., [84], [88], [249]). He first showed that

$\|T_n(b)\|_2 \ge \|P_{\rm circ}(T_n(b))\|_2$, where $P_{\rm circ}(T_n(b))$ is the circulant matrix of Exercise 2. One can straightforwardly verify that

$$\|P_{\rm circ}(T_n(b))\|_2 = \max_{t\in\mathbf{T}_n} |(\sigma_n b)(t)|, \tag{5.57}$$

where $\mathbf{T}_n$ is the group of the nth unit roots. A scaling trick, viz., the equality $\|T_n(b)\|_2 = \|T_n(b_\zeta)\|_2$ with $(b_\zeta)_j := b_j\zeta^j$ ($\zeta \in \mathbf{T}$), eventually gives the desired inequality

$$\begin{aligned}\|T_n(b)\|_2 &= \max_{\zeta\in\mathbf{T}} \|T_n(b_\zeta)\|_2 \ge \max_{\zeta\in\mathbf{T}}\max_{t\in\mathbf{T}_n} |(\sigma_n b_\zeta)(t)|\\ &= \max_{\zeta\in\mathbf{T}}\max_{t\in\mathbf{T}_n} |(\sigma_n b)(\zeta t)| = \|\sigma_n b\|_\infty.\end{aligned}$$

In an attempt to find out who was the first to write down (5.57) explicitly, we looked into the paper [249], and once we saw the first lines of the proof of Lemma 2.1 of [249], we came to understand that our Theorem 5.10 can actually be proved as in Section 5.3.

The results of Sections 5.4 to 5.7 are also taken from paper [58].

Figure 5.1 was done by André Eppler. For Exercises 4 and 5 see [53]. Exercise 8 is from [194], where it is attributed to A. Volberg.

Further results: symbols of minimal norm. Suppose we are given a Toeplitz matrix $T_n(a)$. Since only the Fourier coefficients a_k with $|k| \le n-1$ enter the matrix, we have $T_n(a) = T_n(\varphi)$ for every $\varphi \in L^\infty$ satisfying $\varphi_k = a_k$ for $|k| \le n-1$. Put

$$\nu_n(a) = \inf\{\|\varphi\|_\infty : \varphi_k = a_k \ \text{ for } \ |k| \le n-1\}.$$

It is clear that $\|T_n(a)\|_2 \le \nu_n(a) \le \|a\|_\infty$. Nikolskaya and Farforovskaya [194] showed that actually

$$\|T_n(a)\|_2 \ge \frac{1}{3}\, \nu_n(a).$$

Thus, we can find a symbol $\varphi \in L^\infty$ such that

$$T_n(a) = T_n(\varphi) \quad \text{and} \quad \|T_n(a)\|_2 \le \|\varphi\|_\infty \le 3\|T_n(a)\|_2.$$

From Exercise 8 we learn that if $a(t) = t + t^{-1}$, then $\nu_2(a) \ge \sqrt{2}$, which implies that

$$1 = \|T_2(a)\|_2 \ge c\,\nu_2(a) \ge c\sqrt{2}$$

cannot be true if $c > 1/\sqrt{2}$. We so arrive at the conclusion that the optimal constant in the inequality $\|T_n(a)\|_2 \ge c\,\nu_n(a)$ is in $[1/3,\, 1/\sqrt{2}]$.

Together with $T_n(a) = P_n T(a) P_n$ we may also consider the rectangular Toeplitz matrices

$$T_{n-r,n+r}(a) = P_{n-r} T(a) P_{n+r} \quad (|r| \le n-1).$$

For example, in the case $n = 3$ we have the five matrices

$$T_{5,1}(a) = \begin{pmatrix} a_{-2} \\ a_{-1} \\ a_0 \\ a_1 \\ a_2 \end{pmatrix}, \quad T_{4,2}(a) = \begin{pmatrix} a_{-1} & a_{-2} \\ a_0 & a_{-1} \\ a_1 & a_0 \\ a_2 & a_1 \end{pmatrix},$$

$$T_{3,3}(a) = \begin{pmatrix} a_0 & a_{-1} & a_{-2} \\ a_1 & a_0 & a_{-1} \\ a_2 & a_1 & a_0 \end{pmatrix}, \quad T_{2,4}(a) = \begin{pmatrix} a_1 & a_0 & a_{-1} & a_{-2} \\ a_2 & a_1 & a_0 & a_{-1} \end{pmatrix},$$

$$T_{1,5}(a) = \begin{pmatrix} a_2 & a_1 & a_0 & a_{-1} & a_{-2} \end{pmatrix}.$$

Put

$$\delta_n(a) = \max_{|r| \le n-1} \|T_{n-r,n+r}(a)\|_2.$$

Again it is obvious that $\delta_n(a) \le \nu_n(a) \le \|a\|_\infty$. Bakonyi and Timotin [12] proved the estimate

$$\delta_n(a) \ge \frac{1}{2} \frac{n+2}{n+1} \nu_n(a).$$

Consequently, there is a symbol $\varphi \in L^\infty$ such that

$$T_n(a) = T_n(\varphi) \quad \text{and} \quad \delta_n(a) \le \|\varphi\|_\infty \le 2 \frac{n+1}{n+2} \delta_n(a).$$

Chapter 6

Condition Numbers

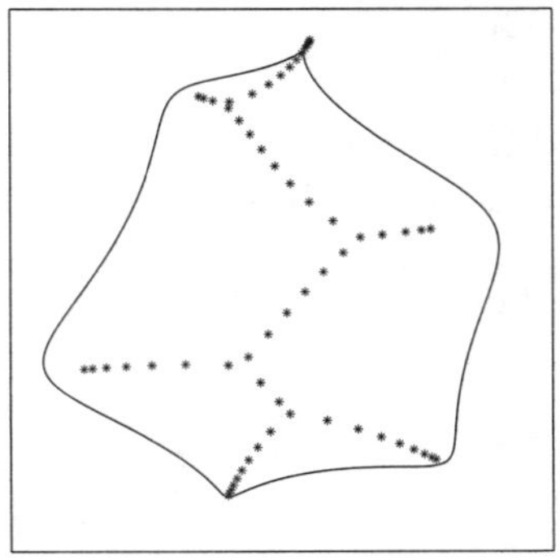

Let B_n be an $n \times n$ matrix. For $1 \le p \le \infty$, we denote by $\kappa_p(B_n)$ the *condition number* of B_n as an operator on ℓ_n^p:

$$\kappa_p(B_n) := \|B_n\|_p \, \|B_n^{-1}\|_p.$$

Throughout what follows we put $\|B_n^{-1}\|_p = \infty$ in case B_n is not invertible.

In this chapter, we study the behavior of the condition numbers of Toeplitz band matrices $T_n(b)$ and of Toeplitz-like matrices $B_n = T_n(b) + P_n K P_n + W_n L W_n$ for large n.

6.1 Asymptotic Inverses of Toeplitz-Like Matrices

As in Section 5.4, we assume that B_n is given by

$$B_n = T_n(b) + P_n K P_n + W_n L W_n, \tag{6.1}$$

where $b(t) = \sum_{j=-s}^{s} b_j t^j$ $(t \in \mathbf{T})$ and where K and L have only a finite number of nonzero entries, that is,

$$P_{n_0} K P_{n_0} = K, \quad P_{n_0} L P_{n_0} = L \tag{6.2}$$

for some $n_0 \in \mathbf{N}$. Theorem 3.15 gives an asymptotic inverse for the pure Toeplitz matrices $T_n(b)$. The purpose of this section is to extend this result to the Toeplitz-like matrices B_n.

We put

$$B = T(b) + K, \quad \widetilde{B} = T(\widetilde{b}) + L, \tag{6.3}$$

and in case B and $\widetilde{B}$ are invertible (which, by Theorem 1.9, implies that b is invertible and that $\operatorname{wind} b = 0$), we set

$$X = B^{-1} - T(b^{-1}), \quad Y = \widetilde{B}^{-1} - T(\widetilde{b}^{-1}). \tag{6.4}$$

Lemma 6.1. *Let $1 \le p \le \infty$ and let α be any number satisfying* (1.23). *If B and $\widetilde{B}$ are invertible on ℓ^p, then $\|Q_n X\|_p = O(e^{-\alpha n})$ and $\|Q_n Y\|_p = O(e^{-\alpha n})$ as $n \to \infty$.*

Proof. From (6.4) we infer that $I = (T(b^{-1}) + X)(T(b) + K)$. This and Proposition 1.3 yield

$$I = X(T(b) + K) + T(b^{-1})T(b) + T(b^{-1})K$$
$$= X(T(b) + K) + I - H(b^{-1})H(\widetilde{b}) + T(b^{-1})K,$$

whence $X = (H(b^{-1})H(\widetilde{b}) - T(b^{-1})K)(T(b) + K)^{-1}$ and thus

$$\|Q_n X\|_p \le \Big(\|Q_n H(b^{-1})\|_p \, \|H(\widetilde{b})\|_p + \|Q_n T(b^{-1})K\|_p\Big) \, \|(T(b) + K)^{-1}\|_p.$$

Taking into account that K has only a finite number of nonzero entries and using Lemma 1.17 it is easy to show that $\|Q_n H(b^{-1})\|_p$ and $\|Q_n T(b^{-1})K\|_p$ are $O(e^{-\alpha n})$, which implies that $\|Q_n X\|_p = O(e^{-\alpha n})$. The proof is analogous for $\|Q_n Y\|_p$. □

Theorem 6.2. *Let $1 \le p \le \infty$, let α satisfy* (1.23), *and suppose B and $\widetilde{B}$ are invertible on ℓ^p. Then the matrices B_n are invertible for all sufficiently large n and*

$$B_n^{-1} = T_n(b^{-1}) + P_n X P_n + W_n Y W_n + E_n,$$

where X and Y are given by (6.4) *and $\|E_n\|_p = O(e^{-\alpha n})$.*

Proof. Put $A_n = T_n(b^{-1}) + P_n X P_n + W_n Y W_n$. Theorem 3.13 tells us that the matrices B_n are invertible for all sufficiently large n and that $\|B_n^{-1}\|_p$ remains bounded as $n \to \infty$. We have $E_n = B_n^{-1} - A_n = B_n^{-1}(P_n - B_n A_n)$, and the theorem will follow as soon as we have shown that $\|P_n - B_n A_n\|_p = O(e^{-\alpha n})$. The matrix $B_n A_n - P_n$ equals

$$P_n(T(b) + K)P_n(T(b^{-1}) + X)P_n - P_n$$
$$+ P_n(T(b) + K)P_n W_n Y W_n + W_n L W_n P_n(T(b^{-1}) + X)P_n + W_n L W_n W_n Y W_n.$$

Let n_0 be the number in (6.2). Since $P_{n_0} W_n = W_{n_0} T(\chi_{-(n-n_0)}) P_n Q_{n-n_0}$ for all $n \ge n_0$, we obtain from (6.2) that

$$P_n K P_n W_n Y W_n = P_{n_0} K P_{n_0} W_n Y W_n = P_{n_0} K W_{n_0} T(\chi_{-(n-n_0)}) P_n Q_{n-n_0} Y W_n,$$

and Lemma 6.1 therefore gives $\|P_n K P_n W_n Y W_n\|_p = O(e^{-\alpha n})$. Analogously one can show that $\|W_n L W_n P_n X P_n\| = O(e^{-\alpha n})$. Thus $B_n A_n - P_n$ is

$$P_n(T(b) + K)P_n(T(b^{-1}) + X)P_n - P_n$$
$$+ P_n T(b) P_n W_n Y W_n + W_n L W_n P_n T(b^{-1}) P_n + W_n L W_n W_n Y W_n + E_n' \tag{6.5}$$

with $\|E_n'\|_p = O(e^{-\alpha n})$. The identities

$$P_n T(b) W_n Y W_n = W_n T(\widetilde{b}) P_n Y W_n,$$
$$W_n L W_n T(b^{-1}) P_n = W_n L P_n T(\widetilde{b}^{-1}) W_n,$$

imply that (6.5) equals

$$
\begin{aligned}
&P_n(T(b)+K)P_n(T(b^{-1})+X)P_n - P_n \\
&+ W_n(T(\widetilde{b})+L)P_n(T(\widetilde{b^{-1}})+Y)W_n - W_nT(\widetilde{b})P_nT(\widetilde{b^{-1}})W_n + E_n',
\end{aligned} \tag{6.6}
$$

and since $(T(b)+K)(T(b^{-1})+X) = (T(\widetilde{b})+L)(T(\widetilde{b^{-1}})+Y) = I$, it results that (6.6) is equal to

$$
\begin{aligned}
&P_n - P_n(T(b)+K)Q_n(T(b^{-1})+X)P_n - P_n \\
&+ P_n - W_n(T(\widetilde{b})+L)Q_n(T(\widetilde{b^{-1}})+Y)W_n - W_nT(\widetilde{b})P_nT(\widetilde{b^{-1}})W_n + E_n'.
\end{aligned} \tag{6.7}
$$

Using (6.2) and Lemma 6.1 we see that (6.7) is

$$
\begin{aligned}
&P_n - P_nT(b)Q_nT(b^{-1})P_n - P_n \\
&+ P_n - W_nT(\widetilde{b})Q_nT(\widetilde{b^{-1}})W_n - W_nT(\widetilde{b})P_nT(\widetilde{b^{-1}})W_n + E_n''
\end{aligned} \tag{6.8}
$$

with $\|E_n''\|_p = O(e^{-\alpha n})$. Finally, taking into account Propositions 1.3 and 3.10 we obtain that (6.8) minus E_n'' is

$$
\begin{aligned}
&P_n - P_nT(b)Q_nT(b^{-1})P_n - W_nT(\widetilde{b})T(\widetilde{b^{-1}})W_n \\
&= P_n - P_nT(b)T(b^{-1})P_n + P_nT(b)P_nT(b^{-1})P_n - W_nT(\widetilde{b})T(\widetilde{b^{-1}})W_n \\
&= P_n - P_n(I - H(b)H(\widetilde{b^{-1}}))P_n + T_n(b)T_n(b^{-1}) - W_n(I - H(\widetilde{b})H(b^{-1}))W_n \\
&= P_n - P_n + P_nH(b)H(\widetilde{b^{-1}})P_n + T_n(b)T_n(b^{-1}) - P_n + W_nH(\widetilde{b})H(b^{-1})W_n \\
&= P_n - P_n + P_n - P_n = 0.
\end{aligned}
$$

Thus, $B_nA_n - P_n = E_n''$. □

6.2 The Limit of the Condition Numbers

Let B_n be of the form (6.1) and define B and $\widetilde{B}$ by (6.3). We know from Theorem 3.13 that $\|B_n^{-1}\|_p \to \infty$ if B or $\widetilde{B}$ is not invertible. So suppose B and $\widetilde{B}$ are invertible and put

$$N_p := \max(\|B^{-1}\|_p, \|\widetilde{B}^{-1}\|_p).$$

Again by Theorem 3.13, the matrices B_n are invertible for all sufficiently large n and $\limsup \|B_n^{-1}\|_p < \infty$. The following theorem shows that in fact the limit $\lim \|B_n^{-1}\|_p$ exists and it identifies this limit.

Theorem 6.3. *If* $1 \le p \le \infty$, *then*

$$\lim_{n\to\infty} \|B_n^{-1}\|_p = N_p. \tag{6.9}$$

Proof. Fix $\varepsilon > 0$. By virtue of Theorem 6.2,

$$B_n^{-1} = T_n(b^{-1}) + P_nXP_n + W_nYW_n + E_n,$$

where X and Y are given by (6.4) and $\|E_n\|_p = O(e^{-\alpha n})$. We can choose a Laurent polynomial a such that $\|b - a^{-1}\|_W < \varepsilon$ and $\|b^{-1} - a\|_W < \varepsilon$. Due to Lemma 6.1, there exist infinite matrices U and V with only finitely many entries such that $\|X - U\|_p < \varepsilon$ and $\|Y - V\|_p < \varepsilon$. Put

$$A_n = T_n(a) + P_n U P_n + W_n V W_n, \quad M_p = \max(\|T(a) + U\|_p, \|T(\widetilde{a}) + V\|_p).$$

Corollary 5.14 implies that

$$\big| \|A_n\|_p - M_p \big| < \varepsilon \tag{6.10}$$

for all sufficiently large n. We have

$$\begin{aligned} &\big| \|B_n^{-1}\|_p - \|A_n\|_p \big| \\ &= \big| \|T_n(b^{-1}) + P_n X P_n + W_n Y W_n + E_n\|_p - \|T_n(a) + P_n U P_n + W_n V W_n\|_p \big| \\ &\le \|T_n(b^{-1} - a)\|_p + \|P_n(X - U)P_n\|_p + \|W_n(Y - V)W_n\|_p + \|E_n\|_p \\ &\le \|b^{-1} - a\|_W + \|X - U\|_p + \|Y - V\|_p + \|E_n\|_p < 4\varepsilon \end{aligned} \tag{6.11}$$

provided n is large enough. Furthermore,

$$\begin{aligned} &| M_p - N_p | \\ &= \big| \max(\|T(a) + U\|_p, \|T(\widetilde{a}) + V\|_p) - \max(\|B^{-1}\|_p, \|\widetilde{B}^{-1}\|_p) \big| \\ &\le \max\big(| \|T(a) + U\|_p - \|B^{-1}\|_p |, | \|T(\widetilde{a}) + V\|_p - \|\widetilde{B}^{-1}\|_p |\big) \\ &\le \max\big(\|T(a) + U - B^{-1}\|_p, \|T(\widetilde{a}) + V - \widetilde{B}^{-1}\|_p\big) \\ &= \max\big(\|T(a - b^{-1}) + U - X\|_p, \|T(\widetilde{a} - \widetilde{b}^{-1}) + V - Y\|_p\big) \\ &\le \max\big(\|a - b^{-1}\|_W + \|U - X\|_p, \|\widetilde{a} - \widetilde{b}^{-1}\|_W + \|V - Y\|_p\big) \\ &< \max(\varepsilon + \varepsilon, \varepsilon + \varepsilon) = 2\varepsilon. \end{aligned} \tag{6.12}$$

Putting (6.10), (6.11), (6.12) together, we arrive at the conclusion that

$$| \|B_n^{-1}\|_p - N_p | < \varepsilon + 4\varepsilon + 2\varepsilon = 7\varepsilon$$

for all sufficiently large n. □

Corollary 6.4. *Let* $1 \le p \le \infty$. *If* $B \ne 0$ *or* $\widetilde{B} \ne 0$, *then*

$$\lim_{n \to \infty} \kappa_p(B_n) = M_p N_p, \tag{6.13}$$

where $M_p = \max(\|B\|_p, \|\widetilde{B}\|_p)$ *and* $N_p = \max(\|B^{-1}\|_p, \|\widetilde{B}^{-1}\|_p)$.

Proof. This is immediate from Corollary 5.14 and Theorem 6.3. (Notice that if $B = \widetilde{B} = 0$, then the assertion reads $\lim 0 = 0 \cdot \infty$.) □

Corollary 6.5. *If* $1 \le p \le \infty$ *and* b *is a Laurent polynomial that does not vanish identically, then*

$$\lim_{n \to \infty} \|T_n^{-1}(b)\|_p = \max(\|T^{-1}(b)\|_p, \|T^{-1}(\widetilde{b})\|_p), \tag{6.14}$$

$$\lim_{n \to \infty} \kappa_p(T_n(b)) = \max(\kappa_p(T(b)), \kappa_p(T(\widetilde{b}))). \tag{6.15}$$

Proof. In the case where $T(b)$ is invertible, formula (6.14) is a special case of Theorem 6.3 and formula (6.15) follows from Corollary 6.4:

$$\begin{aligned}\lim_{n\to\infty} \|T_n(b)\|_p \|T_n^{-1}(b)\|_p &= \|T(b)\|_p \lim_{n\to\infty} \|T_n^{-1}(b)\|_p \\ &= \|T(b)\|_p \max(\|T^{-1}(b)\|_p, \|T^{-1}(\widetilde{b})\|_p) \\ &= \max(\|T(b)\|_p \|T^{-1}(b)\|_p, \|T(b)\|_p \|T^{-1}(\widetilde{b})\|_p) \\ &= \max(\|T(b)\|_p \|T^{-1}(b)\|_p, \|T(\widetilde{b})\|_p \|T^{-1}(\widetilde{b})\|_p),\end{aligned}$$

the equality $\|T(b)\|_p = \|T(\widetilde{b})\|_p$ resulting from Theorem 1.14 for $p = 1$ and $p = \infty$ and from Lemma 5.3 for $1 < p < \infty$. If $T(b)$ is not invertible, then Theorem 3.7 tells us that both sides of (6.14) and (6.15) are infinite. □

If $p = 2$, then $\|T(b)\|_2 = \|T(\widetilde{b})\|_2 = \|b\|_\infty$ and $\|T^{-1}(b)\|_2 = \|T^{-1}(\widetilde{b})\|_2$ (note that $T(\widetilde{b})$ is just the transpose of $T(b)$), whence

$$\lim_{n\to\infty} \|T_n^{-1}(b)\|_2 = \|T^{-1}(b)\|_2, \quad \lim_{n\to\infty} \kappa_2(T_n(b)) = \kappa_2(T(b)). \tag{6.16}$$

As the following example shows, the situation is different for $p \neq 2$.

Example 6.6. Let

$$a(t) = \frac{1}{a_-(t)a_+(t)}, \quad a_-(t) = 4 + 4t^{-1} + 3t^{-2}, \quad a_+(t) = 2 - t.$$

It is easily seen that $a_\pm \in GW_\pm$ (Theorem 1.7). Thus, $T(a)$ is invertible and $T^{-1}(a) = T(a_+)T(a_-)$ due to Proposition 1.4. We have

$$\begin{aligned}T(a_+)T(a_-) &= \begin{pmatrix} 2 & 0 & 0 & \dots \\ -1 & 2 & 0 & \dots \\ 0 & -1 & 2 & \dots \\ \dots & \dots & \dots & \dots \end{pmatrix} \begin{pmatrix} 4 & 4 & 3 & 0 & \dots \\ 0 & 4 & 4 & 3 & \dots \\ 0 & 0 & 4 & 4 & \dots \\ \dots & \dots & \dots & \dots & \dots \end{pmatrix} \\ &= \begin{pmatrix} 8 & 8 & 6 & 0 & 0 & \dots \\ -4 & 4 & 5 & 6 & 0 & \dots \\ 0 & -4 & 4 & 5 & 6 & \dots \\ \dots & \dots & \dots & \dots & \dots & \dots \end{pmatrix}.\end{aligned}$$

This in conjunction with (1.16) shows that

$$\|T^{-1}(a)\|_1 = \max(12, 16, 19) = 19, \quad \|T^{-1}(\widetilde{a})\|_1 = \max(22, 19) = 22.$$

Since $\|T(a)\|_1 = \|T(\widetilde{a})\|_1 = \|a\|_W$, it follows that

$$\|T^{-1}(a)\|_1 < \|T^{-1}(\widetilde{a})\|_1, \quad \kappa_1(T(a)) < \kappa_1(T(\widetilde{a})).$$

If $\|b_n - a\|_W \to 0$, then

$$\|T^{-1}(b_n) - T^{-1}(a)\|_1 \le \|T^{-1}(a)\|_1 \, \|a - b_n\|_W \, \|T^{-1}(b_n)\|_1 \to 0,$$

and hence we can find a Laurent polynomial b such that

$$\|T^{-1}(b)\|_1 < \|T^{-1}(\widetilde{b})\|_1, \quad \kappa_1(T(b)) < \kappa_1(T(\widetilde{b})).$$

By the Riesz-Thorin interpolation theorem, the numbers

$$\log \|T^{-1}(b)\|_p, \quad \log \| T^{-1}(\widetilde{b})\|_p, \quad \log \|T(b)\|_p, \quad \log \|T(\widetilde{b})\|_p$$

are convex functions of $1/p \in [0, 1]$. These functions are therefore continuous, which implies that if $p > 1$ is sufficiently close to 1, then

$$\|T^{-1}(b)\|_p < \|T^{-1}(\widetilde{b})\|_p, \quad \kappa_p(T(b)) < \kappa_p(T(\widetilde{b})).$$

The moral is that, in general, we cannot remove the maxima in formulas (6.14) and (6.15) for $p \neq 2$. □

6.3 Convergence Speed Estimates

Again let $B_n = T_n(b) + P_n K P_n + W_n L W_n$ be as in Section 6.1, set $B = T(b) + K$, $\widetilde{B} = T(\widetilde{b}) + L$, and put

$$N_p = \max(\|B^{-1}\|_p, \|\widetilde{B}^{-1}\|_p), \quad N_p^0 = \|T(b^{-1})\|_p.$$

Theorem 6.7. *Let $1 < p < \infty$ and suppose B and $\widetilde{B}$ are invertible.*

If $N_p > N_p^0$, then there is a $\gamma > 0$ depending only on b, K, L such that

$$\big| \|B_n\|_p - N_p \big| = O(e^{-\gamma\sqrt{n}}). \tag{6.17}$$

If $N_p = N_p^0$, then

$$\big| \|B_n\|_p - N_p \big| = O\left(\frac{\log n}{n}\right). \tag{6.18}$$

Proof. We know from Theorem 6.2 that $B_n^{-1} = T_n(a) + P_n X P_n + W_n Y W_n + E_n$, where $a = b^{-1}$, X and Y are given by (6.4), and $\|E_n\|_p = O(e^{-\alpha n})$. Let $s(n)$ be a sequence of natural numbers such that

$$s(n) \to \infty, \quad \frac{n}{s(n)} \to \infty. \tag{6.19}$$

Put $X_{s(n)} = P_{s(n)} X P_{s(n)}$, $Y_{s(n)} = P_{s(n)} Y P_{s(n)}$, and let c_n denote the $s(n)$th partial sum of the Fourier series of a. Set

$$A_n = T_n(c_n) + P_n X_{s(n)} P_n + W_n Y_{s(n)} W_n. \tag{6.20}$$

Lemma 1.17 implies that

$$\|a - c_n\|_W = O(\varrho^{s(n)}) \text{ with some } \varrho < 1. \tag{6.21}$$

We have

$$
\begin{aligned}
&\big|\, \|B_n^{-1}\|_p - \|A_n\|_p \,\big| \le \|B_n^{-1} - A_n\|_p \\
&\le \|T_n(a - c_n)\|_p + \|P_n(X - X_{s(n)})P_n\|_p + \|W_n(Y - Y_{s(n)})W_n\|_p + \|E_n\|_p \\
&\le \|a - c_{s(n)}\|_W + \|X - X_{s(n)}\|_p + \|Y - Y_{s(n)}\|_p + \|E_n\|_p. \qquad (6.22)
\end{aligned}
$$

By Lemma 6.1, $\|X - P_s X P_s\|_p = O(\sigma^s)$ and $\|Y - P_s Y P_s\|_p = O(\sigma^s)$ with $\sigma = e^{-\alpha} < 1$ as $s \to \infty$. This and (6.17) show that (6.22) is

$$O(\varrho^{s(n)}) + O(\sigma^{s(n)}) + O(\sigma^{s(n)}) + O(\sigma^{s(n)}) = O(\tau^{s(n)}),$$

where $\max(\varrho, \sigma) < \tau < 1$. Let

$$M_p(s(n)) = \max(\|T(c_n) + X_{s(n)}\|_p, \|T(\widetilde{c}_n) + Y_{s(n)}\|_p), \quad M_p^0(s(n)) = \|T(c_n)\|_p.$$

Then $|\, M_p(s(n)) - N_p \,|$ does not exceed

$$
\begin{aligned}
&\max\Big(\big|\, \|T(c_n) + X_{s(n)}\|_p - \|T(a) + X\|_p \,\big|, \big|\, \|T(\widetilde{c}_n) + Y_{s(n)}\|_p - \|T(\widetilde{a}) + Y\|_p \,\big|\Big) \\
&\le \max\Big(\|T(c_n - a) + X_{s(n)} - X\|_p, \|T(\widetilde{c}_n - \widetilde{a}) + Y_{s(n)} - Y\|_p\Big),
\end{aligned}
$$

and again taking into account (6.21) and Lemma 6.1 we see that this is

$$O(\varrho^{s(n)}) + O(\sigma^{s(n)}) = O(\tau^{s(n)}).$$

Furthermore, also by (6.21),

$$
\begin{aligned}
&\big|\, M_p^0(s(n)) - N_p^0 \,\big| \\
&= \big|\, \|T(c_n)\|_p - \|T(a)\|_p \,\big| \le \|T(c_n - a)\|_p = O(\varrho^{s(n)}) = O(\tau^{s(n)}).
\end{aligned}
$$

In summary, at the present moment we have shown that

$$\big|\, \|B_n^{-1}\|_p - N_p \,\big| \le \big|\, \|A_n\|_p - M_p(s(n)) \,\big| + O(\tau^{s(n)}) \qquad (6.23)$$

and that

$$M_p(s(n)) = N_p + O(\tau^{s(n)}), \quad M_p^0(s(n)) = N_p^0 + O(\tau^{s(n)}). \qquad (6.24)$$

Now suppose that $N_p > N_p^0$. Then, by (6.24), $M_p(s(n)) > M_p^0(s(n))$ for all sufficiently large n, and Theorem 5.16 therefore gives

$$
\begin{aligned}
&\big|\, \|A_n\|_p - M_p(s(n)) \,\big| \\
&\le 2s(n) M_p(s(n)) \left(\frac{M_p(s(n))}{M_p^0(s(n))}\right)^{(6s(n)+1)/(4s(n))} \left(\frac{M_p^0(s(n))}{M_p(s(n))}\right)^{n/(4s(n))} \\
&\le 2s(n) M_p(s(n)) \left(\frac{M_p(s(n))}{M_p^0(s(n))}\right)^{2} \left(\frac{M_p^0(s(n))}{M_p(s(n))}\right)^{n/(4s(n))} . \qquad (6.25)
\end{aligned}
$$

Choose any $\varepsilon > 0$ so that $(N_p^0 + \varepsilon)/(N_p - \varepsilon) \in (0, 1)$ and assume that $\tau \in (0, 1)$ is larger than $(N_p^0 + \varepsilon)/(N_p - \varepsilon)$. If n is sufficiently large, then (6.25) is at most

$$2s(n)(N_p + \varepsilon)\left(\frac{N_p + \varepsilon}{N_p^0 - \varepsilon}\right)^2 \left(\frac{N_p^0 + \varepsilon}{N_p - \varepsilon}\right)^{n/(4s(n))} = O\left(s(n)\tau^{n/(4s(n))}\right).$$

This in conjunction with (6.23) yields the estimate

$$\left|\, \|B_n^{-1}\|_p - N_p \,\right| = O\left(s(n)\tau^{n/(4s(n))}\right).$$

Letting $s(n) = [\sqrt{n}\,]$, we obtain (6.17) with any $\gamma > 0$ such that $\tau^{1/4} < e^{-\gamma}$.

Now assume $N_p = N_p^0$. Then $M_p(s(n)) \ge M_p^0(s(n))$ for all n by virtue of Lemma 6.10. From Theorem 5.13 and (6.24) we obtain

$$\|A_n\|_p - M_p(s(n)) \le \frac{8}{p} M_p(s(n)) \frac{s(n)}{n} = O\left(\frac{s(n)}{n}\right),$$

while the lower estimate of Theorem 5.18 and (6.24) give

$$\begin{aligned} &-\|A_n\|_p + M_p(s(n)) \\ &\le \frac{81}{p} M_p^0(s(n)) \frac{s(n)}{n} + \left| M_p(s(n)) - M_p^0(s(n)) \right| \\ &= O\left(\frac{s(n)}{n}\right) + O(\tau^{s(n)}) \end{aligned}$$

(to get the $O(\tau^{s(n)})$ we used (6.24) and the equality $N_p = N_p^0$). Choose $\gamma > 0$ and $\alpha > 0$ so that $\tau < e^{-\gamma}$ and $\gamma\alpha > 1$. Then put $s(n) = [\alpha \log n]$. Clearly, (6.19) is satisfied. As

$$O\left(\frac{s(n)}{n}\right) = O\left(\frac{\log n}{n}\right),$$

$$O(\tau^{s(n)}) = O(e^{-\gamma\alpha \log n}) = O(n^{-\gamma\alpha}) = O\left(\frac{\log n}{n}\right),$$

we get (6.18) from (6.23). □

Theorem 6.8. *Let $p = 1$ or $p = \infty$ and suppose B and $\widetilde{B}$ are invertible. Let further $\alpha > 0$ be any number satisfying* (1.23). *Then*

$$\left|\, \|B_n\|_p - N_p \,\right| = O(e^{-\alpha n/2}). \tag{6.26}$$

Proof. Write B_n^{-1} as in the proof of Theorem 6.7, put $s(n) = [n/2] - 1$, and define A_n by (6.20). Proceeding in the way we derived (6.22) and using Lemmas 1.17 and 6.1, we get

$$\begin{aligned} &\left|\, \|B_n^{-1}\|_p - \|A_n\|_p \,\right| \\ &\le \|a - c_n\|_W + \|X - X_{s(n)}\|_p + \|Y - Y_{s(n)}\|_p + \|E_n\|_p \\ &= O(e^{-\alpha s(n)}) = O(e^{-\alpha n/2}). \end{aligned} \tag{6.27}$$

Since $2s(n)+1 \le n$, we may employ (5.27) and (5.28) to conclude that

$$\|A_n\|_p = \max\left(\|T(c_n)+X_{s(n)}\|_p, \|T(\widetilde{c}_n)+Y_{s(n)}\|_p\right),$$

whence, again by Lemmas 1.17 and 6.1,

$$\begin{aligned} &\big|\, \|A_n\|_p - N_p \,\big| \\ &\le \max\Big(\|c_n - a\|_W + \|X - X_{s(n)}\|_p, \|\widetilde{c}_n - \widetilde{a}\|_W + \|Y - Y_{s(n)}\|_p\Big) \\ &= O(e^{-\alpha s(n)}) = O(e^{-\alpha n/2}). \end{aligned} \tag{6.28}$$

Combining (6.27) and (6.28) we arrive at the asserted estimate (6.26). □

Recall that

$$M_p = \max(\|B\|_p, \|\widetilde{B}\|_p), \quad M_p^0 = \|T(b)\|_p \tag{6.29}$$

$$N_p = \max(\|B^{-1}\|_p, \|\widetilde{B}^{-1}\|_p), \quad N_p^0 = \|T(b^{-1})\|_p. \tag{6.30}$$

Theorem 6.9. *Suppose B and $\widetilde{B}$ are invertible on ℓ^p.*

If $1 < p < \infty$, $M_p > M_p^0$, and $N_p > N_p^0$, then there is a constant $\gamma > 0$ (depending on p, b, K, L) such that

$$\big|\, \kappa_p(B_n) - M_pN_p \,\big| = O(e^{-\gamma\sqrt{n}}). \tag{6.31}$$

If $1 < p < \infty$ and if $M_p = M_p^0$ or $N_p = N_p^0$, then

$$\big|\, \kappa_p(B_n) - M_pN_p \,\big| = O\left(\frac{\log n}{n}\right). \tag{6.32}$$

If $p = 1$ or $p = \infty$, then

$$\big|\, \kappa_p(B_n) - M_pN_p \,\big| = O(e^{-\alpha n/2}), \tag{6.33}$$

where $\alpha > 0$ satisfies (1.23).

Proof. Estimate (6.31) follows from (5.51) and Theorem 6.7, estimate (6.32) is a consequence of (5.51), (5.52), and Theorem 6.7, and equality (6.33) results from (5.27), (5.28), and Theorem 6.8. □

6.4 Generic and Exceptional Cases

Let $B_n = T_n(b)+P_nKP_n+W_nLW_n$ be as in Section 6.1, put $B = T(b)+K$, $\widetilde{B} = T(\widetilde{b})+L$, and define M_p, M_p^0, N_p, N_p^0 by (6.29) and (6.30).

Lemma 6.10. *Let $1 \le p \le \infty$. We always have $M_p^0 \le M_p$ and $N_p^0 \le N_p$.*

Proof. The inequality $M_p^0 \le M_p$ follows from Lemma 5.12 for $1 < p < \infty$ and from (1.16) for $p = 1$ and $p = \infty$. To prove that $N_p^0 \le N_p$, note first that $B^{-1} = T(b^{-1}) + X$

by (6.4). Hence $Q_nB^{-1}Q_n = Q_nT(b^{-1})Q_n + Q_nXQ_n$, and since $\|Q_nT(b^{-1})Q_n\|_p = \|T(b^{-1})\|_p$ and $\|Q_nXQ_n\|_p \to 0$ (Lemma 6.1), it follows that $\lim \|Q_nB^{-1}Q_n\|_p = \|T(b^{-1})\|_p$. As $\|B^{-1}\|_p \ge \|Q_nB^{-1}Q_n\|_p$, we conclude that $\|B^{-1}\|_p \ge \|T(b^{-1})\|_p$. Analogously, $\|\widetilde{B}^{-1}\|_p \ge \|T(\widetilde{b}^{-1})\|_p = \|T(b^{-1})\|_p$. □

In Chapter 5, we observed that the norm $\|B_n\|_p$ converges exponentially fast to M_p if $M_p > M_p^0$, while the convergence may be slow in the case $M_p = M_p^0$. We also proved that the strict inequality $M_2 > M_2^0$ represents the generic case and that we have the (exceptional) equality in the case where $B_n = T_n(b)$ is a pure Toeplitz matrix. The results of Section 6.3 show that the same phenomenon is encountered when treating the norms of inverses: $\|B_n^{-1}\|_p$ converges to N_p very fast if $N_p > N_p^0$, whereas this convergence may be slow provided $N_p = N_p^0$. As the following proposition reveals, at least for $p = 2$ the strict inequality $N_2 > N_2^0$ is the generic case.

Define $\mathcal{P} \times \mathcal{X} \times \mathcal{X}$ as in the paragraph before Proposition 5.17 and let $\mathcal{P}_0$ be the set of all Laurent polynomials without zeros on $\mathbf{T}$.

Proposition 6.11. *The set of all $(b, K, L) \in \mathcal{P}_0 \times \mathcal{X} \times \mathcal{X}$ for which*

$$\max(\|(T(b) + K)^{-1}\|_2, \|(T(\widetilde{b}) + L)^{-1}\|_2) > \|T(b^{-1})\|_2$$

is an open and dense subset of $\mathcal{P}_0 \times \mathcal{X} \times \mathcal{X}$.

Proof. Clearly, the set under consideration is open. It remains to show that

$$\{(b, K) \in \mathcal{P}_0 \times \mathcal{X} : \|(T(b) + K)^{-1}\|_2 > \|T(b^{-1})\|_2\}$$

is dense in $\mathcal{P}_0 \times \mathcal{X}$. So let $b \in \mathcal{P}_0$ and $K \in \mathcal{X}$ and suppose $\|(T(b)+K)^{-1}\|_2 = \|T(b^{-1})\|_2 = \|b^{-1}\|_\infty$. Fix a number $\varepsilon > 0$. We know from (6.4) and Lemma 6.1 that $(T(b) + K)^{-1} = T(b^{-1})+X$ with a compact operator X. If $\lambda \in \mathbf{C}$ is sufficiently small, then $T(b^{-1})+X+\lambda P_n$ is invertible and

$$\|(T(b^{-1}) + X + \lambda P_n)^{-1}\|_2 \le \frac{\|(T(b^{-1}) + X)^{-1}\|_2}{1 - \|(T(b^{-1}) + X)^{-1}\|_2\,|\lambda|}.$$

Consequently,

$$\|(T(b^{-1}) + X + \lambda P_n)^{-1} - (T(b^{-1}) + X)^{-1}\|_2 < \varepsilon/2 \tag{6.34}$$

for all $n \ge 1$ whenever $|\lambda|$ is small enough. Now, as in the proof of Proposition 5.17, choose a $\lambda \in \mathbf{C}$ of sufficiently small absolute value so that $\|b^{-1}+\lambda\|_\infty > \|b^{-1}\|_\infty$, $T(b^{-1})+X+\lambda P_n$ is invertible, and (6.34) holds. It follows, again as in the proof of Proposition 5.17, that $\|T(b^{-1}) + X + \lambda P_n\|_2 > \|T(b^{-1})\|_2$ for all sufficiently large n. Put

$$K_n = (T(b^{-1}) + X + \lambda P_n)^{-1} - T(b).$$

By construction, $\|(T(b) + K_n)^{-1}\|_2 > \|T(b^{-1})\|_2$. Since K_n is compact together with $X + \lambda P_n$, the operators $P_mK_nP_m \in \mathcal{X}$ converge uniformly to K_n as $m \to \infty$ (Proposition 3.2). Hence $\|(T(b) + P_mK_nP_m)^{-1}\| > \|T(b^{-1})\|_2$ for all sufficiently large m. If m is large enough, then $\|P_mK_nP_m - K_n\|_2 < \varepsilon/2$. This together with (6.34) shows that

$$\|K_n - K\|_2 = \|(T(b^{-1}) + X + \lambda P_n)^{-1} - (T(b^{-1}) + X)^{-1}\|_2 < \varepsilon,$$

which completes the proof. □

Combining Propositions 5.17 and 6.11, we arrive at the conclusion that, at least for $p = 2$, the fast convergence (6.31) of the condition numbers is generic, while the slow convergence (6.32) occurs in exceptional cases only.

6.5 Norms of Inverses of Pure Toeplitz Matrices

We now turn to pure Toeplitz matrices. Throughout what follows, b is a Laurent polynomial. In the case $B_n = T_n(b)$, we have

$$M_p = \|T(b)\|_p, \quad M_p^0 = \|T(b)\|_p,$$
$$N_p = \max\left(\|T^{-1}(b)\|_p, \|T^{-1}(\widetilde{b})\|_p\right), \quad N_p^0 = \|T(b^{-1})\|_p.$$

Corollary 6.12. *Let b be a Laurent polynomial and suppose $T(b)$ is invertible on the space ℓ^p.*

If $1 < p < \infty$ and $\|T^{-1}(b)\|_p > \|T(b^{-1})\|_p$, then there exists a $\gamma > 0$ (depending only on p and b) such that

$$\left| \|T_n^{-1}(b)\|_p - \max\left(\|T^{-1}(b)\|_p, \|T^{-1}(\widetilde{b})\|_p\right) \right| = O(e^{-\gamma\sqrt{n}}). \tag{6.35}$$

If $1 < p < \infty$ and $\|T^{-1}(b)\|_p = \|T(b^{-1})\|_p$, then

$$\left| \|T_n^{-1}(b)\|_p - \max\left(\|T^{-1}(b)\|_p, \|T^{-1}(\widetilde{b})\|_p\right) \right| = O\left(\frac{\log n}{n}\right). \tag{6.36}$$

If $p = 1$ or $p = \infty$, then

$$\left| \|T_n^{-1}(b)\|_p - \max\left(\|T^{-1}(b)\|_p, \|T^{-1}(\widetilde{b})\|_p\right) \right| = O(e^{-\alpha n/2}), \tag{6.37}$$

where $\alpha > 0$ is any number satisfying (1.23).

Proof. This is Theorems 6.7 and 6.8 in the special case $B_n = T_n(b)$. □

We remark that for $p = 2$ the number $\max(\|T^{-1}(b)\|_p, \|T^{-1}(\widetilde{b})\|_p)$ can be replaced by $\|T^{-1}(b)\|_2$. Thus, $N_2 = \|T^{-1}(b)\|_2$ and $N_2^0 = \|T(b^{-1})\|_2$.

Corollary 6.12 motivates the discussion of the question of when in $\|T^{-1}(b)\|_p \geq \|T(b^{-1})\|_p$ equality or strict inequality holds. This section gives partial answers to this question in the case $p = 2$.

Proposition 6.13. *Let $T(b)$ be invertible and let $b = b_- b_+$ be a Wiener-Hopf factorization. If $|b_-|$ and $|b_+|$ attain their minimum on $\mathbf{T}$ at the same point,*

$$\min_{t\in\mathbf{T}} |b_-(t)| = |b_-(t_0)|, \quad \min_{t\in\mathbf{T}} |b_+(t)| = |b_+(t_0)|, \tag{6.38}$$

then $N_2 = N_2^0 = 1/|b(t_0)|$.

Proof. By (6.38),

$$\|T^{-1}(b)\|_2 = \|T(b_+^{-1})T(b_-^{-1})\|_2 \leq \|b_+^{-1}\|_\infty \|b_-^{-1}\|_\infty$$
$$= |b_+^{-1}(t_0)|\,|b_-^{-1}(t_0)| = |b^{-1}(t_0)| \leq \|b^{-1}\|_\infty = \|T(b^{-1})\|_2.$$

Since always $\|T(b^{-1})\|_2 \leq \|T^{-1}(b)\|_2$ due to Lemma 6.10, we arrive at the assertion. □

Corollary 6.14. *Suppose $T(b)$ is invertible. If $T(b)$ is Hermitian or triangular, then $N_2 = N_2^0$. Moreover, if $t_0 \in \mathbf{T}$ is any point at which $|b|$ attains its minimum on $\mathbf{T}$, then $N_2 = N_2^0 = 1/|b(t_0)|$.*

Proof. Let $T(b)$ be Hermitian. Then b is real valued, and the invertibility of $T(b)$ implies that $b(t) \geq \varepsilon > 0$ or $b(t) \leq -\varepsilon < 0$ for all $t \in \mathbf{T}$. In the first case, $\log b$ is a real-valued function in W. Therefore $(\log b)_{-k} = \overline{(\log b)_k}$, whence

$$b_-(t) := \exp\left(\frac{1}{2}(\log b)_0 + \sum_{k=1}^{\infty}(\log b)_{-k}t^{-k}\right)$$

$$= \exp\left(\frac{1}{2}(\log b)_0 + \sum_{k=1}^{\infty}\overline{(\log b)_k t^k}\right) =: \overline{b_+(t)}.$$

The factorization $b = b_- b_+$ is a Wiener-Hopf factorization. Since $|b_+| = |b_-|$, condition (6.38) is satisfied for some $t_0 \in \mathbf{T}$. Thus, $N_2 = N_2^0 = 1/|b(t_0)|$ by Proposition 6.13. If $b(t) \leq -\varepsilon < 0$, we can apply the preceding argument to $-b$.

If $T(b)$ is triangular, then $T^{-1}(b) = T(b^{-1})$, and this yields all assertions. □

Theorem 6.15. *Let $\mathcal{P}_{00}$ be the set of all Laurent polynomials that have no zeros on $\mathbf{T}$ and whose winding number is zero. Equip $\mathcal{P}_{00}$ with the L^∞ metric. The set of all $b \in \mathcal{P}_{00}$ for which $N_2 = \|T^{-1}(b)\|_2 > \|T(b^{-1})\|_2 = N_2^0$ is an open and dense subset of $\mathcal{P}_{00}$.*

Proof. It is clear that the set is open. To show that it is dense, pick $b \in \mathcal{P}_{00}$ and $\varepsilon > 0$. We construct a $c \in \mathcal{P}_{00}$ such that $\|c - b\|_\infty < \varepsilon$ and $\|T^{-1}(c)\|_2 > \|T(c^{-1})\|_2$. Since $T(\mu b(t/t_0)) = \mu \operatorname{diag}(1, t_0, t_0^2, \ldots)\, T(b) \operatorname{diag}(1, t_0^{-1}, t_0^{-2}, \ldots)$, we may without loss of generality assume that $\min_{t\in\mathbf{T}} |b(t)| = 1$ and $b(1) = 1$. Let $0 < \delta < 1/4$ and consider $c = b\varphi_\delta^- \varphi_\delta^+$ where

$$\varphi_\delta^-(t) = \frac{1 - 2\delta + 2\delta t^{-n}}{1 - \delta + \delta t^{-n}}, \quad \varphi_\delta^+(t) = \frac{1 - \delta + \delta t^n}{1 - 2\delta + 2\delta t^n}.$$

Clearly, $\|c - b\|_\infty < \varepsilon$ if only $\delta > 0$ is sufficiently small. It is easy to see that $|\varphi_\delta^-(t)\varphi_\delta^+(t)| = 1$ for all $t \in \mathbf{T}$ and that $\varphi_\delta^-(1) = \varphi_\delta^+(1) = 1$. Since $\|T(c^{-1})\|_2 = \|c^{-1}\|_\infty = 1$, we are left with proving that $\|T^{-1}(c)\|_2 > 1$.

Let $b = b_- b_+$ be a Wiener-Hopf factorization with $b_-(1) = b_+(1) = 1$. Then

$$c = c_- c_+, \quad c_- := b_- \varphi_\delta^-, \quad c_+ := b_+ \varphi_\delta^+$$

is a Wiener-Hopf factorization of c. Given $z \in \mathbf{C}$ such that $|z| > 1$, put

$$f_z(t) = \sqrt{1 - \frac{1}{|z|^2}}\left(1 - \frac{t}{z}\right)^{-1}.$$

The sequence of the Fourier coefficients of f_z is

$$x_z := \sqrt{1 - \frac{1}{|z|^2}}\left\{\ldots, 0, 0, 1, \frac{1}{z}, \frac{1}{z^2}, \ldots\right\}.$$

Thus, we may think of x_z as a unit vector in ℓ^2. It is readily seen that $T(d_-)x_z = d_-(z)x_z$ for every $d_- \in W_-$. This implies that

$$\|T^{-1}(c)x_z\|_2^2 = \|T(c_+^{-1})T(c_-^{-1})x_z\|_2^2 = |c_-^{-1}(z)|^2\|T(c_+^{-1})x_z\|_2^2.$$

Using the analogue of Lemma 4.7 for infinite matrices, we get

$$\begin{aligned}
&\|T(c_+^{-1})x_z\|_2^2 = \|c_+^{-1} f_z\|_2^2 \\
&= \|(c_+^{-1} - c_+^{-1}(1/\overline{z}))f_z - c_+^{-1}(1/\overline{z})f_z\|_2^2 \\
&= \|(c_+^{-1} - c_+^{-1}(1/\overline{z}))f_z\|_2^2 + |c_+^{-1}(1/\overline{z})|^2 \\
&\quad + 2\left(1 - \frac{1}{|z|^2}\right)\operatorname{Re}\int_0^{2\pi} \frac{(c_+^{-1}(e^{i\theta}) - c_+^{-1}(1/\overline{z}))\,\overline{c_+^{-1}(1/\overline{z})}}{(1-e^{i\theta}/z)(1-e^{-i\theta}/\overline{z})}\frac{d\theta}{2\pi}.
\end{aligned}$$

The integral is the 0th Fourier coefficient of its integrand, and taking into account that

$$\begin{aligned}
&\left[\frac{a_+(e^{i\theta})}{1-e^{i\theta}/\overline{z}}\right]_0 = \left[(a_0 + a_1e^{i\theta} - a_2e^{2i\theta} + \cdots)\left(1 + \frac{e^{-i\theta}}{\overline{z}} + \frac{e^{-2i\theta}}{\overline{z}^2} + \cdots\right)\right]_0 \\
&= a_0 + \frac{a_1}{\overline{z}} + \frac{a_2}{\overline{z}^2} + \cdots = a_+(1/\overline{z}),
\end{aligned}$$

we see that the integral is zero. Thus,

$$\|T^{-1}(c)x_z\|_2^2 = |c_-^{-1}(z)c_+^{-1}(1/\overline{z})|^2 + |c_-^{-1}(z)|^2\,\|(c_+^{-1} - c_+^{-1}(1/\overline{z}))f_z\|_2^2. \tag{6.39}$$

Now put $n = [1/\delta^3]$ and $z = 1 + \delta^3$. We have

$$\frac{1}{\varphi_\delta^-(z)} = \frac{1}{\varphi_\delta^-(1+\delta^3)} = 1 + \frac{\delta(1-(1+\delta^3)^{-n})}{1-2\delta(1-(1+\delta^3)^{-n})}$$

and since $(1+1/n)^{-n} = e^{-1} + O(1/n)$ and hence $(1+\delta^3)^{-n} = e^{-1} + O(\delta^3)$, we obtain that

$$\frac{1}{\varphi_\delta^-(z)} = 1 + \frac{\delta(1-e^{-1})}{1-2\delta(1-e^{-1})} + O(\delta^4) = 1 + \delta(1-e^{-1}) + O(\delta^2). \tag{6.40}$$

Analogously,

$$\frac{1}{\varphi_\delta^+(1/z)} = 1 - \frac{\delta(1-e^{-1})}{1-\delta(1-e^{-1})} + O(\delta^4) = 1 - \delta(1-e^{-1}) + O(\delta^2). \tag{6.41}$$

It is obvious that

$$\varphi_\delta^-(z)\varphi_\delta^+(1/z) = 1. \tag{6.42}$$

Since $z = \overline{z}$, equalities (6.39) and (6.42) imply that

$$\|T^{-1}(c)x_z\|_2^2 = |b_-^{-1}(z)b_+^{-1}(1/z)|^2 + |b_-^{-1}(z)|^2\,|1/\varphi_\delta^-(z)|^2\,\|(c_+^{-1} - c_+^{-1}(1/z))f_z\|_2^2.$$

The formulas

$$b_-(z) = b_-(1+\delta^3) = 1 + O(\delta^3), \quad b_+(1/z) = b_+(1/(1+\delta^3)) = 1 + O(\delta^3)$$

together with (6.40) therefore give

$$\|T^{-1}(c)x_z\|_2^2 = 1 + O(\delta^3) + (1+O(\delta^3))(1+O(\delta))\,\|(c_+^{-1} - c_+^{-1}(1/z))f_z\|_2^2. \tag{6.43}$$

Further,

$$\begin{aligned}
&\|(c_+^{-1} - c_+^{-1}(1/z))f_z\|_2 = \|((\varphi_\delta^+)^{-1}b_+^{-1} - (\varphi_\delta^+)^{-1}(1/z)b_+^{-1}(1/z))f_z\|_2 \\
&\geq \|((\varphi_\delta^+)^{-1} - (\varphi_\delta^+)^{-1}(1/z))b_+^{-1}f_z\|_2 - |(\varphi_\delta^+)^{-1}(1/z)|\,\|(b_+^{-1} - b_+^{-1}(1/z))f_z\|_2 \\
&=: A - B.
\end{aligned}$$

We have

$$\begin{aligned}
B^2 &= |(\varphi_\delta^+)^{-1}(1/z)|^2\left(1 - \frac{1}{z^2}\right)\int_{\mathbf{T}} \frac{|b_+^{-1}(t) - b_+^{-1}(1/z)|^2}{|1-t/z|^2}\frac{|dt|}{2\pi} \\
&= |(\varphi_\delta^+)^{-1}(1/z)|^2\left(1 - \frac{1}{z^2}\right)\int_{\mathbf{T}} \frac{|b_+^{-1}(t) - b_+^{-1}(1/z)|^2}{|t-1/z|^2}\frac{|dt|}{2\pi} \\
&= |(\varphi_\delta^+)^{-1}(1/z)|^2\left(1 - \frac{1}{z^2}\right)\int_{\mathbf{T}} |g_z(t)|^2\frac{|dt|}{2\pi}
\end{aligned}$$

with the function

$$g_z(t) = \frac{b_+^{-1}(t) - b_+^{-1}(1/z)}{t - 1/z} = \sum_{j=1}^{\infty} h_j(t-1/z)^j.$$

The function g_z is analytic in some open disk containing $\mathbf{D}$. Thus, there is a constant $M < \infty$ such that $\|g_z\| \leq M$ for all z sufficiently close to 1. Since, by (6.41), $|(\varphi_\delta^+)^{-1}(1/z)|^2 = 1 + O(\delta)$ and, obviously, $1 - 1/z^2 = O(\delta^3)$, it follows that

$$B \leq (1+O(\delta))\,O(\delta^{3/2})\,M = O(\delta^{3/2}).$$

On the other hand,

$$\begin{aligned}
A^2 &= \left(1 - \frac{1}{z^2}\right)\int_{\mathbf{T}} \frac{|(\varphi_\delta^+)^{-1}(t) - (\varphi_\delta^+)^{-1}(1/z)|^2\,|b_+^{-1}(t)|^2}{|1-t/z|^2}\frac{|dt|}{2\pi} \\
&\geq \gamma^2\left(1 - \frac{1}{z^2}\right)\int_{\mathbf{T}} \frac{|(\varphi_\delta^+)^{-1}(t) - (\varphi_\delta^+)^{-1}(1/z)|^2}{|1-t/z|^2}\frac{|dt|}{2\pi},
\end{aligned}$$

where $\gamma = \min_{t\in\mathbf{T}} |b_+^{-1}(t)|$. Taking into account (6.41) we get

$$\begin{aligned}
&(\varphi_\delta^+)^{-1}(t) - (\varphi_\delta^+)^{-1}(1/z) = 1 - \frac{\delta(1-t^n)}{1-\delta(1-t^n)} - 1 + \frac{\delta(1-1/z^n)}{1-\delta(1-1/z^n)} \\
&= \delta\left(\frac{1-t^n}{1-\delta(1-t^n)} - \frac{1-e^{-1}}{1-\delta(1-e^{-1})} + O(\delta^3)\right) \\
&= \delta(1 - t^n - (1-e^{-1}) + O(\delta)) \\
&= \delta(e^{-1} - t^n + O(\delta))
\end{aligned}$$

and hence arrive at the estimate

$$\begin{aligned} A^2 &\geq \gamma^2 \left(1 - \frac{1}{z^2}\right) \delta^2 \int_{\mathbf{T}} \frac{(|t^n - e^{-1}| + O(\delta))^2}{|1 - t/z|^2} \frac{|dt|}{2\pi} \\ &\geq \gamma^2 \left(1 - \frac{1}{z^2}\right) \delta^2 \int_{\mathbf{T}} \frac{(1 - e^{-1} + O(\delta))^2}{|1 - t/z|^2} \frac{|dt|}{2\pi} \\ &= \gamma^2 \delta^2 (1 - e^{-1} + O(\delta))^2 = \gamma^2 (1 - e^{-1})^2 \delta^2 + O(\delta^3). \end{aligned}$$

In summary, (6.43) yields

$$\begin{aligned} \|T^{-1}(c)x_z\|^2 &\geq 1 + O(\delta^3) + (1 + O(\delta))(A - B)^2 \\ &\geq 1 + O(\delta^3) + (1 + O(\delta))(\gamma(1 - e^{-1})\delta + O(\delta^{3/2}))^2 \\ &= 1 + \gamma^2 (1 - e^{-1})^2 \delta^2 + O(\delta^{5/2}), \end{aligned}$$

and this is certainly greater than 1 if only $\delta > 0$ is sufficiently small. □

By virtue of Theorem 6.15, we may say that the strict inequality $N_2 > N_2^0$ is the generic case, whereas the equality $N_2 = N_2^0$ represents the exceptional case. Corollary 6.14 tells us that we are in the exceptional case whenever $T(b)$ is Hermitian or triangular.

Estimates (6.35) and (6.36) result from our techniques, and it may be that these estimates can be improved. We conjecture that Corollary 6.12 remains valid with the right-hand sides of (6.35) and (6.36) replaced by $O(e^{-\gamma n})$ and $O(1/n^2)$, respectively. However, as the following example shows, the gap between (6.35) and (6.36) is essential and cannot be removed.

Example 6.16. Let $b(t) = 3 - t - t^{-1}$ $(t \in \mathbf{T})$. By Theorem 2.4, the eigenvalues of the Hermitian matrix $T_n(b)$ are

$$\lambda_j(T_n(b)) = 3 + 2\cos\frac{j\pi}{n+1} \quad (j = 1, \ldots, n).$$

This implies that

$$\|T_n^{-1}(b)\|_2 = \frac{1}{\lambda_n(T_n(b))} = 1 - \frac{\pi^2}{n^2} + O\left(\frac{1}{n^3}\right),$$

and it also shows that $\|T_n^{-1}(b)\|_2 - \|T^{-1}(b)\|_2 = \|T_n^{-1}(b)\|_2 - 1$ cannot decay faster than $O(1/n^2)$. □

Estimate (6.36) is universal in the sense that it does not depend on any further information about b. In case we know more about b, we can improve (6.36).

Lemma 6.17. *Let b be a Laurent polynomial for which $T(b)$ is invertible. Further, let $t_0 \in \mathbf{T}$ be any point at which $|b|$ attains its minimum and suppose that*

$$|b(t) - b(t_0)| \leq D|t - t_0|^\gamma \text{ for all } t \in \mathbf{T},$$

where $D \in (0, \infty)$ and $\gamma \in \mathbf{N}$. Then

$$\|T_n^{-1}(b)\|_2 \geq \|T(b^{-1})\|_2 - \frac{C}{n^\gamma} \text{ for all } n \geq 1$$

with some constant $C \in (0, \infty)$.

Proof. Without loss of generality assume that $t_0 = 1$. Put $j = \gamma + 1$ and let m be the nonnegative integer for which $mj < n \le (m+1)j$. Define $p_m^j \in \mathcal{P}_{mj+1}^+$ by (4.20), (4.21). Clearly, $p_m^j \in \mathcal{P}_n^+$. To estimate $\|T_n^{-1}(b)\|_2$, we make use of Lemma 4.9. We have $\|bp_m^j\|_2 \le |b(1)|\,\|p_m^j\|_2 + \|(b-b(1))p_m^j\|_2$, and

$$\begin{aligned}\|(b-b(1))p_m^j\|_2^2 &= \int_{-\pi}^{\pi} |b(e^{i\theta}) - b(1)|^2 \left(\frac{\sin((m+1)\theta/2)}{\sin(\theta/2)}\right)^{2j} \frac{d\theta}{2\pi}\\ &= O\left(\int_{|\theta|<1/(m+1)} |\theta|^{2\gamma}(m+1)^{2j}\frac{d\theta}{2\pi} + \int_{|\theta|>1/(m+1)} |\theta|^{2\gamma}|\theta|^{-2j}\frac{d\theta}{2\pi}\right)\\ &= O\left(\frac{(m+1)^{2j}}{(m+1)^{2\gamma+1}} + \frac{(m+1)^{2\gamma+1}}{(m+1)^{2j}}\right) = O\,(m+1) = O(m).\end{aligned}$$

Thus, by Lemma 4.10,

$$\frac{\|bp_m^j\|_2}{\|p_m^j\|_2} \le |b(1)| + O\left(\frac{m^{1/2}}{m^{j-1/2}}\right) = |b(1)| + O\left(\frac{1}{m^\gamma}\right).$$

From Lemma 4.9 we now obtain that $\|T_n^{-1}(b)\|_2$ is at least

$$\frac{1}{|b(1)| + O(1/m^\gamma)} = \frac{1}{|b(1)| + O(1/n^\gamma)} \ge \frac{1}{|b(1)|} - O\left(\frac{1}{n^\gamma}\right) \ge \frac{1}{|b(1)|} - C\frac{1}{n^\gamma},$$

and since $\|T(b^{-1})\|_2 = \|b^{-1}\|_\infty = 1/\min|b| = 1/|b(1)|$, we arrive at the assertion. □

Theorem 6.18. *Let b be a Laurent polynomial and suppose $T(b)$ is invertible. Let $t_0 \in \mathbf{T}$ and assume*

$$|b(t_0)| = \min_{t\in\mathbf{T}} |b(t)|, \quad |b(t) - b(t_0)| \le D|t - t_0|^\gamma \quad (t \in \mathbf{T}).$$

If $T(b)$ is Hermitian or triangular, then $N_2 = \|T(b^{-1})\|_2 = 1/|b(t_0)|$ and

$$N_2\left(1 - \frac{C}{n^\gamma}\right) \le \|T_n^{-1}(b)\|_2 \le N_2 \text{ for all } n \ge 1, \tag{6.44}$$

where $C \in (0,\infty)$ is some constant independent of n.

Proof. The equalities $N_2 = \|T(b^{-1})\|_2 = 1/|b(t_0)|$ follow from Corollary 6.14, and the lower estimate in (6.44) results from Lemma 6.17. To prove the upper estimate of (6.44), assume first that $T(b)$ is Hermitian. Then $b(t) \ge |b(t_0)| > 0$ or $b(t) \le -|b(t_0)| < 0$ for all $t \in \mathbf{T}$. We consider the first case (the second case can be reduced to the first case by replacing b with $-b$). Let $\lambda_{\min}$ be the smallest eigenvalue of $T_n(b)$ and let x be an eigenvector such that $\|x\|_2 = 1$. Define $f \in \mathcal{P}_n^+$ by (4.15). Then, by (4.16),

$$\lambda_{\min} = (T_n(b)x, x) = \frac{1}{2\pi}\int_0^{2\pi} b(e^{i\theta})|f(e^{i\theta})|^2 d\theta \ge |b(t_0)|,$$

whence $\|T_n^{-1}(b)\|_2 = 1/\lambda_{\min} \le 1/|b(t_0)| = N_2$. This is the upper estimate of (6.44) for Hermitian matrices. If $T(b)$ is triangular, we have $T_n^{-1}(b) = T_n(b^{-1})$, which implies that $\|T_n^{-1}(b)\|_2 = \|T_n(b^{-1})\|_2 \le \|b^{-1}\|_\infty = 1/|b(t_0)| = N_2$. □

6.6 Condition Numbers of Pure Toeplitz Matrices

Combining our results on the converge speed of norms and norms of inverses, we can establish a result on the convergence speed for condition numbers.

Corollary 6.19. *Let b be a Laurent polynomial and let $T(b)$ be invertible on the space ℓ^p.*
If $1 < p < \infty$, then

$$\left|\,\kappa_p(T_n(b)) - \max\left(\kappa_p(T(b)), \kappa_p(T(\widetilde{b}))\right)\right| = O\left(\frac{\log n}{n}\right). \tag{6.45}$$

If $p = 1$ or $p = \infty$, then

$$\left|\,\kappa_p(T_n(b)) - \max\left(\kappa_p(T(b)), \kappa_p(T(\widetilde{b}))\right)\right| = O(e^{-\alpha n/2}), \tag{6.46}$$

where $\alpha > 0$ is subject to (1.23).

Proof. In the case at hand, $M_p N_p = \max(\kappa_p(T(b)), \kappa_p(T(\widetilde{b})))$. Thus, (6.32) gives (6.45), while (6.33) yields (6.46). □

Note that in the case $p = 2$ the equality $\max\left(\kappa_2(T(b)), \kappa_2(T(\widetilde{b}))\right) = \kappa_2(T(b))$ holds.

We emphasize that generically $\|T_n^{-1}(b)\|_p$ converges to N_p very fast, namely, at least of the order $O(e^{-\gamma\sqrt{n}})$. The generically slow convergence of $\kappa_p(T_n(b))$ to $M_p N_p$ is caused by the generically slow convergence of $\|T_n(b)\|_p$ to M_p.

As in the case of the norms of the inverses, we believe that (6.45) can be sharpened. However, the following result shows that generically we cannot expect more than polynomially fast convergence.

Proposition 6.20. *The set of all Laurent polynomials $b \in \mathcal{P}_{00}$ for which there exist constants $2\gamma \in \{2, 4, 6, \dots\}$ and $\mu > 0$ such that*

$$|\,\kappa_2(T_n(b)) - \kappa_2(T(b))\,| \ge \frac{\mu}{n^{2\gamma}} \quad \textit{for all } n \ge 1$$

is an open and dense subset of $\mathcal{P}_{00}$.

Proof. Let

$$E_1 = \{b \in \mathcal{P}_{00} : |b| \text{ is not constant}\}, \quad E_2 = \{b \in \mathcal{P}_{00} : \|T^{-1}(b)\|_2 > \|T(b^{-1})\|_2\}.$$

Proposition 5.6 implies that E_1 is open and dense, and Theorem 6.15 says that E_2 is open and dense. Hence, $E_1 \cap E_2$ is also an open and dense subset of $\mathcal{P}_{00}$. If $b \in E_1 \cap E_2$, then

$$|\,\|T_n(b)\|_2 - \|T(b)\|_2\,| \ge \frac{c}{n^{2\gamma}}, \quad \left|\,\|T_n^{-1}(b)\|_2 - \|T^{-1}(b)\|_2\,\right| \le d e^{-\delta\sqrt{n}}$$

by Theorem 5.8 and (6.35), whence

$$\begin{aligned}
&|\,\kappa_2(T_n(b)) - \kappa_2(T(b))\,| = \left|\,\|T_n(b)\|_2\,\|T_n^{-1}(b)\|_2 - \|T(b)\|_2\,\|T^{-1}(b)\|_2\,\right| \\
&\ge \|T^{-1}(b)\|_2 \left|\,\|T_n(b)\|_2 - \|T(b)\|_2\,\right| - \|T_n(b)\|_2 \left|\,\|T_n^{-1}(b)\|_2 - \|T^{-1}(b)\|_2\,\right| \\
&\ge \|T^{-1}(b)\|_2\, c n^{-2\gamma n} - \|T_n(b)\|_2\, d e^{-\delta\sqrt{n}} \ge \mu n^{-2\gamma}
\end{aligned}$$

for some $\mu > 0$. □

6.7 Conclusions

In Chapters 5 and 6, we proved several estimates that make precise the following insights:

(a) *Within the class of Toeplitz-like matrices, fast convergence of the norms, of the norms of the inverses, and of the condition numbers is generic.*

(b) *Within the class of pure Toeplitz matrices, norms converge generically slow, norms of inverses converge generically fast, and condition numbers converge generically slow.*

Exercises

1. Prove that there exist $b, c \in \mathcal{P}$ such that

$$\|T(b)T(c)\|_2 < \|T(c)T(b)\|_2.$$

2. Let $K = \operatorname{diag}(0, -3/4, 0, 0, \dots)$ and $L = \operatorname{diag}(2, -1/2, 0, 0, \dots)$. Then for $n \geq 4$,

$$A_n := I + P_n K P_n + W_n L W_n = \operatorname{diag}\Big(1, \frac{1}{4}, \underbrace{1, \dots, 1}_{n-4}, \frac{1}{3}, 3\Big).$$

Let A and $\widetilde{A}$ denote the strong limits of A_n and $W_n A_n W_n$. Show that

$$\lim_{n\to\infty} \kappa_2(A_n) = 12, \quad \kappa_2(A) = 4, \quad \max(\kappa_2(A), \kappa_2(\widetilde{A})) = 6.$$

3. Show that there exist $b \in \mathcal{P}$ such that $\kappa_2(T_n(b)) \to \infty$ as $n \to \infty$ although $\|T_n(b)\|_2 \leq 2$ and $\det T_n(b) = 1$ for all $n \geq 1$.

4. Let $\{a_n\}_{n=0}^{\infty}$ be a sequence of complex numbers satisfying $|a_n| = O(1/n^2)$ and put

$$A_n = \begin{pmatrix} 0 & 0 & \dots & 0 & a_0 \\ a_0 & 0 & \dots & 0 & a_1 \\ a_1 & a_0 & \dots & 0 & a_2 \\ \vdots & \vdots & \ddots & \vdots & \vdots \\ a_n & a_{n-1} & \dots & a_0 & a_{n+1} \end{pmatrix}.$$

Fix $p \in [1, \infty]$. Show that $\kappa_p(A_n)$ remains bounded as $n \to \infty$ if and only if $\sum_{n=0}^{\infty} a_n z^n \neq 0$ for $|z| \leq 1$.

5. Find four Laurent polynomials a, b, c, d such that $ad - bc$ has no zeros on $\mathbf{T}$ and winding number zero but

$$\kappa_p \begin{pmatrix} T_n(a) & T_n(b) \\ T_n(c) & T_n(d) \end{pmatrix} \to \infty \quad \text{as} \quad n \to \infty$$

for every $p \in [1, \infty]$.

6. Let $b \in \mathcal{P}$ and suppose $b \geq 0$ on $\mathbf{T}$ and $G(b) = 1$. Prove that $\kappa_2(T_n(b))/D_n(b)$ converges to zero if b has at least three distinct zeros on $\mathbf{T}$. What can be said about the ratio $\kappa_2(T_n(b))/D_n(b)$ if b has no, exactly one, or exactly two distinct zeros on $\mathbf{T}$?

7. Let $p(z) = p_0 + \cdots + p_{n-1}z^{n-1} + z^n$ ($p_0 \neq 0$) and suppose we know that $p(z)$ has m zeros in $\{z \in \mathbf{C} : |z| < r < 1\}$ and $n-m$ zeros in $\{z \in \mathbf{C} : |z| > R > 1\}$. We want a polynomial factorization $p(z) = v(z)\ell(z)$ such that all zeros of $v(z)$ and $\ell(z)$ are of modulus less than r and greater than R, respectively. To estimate the conditioning of this problem, let $\widetilde{p}(z)$ be another polynomial of the same form as $p(z)$ and denote by $\widetilde{p}(z) = \widetilde{v}(z)\widetilde{\ell}(z)$ the corresponding factorization.

   (a) Prove that the coefficients of ℓ are the solution x of the equation $T(a^{-1})x = e_1$, where $a(t) = t^{-m}p(t)$ and $e_1 = \{1, 0, 0, \dots\}$.

   (b) Prove that if $\|p - \widetilde{p}\|_\infty \leq \varepsilon \, \|p\|_\infty$, then

$$\frac{\|\widetilde{\ell} - \ell\|_2}{\|\ell\|_2} \leq \varepsilon \, \kappa_2(T(a^{-1})) \, \|p\|_\infty \, \|p^{-1}\|_\infty + O(\varepsilon^2) \quad \text{as} \quad \varepsilon \to 0,$$

   where $\|\cdot\|_2$ and $\|\cdot\|_\infty$ are the norms in $L^2(\mathbf{T})$ and $L^\infty(\mathbf{T})$.

8. Let A_n be an $n \times n$ matrix with distinct eigenvalues $\lambda_1, \dots, \lambda_n$. Choose nonzero vectors x_j and y_j such that $A_n x_j = \lambda_j x_j$ and $A_n^* y_j = \overline{\lambda}_j y_j$ ($j = 1, \dots, n$). The *instability index* $i(A_n)$ is defined as

$$i(A_n) = \max_{1 \leq j \leq n} \frac{\|x_j\|_2 \|y_j\|_2}{|(x_j, y_j)|}.$$

   (a) Prove the following implications:

$$\begin{aligned} &V_n^{-1} A_n V_n \text{ is diagonal for some } V_n \text{ with } \kappa_2(V_n) \leq k \\ &\Longrightarrow \ \|(A_n - \lambda I)^{-1}\|_2 \leq k/\text{dist}\,(\lambda, \text{sp}\, A_n) \text{ for all } \lambda \notin \text{sp}\, A_n \\ &\Longrightarrow \ i(A_n) \leq k. \end{aligned}$$

   (b) Show that if

$$A_n = \begin{pmatrix} f_0 & f_{n-1}a^{-1} & \dots & f_1 a^{-(n-1)} \\ f_1 a & f_0 & \dots & f_2 a^{-(n-2)} \\ \vdots & \vdots & & \vdots \\ f_{n-1}a^{n-1} & f_{n-2}a^{n-2} & \dots & f_0 \end{pmatrix}$$

   with $a > 1$, then

$$i(A_n) = \frac{a}{(a^2 - 1)n}(a^n - a^{-n}).$$

Notes

Corollary 6.5 was established in [34] for $p = 2$ and in [146] for $p = 1$. Example 6.6 is from [146]. All other results of this chapter are taken from the papers [49], [59], [62]. We remark that Theorem 6.15 is stated in [59] but that the proof given there is incorrect. We thank Alexander Rogozhin for bringing this to our attention. The proof presented here is new.

The result of Exercise 7 was established in [28]. Exercise 8 is from [95]. Note that what is called the "instability index" in [95] and here is also known as the maximum of the eigenvalue condition numbers (see [298] and [137]).

Chapter 7

Substitutes for the Spectrum

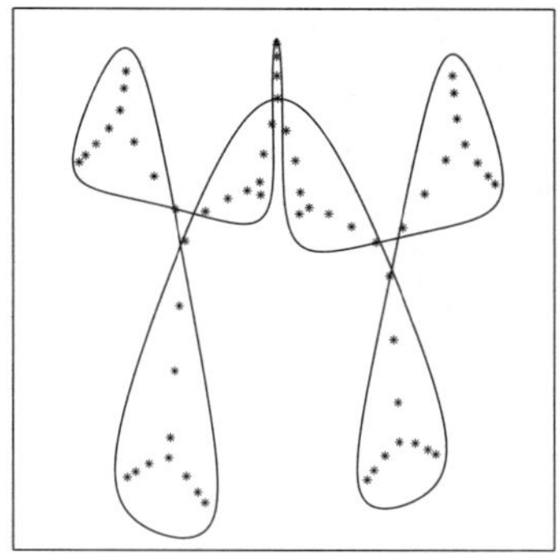

As will be seen in Chapter 11, the spectrum $\operatorname{sp} T_n(b)$ need not mimic $\operatorname{sp} T(b)$ as n goes to infinity. In contrast to this, pseudospectra and numerical ranges behave as nicely as we could ever expect. These sets are the concern of the present chapter.

7.1 Pseudospectra

For $\varepsilon > 0$, the *ε-pseudospectrum* of an operator $A \in \mathcal{B}(X)$ on a Banach space X is defined by

$$\operatorname{sp}_\varepsilon A = \left\{\lambda \in \mathbf{C} : \|(A - \lambda I)^{-1}\|_{\mathcal{B}(X)} \geq 1/\varepsilon\right\}. \tag{7.1}$$

Here we put $\|(A - \lambda I)^{-1}\|_{\mathcal{B}(X)} = \infty$ if $A - \lambda I$ is not invertible. Thus, the usual spectrum $\operatorname{sp} A$ is always a subset of $\operatorname{sp}_\varepsilon A$. Clearly, $\operatorname{sp}_\varepsilon A$ depends on X. If A acts on ℓ^p or ℓ^p_n, we denote the ε-pseudospectrum of A by $\operatorname{sp}_\varepsilon^{(p)} A$.

Definition (7.1) admits several modifications and generalizations. An important generalization, which is motivated by plenty of applications, is the so-called *structured pseudospectrum*. In this context we are given three operators $A, B, C \in \mathcal{B}(X)$ and we define

$$\operatorname{sp}_\varepsilon^{B,C} A = \operatorname{sp} A \cup \left\{\lambda \notin \operatorname{sp} A : \|C(A - \lambda I)^{-1} B\|_{\mathcal{B}(X)} \geq 1/\varepsilon\right\}. \tag{7.2}$$

Evidently, $\operatorname{sp}_\varepsilon A$ is nothing but $\operatorname{sp}_\varepsilon^{I,I} A$. Furthermore, some authors prefer (7.1) and (7.2) with "$\geq$" replaced by "$>$". Theorem 7.2 and its Corollary 7.3 will give alternative descriptions of the sets (7.1) and (7.2) in the Hilbert space case.

Lemma 7.1. *If M and N are linear operators, then $I + MN$ is invertible if and only if $I + NM$ is invertible, in which case*

$$(I + MN)^{-1} = I - M(I + NM)^{-1}N.$$

Proof. Simply check that

$$(I + MN)(I - M(I + NM)^{-1}N) = (I - M(I + NM)^{-1}N)(I + MN) = I. \qquad \square$$

Theorem 7.2. *If H is a Hilbert space, $A, B, C \in \mathcal{B}(H)$, and $\varepsilon > 0$, then*

$$\bigcup_{\|K\| \le \varepsilon} \operatorname{sp}(A + BKC) = \operatorname{sp} A \cup \left\{\lambda \notin \operatorname{sp} A : \|C(A - \lambda I)^{-1}B\| \ge 1/\varepsilon\right\}, \tag{7.3}$$

$$\bigcup_{\|K\| < \varepsilon} \operatorname{sp}(A + BKC) = \operatorname{sp} A \cup \left\{\lambda \notin \operatorname{sp} A : \|C(A - \lambda I)^{-1}B\| > 1/\varepsilon\right\}, \tag{7.4}$$

the union taken over all $K \in \mathcal{B}(H)$ with the given norm constraint.

Proof. Suppose A is invertible and $\|CA^{-1}B\| \le 1/\varepsilon$ (respectively, $\|CA^{-1}B\| < 1/\varepsilon$) and $\|K\| < \varepsilon$ (respectively, $\|K\| \le \varepsilon$). In either case, $\|CA^{-1}BK\| < 1$, which implies that $I + CA^{-1}BK$ is invertible. Using Lemma 7.1 with $N = C$ and $M = A^{-1}BK$, we see that $I + A^{-1}BKC$ and thus also $A + BKC = A(I + A^{-1}BKC)$ are invertible. This proves that the left-hand sides of (7.3) and (7.4) are contained in the corresponding right-hand sides.

To show that the right-hand sides of (7.3) and (7.4) are subsets of the corresponding left-hand sides, it suffices to prove that if A is invertible and $\|CA^{-1}B\| = 1/\delta$, then there exists an operator K such that $\|K\| = \delta$ and $A + BKC$ is not invertible. So assume that A is invertible and $\|CA^{-1}B\| = 1/\delta$. Clearly, $A + BKC = A(I + A^{-1}BKC)$ is invertible if and only if so is $I + A^{-1}BKC$, which, by Lemma 7.1 with $M = A^{-1}B$ and $N = KC$, is equivalent to the invertibility of $I + KCA^{-1}B$. Abbreviate $CA^{-1}B$ to S and put $K = -\delta^2 S^*$. Then $\|K\| = \delta$ and $I + KCA^{-1}B = I - \delta^2 S^* S$. The spectral radius of the positive semi-definite selfadjoint operator S^*S coincides with its norm, that is, with $\|S^*S\| = \|S\|^2 = 1/\delta^2$. It follows that $1/\delta^2 \in \operatorname{sp} S^*S$. The spectral mapping theorem therefore implies that $0 \in \operatorname{sp}(I - \delta^2 S^* S)$. □

Corollary 7.3. *Let H be a Hilbert space and $A \in \mathcal{B}(H)$. Then for every $\varepsilon > 0$,*

$$\operatorname{sp}_\varepsilon A = \bigcup_{\|K\| \le \varepsilon} \operatorname{sp}(A + K),$$

the union taken over all $K \in \mathcal{B}(H)$ of norm at most ε.

Proof. This is (7.3) with $B = C = I$. □

The following result sharpens (7.4).

Theorem 7.4. *Let X be a Banach space, let A, B, C be operators in $\mathcal{B}(X)$, and let $\varepsilon > 0$. Then*

$$\operatorname{sp} A \cup \{\lambda \notin \operatorname{sp} A : \|C(A - \lambda I)^{-1}B\| > 1/\varepsilon\} \tag{7.5}$$

$$= \bigcup_{\|K\| < \varepsilon} \operatorname{sp}(A + BKC) \tag{7.6}$$

$$= \bigcup_{\|K\| < \varepsilon,\ \operatorname{rank} K = 1} \operatorname{sp}(A + BKC). \tag{7.7}$$

Proof. The first part of the proof of Theorem 7.2 can literally be used to show that (7.6) is a subset of (7.5).

It remains to prove that (7.5) is contained in (7.7). This will follow as soon as we have shown that if A is invertible and $\|CA^{-1}B\| > 1/\varepsilon$, then there exists a rank-one operator K such that $\|K\| < \varepsilon$ and $A + BKC$ is not invertible. So assume A is invertible and $\|CA^{-1}B\| > 1/\varepsilon$. Then we can find a $u \in X$ such that $\|u\| = 1$ and $\|CA^{-1}Bu\| > 1/\varepsilon$. Thus, $\|CA^{-1}Bu\| = 1/\delta$ with $\delta < \varepsilon$. By the Hahn-Banach theorem, there is a functional $\varphi \in X^*$ such that $\|\varphi\| = 1$ and $\varphi(CA^{-1}Bu) = \|CA^{-1}Bu\| = 1/\delta$. Let $K \in \mathcal{B}(X)$ be the rank-one operator defined by $Kx = -\delta\varphi(x)u$. Clearly, $\|K\| \le \delta < \varepsilon$. Furthermore,

$$\begin{aligned} BKCA^{-1}Bu &= B(-\delta\varphi(CA^{-1}Bu)u) \\ &= -\delta\varphi(CA^{-1}Bu)Bu = -\delta(1/\delta)Bu = -Bu. \end{aligned} \tag{7.8}$$

Put $y = A^{-1}Bu$. If $y = 0$, then $CA^{-1}Bu = Cy = 0$, which contradicts the assumption $\|CA^{-1}Bu\| = 1/\delta > 0$. Consequently, $y \ne 0$. From (7.8) we see that $BKCy = -Bu = -Ay$, whence $(A + BKC)y = 0$. This implies that $A + BKC$ is not invertible. □

We have not been able to prove Theorem 7.4 with strict inequalities replaced by nonstrict inequalities. However, one can show that if X is a Banach space, $A, B, C \in \mathcal{B}(X)$, $\varepsilon > 0$, and at least one of the operators B or C is compact, then

$$\begin{aligned} &\operatorname{sp} A \cup \{\lambda \notin \operatorname{sp} A : \|C(A - \lambda I)^{-1}B\| \ge 1/\varepsilon\} \\ &= \bigcup_{\|K\| \le \varepsilon} \operatorname{sp}(A + BKC) = \bigcup_{\|K\| \le \varepsilon,\ \operatorname{rank} K = 1} \operatorname{sp}(A + BKC). \end{aligned}$$

To see this, assume that A is invertible and $\|CA^{-1}B\| = 1/\delta$. Since $CA^{-1}B$ is compact, there exists a $u \in X$ such that $\|u\| = 1$ and $\|CA^{-1}Bu\| = 1/\delta$. The rest of the proof is as in the proof of Theorem 7.4.

7.2 Norm of the Resolvent

In this section we show that the norm of the resolvent of a bounded operator on ℓ^p $(1 < p < \infty)$ cannot be locally constant. It should be noted that such a result is not true for arbitrary analytic operator-valued functions. To see this, consider the function

$$A : \mathbf{C} \to \mathcal{B}(\ell_2^p),\ \lambda \mapsto \begin{pmatrix} \lambda & 0 \\ 0 & 1 \end{pmatrix}.$$

Obviously, $\|A(\lambda)\|_p = \max(|\lambda|, 1)$ and thus $\|A(\lambda)\|_p = 1$ for all λ in the unit disk.

Theorem 7.5 (Daniluk). *Let H be a Hilbert space and $A \in \mathcal{B}(H)$. Suppose that $A - \lambda I$ is invertible for all λ in some open subset U of $\mathbf{C}$. If there is an $M < \infty$ such that $\|(A - \lambda I)^{-1}\| \le M < \infty$ for all $\lambda \in U$, then $\|(A - \lambda I)^{-1}\| < M$ for all $\lambda \in U$.*

Proof. A little thought reveals that what we must show is the following: If U is an open subset of $\mathbf{C}$ containing the origin and $\|(A - \lambda I)^{-1}\| \le M$ for all $\lambda \in U$, then $\|A^{-1}\| < M$.

To prove this, assume the contrary, i.e., let $\|A^{-1}\| = M$. There is a sufficiently small $r > 0$ such that

$$(A - \lambda I)^{-1} = \sum_{j=0}^{\infty} \lambda^j A^{-j-1} \text{ for } |\lambda| = r.$$

Given $f \in H$, we therefore get

$$\|(A - \lambda I)^{-1} f\|^2 = \sum_{j,k \geq 0} \lambda^j \overline{\lambda}^k (A^{-j-1} f, A^{-k-1} f)$$

whenever $\lambda = re^{i\varphi}$. Integrating the last equality we obtain

$$\frac{1}{2\pi} \int_0^{2\pi} \|(A - re^{i\varphi} I)^{-1} f\|^2 d\varphi = \sum_{j=0}^{\infty} r^{2j} \|A^{-j-1} f\|^2,$$

and since $\|(A - re^{i\varphi} I)^{-1} f\| \leq M\|f\|$, it follows that $\|A^{-1} f\|^2 + r^2 \|A^{-2} f\|^2 \leq M^2 \|f\|^2$. Now pick an arbitrary $\varepsilon > 0$ and choose an $f_\varepsilon \in H$ such that $\|f_\varepsilon\| = 1$ and $\|A^{-1} f_\varepsilon\|^2 > M^2 - \varepsilon$. Then $M^2 - \varepsilon + r^2 \|A^{-2} f_\varepsilon\|^2 < M^2$, whence $1 = \|f_\varepsilon\|^2 \leq \|A^2\|^2 \|A^{-2} f_\varepsilon\|^2 < \varepsilon r^{-2} \|A^2\|^2$, which is impossible if $\varepsilon > 0$ is small enough. □

The following result extends Theorem 7.5 to operators on ℓ^p for $1 < p < \infty$.

Theorem 7.6. *Let $1 < p < \infty$ and let $A \in \mathcal{B}(\ell^p)$. If $A - \lambda I$ is invertible for all λ in some open set $U \subset \mathbf{C}$ and $\|(A - \lambda I)^{-1}\|_p \leq M < \infty$ for $\lambda \in U$, then $\|(A - \lambda I)^{-1}\|_p < M$ for $\lambda \in U$.*

Proof. We may without loss of generality assume that $p \geq 2$; otherwise we can pass to adjoint operators. Again it suffices to show that $\|A^{-1}\|_p < M$ provided U contains the origin. Assume the contrary, that is, let $\|A^{-1}\|_p = M$. There is an $r > 0$ such that

$$(A - \lambda I)^{-1} = \sum_{j=0}^{\infty} \lambda^j A^{-j-1}$$

for all $\lambda = re^{i\varphi}$. Hence, for every $x \in \ell^p$,

$$\|(A - \lambda I)^{-1} x\|_p^p = \sum_{n=0}^{\infty} \left| \sum_{j=0}^{\infty} \lambda^j (A^{-j-1} x)_n \right|^p = \sum_{n=0}^{\infty} \left| \sum_{j=0}^{\infty} r^j e^{ij\varphi} (A^{-j-1} x)_n \right|^{2p/2}$$

$$= \sum_{n=0}^{\infty} \left| \left(\sum_{j=0}^{\infty} r^j e^{ij\varphi} (A^{-j-1} x)_n \right) \left(\sum_{k=0}^{\infty} r^k e^{-ik\varphi} \overline{(A^{-k-1} x)_n} \right) \right|^{p/2}$$

$$= \sum_{n=0}^{\infty} \left| C(r, n) + \sum_{\ell=1}^{\infty} B_\ell(r, \varphi, n) \right|^{p/2}, \tag{7.9}$$

where

$$C(r,n) = \sum_{j=0}^{\infty} r^{2j} |(A^{-j-1}x)_n|^2,$$

$$B_\ell(r,\varphi,n) = 2\sum_{k=0}^{\infty} r^{\ell+2k} \operatorname{Re}\left(e^{i\ell\varphi}(A^{-\ell-k-1}x)_n \overline{(A^{-k-1}x)_n}\right).$$

For $m = 0, 1, 2, \ldots$, put

$$I_m(r,\varphi) = \sum_{n=0}^{\infty} \left| C(r,n) + \sum_{\ell=1}^{\infty} B_{2^m\ell}(r,\varphi,n) \right|^{p/2}.$$

Clearly,

$$\lim_{m\to\infty} I_m(r,\varphi) = \sum_{n=0}^{\infty} |C(r,n)|^{p/2}. \tag{7.10}$$

We now apply the inequality

$$|a|^{p/2} \le \frac{1}{2}\left(|a+b|^{p/2} + |a-b|^{p/2}\right) \tag{7.11}$$

to

$$a = C(r,n) + \sum_{\ell=1}^{\infty} B_{2\ell}(r,\varphi,n), \quad b = \sum_{\ell=1}^{\infty} B_{2\ell-1}(r,\varphi,n)$$

and sum up the results for $n = 0, 1, 2, \ldots$. Taking into account that $\sum_{n=0}^{\infty} |a-b|^{p/2}$ is nothing but $I_0(r,\varphi+\pi)$, we get

$$I_1(r,\varphi) \le \frac{1}{2}\Big(I_0(r,\varphi) + I_0(r,\varphi+\pi)\Big). \tag{7.12}$$

Letting

$$a = C(r,n) + \sum_{\ell=1}^{\infty} B_{4\ell}(r,\varphi,n), \quad b = \sum_{\ell=1}^{\infty} B_{4\ell-2}(r,\varphi,n)$$

in (7.11), we analogously obtain that

$$I_2(r,\varphi) \le \frac{1}{2}\Big(I_1(r,\varphi) + I_1(r,\varphi+\pi/2)\Big). \tag{7.13}$$

Combining (7.12) and (7.13) we arrive at the inequality

$$\begin{aligned} I_2(r,\varphi) &\le \frac{1}{2}\Big(I_1(r,\varphi) + I_1(r,\varphi+\pi/2)\Big) \\ &\le \frac{1}{4}\Big(I_0(r,\varphi) + I_0(r,\varphi+\pi/2) + I_0(r,\varphi+\pi) + I_0(r,\varphi+3\pi/2)\Big). \end{aligned} \tag{7.14}$$

In the same way we see that

$$I_3(r,\varphi) \le \frac{1}{2}\Big(I_2(r,\varphi) + I_2(r,\varphi+\pi/4)\Big),$$

which together with (7.14) gives

$$I_3(r,\varphi) \le \frac{1}{8}\sum_{k=0}^{7} I_0\left(r,\varphi+\frac{k\pi}{4}\right).$$

Continuing this procedure we get

$$I_m(r,\varphi) \le \frac{1}{2^m}\sum_{k=0}^{2^m-1} I_0\left(r,\varphi+\frac{k\pi}{2^{m-1}}\right) \tag{7.15}$$

for every $m \ge 0$.

Now put $\varphi = 0$ in (7.15) and pass to the limit $m \to \infty$. The limit of the left-hand side is given by (7.10). The right-hand side is an integral sum and hence

$$\lim_{m\to\infty}\frac{1}{2^m}\sum_{k=0}^{2^m-1} I_0\left(r,\frac{k\pi}{2^{m-1}}\right) = \lim_{m\to\infty}\frac{1}{2\pi}\frac{2\pi}{2^m}\sum_{k=0}^{\infty} I_0\left(r,\frac{2k\pi}{2^m}\right) = \frac{1}{2\pi}\int_0^{2\pi} I_0(r,\varphi)d\varphi.$$

Thus,

$$\sum_{n=0}^{\infty}|C(r,n)|^{p/2} \le \frac{1}{2\pi}\int_0^{2\pi} I_0(r,\varphi)d\varphi.$$

Since $\|(A-\lambda I)^{-1}\|_p \le M$, we have $I_0(r,\varphi) = \|(A-re^{i\varphi}I)^{-1}x\|_p^p \le M^p\|x\|_p^p$. Consequently,

$$\sum_{n=0}^{\infty}|C(r,n)|^{p/2} \le M^p\|x\|_p^p. \tag{7.16}$$

Because

$$\sum_{n=0}^{\infty}|C(r,n)|^{p/2} \ge \sum_{n=0}^{\infty}\Big|\,|(A^{-1}x)_n|^2 + r^2\,|(A^{-2}x)_n|^2\,\Big|^{p/2}$$

$$\ge \sum_{n=0}^{\infty}|(A^{-1}x)_n|^p + r^p\sum_{n=0}^{\infty}|(A^{-2}x)_n|^p = \|A^{-1}x\|_p^p + r^p\|A^{-2}x\|_p^p$$

(here we used the inequality $(|a|+|b|)^{p/2} \ge |a|^{p/2} + |b|^{p/2}$), we deduce from (7.16) that

$$\|A^{-1}x\|_p^p + r^p\|A^{-2}x\|_p^p \le M^p\|x\|_p^p. \tag{7.17}$$

Finally, let $\varepsilon > 0$ and choose $x_\varepsilon \in \ell^p$ so that $\|x_\varepsilon\|_p = 1$ and $\|A^{-1}x\|_p^p > M^p - \varepsilon$. Then (7.17) yields $M^p - \varepsilon + r^p\|A^{-2}x_\varepsilon\|_p^p < M^p$, and this implies that

$$1 = \|x_\varepsilon\|_p^p \le \|A^2\|_p^p\|A^{-2}x_\varepsilon\|_p^p < \varepsilon r^{-p}\|A^2\|_p^p.$$

This inequality is impossible if $\varepsilon > 0$ is sufficiently small. Thus, our assumption $\|A^{-1}\|_p = M$ must be false. □

7.3 Limits of Pseudospectra

Let $\{M_n\}_{n=1}^{\infty}$ be a sequence of sets $M_n \subset \mathbf{C}$. We define

$$\liminf_{n\to\infty} M_n$$

as the set of all $\lambda \in \mathbf{C}$ for which there are $\lambda_1 \in M_1, \lambda_2 \in M_2, \dots$ such that $\lambda_n \to \lambda$, and we let

$$\limsup_{n\to\infty} M_n$$

denote the set of all $\lambda \in \mathbf{C}$ for which there exist $n_1 < n_2 < \cdots$ and $\lambda_{n_k} \in M_{n_k}$ such that $\lambda_{n_k} \to \lambda$. In other words, $\lambda \in \liminf M_n$ if and only if λ is the limit of some sequence $\{\lambda_n\}_{n=1}^{\infty}$ with $\lambda_n \in M_n$, while $\lambda \in \limsup M_n$ if and only if λ is a partial limit of such a sequence. We remark that if M and the members of the sequence $\{M_n\}$ are nonempty compact subsets of $\mathbf{C}$, then

$$\liminf_{n\to\infty} M_n = \limsup_{n\to\infty} M_n = M$$

if and only if M_n converges to M in the Hausdorff metric, which means that $d(M_n, M) \to 0$ with

$$d(A, B) := \max\left(\max_{a\in A} \operatorname{dist}(a, B), \max_{b\in B} \operatorname{dist}(b, A)\right).$$

This result is due to Hausdorff. Proofs can be found in [149, Sections 3.1.1 and 3.1.2] and [153, Section 2.8].

Theorem 7.7. *Let b be a Laurent polynomial. Then for every $\varepsilon > 0$ and every $p \in (1, \infty)$,*

$$\liminf_{n\to\infty} \operatorname{sp}_\varepsilon^{(p)} T_n(b) = \limsup_{n\to\infty} \operatorname{sp}_\varepsilon^{(p)} T_n(b) = \operatorname{sp}_\varepsilon^{(p)} T(b) \cup \operatorname{sp}_\varepsilon^{(p)} T(\widetilde{b}). \tag{7.18}$$

Proof. We first show that

$$\operatorname{sp}_\varepsilon^{(p)} T(b) \subset \liminf_{n\to\infty} \operatorname{sp}_\varepsilon^{(p)} T_n(b). \tag{7.19}$$

If $\lambda \in \operatorname{sp} T(b)$, then $\|T_n^{-1}(b-\lambda)\|_p \to \infty$ by virtue of Lemma 3.4. Thus, we have $\|T_n^{-1}(b-\lambda)\|_p \geq 1/\varepsilon$ for all $n \geq n_0$, which implies that $\lambda \in \operatorname{sp}_\varepsilon^{(p)} T_n(b)$ for all $n \geq n_0$. Consequently, λ belongs to $\liminf \operatorname{sp}_\varepsilon^{(p)} T_n(b)$.

Now suppose that $\lambda \in \operatorname{sp}_\varepsilon^{(p)} T(b) \setminus \operatorname{sp} T(b)$. Then $\|T^{-1}(b-\lambda)\|_p \geq 1/\varepsilon$. Let $U \subset \mathbf{C}$ be any open neighborhood of λ. From Theorem 7.6 we deduce that there is a point $\mu \in U$ such that $\|T^{-1}(b-\mu)\|_p > 1/\varepsilon$. Hence, we can find a natural number k_0 such that

$$\|T^{-1}(b-\mu)\|_p \geq \frac{1}{\varepsilon - 1/k} \quad \text{for all } k \geq k_0.$$

As U was arbitrary, it follows that there exists a sequence $\mu_1, \mu_2, \dots$ such that $\mu_k \in \operatorname{sp}_{\varepsilon-1/k}^{(p)} T(b)$ and $\mu_k \to \lambda$. For every invertible operator $A \in \mathcal{B}(\ell^p)$,

$$\|A^{-1}\|_p = \sup_{x\neq 0} \frac{\|A^{-1}x\|_p}{\|x\|_p} = \sup_{y\neq 0} \frac{\|y\|_p}{\|Ay\|_p} = \left(\inf_{y\neq 0} \frac{\|Ay\|_p}{\|y\|_p}\right)^{-1}. \tag{7.20}$$

Since $\|T^{-1}(b-\mu_k)\|_p \geq 1/(\varepsilon - 1/k)$, it results that

$$\inf_{\|y\|_p=1} \|T(b-\mu_k)y\|_p \leq \varepsilon - 1/k.$$

Thus, there are $y_k \in \ell^p$ such that $\|y_k\|_p = 1$ and $\|T(b-\mu_k)y_k\|_p < \varepsilon - 1/(2k)$. Clearly, $\|T_n(b-\mu_k)P_n y_k\|_p \to \|T(b-\mu_k)y_k\|_p$ and $\|P_n y_k\|_p \to \|y_k\|_p = 1$ as $n \to \infty$. Hence,

$$\frac{\|T_n(b-\mu_k)P_n y_k\|_p}{\|P_n y_k\|_p} < \varepsilon - 1/(3k)$$

for all $n > n_0(k)$. Again invoking (7.20) we see that

$$\|T_n^{-1}(b-\mu_k)\|_p > (\varepsilon - 1/(3k))^{-1} > 1/\varepsilon$$

and thus $\mu_k \in \mathrm{sp}_\varepsilon^{(p)} T_n(b)$ for all $n > n_0(k)$. This implies that $\lambda = \lim \mu_k$ belongs to $\liminf \mathrm{sp}_\varepsilon^{(p)} T_n(b)$. At this point the proof of (7.19) is complete.

Repeating the above reasoning with $\widetilde{b}$ in place of b we get the inclusion

$$\mathrm{sp}_\varepsilon^{(p)} T(\widetilde{b}) \subset \liminf_{n\to\infty} \mathrm{sp}_\varepsilon^{(p)} T_n(\widetilde{b}). \tag{7.21}$$

As $T_n(\widetilde{b}-\lambda) = W_n T_n(b-\lambda) W_n$ and W_n is an isometry of ℓ_n^p, it is clear that $\mathrm{sp}_\varepsilon^{(p)} T_n(\widetilde{b}) = \mathrm{sp}_\varepsilon^{(p)} T_n(b)$. Thus, in (7.21) we may replace the $T_n(\widetilde{b})$ on the right by $T_n(b)$, which in conjunction with (7.19) proves that $\mathrm{sp}_\varepsilon^{(p)} T(b) \cup \mathrm{sp}_\varepsilon^{(p)} T(\widetilde{b})$ is contained in $\liminf \mathrm{sp}_\varepsilon^{(p)} T_n(b)$.

We are left to prove the inclusion

$$\limsup_{n\to\infty} \mathrm{sp}_\varepsilon^{(p)} T_n(b) \subset \mathrm{sp}_\varepsilon^{(p)} T(b) \cup \mathrm{sp}_\varepsilon^{(p)} T(\widetilde{b}). \tag{7.22}$$

So let $\lambda \notin \mathrm{sp}_\varepsilon^{(p)} T(b) \cup \mathrm{sp}_\varepsilon^{(p)} T(\widetilde{b})$. Then $\|T^{-1}(b-\lambda)\|_p < 1/\varepsilon$ and $\|T^{-1}(\widetilde{b}-\lambda)\|_p < 1/\varepsilon$, whence, by Theorem 6.3,

$$\|T_n^{-1}(b-\lambda)\|_p < 1/\varepsilon - \delta < 1/\varepsilon \quad \text{for all } n \geq n_0 \tag{7.23}$$

with some $\delta > 0$. If $|\mu - \lambda|$ is sufficiently small, then $T_n(b-\mu)$ is invertible together with $T_n(b-\lambda)$, and we have, from the first resolvent identity,

$$\|T_n^{-1}(b-\mu)\|_p \leq \frac{\|T_n^{-1}(b-\lambda)\|_p}{1 - |\mu-\lambda|\, \|T_n^{-1}(b-\lambda)\|_p}. \tag{7.24}$$

Let $|\mu - \lambda| < \varepsilon\delta(1/\varepsilon - \delta)^{-1}$. In this case (7.23) and (7.24) give

$$\|T_n^{-1}(b-\mu)\|_p < \frac{1/\varepsilon - \delta}{1 - \varepsilon\delta(1/\varepsilon-\delta)^{-1}(1/\varepsilon - \delta)} = \frac{1}{\varepsilon}.$$

Thus, $\mu \notin \mathrm{sp}_\varepsilon^{(p)} T_n(b)$ for $n \geq n_0$. This shows that λ cannot belong to the left-hand side of (7.22). □

Corollary 7.8. *If b is a Laurent polynomial and $\varepsilon > 0$, then*

$$\liminf_{n\to\infty} \mathrm{sp}_\varepsilon^{(2)} T_n(b) = \limsup_{n\to\infty} \mathrm{sp}_\varepsilon^{(2)} T_n(b) = \mathrm{sp}_\varepsilon^{(2)} T(b).$$

Proof. Since $T(\widetilde{b})$ is simply the transpose of $T(b)$, the two norms $\|T^{-1}(b-\lambda)\|_2$ and $\|T^{-1}(\widetilde{b}-\lambda)\|_2$ coincide. Therefore $\mathrm{sp}_\varepsilon^{(2)} T(b) = \mathrm{sp}_\varepsilon^{(2)} T(\widetilde{b})$. The assertion is now immediate from Theorem 7.7. □

Figures 7.1 and 7.2 show an example with the symbol $b(t) = -(6-13i)t + (5-4i)t^2 + 3t^3 - (4+3i)t^{-2} + 3t^{-3}$.

7.4 Pseudospectra of Infinite Toeplitz Matrices

For every operator $A \in \mathcal{B}(\ell^2)$ the inequality

$$1/\mathrm{dist}\,(\lambda, \mathrm{sp}\, A) \le \|(A-\lambda I)^{-1}\|_2 \tag{7.25}$$

holds. This implies that $\|(A-\lambda I)^{-1}\|_2 \ge 1/\varepsilon$ whenever $\mathrm{dist}\,(\lambda, \mathrm{sp}\, A) \le \varepsilon$ and hence yields the universal lower estimate

$$\mathrm{sp}\, A + \varepsilon \overline{\mathbf{D}} \subset \mathrm{sp}_\varepsilon^{(2)} A.$$

For Toeplitz operators, Theorem 4.29 gives

$$\|(T(a)-\lambda I)^{-1}\|_2 = \|T^{-1}(a-\lambda)\|_2 \le 1/\mathrm{dist}\,(\lambda, \mathrm{conv}\, \mathcal{R}(a)). \tag{7.26}$$

Consequently, if $\mathrm{dist}\,(\lambda, \mathrm{conv}\, \mathcal{R}(a)) > \varepsilon$ then $\|T^{-1}(a-\lambda)\|_2 < 1/\varepsilon$ and λ cannot belong to $\mathrm{sp}_\varepsilon^{(2)} T(a)$. We therefore arrive at the upper estimate

$$\mathrm{sp}_\varepsilon^{(2)} T(a) \subset \mathrm{conv}\, \mathcal{R}(a) + \varepsilon \overline{\mathbf{D}}.$$

Given $a \in W$, let $V(a)$ be the set of all $\lambda \in \mathbf{C} \setminus \mathrm{sp}\, T(a)$ for which

$$\mathrm{dist}\,(\lambda, \mathrm{sp}\, T(a)) = \mathrm{dist}\,(\lambda, \mathrm{conv}\, \mathcal{R}(a)).$$

From (7.25) and (7.26) we infer that if $\lambda \in V(a)$, then

$$\|T^{-1}(a-\lambda)\|_2 = 1/\mathrm{dist}\,(\lambda, \mathrm{conv}\, \mathcal{R}(a))$$

and hence in $V(a)$ the level curves $\|T^{-1}(a-\lambda)\|_2 = 1/\varepsilon$ coincide with the curves $\mathrm{dist}\,(\lambda, \mathrm{conv}\, \mathcal{R}(a)) = \varepsilon$. If $V(a) = \mathbf{C} \setminus \mathrm{sp}\, T(a)$, or equivalently, if $\mathrm{sp}\, T(a)$ is a convex set, then

$$\mathrm{sp}_\varepsilon^{(2)} T(a) = \mathrm{conv}\, \mathcal{R}(a) + \varepsilon \overline{\mathbf{D}}. \tag{7.27}$$

Equality (7.27) is particularly true for tridiagonal Toeplitz matrices, in which case $\mathrm{sp}\, T(a) = \mathrm{conv}\, \mathcal{R}(a)$ is an ellipse.

7.5 Numerical Range

Let X be a Banach space and put

$$\Pi(X) = \{(f, x) \in X^* \times X : \|f\| = 1,\ \|x\| = 1,\ f(x) = 1\}.$$

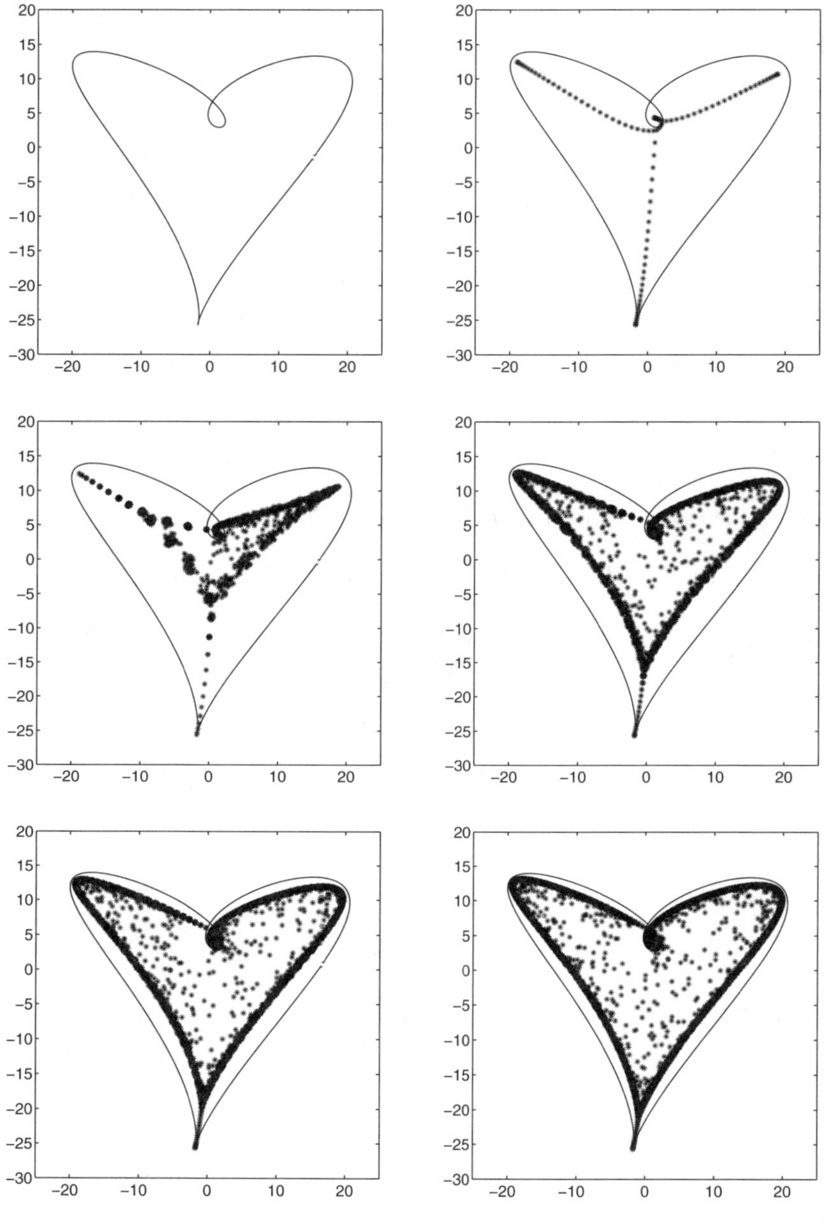

Figure 7.1. *In the two top pictures we see* $b(\mathbf{T})$ *(left) and* $b(\mathbf{T})$ *together with the* 100 *eigenvalues of* $T_{100}(b)$ *(right). The other four pictures indicate* $\text{sp}_\varepsilon^{(2)} T_n(b)$ *for* $\varepsilon = 1/100$ *and* $n = 50, 100, 150, 200$. *Each picture shows the superposition of the spectra* $\text{sp}\,(T_n(b) + E)$ *for* 50 *randomly chosen matrices* E *with* $\|E\|_2 = \varepsilon$.

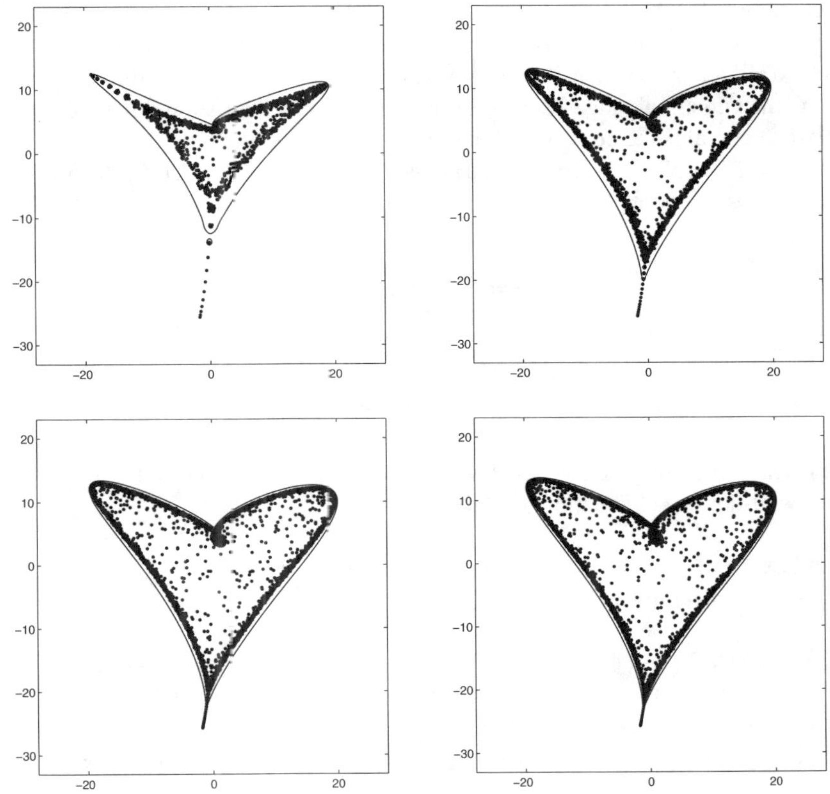

Figure 7.2. *These pictures were done by Mark Embree. In contrast to the lower four pictures of Figure 7.1, the solid curves are the boundaries of the pseudospectra* $\mathrm{sp}_\varepsilon^{(2)} T_n(b)$ *for* $\varepsilon = 1/100$ *and* $n = 50, 100, 150, 200$. *These curves were determined with the help of Tom Wright's package* [299].

The (spatial) *numerical range* (*=field of values*) $\mathcal{H}_X(A)$ of an operator $A \in \mathcal{B}(X)$ is defined as

$$\mathcal{H}_X(A) = \{f(Ax) : (f, x) \in \Pi(X)\}.$$

If $X = H$ is a Hilbert space, we can identify the dual space X^* with H and, accordingly,

$$\Pi(H) = \{(y, x) \in H \times H : \|y\| = 1, \|x\| = 1, (y, x) = 1\}.$$

Since equality holds in the Cauchy-Schwarz inequality $|(y, x)| \le \|y\|\,\|x\|$ if and only if y and x are linearly dependent, we see that actually

$$\Pi(H) = \{(x, x) \in H \times H : \|x\| = 1\}.$$

This implies that in the Hilbert space case the numerical range may also be defined by

$$\mathcal{H}_H(A) = \{(Ax, x) : \|x\| = 1\}.$$

It is well known that $\mathcal{H}_X(A)$ is always a bounded and connected set whose closure contains the spectrum of A: $\mathrm{sp}\, A \subset \mathrm{clos}\, \mathcal{H}_X(A)$. The Toeplitz-Hausdorff-Stone theorem says that if $X = H$ is a Hilbert space, then $\mathcal{H}_H(A)$ is necessarily convex. For finite-dimensional Banach spaces X, the numerical range $\mathcal{H}_X(A)$ is obviously closed. This shows that if $X = H = \ell_n^2$, then $\mathcal{H}_H(A)$ contains the convex hull of the eigenvalues of A.

If $X = \ell^p$ or $X = \ell_n^p$, we denote $\mathcal{H}_X(A)$ by $\mathcal{H}_p(A)$. If $A \in \mathcal{B}(\ell^p)$, then $P_n A P_n$ may be thought of as an operator on ℓ_n^p. The purpose of this section is to show that if $1 < p < \infty$, then always

$$\liminf_{n\to\infty} \mathcal{H}_p(P_n A P_n) = \limsup_{n\to\infty} \mathcal{H}_p(P_n A P_n) = \mathrm{clos}\, \mathcal{H}_p(A).$$

In particular,

$$\liminf_{n\to\infty} \mathcal{H}_p(T_n(b)) = \limsup_{n\to\infty} \mathcal{H}_p(T_n(b)) = \mathrm{clos}\, \mathcal{H}_p(T(b)).$$

The dual space of ℓ^p may be identified with ℓ^q $(1/p + 1/q = 1)$. Thus, for f in $(\ell^p)^* = \ell^q$, $P_n f$ is a well-defined element of $(\ell^p)^* = \ell^q$.

Lemma 7.9. *Let $1 < p < \infty$. If $(f, x) \in \Pi(\ell^p)$, then $(P_n f/\|P_n f\|_q, P_n x/\|P_n x\|_p)$ is in $\Pi(\ell_n^p)$ for all sufficiently large n.*

Proof. Since $\|P_n f\|_q \to \|f\|_q = 1$ and $\|P_n x\|_p \to \|x\|_p = 1$, it follows that $\|P_n f\|_q \neq 0$ and $\|P_n x\|_p \neq 0$ for all n large enough. Put $f_n = P_n f/\|P_n f\|_q$ and $x_n = P_n x/\|P_n x\|_p$. We are left with showing that $f_n(x_n) = 1$.

Let $x = \{r_0 e^{i\varphi_0}, r_1 e^{i\varphi_1}, \ldots\}$ with $0 \le r_j < \infty$ and $0 \le \varphi_j < 2\pi$. By assumption,

$$\|x\|_p = (r_0^p + r_1^p + \cdots)^{1/p} = 1.$$

Set $g = \{r_0^{p/q} e^{-i\varphi_0}, r_1^{p/q} e^{-i\varphi_1}, \ldots\}$. Then

$$\|g\|_q = (r_0^p + r_1^p + \cdots)^{1/q} = 1$$

and

$$g(x) = r_0^{p/q} r_0 + r_1^{p/q} r_1 + \cdots = r_0^p + r_1^p + \cdots = 1.$$

Thus, $(g, x) \in \Pi(\ell^p)$.

The space ℓ^q is uniformly convex. This means that if $\|h_1\|_q = \|h_2\|_q = 1$ and $\|h_1 + h_2\|_q = 2$, then $h_1 = h_2$. Since $\|f\|_q = \|g\|_q = 1$ and

$$\|f + g\|_q \ge |f(x) + g(x)| = |1 + 1| = 2,$$

we arrive at the conclusion that $f = g$. Consequently,

$$\begin{aligned}
&\|P_n f\|_q \|P_n x\|_p = \|P_n g\|_q \|P_n x\|_p \\
&= \left(r_0^p + r_1^p + \cdots + r_{n-1}^p\right)^{1/q} \left(r_0^p + r_1^p + \cdots + r_{n-1}^p\right)^{1/p} \\
&= r_0^p + r_1^p + \cdots + r_{n-1}^p \\
&= r_0^{p/q} r_0 + r_1^{p/q} r_1 + \cdots + r_{n-1}^{p/q} r_{n-1} = (P_n g)(P_n x) = (P_n f)(P_n x),
\end{aligned}$$

which is equivalent to the desired equality $f_n(x_n) = 1$. □

Theorem 7.10 (Roch). *Let* $1 < p < \infty$ *and* $A \in \mathcal{B}(\ell^p)$. *Then*

$$\liminf_{n\to\infty} \mathcal{H}_p(P_nAP_n) = \limsup_{n\to\infty} \mathcal{H}_p(P_nAP_n) = \operatorname{clos} \mathcal{H}_p(A).$$

Proof. On regarding ℓ_n^p as a subspace of ℓ_m^p for $n \le m$, we have

$$\Pi(\ell_n^p) \subset \Pi(\ell_m^p) \subset \Pi(\ell^p),$$

whence

$$\mathcal{H}_p(P_nAP_n) \subset \mathcal{H}_p(P_mAP_m) \subset \mathcal{H}_p(A).$$

This shows that

$$\limsup_{n\to\infty} \mathcal{H}_p(P_nAP_n) \subset \operatorname{clos} \mathcal{H}_p(A).$$

To prove the reverse inclusion, let $(f, x) \in \Pi(\ell^p)$ and define $(f_n, x_n) \in \Pi(\ell_n^p)$ as in (the proof of) Lemma 7.9. Since $\|f_n - f\|_q \to 0$ and $\|x_n - x\|_p \to 0$, we obtain that

$$f(Ax) = \lim_{n\to\infty} f_n(Ax_n) = \lim_{n\to\infty} f_n(P_nAP_nx_n).$$

From Lemma 7.9 we infer that $(f_n, x_n) \in \Pi(\ell_n^p)$. Thus,

$$\mathcal{H}_p(A) \subset \liminf_{n\to\infty} \mathcal{H}_p(P_nAP_n),$$

and since limiting sets are always closed, it results that

$$\operatorname{clos} \mathcal{H}_p(A) \subset \liminf_{n\to\infty} \mathcal{H}_p(P_nAP_n). \qquad \square$$

In the case $p = 2$ and $A = T(a)$, the limit in Theorem 7.10 is known.

Theorem 7.11. *If* $a \in W$, *then*

$$\operatorname{clos} \mathcal{H}_2(T(a)) = \operatorname{conv} \operatorname{sp} T(a) = \operatorname{conv} a(\mathbf{T}).$$

Proof. Let $M(a) : L^2 \to L^2$ be the operator of multiplication by a. From Section 1.6 we know that $\Phi^{-1}T(a)\Phi = PM(a)|H^2$, where P is the orthogonal projection of L^2 onto H^2. This implies that

$$\begin{aligned} \mathcal{H}_2(T(a)) &= \mathcal{H}_{H^2}(\Phi^{-1}T(a)\Phi) = \mathcal{H}_{H^2}(PM(a)|H^2) \\ &= \{(PM(a)f, f) : f \in H^2, \|f\|_2 = 1\} \\ &= \{(M(a)f, f) : f \in H^2, \|f\|_2 = 1\} \quad (\text{since } P^*f = Pf = f) \\ &\subset \{(M(a)g, g) : g \in L^2, \|g\|_2 = 1\} = \mathcal{H}_{L^2}(M(a)). \end{aligned} \tag{7.28}$$

The closure of the numerical range of a normal operator is the convex hull of its spectrum (see, e.g., [150, Problem 171]). As $M(a)$ is normal, we deduce that

$$\operatorname{clos} \mathcal{H}_{L^2}(M(a)) = \operatorname{conv} \operatorname{sp} M(a) = \operatorname{conv} a(\mathbf{T}). \tag{7.29}$$

Consequently,

$$\begin{aligned} \operatorname{clos} \mathcal{H}_2(T(a)) &\subset \operatorname{clos} \mathcal{H}_{L^2}(M(a)) \quad \text{(by (7.28))} \\ &= \operatorname{conv} a(\mathbf{T}) \quad \text{(by (7.29))} \\ &\subset \operatorname{conv} \operatorname{sp} T(a) \quad \text{(by Corollary 1.12)} \\ &\subset \operatorname{clos} \mathcal{H}_2(T(a)) \quad \text{(since always } \operatorname{sp} A \subset \operatorname{clos} \mathcal{H}_X(A)), \end{aligned}$$

which gives the assertion. □

7.6 Collective Perturbations

Let $\mathbf{G}$ be the collection of all sequences $\{G_n\}_{n=1}^\infty$ of complex $n \times n$ matrices G_n such that $\|G_n\|_2 \to 0$ as $n \to \infty$.

Theorem 7.12 (Roch). *If $a \in W$ then*

$$\bigcup_{\{G_n\} \in \mathbf{G}} \limsup_{n \to \infty} \operatorname{sp} (T_n(a) + G_n) = \operatorname{sp} T(a).$$

Proof. Let $\lambda \notin \operatorname{sp} T(a)$. Then, by Theorem 3.7, $\{T_n(a - \lambda)\}$ is a stable sequence:

$$\limsup_{n \to \infty} \|T_n^{-1}(a - \lambda)\|_2 < \infty.$$

It follows that if $\|G_n\|_2 \to 0$ and μ is in some sufficiently small open neighborhood U of λ, then

$$\limsup_{n \to \infty} \| \, (T_n(a - \mu) + G_n)^{-1} \, \|_2 < \infty.$$

This implies that $U \cap \operatorname{sp} (T_n(a) + G_n) = \emptyset$ for all sufficiently large n, whence

$$\lambda \notin \limsup_{n \to \infty} \operatorname{sp} (T_n(a) + G_n).$$

Now take $\lambda \in \operatorname{sp} T(a)$. By virtue of Theorem 3.7, $\{T_n(a - \lambda)\}$ is not stable. If $T_{n_k}(a - \lambda)$ is not invertible for infinitely many n_k, then

$$\lambda \in \limsup_{n \to \infty} \operatorname{sp} T_n(a).$$

So assume $T_n(a - \lambda)$ is invertible for all $n \ge n_0$ but

$$\limsup_{n \to \infty} \|T_n^{-1}(a - \lambda)\|_2 = \infty.$$

There are $n_1 < n_2 < n_3 < \cdots$ and $x_{n_k} \in \ell^2_{n_k}$ such that

$$\|x_{n_k}\|_2 = 1 \text{ and } \|T_{n_k}^{-1}(a-\lambda)x_{n_k}\|_2 \geq k.$$

Put $y_{n_k} = T_{n_k}^{-1}(a-\lambda)x_{n_k}$ and let G_{n_k} be the matrix of the linear operator

$$G_{n_k} : \ell^2_{n_k} \to \ell^2_{n_k}, \ z \mapsto -\frac{(z, y_{n_k})x_{n_k}}{\|y_{n_k}\|_2^2}.$$

Obviously,

$$\|G_{n_k}\|_2 = \sup_{\|z\|_2=1} \frac{|(z, y_{n_k})| \, \|x_{n_k}\|_2}{\|y_{n_k}\|_2^2} \leq \frac{\|x_{n_k}\|_2}{\|y_{n_k}\|_2} \leq \frac{1}{k}.$$

Let $G_n = 0$ for $n \in \mathbf{N} \setminus \{n_1, n_2, \dots\}$ Then $\{G_n\} \in \mathbf{G}$. Since

$$\left(T_{n_k}(a) + G_{n_k} - \lambda I\right) y_{n_k} = T_{n_k}(a-\lambda)y_{n_k} + G_{n_k}y_{n_k} = x_{n_k} - x_{n_k} = 0,$$

it follows that $T_{n_k} + G_{n_k} - \lambda I$ is not invertible. Hence

$$\lambda \in \limsup_{n\to\infty} \operatorname{sp}(T_n(a) + G_n). \qquad \square$$

Exercises

1. Let K be a compact operator on ℓ^2. Prove that

$$\liminf_{n\to\infty} \operatorname{sp}(I + P_n K P_n) = \limsup_{n\to\infty} \operatorname{sp}(I + P_n K P_n) = \operatorname{sp}(I + K).$$

2. Let a and b satisfy $0 < a \leq b \leq \infty$. Show that there exists a selfadjoint operator $A \in \mathcal{B}(\ell^2)$ such that

$$\|A^{-1}\|_2 = a \quad \text{and} \quad \limsup_{n\to\infty} \|(P_n A P_n)^{-1} P_n\|_2 = b.$$

3. Let A be a selfadjoint operator on ℓ^2 and suppose that $\operatorname{sp}_{\text{ess}} A$ is a connected set. Put $A_n = P_n A P_n | \operatorname{Im} P_n$. Prove that

$$\liminf_{n\to\infty} \operatorname{sp} A_n = \limsup_{n\to\infty} \operatorname{sp} A_n = \operatorname{sp} A,$$
$$\lim_{n\to\infty} \|(A_n - \lambda I)^{-1}\|_2 = \|(A - \lambda I)^{-1}\|_2 \quad (\lambda \in \mathbf{C} \setminus \operatorname{sp} A).$$

4. For $m \geq 0$, let $A : \ell^2 \to \ell^2$ be the operator

$$A : \{x_0, x_1, x_2, \dots\} \mapsto \{x_m, x_{m+1}, \dots\}.$$

Prove that $\operatorname{sp}_\varepsilon^{(2)} A = (1+\varepsilon)\overline{\mathbf{D}}$.

5. Let $m \geq 2$ and $b(t) = t + t^{-m}/m$. Find the set $V(b)$ and show that

$$\{\lambda \in V(b) : \|T^{-1}(b - \lambda)\|_2 = 1/\varepsilon\}$$

is the union of $n + 1$ pure circular arcs of curvature $1/\varepsilon$.

6. Let A_n be an $n \times n$ matrix and put

$$R_i'(A_n) = \left(\sum_{j=1}^{n} |a_{ij}| \right) - |a_{ii}|.$$

Show that

$$\mathrm{sp}_\varepsilon^{(p)}(A_n) \subset \bigcup_{i=1}^{n} \{\lambda \in \mathbf{C} : |\lambda - a_{ii}| \leq R_i'(A_n) + \varepsilon n\}.$$

7. Let $\varepsilon > 0$. Show that there exist Laurent polynomials b and c such that

$$\mathrm{sp}\, T(b) + \varepsilon \overline{\mathbf{D}} \neq \mathrm{sp}_\varepsilon^{(2)} T(b), \quad \mathrm{sp}_\varepsilon^{(2)} T(c) \neq \mathrm{conv}\, \mathcal{R}(c) + \varepsilon \overline{\mathbf{D}}.$$

8. Show that there exist finite matrices A and B such that

$$\|(\lambda I - A)^{-1}\|_2 = \|(\lambda I - B)^{-1}\|_2 \text{ for all } \lambda \in \mathbf{C}$$

but $\|p(A)\|_2 \neq \|p(B)\|_2$ for some polynomial p.

9. Show that the numerical range is robust in the following sense: If $E \in \mathcal{B}(X)$ and $\|E\| \leq \varepsilon$, then $\mathcal{H}_X(A + E) \subset \mathcal{H}_X(A) + \varepsilon \overline{\mathbf{D}}$.

10. Let A be an $n \times n$ matrix. The numbers

$$\alpha_{\mathcal{H}}(A) = \max_{\lambda \in \mathcal{H}_2(A)} \mathrm{Re}\, \lambda, \quad \alpha(A) = \max_{\lambda \in \mathrm{sp}\, A} \mathrm{Re}\, \lambda$$

are called the numerical and spectral abscissas of A, respectively. Prove that

$$\lim_{t \to 0+0} \frac{d}{dt} \log \|e^{tA}\|_2 = \alpha_{\mathcal{H}}(A), \quad \lim_{t \to +\infty} \frac{d}{dt} \log \|e^{tA}\|_2 = \alpha(A).$$

11. Let $a, b \in \mathcal{P} \setminus \{0\}$ and suppose that $a_0 = b_0 = 0$. Prove that the equality $T_n(a)T_n(b) = 0$ is impossible for $n \leq 3$ but possible for $n \geq 4$.

Notes

Embree and Trefethen's Web page [110] and book [275] are inexhaustible sources on all aspects of pseudospectra. The first chapter of [275] is on eigenvalues and it ends as follows: "In the highly nonnormal case, vivid though the image may be, the location of the eigenvalues may be as fragile an indicator of underlying character as the hair color of a Hollywood actor.

We shall see that pseudospectra provide equally compelling images that may capture the spirit underneath more robustly."

"In summary, eigenvalues and eigenfunctions have a distinguished history of application throughout the mathematical sciences; we could not get around without them. Their clearest successes, however, are associated with problems that involve well-behaved systems of eigenvectors, which in most contexts means matrices or operators that are normal or nearly so. This class encompasses the majority of applications, but not all of them. For nonnormal problems, the record is less clear, and even the conceptual significance of eigenvalues is open to question."

As for the history of pseudospectra, we take the liberty of citing Trefethen and Embree [275] again. "These data suggest that the notion of pseudospectra has been invented at least five times:

J. M. Varah	1967	r-approximate eigenvalue
	1979	ε-spectrum
H. J. Landau	1975	ε-approximate eigenvalues
S. K. Godunov et al.	1975	spectral portrait
L. N. Trefethen	1990	ε-pseudospectrum
D. Hinrichsen and A. J. Pritchard	1992	spectral value set

One should not trust this table too far, however, as even recent history is notoriously hard to pin down. It is entirely possible that Godunov or Wilkinson thought about pseudospectra in the 1960s, and indeed von Neumann may have thought about them in the 1930s. Nor were others such as Dunford and Schwartz, Gohberg, Halmos, Kato, Keldysch, or Kreiss far away."

The infinite Toeplitz matrix $T(b)$ is normal if and only if the essential range of b is a line segment in the complex plane [77]. Consequently, infinite Toeplitz matrices are typically nonnormal and hence pseudospectra are expected to tell us more about them than spectra. The pioneering work on pseudospectra of Toeplitz matrices is the paper [219] by Reichel and Trefethen. This paper was the source of inspiration for one of the authors' paper [34] and thus for investigations that have essentially resulted in large parts of the present book.

For $B = C = I$, that is, in the unstructured case, Theorems 7.2 and 7.4 are in principle already in [269], [270]. In the structured case, these theorems are due to Hinrichsen, Kelb, Pritchard, and Gallestey [124], [125], [162], [163]. Section 7.1 is based on ideas of [125] and follows our paper [50].

The question whether the resolvent norm $\|(A - \lambda I)^{-1}\|$ may be locally constant arose in connection with [34]. One of the authors (A. B.) posed this question as an open problem at a Banach semester in Warsaw in 1994, and a few weeks later, Andrzej Daniluk of Cracow was able to solve the problem. The proof of Theorem 7.5 is due to him. Theorem 7.6 was established in [62].

Corollary 7.8, that is, Theorem 7.7 for $p = 2$, is due to Landau [185], [186], [187] and Reichel and Trefethen [219]. The first clean proof of this result was given in [34]. For general $p \in (1, \infty)$, Theorem 7.7 was proved in [62].

The (Hilbert space) numerical range $\mathcal{H}_H(A)$ was introduced by Otto Toeplitz [268]. For more on numerical ranges, in particular for proofs of the properties quoted in the text, we refer to the books [30], [148], [150], [167], [275]. Theorem 7.10 was established by

Roch [220]. Theorem 7.12 is also Roch's; it appeared first in [149, Theorem 3.19]. Theorem 7.11 and the proof given here are due to Halmos [150]. This theorem gives us the closure of $\mathcal{H}_2(T(b))$. The set $\mathcal{H}_2(T(b))$ itself was determined by Klein [180]. There are two theorems in [180]. Theorem 1 says that if the Toeplitz operator has a nonconstant symbol and is normal so that the spectrum is a closed interval $[\gamma, \delta] \subset \mathbf{C}$, then the numerical range is the corresponding open interval (γ, δ). Theorem 2 states that if the Toeplitz operator is not normal, then its numerical range is the interior of the convex hull of its spectrum. We remark that Halmos and Klein's results are actually true for arbitrary $b \in L^\infty$.

We will say more on the numerical range of finite Toeplitz matrices in the notes to Chapter 8.

Exercise 6 is from [111]. For Exercise 7 see [71]. Exercise 8 is a result by Greenbaum and Trefethen [144] (and can also be found in [275]). A solution to Exercise 10, which shows that $\alpha_{\mathcal{H}}(A)$ and $\alpha(A)$ give the initial and final slope of the curve $t \mapsto \log \|e^{tA}\|_2$, is in [275], for example. Exercise 11 is from [147].

Further results: convergence speed. Let b be a Laurent polynomial. If λ is in $\mathbf{C} \setminus b(\mathbf{T})$ and wind $(b, \lambda) \neq 0$, then $\|T_n^{-1}(b - \lambda)\|_2$ goes to infinity at least exponentially due to Theorem 4.1. This implies that the inequality $\|T_n^{-1}(b - \lambda)\|_2 \geq 1/\varepsilon$ is already satisfied for n's of moderate size, and consequently, the convergence of $\mathrm{sp}_\varepsilon^{(2)} T_n(b)$ to $\mathrm{sp}_\varepsilon^{(2)} T(b)$, which is ensured by Corollary 7.8, is very fast. In [34], it is shown that Corollary 7.8 remains true for dense Toeplitz matrices provided the symbol b is piecewise continuous. It was observed in [45] that in the case of piecewise continuous symbols the convergence of $\mathrm{sp}_\varepsilon^{(2)} T_n(b)$ to $\mathrm{sp}_\varepsilon^{(2)} T(b)$ may be spectacularly slow, which has its reason in the fact that $\|T_n^{-1}(b - \lambda)\|_2$ may grow only polynomially. The main result of our paper [54] says that such a slow convergence of pseudospectra is generic even within the class of continuous symbols. In [54], we proved the following. Let $b \in C^2$ and let $\lambda \in \mathbf{C}$ be a point whose winding number with respect to $b(\mathbf{T})$ is -1 (respectively, 1). Then $\|T_n^{-1}(b - \lambda)\|_2$ increases faster than every polynomial,

$$\lim_{n\to\infty} \|T_n^{-1}(b - \lambda)\|_2 \, n^{-\beta} = \infty \quad \text{for each} \quad \beta > 0,$$

if and only if Pb (respectively, Qb) is in C^∞. Here $(Pb)(t) := \sum_{j=0}^{\infty} b_j t^j$ and $(Qb)(t) := \sum_{j=-\infty}^{-1} b_j t^j$ for $b(t) = \sum_{j=-\infty}^{\infty} b_j t^j$.

Further results: operator polynomials. Roch [221] considered the polynomials

$$L_n(\lambda) = T_n(b_0) + T_n(b_1)\lambda + \cdots + T_n(b_k)\lambda^k,$$
$$L_\infty(\lambda) = T(b_0) + T(b_1)\lambda + \cdots + T(b_k)\lambda^k,$$

thought of as acting on ℓ_n^2 and ℓ^2, respectively, and proved that if $T(b_k)$ is invertible, then

$$\liminf_{n\to\infty} \{\lambda \in \mathbf{C} : \|L_n^{-1}(\lambda)\|_2 \geq 1/\varepsilon\} = \limsup_{n\to\infty} \{\lambda \in \mathbf{C} : \|L_n^{-1}(\lambda)\|_2 \geq 1/\varepsilon\}$$
$$= \{\lambda \in \mathbf{C} : \|L_\infty^{-1}(\lambda)\|_2 \geq 1/\varepsilon\}$$

for each $\varepsilon > 0$.

Further results: higher order relative spectra. Let $A \in \mathcal{B}(\ell^2)$ and let L be a closed subspace of ℓ^2. We denote by P_L the orthogonal projection of ℓ^2 onto L. For a natural number k, the kth order spectrum $\mathrm{sp}_k(A, L)$ of A relative to L is defined as the set of all $\lambda \in \mathbf{C}$ for which the compression $P_L(A - \lambda I)^k P_L|L$ is not invertible on L. This definition is due to Brian Davies [94], who suggested that second order spectra might be useful for the approximate computation of spectra of self-adjoint operators.

Shargorodsky's paper [252] is devoted to the geometry of $\mathrm{sp}_k(A, L)$ for fixed L and to the limiting behavior of $\mathrm{sp}_k(A, L_n)$ as P_{L_n} converges strongly to the identity operator. One main result is a purely geometric description of the minimal set $\mathcal{Q}_k(K)$ with the property that $\mathrm{sp}_k(A, L) \subset \mathcal{Q}_k(K)$ whenever A is a normal operator with $\mathrm{sp}\, A \subset K$. Let, for example, K be a compact subset of $\mathbf{R}$. Put $a = \min K$ and $b = \max K$. The set $(a, b) \setminus K$ is an at most countable union of open intervals, that is, of the form $\cup_j (a_j, b_j)$. Let $B(c_1, c_2)$ denote the closed disk with diameter $[c_1, c_2]$. Then

$$\mathcal{Q}_2(K) = B(a, b) \setminus \bigcup_j \mathrm{Int}\, B(a_j, b_j),$$

where Int stands for the interior points.

Another remarkable result of [252] states that if A is normal, then

$$\bigcup_{\{L_n\}} \limsup_{n\to\infty} \mathrm{sp}_k(A, L_n) = \mathrm{sp}\, A \cup \mathcal{Q}_k(\mathrm{sp}_{\mathrm{ess}} A),$$

where $\mathrm{sp}_{\mathrm{ess}} A$ is the essential spectrum of A and the union is over all sequences $\{L_n\}$ for which P_{L_n} converges strongly to the identity operator. As a consequence, Shargorodsky obtains that if k is even, then

$$\bigcup_{\{L_n\}} \limsup_{n\to\infty} \mathrm{sp}_k(A, L_n) \cap \mathbf{R} = \mathrm{sp}\, A$$

for every selfadjoint operator A. (In the last two equalities, lim sup may be replaced by lim inf.) This shows that, in contrast to usual spectra, even order relative spectra do not deliver spurious points in the gaps of $\mathrm{sp}_{\mathrm{ess}} A$ when employing a projection method for the approximate computation of $\mathrm{sp}\, A$. For an arbitrary bounded operator A, the estimate

$$\bigcup_{\{L_n\}} \limsup_{n\to\infty} \mathrm{sp}_k(A, L_n) \subset \mathrm{sp}\, A \,\cup\, \frac{\|A\|_{\mathrm{ess}}}{\sin(\pi/(2k))}\,\overline{\mathbf{D}}$$

is proved in [252].

Further results: normal finite Toeplitz matrices. The characterization of finite normal Toeplitz matrices is discussed in [114], [126], [147], [169], [170], [171], [172]. The approach of Gu and Patton [147] is especially elegant and is applicable to the more general problem of determining all $n \times n$ Toeplitz matrices A, B, C, D such that $AB - CD$ is again Toeplitz or zero. For example, [147] contains a simple proof of the following result: The matrix $T_n(a) = (a_{j-k})_{j,k=1}^n$ is normal if and only if there is a $\lambda \in \mathbf{T}$ such that either $a_j = \lambda a_{-(n-j)}$ for $1 \le j \le n-1$ or $a_j = \lambda \overline{a_{-j}}$ for $1 \le j \le n-1$. Note that if $a_j = \lambda \overline{a_{-j}}$

with $\lambda \in \mathbf{T}$ for $1 \le j \le n-1$, then $a(\mathbf{T})$ is a line segment. Indeed, choose $\mu \in \mathbf{T}$ so that $\mu^2 = \lambda$. Then, for $t \in \mathbf{T}$,

$$\begin{aligned} a(t) &= a_0 + \sum_{j=1}^{n} \left(\lambda \overline{a_{-j}}\, t^j + a_{-j} t^{-j}\right) \\ &= a_0 + \mu \sum_{j=1}^{n} \left(\mu \overline{a_{-j}}\, t^j + \overline{\mu} a_{-j} t^{-j}\right) = a_0 + 2\mu \sum_{j=1}^{n} \mathrm{Re}\, \left(\mu \overline{a_{-j}}\, t^j\right). \end{aligned}$$

In the other case, $a_j = \lambda\, a_{-(n-j)}$ with $\lambda \in \mathbf{T}$ for $1 \le j \le n-1$, the range $a(\mathbf{T})$ need not to be a line segment (consider, for instance, $n = 3$ and $a(t) = t^{-1} + it^2$).

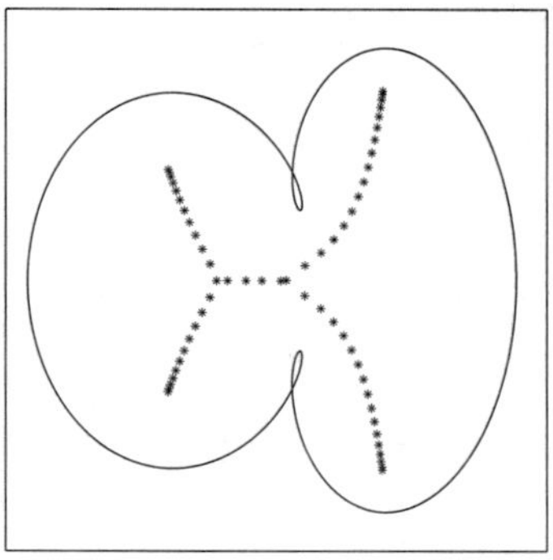

Chapter 8

Transient Behavior

Let A_n be a complex $n \times n$ matrix. The behavior of the norms $\|A_n^k\|$ is of considerable interest in connection with several problems. We specify $\|\cdot\|$ to be the spectral norm $\|\cdot\|_2$. The norms $\|A_n^k\|_2$ converge to zero as $k \to \infty$ if and only if $\operatorname{rad} A_n < 1$, where $\operatorname{rad} A_n$ denotes the spectral radius of A_n. However, sole knowledge of the spectral radius or even of all the eigenvalues of A_n does not tell us whether the norms $\|A_n^k\|_2$ run through a *critical transient phase*, that is, whether there are k for which $\|A_n^k\|_2$ becomes very large, before eventually decaying exponentially to zero. In this chapter we embark on this problem in the case where A_n is a Toeplitz band matrix.

8.1 The General Message

So what can be said about the norms $\|T_n^k(b)\|_2 := \|(T_n(b))^k\|_2$? The computer is expected to give a reliable answer in the case where n is small. We therefore assume that n is large. The message of this chapter is that $\|T_n^k(b)\|_2$ has *critical behavior* as k increases (which means that the norms $\|T_n^k(b)\|_2$ become large before decaying to zero or that the norms $\|T_n^k(b)\|_2$ go to infinity) if and only if $\|b\|_\infty > 1$. In other words, to find out whether $\|T_n^k(b)\|_2$ shows critical transient or limiting behavior, we need only look whether the L^∞ norm of the symbol is greater than one. Notice that it is much more difficult to decide whether $\operatorname{rad} T_n(b)$ or $\limsup_{n\to\infty} \operatorname{rad} T_n(b)$ is smaller than one (see Sections 10.3 and 10.4).

Since $T_n(b) = P_n T(b) P_n | \operatorname{Im} P_n$, it follows at once that

$$\|T_n^k(b)\|_2 \le \|T_n(b)\|_2^k \le \|T(b)\|_2^k = \|b\|_\infty^k \tag{8.1}$$

for all n and k. Thus, for the norms of powers of Toeplitz matrices we have the simple universal upper estimate (8.1). In particular, if $\|b\|_\infty \le 1$, then there is no critical behavior. The following theorem shows that (8.1) contracts to an equality in the $n \to \infty$ limit. This theorem may be viewed as an argument in support of the statement that $\|T_n^k(b)\|$ shows critical behavior whenever $\|b\|_\infty > 1$.

Theorem 8.1. *If $b \in W$, then*

$$\lim_{n\to\infty} \|T_n^k(b)\|_2 = \|b\|_\infty^k$$

for each natural number k.

Proof. This follows from Corollary 5.14. Here is a direct proof. For fixed k, the operators $T_n^k(b)$ converge strongly to $T^k(b)$ as $n \to \infty$. Hence, by the Banach-Steinhaus theorem,

$$\liminf_{n\to\infty} \|T_n^k(b)\|_2 \geq \|T^k(b)\|_2. \tag{8.2}$$

By (8.1) and (8.2), we are left with proving that $\|T^k(b)\|_2 \geq \|b\|_\infty^k$. But $\|T^k(b)\|_2 \geq \operatorname{rad} T^k(b) = (\operatorname{rad} T(b))^k$, and Corollary 1.12 says that the spectrum of $T(b)$ contains the range of b, whence $\operatorname{rad} T(b) \geq \|b\|_\infty$. □

The problem studied here amounts to looking for peaks of the "surface" $(k, n) \mapsto \|T_n^k(b)\|_2$ along the lines $n =$ constant, while Theorem 8.1 concerns the behavior of $\|T_n^k(b)\|_2$ along the lines $k =$ constant. We will return to this question later. For the moment, let us consider an example.

Example 8.2. The simplest nontrivial Toeplitz matrix is the Jordan block

$$J_n(\lambda) = \begin{pmatrix} \lambda & 0 & 0 & \dots & 0 \\ 1 & \lambda & 0 & \dots & 0 \\ 0 & 1 & \lambda & \dots & 0 \\ \vdots & \vdots & \vdots & \ddots & \vdots \\ 0 & 0 & 0 & \dots & \lambda \end{pmatrix}. \tag{8.3}$$

Clearly, $J_n(\lambda) = T_n(b)$ with $b(t) = \lambda + t$ ($t \in \mathbf{T}$). Figure 8.1 indicates the shape of the "surface" $(k, n) \mapsto \|J_n^k(0.8)\|_2$. We will omit the quotation marks in the following and will simply speak of the *norm surface*. We will also silently identify the surface with its "map" in the k, n plane; that is, we will not distinguish the point $(k, n, f(k, n))$ on the surface from its projection (k, n) in the plane. The spectral radius of $J_n(0.8)$ equals 0.8, while the L^∞ norm of the symbol is $\|b\|_\infty = 1.8$. We see that the surface has a *lowland* (say below the curve $\|J_n^k(0.8)\|_2 = 10^{-6}$, a *steep region* (say between the curves $\|J_n^k(0.8)\|_2 = 10^{-6}$ and $\|J_n^k(0.8)\|_2 = 10^2$), and a part where it grows enormously (say above the curve $\|J_n^k(0.8)\|_2 = 10^2$). We call the last part the *sky region*. If n is fixed and k increases, we move horizontally in Figure 8.1. On the surface, we will soon be in the sky region, then step down the steep region, and finally be caught in the lowland forever. Thus, we have the critical transient phase shown in the left picture of Figure 8.2. On the other hand, if we fix k and let n increase, then this corresponds to a vertical movement in Figure 8.1. This time we will fairly quickly reach the sky region and move at nearly constant height for the rest of the journey, as shown the right picture of Figure 8.2. Note that in the right picture of Figure 8.2 the norms converge to the limit $1.8^{20} = 12.75 \cdot 10^4$ and that the norms are already very close to this limit beginning with n between 20 and 40. □

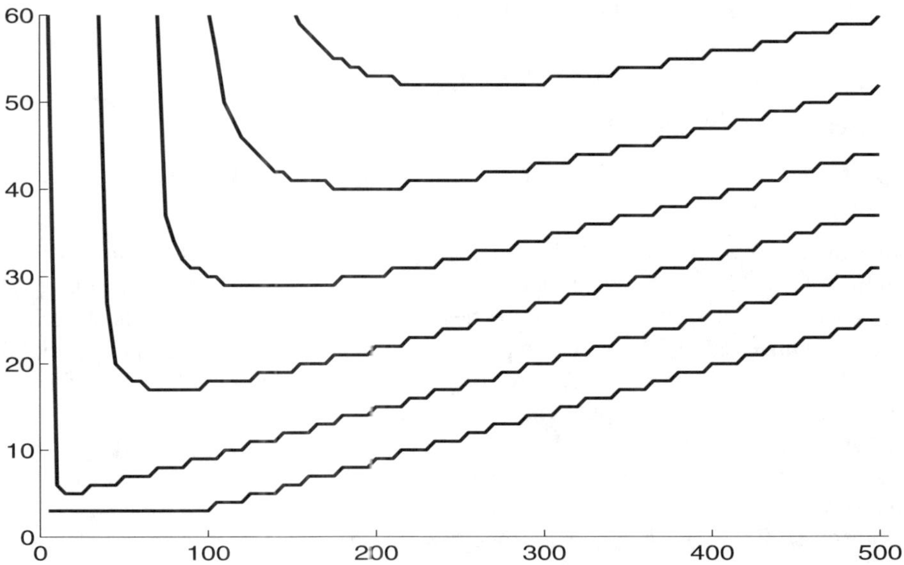

Figure 8.1. *Level curves* $\|J_n^k(0.8)\|_2 = h$ *for* $h = 10^{-6}, 10^2, 10^{10}, 10^{18}, 10^{26}, 10^{34}$ *(the lower curve corresponds to* $h = 10^{-6}$*, the upper to* 10^{34}*). We took* $n = 3, 4, 5, \dots, 60$ *and* $k = 5, 10, 15, \dots, 500$.

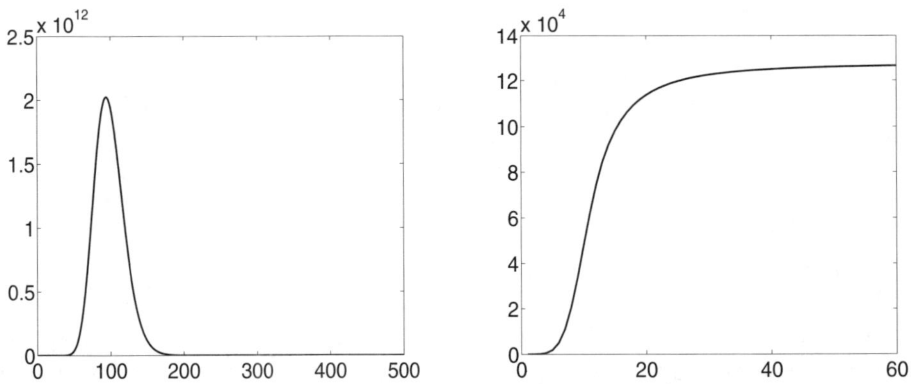

Figure 8.2. *Movement on the surface* $(k, n) \mapsto \|J_n^k(0.8)\|_2$ *along the horizontal line* $n = 20$ *(left) and along the vertical line* $k = 20$ *(right).*

8.2 Polynomial Numerical Hulls

Let A_n be an $n \times n$ matrix. The *polynomial numerical hull* $G_k(A_n)$ *of degree* k is defined as

$$G_k(A_n) = \{z \in \mathbf{C} : \ p(z)| \le \|p(A_n)\|_2 \ \text{ for all } \ p \in \mathcal{P}_{k+1}^+\},$$

where $\mathcal{P}_{k+1}^{+}$ is the set of all polynomials of the form

$$p(z) = p_0 + p_1 z + \cdots + p_k z^k.$$

The objective of $G_k(A_n)$ is to employ the obvious inequality

$$\|p(A_n)\|_2 \geq \max_{z \in G_k(A_n)} |p(z)|$$

in order to get a lower estimate for $\|p(A_n)\|_2$. The sets $G_k(A_n)$ were introduced by Olavi Nevanlinna [192], [193] and were independently discovered by Anne Greenbaum [141], [142]. Their works describe various properties of polynomial numerical hulls. In general, polynomial numerical hulls can be computed only numerically (see, e.g., [141]), and the development of algorithms and software for polynomial numerical hulls is still far away from the advanced level of the pseudospecta counterpart [299], [300].

Faber, Greenbaum, and Marshall [112] obtained remarkably precise results on the polynomial numerical hulls of Jordan blocks. Let $J_n(\lambda)$ be the Jordan block (8.3). If k is greater than the degree of the minimal polynomial of A_n, then $G_k(A_n)$ collapses to the spectrum of A_n. Thus, $G_k(J_n(\lambda)) = \{\lambda\}$ for $k \geq n$. In [112], it is shown that if $1 \leq k \leq n-1$, then $G_k(J_n(\lambda))$ is a closed disk with the center λ whose radius $\varrho_{n,k}$ satisfies

$$\cos \frac{\pi}{n+1} = \varrho_{1,n} \geq \varrho_{k,n} \geq \varrho_{n-1,n} = 1 - \frac{\log(2n)}{n} + \frac{\log(\log(2n))}{n} + o\left(\frac{1}{n}\right).$$

Consequently,

$$\|J_n^k(\lambda)\|_2 \geq \max_{|\zeta - \lambda| \leq \varrho_{n,k}} |\zeta|^k = (|\lambda| + \varrho_{n,k})^k.$$

It is also shown in [112] that $\varrho_{n-1,n}$ is greater than or equal to the positive root of $2r^n + r - 1 = 0$, which implies that

$$\varrho_{n-1,n} > 1 - \frac{\log(2n)}{n} \quad \text{for all} \quad n.$$

This yields, for instance, $\|J_{100}^{50}(0.8)\|_2 \geq 10^{12.11}$. MATLAB gives that $\|J_{100}^{50}(0.8)\|_2$ equals $10^{12.76}$.

We remark that a matrix with a norm that is gigantic in comparison with the matrix dimension must have a gigantic entry. For $n > k$, the ℓth entry of the first column of the lower-triangular Toeplitz matrix $J_n^k(\lambda)$ is $\binom{k}{\ell}\lambda^{k-\ell}$. Taking $\ell = [k/(1 + |\lambda|)]$, where $[\cdot]$ denotes the integer part, we get

$$\|J_n^k(\lambda)\|_2 \geq \binom{k}{[k/(1+|\lambda|)]} |\lambda|^{k-[k/(1+|\lambda|)]}.$$

This simple observation delivers

$$\|J_{100}^{50}(0.8)\|_2 \geq \binom{50}{27} 0.8^{23} = 10^{11.80},$$

which is fairly good.

8.3 The Pseudospectra Perspective

The role of pseudospectra in connection with the norms of powers of matrices and operators is as follows: norms of powers can be related to the resolvent norm, and pseudospectra decode information about the resolvent norm in a visual manner. A relation between resolvent and power norms is established by the Kreiss matrix theorem. Let A_n be an $n \times n$ matrix with rad $A_n \le 1$. For λ outside the closed unit disk $\overline{\mathbf{D}}$, put $R_n(\lambda) = \|(A_n - \lambda I)^{-1}\|_2$. The **Kreiss matrix theorem** says that

$$\sup_{|\lambda|>1} (|\lambda| - 1) R_n(\lambda) \le \max_{k \ge 0} \|A_n^k\|_2 \le en \sup_{|\lambda|>1} (|\lambda| - 1) R_n(\lambda).$$

We refer the reader to [289] for a delightful discussion of the theorem. The "easy half" of the Kreiss matrix theorem is the lower estimate. It implies that if we pick a point $\lambda \in \mathbf{C}$ with $\varrho := |\lambda| > 1$, then the maximum of the norms $\|A_n^k\|_2$ $(k \ge 0)$ is at least $(\varrho - 1)R_n(\lambda)$. Thus, the maximum is seen to be large whenever we can find a λ outside $\overline{\mathbf{D}}$ with large resolvent norm.

The following theorem and its proof are due to Nick Trefethen [272]. This theorem is similar to but nevertheless slightly different from the lower estimate of the Kreiss matrix theorem.

Theorem 8.3. *Let A be a bounded linear operator and let $\lambda \in \mathbf{C} \setminus \operatorname{sp} A$. Put $\varrho = |\lambda|$ and $R(\lambda) = \|(A - \lambda I)^{-1}\|_2$. If $R(\lambda) > 1/\varrho$, then*

$$\max_{1 \le j \le k} \|A^j\|_2 \ge \varrho^k \bigg/ \left(1 + \frac{\varrho^k - 1}{\varrho - 1} \frac{1}{\varrho R(\lambda) - 1}\right). \tag{8.4}$$

Proof. Put $M_k = \max_{1 \le j \le k} \|A^j\|_2$. By assumption, $\varrho R(\lambda) - 1$ is positive, so the term in the large brackets of (8.4) is at least 1 and the right-hand side of (8.4) is thus at most ϱ^k. If ϱ does not exceed the spectral radius of A, then (8.4) is accordingly trivial. We may therefore assume that ϱ is larger than the spectral radius of A. In this case we have

$$\lambda(\lambda I - A)^{-1} = (I - \lambda^{-1}A)^{-1} = I + (\lambda^{-1}A) + (\lambda^{-1}A)^2 + (\lambda^{-1}A)^3 + \cdots$$

and the corresponding bound

$$\varrho R(\lambda) \le 1 + \varrho^{-1}\|A\|_2 + \varrho^{-2}\|A^2\|_2 + \varrho^{-3}\|A^3\|_2 + \cdots. \tag{8.5}$$

Let us take the case $k = 2$ for illustration. Clearly,

$$\|A\|_2, \|A^2\|_2 \le M_2, \quad \|A^3\|_2, \|A^4\|_2 \le M_2^2, \quad \|A^5\|_2, \|A^6\|_2 \le M_2^3,$$

and so on. Grouping the terms in (8.5) accordingly into pairs gives

$$\varrho R(\lambda) \le 1 + (\varrho^{-2}M_2)(1 + \varrho) + (\varrho^{-2}M_2)^2(1 + \varrho) + \cdots.$$

If $M_2 \ge \varrho^2$, then (8.4) is true, so we may assume that $\varrho^{-2}M_2 < 1$. In this case

$$\varrho R(\lambda) \le 1 + \frac{\varrho^{-2}M_2(1 + \varrho)}{1 - \varrho^{-2}M_2} = 1 + \frac{\varrho + 1}{\varrho^2/M_2 - 1}.$$

For general k, we obtain similarly that $\varrho R(\lambda)$ is at most

$$\begin{aligned}&1+(\varrho^{-k}M_k)(1+\cdots+\varrho^{k-1})+(\varrho^{-k}M_k)^2(1+\cdots+\varrho^{k-1})+\cdots\\&=1+\frac{1+\varrho+\cdots+\varrho^{k-1}}{\varrho^k/M_k-1},\end{aligned}$$

that is,

$$\varrho R(\lambda)-1\le\frac{\varrho^k-1}{\varrho-1}\,\frac{1}{\varrho^k/M_k-1}.$$

It follows that

$$\frac{1}{\varrho R(\lambda)-1}\ge\frac{\varrho-1}{\varrho^k-1}\left(\frac{\varrho^k}{M_k}-1\right),$$

which implies that

$$\frac{\varrho^k}{M_k}-1\le\frac{\varrho^k-1}{\varrho-1}\,\frac{1}{\varrho R(\lambda)-1}.$$

This is (8.4). □

Thus, if sp A is known to be a subset of the closed unit disk $\overline{\mathbf{D}}$ and if there is an $\varepsilon\le 1$ such that the pseudospectrum $\mathrm{sp}_\varepsilon A$ contains points outside $\overline{\mathbf{D}}$, then, by (7.1), each point $\lambda\in\mathrm{sp}_\varepsilon A\setminus\overline{\mathbf{D}}$ yields an estimate

$$\max_{1\le j\le k}\|A^j\|_2\ge\varrho^k\Big/\left(1+\frac{\varrho^k-1}{\varrho-1}\,\frac{1}{\varrho/\varepsilon-1}\right)\tag{8.6}$$

with $\varrho=|\lambda|$. Different choices of λ give different values of the right-hand sides of (8.4) and (8.6), and we want these right-hand sides to be as large as possible. Tom Wright [299], [300] has implemented the search for an optimal λ in EigTool.

Recall that $\mathcal{P}$ denotes the set of all Laurent polynomials. Fix $b\in\mathcal{P}$ and consider $T_n(b)$. We put

$$M_k(n)=\max_{1\le j\le k}\|T_n^j(b)\|_2,\quad M(n)=\lim_{k\to\infty}M_k(n).$$

Clearly, $M(n)=\sup_{k\ge1}M_k(n)$; that is, $M(n)$ is the height of the highest peak of $\|T_n^k(b)\|_2$ as k ranges over $\mathbf{N}$. Thus, the powers of $T_n(b)$ go to infinity if and only if $M(n)=\infty$ and they have a critical transient behavior before decaying to zero if and only if $M(n)$ is large but finite. We assume that the plane Lebesgue measure of sp $T(b)$ is nonzero. Equivalently, we assume that there exist points in the plane that are encircled by the (naturally oriented) curve $b(\mathbf{T})$ with nonzero winding number. Suppose $\|b\|_\infty>1$. Then there exist points $\lambda\in\mathbf{C}\setminus b(\mathbf{T})$ such that $\varrho=|\lambda|>1$ and the winding number of b about λ is nonzero. From Theorem 4.1 we know that

$$R_n(\lambda):=\|(T_n(b)-\lambda I)^{-1}\|_2=\|T_n^{-1}(b-\lambda)\|_2$$

increases at least exponentially, i.e., there exist positive constants $C = C(b, \lambda)$ and $\beta = \beta(b, \lambda)$ such that

$$R_n(\lambda) \geq Ce^{\beta n} \tag{8.7}$$

for all $n \geq 1$. Consequently, $R_n(\lambda) > 1$ for all sufficiently large n and we obtain from Theorem 8.3 that

$$M_k(n) \geq \varrho^k \bigg/ \left(1 + \frac{\varrho^k - 1}{\varrho - 1} \frac{1}{\varrho R_n(\lambda) - 1}\right),$$

and

$$M(n) \geq (\varrho - 1)(\varrho R_n(\lambda) - 1). \tag{8.8}$$

This in conjunction with (8.7) gives

$$M(n) \geq (\varrho - 1)(\varrho C e^{\beta n} - 1), \tag{8.9}$$

that is, *there must be critical transient or limiting behavior of* $\|T_n^k(b)\|_2$ *for all n larger than a moderately sized* n_0. We will now say something about the constants in (8.7) and (8.9).

We denote by $\mathcal{P}_r$ $(r \geq 1)$ the set of all Laurent polynomials of degree at most r, that is, the set of all functions of the form

$$b(t) = \sum_{j=-r}^{r} b_j t^j. \tag{8.10}$$

Let $b \in \mathcal{P}_r$ and suppose $b_r \neq 0$. Assume $\|b\|_\infty > 1$ and choose λ as above. For the sake of definiteness, let wind $(b, \lambda) = -\kappa \leq -1$. We have

$$t^r(b(t) - \lambda) = b_r \prod_{i=1}^{s}(t - \mu_i) \prod_{i=s+1}^{2r}(t - \delta_i)$$

with $|\mu_i| > 1$ and $|\delta_i| < 1$ (recall Section 1.4). It follows that

$$b(t) - \lambda = b_r t^{-\kappa} \prod_{i=1}^{s}(t - \mu_i) \prod_{i=s+1}^{2r}\left(1 - \frac{\delta_j}{t}\right),$$

where $\kappa = s - r$ (≥ 1). Put

$$d(t) = \prod_{i=1}^{s}(t - \mu_i)^{-1} =: \sum_{j=0}^{\infty} d_j t^j \quad (t \in \mathbf{T}), \tag{8.11}$$

$$\|P_n d\|_2^2 = \sum_{j=0}^{n-1} |d_j|^2, \quad \|Q_n d\|_2^2 = \sum_{j=n}^{\infty} |d_j|^2. \tag{8.12}$$

Theorem 8.4. *For all $n \geq 1$,*

$$R_n(\lambda) \geq \frac{\|P_n d\|_2}{\|b - \lambda\|_\infty \|Q_n d\|_2}.$$

Proof. Put $c = b - \lambda$. With the notation $\chi_k(t) = t^k$, we have $c = \chi_{-\kappa} c_- c_+$, where

$$c_-(t) = \prod_{i=s+1}^{2r} \left(1 - \frac{\delta_i}{t}\right), \quad c_+(t) = b_r \prod_{i=1}^{s} (t - \mu_i).$$

Define $x^{(n)} \in \mathbf{C}^n$ and $x \in \ell^2$ by $x^{(n)} = (d_0, \dots, d_{n-1})$ and $x = \{d_0, d_1, \dots\}$. Since $T(c) = T(c_-)T(\chi_{-\kappa})T(c_+)$, we obtain

$$T_n(c)x^{(n)} = P_n T(c_-)T(\chi_{-\kappa})T(c_+)x^{(n)}. \tag{8.13}$$

Because $\kappa \geq 1$, we have $P_n T(c_-)T(\chi_{-\kappa})e_0 = 0$, where $e_0 = \{1, 0, 0, \dots\}$. As $T(c_+)x = e_0$, it follows that $P_n T(c_-)T(\chi_{-\kappa})T(c_+)x = 0$. This equality and (8.13) give

$$T_n(c)x^{(n)} = P_n T(c_-)T(\chi_{-\kappa})T(c_+)(x^{(n)} - x) = P_n T(c)(x^{(n)} - x).$$

Taking into account that $\|x^{(n)} - x\|_2 = \|Q_n d\|_2$, we arrive at the estimate

$$\|T_n(c)x^{(n)}\|_2 \leq \|T(c)\|_2 \, \|Q_b d\|_2 = \|c\|_\infty \|Q_n d\|_2,$$

and since $\|x^{(n)}\|_2 = \|P_n d\|_2$, it results that

$$\|T_n^{-1}(c)\|_2 \geq \frac{\|x^{(n)}\|_2}{\|T_n(c)x^{(n)}\|_2} \geq \frac{\|P_n d\|_2}{\|c\|_\infty \|Q_n d\|_2}. \qquad \square$$

For the sake of simplicity, assume that the zeros $\mu_1, \dots, \mu_s$ are distinct and that $|\mu_1| < |\mu_i|$ for all $i \geq 2$. Decomposition into partial fractions gives

$$d(t) = \sum_{i=1}^{s} \frac{A_i}{1 - t/\mu_i} = \sum_{j=0}^{\infty} \sum_{i=1}^{s} \frac{A_i}{\mu_i^j} t^j, \tag{8.14}$$

with explicitly available constants $A_1, \dots, A_s$. Theorem 8.4 in conjunction with the estimates

$$\|P_n d\|_2 \geq |d_0| = \left| \sum_{i=1}^{s} A_i \right| = \prod_{i=1}^{s} \frac{1}{|\mu_i|}$$

and

$$\|Q_n d\|_2^2 = \sum_{j=n}^{\infty} \left| \sum_{i=1}^{s} \frac{A_i}{\mu_i^j} \right|^2 \leq s \sum_{i=1}^{s} \sum_{j=n}^{\infty} \frac{|A_i|^2}{|\mu_i|^{2j}} = \sum_{i=1}^{s} \frac{|A_i|^2}{|\mu_i|^{2n}} \frac{s}{1 - 1/|\mu_i|^2}$$

$$=: \sum_{i=1}^{s} \frac{B_i}{|\mu_i|^{2n}} = \frac{B_1}{|\mu_1|^{2n}} \left(1 + \sum_{i=2}^{s} \frac{B_i}{B_1} \left| \frac{\mu_1}{\mu_i} \right|^{2n} \right)$$

yields

$$R_n(\lambda)^2 \geq \frac{|\mu_1|^{2n}}{B_1 \|b - \lambda\|_\infty^2 \prod_{i=1}^{s} |\mu_i|^2} \frac{1}{1 + \sum_{i=2}^{s} (B_i/B_1)|\mu_1/\mu_i|^{2n}}. \tag{8.15}$$

In practice, we could try computing $R_n(\lambda)$ directly via the MATLAB commands $R_n(\lambda) = \texttt{norm(inv(}T_n(b-\lambda)\texttt{))}$ or $R_n(\lambda) = \texttt{1/min(svd(}T_n(b-\lambda)\texttt{))}$. However, as (8.7) shows, this is an ill-conditioned problem for large n. Estimate (8.15) is more reliable. It first of all shows that (8.7) and (8.9) are true with

$$\beta = \log|\mu_1|.$$

To get the constants contained in (8.15) we may proceed as follows. We first determine the zeros $\mu_1, \ldots, \mu_s$, we then use MATLAB's `residue` command to find the numbers A_i in formula (8.14), and finally we put $B_i = s|A_i|^2/(1-1/|\mu_i|^2)$. Note that s is in general much smaller than n so that possible numerical instabilities are no longer caused by large matrix dimensions but at most by unfortunate location of the zeros $\mu_1, \ldots, \mu_s$. Notice also that (8.11) implies that the numbers $d_0, \ldots, d_{s-1}$ are the entries of the first column of the lower-triangular Toeplitz matrix $T_s(d)$. Thus, alternatively we could solve the $s \times s$ system

$$T_s\left(\prod_{i=1}^{s}(t-\mu_i)\right)(\,d_0\;\ldots\;d_{s-1}\,)^\top = (\,1\;0\;\ldots\;0\,)^\top$$

to obtain $d_0, \ldots, d_{s-1}$ and then, taking into account (8.14), find the constants $A_1, \ldots, A_s$ as the solutions of the $s \times s$ Vandermonde system

$$\begin{pmatrix} 1 & 1 & \ldots & 1 \\ 1/\mu_1 & 1/\mu_2 & \ldots & 1/\mu_s \\ \vdots & \vdots & & \vdots \\ 1/\mu_1^{s-1} & 1/\mu_2^{s-1} & \ldots & 1/\mu_s^{s-1} \end{pmatrix} \begin{pmatrix} A_1 \\ A_2 \\ \vdots \\ A_s \end{pmatrix} = \begin{pmatrix} d_0 \\ d_1 \\ \vdots \\ d_{s-1} \end{pmatrix}.$$

Here is an example that can be done by hand.

Example 8.5. Let $b(t) = t^{-1} + \alpha^2 t$ with $0 < \alpha < 1/2$. The range $b(\mathbf{T})$ is the ellipse

$$\frac{x^2}{(1+\alpha^2)^2} + \frac{y^2}{(1-\alpha^2)^2} = 1$$

and the eigenvalues of the matrix $T_n(b)$ are densely spread over the interval $(-2\alpha, 2\alpha)$ between the foci of the ellipse. The spectral radius is

$$\text{rad}\,(T_n(b)) = 2\alpha \cos\frac{\pi}{n+1}.$$

This is smaller than 1 but may be close to 1. The norm of the symbol is $\|b\|_\infty = 1+\alpha^2 > 1$. Fix $\lambda = \varrho \in (1, 1+\alpha^2)$. The zeros of $t(b(t)-\varrho)$ are

$$\mu_1 = \frac{\varrho - \sqrt{\varrho^2 - 4\alpha^2}}{2\alpha^2}, \quad \mu_2 = \frac{\varrho + \sqrt{\varrho^2 - 4\alpha^2}}{2\alpha^2}.$$

The numbers μ_1 and μ_2 are greater than 1, and we have

$$A_1 = \frac{1}{\mu_1 - \mu_2}, \quad A_2 = \frac{1}{\mu_2 - \mu_1}, \quad B_1 = \frac{2A_1^2}{1 - 1/\mu_1^2}, \quad B_2 = \frac{2A_2^2}{1 - 1/\mu_2^2}.$$

Thus, (8.15) becomes

$$R_n(\lambda)^2 \geq \frac{\alpha^4 \mu_1^{2n}}{B_1(\varrho+1+\alpha^2)^2} \Bigg/ \left(1+\frac{B_2}{B_1}\left(\frac{\mu_1}{\mu_2}\right)^{2n}\right).$$

Here are a few concrete samples. In each case we picked $\lambda = \varrho = 1.01$. Note that (8.8) with $\varrho = 1.01$ gives

$$M(n) \geq 0.1(1.01\, R_n(1.01) - 1),$$

which is almost the same as the Kreiss estimate $M(n) \geq 0.1 R_n(1.01)$.

$\boldsymbol{\alpha = 0.2}$. We have $\mu_1 = 1.0323$, $\mu_2 = 24.2177$,

$$R_n(1.01) \geq 10^{0.0138\,n - 1.0863} =: E_{0.2}(n).$$

In particular, $E_{0.2}(1000) = 10^{12.70}$. MATLAB result: $R_{1000}(1.01) = 10^{15.03}$.

$\boldsymbol{\alpha = 0.4}$. Now $\mu_1 = 1.2296$, $\mu_2 = 5.0829$,

$$R_n(1.01) \geq 10^{0.0897\,n - 0.9316} =: E_{0.4}(n).$$

We get $E_{0.4}(100) = 10^{8.04}$ and $E_{0.4}(1000) = 10^{88.77}$, while the MATLAB results are $R_{100}(1.01) = 10^{9.48}$ and $R_{1000}(1.01) = 10^{90.28}$.

$\boldsymbol{\alpha = 0.49}$. This time $\mu_1 = 1.5945$, $\mu_2 = 2.6121$,

$$R_n(1.01) \geq 10^{0.2026\,n - 1.2233} =: E_{0.49}(n).$$

We obtain

$$E_{0.49}(50) = 10^{8.91}, \quad E_{0.49}(100) = 10^{19.04},$$
$$E_{0.49}(1000) = 10^{201.38}, \quad E_{0.49}(2000) = 10^{403.98},$$

and MATLAB delivers

$$R_{50}(1.01) = 10^{10.22}, \quad R_{100}(1.01) = 10^{20.35}, \quad R_{1000}(1.01) = 10^{202.71};$$

the last two values are delivered with a warning. MATLAB returns a warning without a value for $n = 2000$. □

8.4 A Triangular Example

The next question after Theorem 8.1 concerns the convergence speed. Let $b \in \mathcal{P}$ be of the form $b(t) = \sum_{j\geq 0} b_j t^j$, that is, suppose $T(b)$ is lower triangular. Assume also that $|b|$ is not constant on $\mathbf{T}$. We have $T_n^k(b) = T_n(b^k)$, and hence we may employ Theorem 5.8 with b replaced by b^k. It results that there are constants $c_k, d_k \in (0, \infty)$ and $\gamma \in \{1, 2, \dots\}$ such that

$$\|b\|_\infty^k - \frac{d_k}{n^{2\gamma}} \leq \|T_n^k(b)\|_2 \leq \|b\|_\infty^k - \frac{c_k}{n^{2\gamma}} \tag{8.16}$$

for all $n \geq 1$. However, the constants c_k and d_k may be large if k is large. We remark that the constants C of inequalities like

$$\|b\|_\infty - |b(t)| \leq C|t - t_0|^{2\gamma} \tag{8.17}$$

enter the d_2 of (5.14). For the powers of b, we obtain from (8.17) something like

$$\|b\|_\infty^k - |b^k(t)| \lessapprox Ck\|b\|_\infty^k |t - t_0|^{2\gamma},$$

and since $Ck\|b\|_\infty^k$ is large whenever $\|b\|_\infty > 1$ and k is large, we can expect that d_k is also large.

Let us consider a concrete example. Take

$$b(t) = 10\left(-\frac{1}{16} - \frac{1}{4}t + t^2 - \frac{1}{4}t^3 - \frac{1}{16}t^4\right) = 10t^2\left(\frac{11}{8} - \frac{1}{16}|1 - t|^4\right).$$

In this case $2\gamma = 4$. The spectral radius of $T_n(b)$ is about 0.63 and $\|b\|_\infty$ equals $110/8 = 13.75$. The level curves of the norm surface are shown in Figure 8.3 and we clearly see the expected critical transient behavior. An interesting feature of Figure 8.3 is the indents of the level curves. These indents correspond to vertical valleys in the surface. For instance, along the horizontal line $n = 18$ we have Figure 8.4. The interesting piece of Figure 8.4 is between $k = 15$ and $k = 40$. If we move vertically along one of the lines $k = 15$, $k = 20$, $\ldots$, $k = 40$, we obtain Figure 8.5. Figure 8.6 reveals some mild turbulence in the steep region, but Figure 8.5 convincingly shows that in the sky region everything goes smoothly. In particular, the speed of the convergence of $\|T_n^k(b)\|_2$ to $\|T^k(b)\|_2 = \|b\|_\infty^k$ is not affected by small fluctuations of the exponent k.

8.5 Gauss-Seidel for Large Toeplitz Matrices

To solve the $n \times n$ system $C_n x = y$, one decomposes C_n into a sum $C_n = L_n + U_n$ of a lower-triangular matrix L_n and an upper-triangular matrix U_n with zeros on the main diagonal. If L_n is invertible, then the system $C_n x = y$ is equivalent to the system

$$x = -L_n^{-1}U_n x + L_n^{-1}y. \tag{8.18}$$

The *Gauss-Seidel iteration* consists of choosing an initial x_0 and computing the iterations by

$$x_{k+1} = -L_n^{-1}U_n x_k + L_n^{-1}y. \tag{8.19}$$

Sometimes (8.18) and (8.19) are written in the form

$$x = x + L_n^{-1}(y - C_n x), \quad x_{k+1} = x_k + L_n^{-1}(y - C_n x_k).$$

We have $x_k - x = (-L_n^{-1}U_n)^k(x_0 - x)$. Thus, the iteration matrix is $A_n := -L_n^{-1}U_n = I - L_n^{-1}C_n$, and the iteration converges whenever $\operatorname{rad} A_n < 1$. The problem is whether a critical transient behavior of the norms $\|A_n^k\|$ may garble the solution.

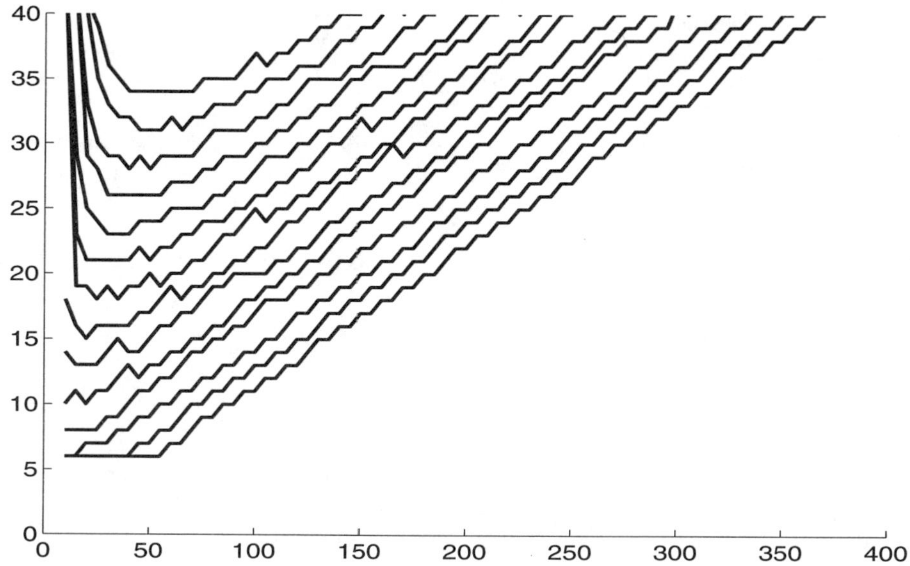

Figure 8.3. *The symbol is* $b(t) = 10\,(-\frac{1}{16} - \frac{1}{4}t + t^2 - \frac{1}{4}t^3 - \frac{1}{16}t^4)$. *The plot shows the level curves* $\|T_n^k(b)\|_2 = h$ *for* $h = 10^{-2}, 1, 10^2, 10^4, \ldots, 10^{24}$ *(the lower curve corresponds to* $h = 10^{-2}$, *the upper to* 10^{24}*). We took* $6 \le n \le 40$ *and* $k = 10, 15, 20, \ldots, 400$.

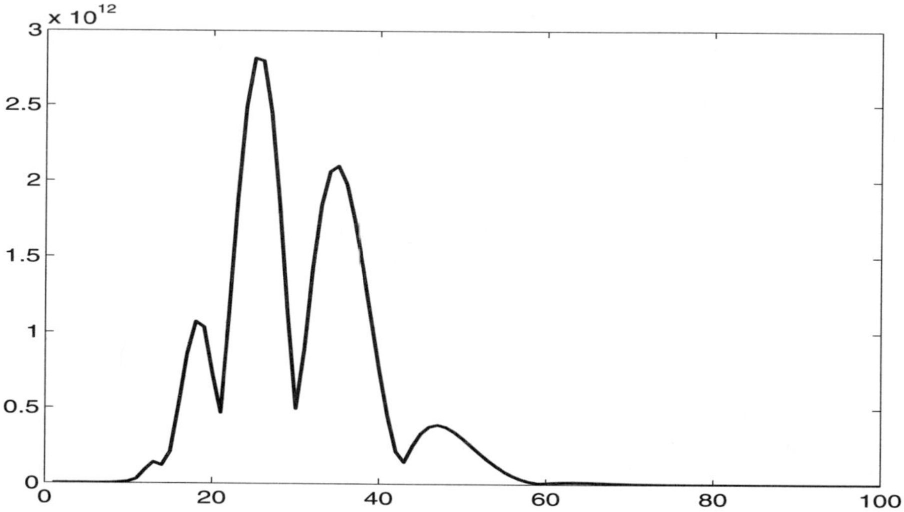

Figure 8.4. *The symbol is as in Figure* 8.3. *The picture shows the norms* $\|T_{18}^k(b)\|_2$ *for* $1 \le k \le 100$.

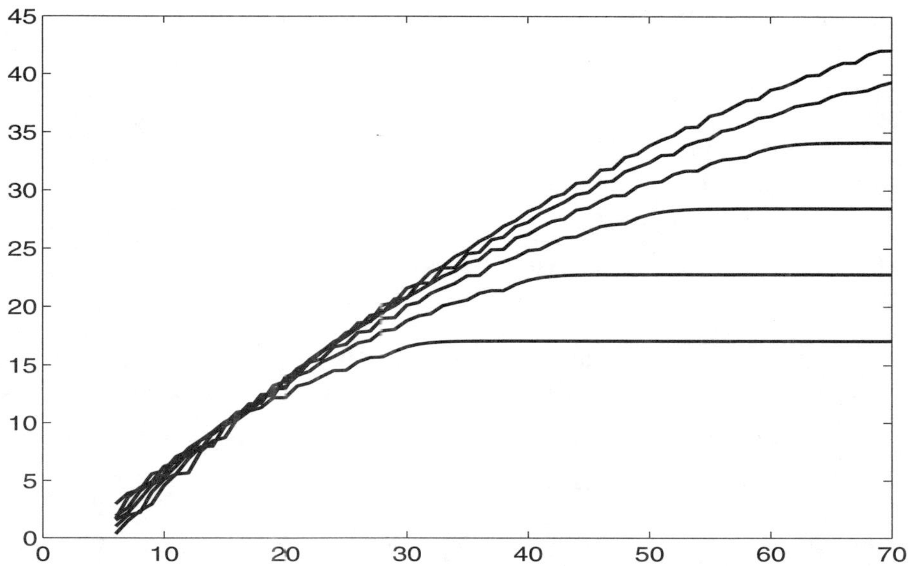

Figure 8.5. *The symbol is again as in Figure* 8.3. *We see the values of* $\log_{10} \|T_n^k(b)\|_2$ *for* $6 \le n \le 70$ *and* $k = 15, 20, 25, \ldots, 40$. *Eventually higher curves correspond to higher values of k.*

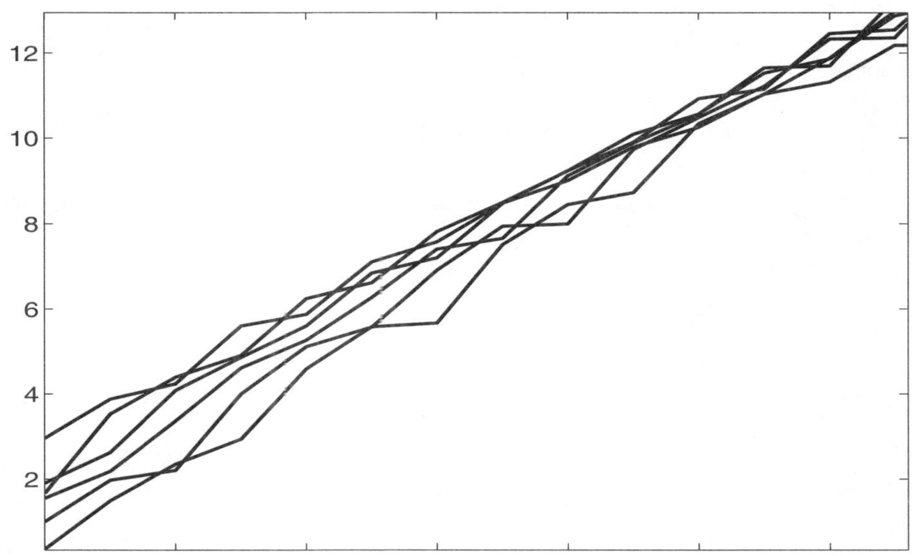

Figure 8.6. *A close-up of Figure* 8.5.

Now suppose $C_n = T_n(c)$ is a Toeplitz matrix and let us for the sake of simplicity assume that

$$c(t) = \sum_{j=-\infty}^{\infty} c_j t^j, \quad \text{where} \quad \sum_{j=-\infty}^{\infty} |c_j| < \infty.$$

We write $C_n = L_n + U_n = T_n(c_+) + T_n(c_-)$ with

$$c_+(t) = \sum_{j=0}^{\infty} c_j t^j, \quad c_-(t) = \sum_{j=-\infty}^{-1} c_j t^j.$$

If $c_+(z) \neq 0$ for $|z| \leq 1$, then the inverse of $T_n(c_+)$ is $T_n(c_+^{-1})$ and the iteration matrix becomes $A_n = -T_n(c_+^{-1})T_n(c_-)$. The spectrum of A_n is the set of all complex numbers λ for which

$$-T_n(c_+^{-1})T_n(c_-) - \lambda I = -T_n(c_+^{-1})T_n(c_- + \lambda c_+)$$

is not invertible. Since $T_n(c_+^{-1})$ is invertible by assumption, we have to look for the λ's for which $T_n(c_- + \lambda c_+)$ is not invertible.

By the Banach-Steinhaus theorem,

$$\liminf_{n\to\infty} \|A_n^k\|_2 = \liminf_{n\to\infty} \|(T_n(c_+^{-1})T_n(c_-))^k\|_2 \geq \|(T(c_+^{-1})T(c_-))^k\|_2.$$

Propositions 1.2 and 1.3 imply that $(T(c_+^{-1})T(c_-))^k$ is $T(c_+^{-k}c_-^k)$ plus a compact operator. The function $c_+^{-k}c_-^k$ is not a Laurent polynomial, but it lies in the Wiener algebra, and the argument of the proof of Lemma 5.12 remains valid for symbols in the Wiener algebra. This shows that $\|(T(c_+^{-1})T(c_-))^k\|_2 \geq \|c_+^{-k}c_-^k\|_\infty = \|c_+^{-1}c_-\|_\infty^k$. Consequently,

$$\liminf_{n\to\infty} \|A_n^k\|_2 \geq \|c_+^{-1}c_-\|_\infty^k. \tag{8.20}$$

Thus, if n is large enough, a critical transient phase will certainly occur in the case where $\|c_+^{-1}c_-\|_\infty > 1$.

If $a(t) = a_{-1}t^{-1} + a_0 + a_1 t$, then the eigenvalues of $T_n(a)$ are known to be

$$a_0 + 2\sqrt{a_1 a_{-1}}\, \cos\frac{\pi j}{n+1} \quad (j = 1, \ldots, n)$$

(Theorem 2.4). Hence, in case c is a trinomial, $c(t) = c_{-1}t^{-1} + c_0 + c_1 t$, the matrix $T_n(c_- + \lambda c_+)$ has the eigenvalues

$$\lambda c_0 + 2\sqrt{c_1 c_{-1}}\, \sqrt{\lambda}\, \cos\frac{\pi j}{n+1} \quad (j = 1, \ldots, n).$$

It follows that

$$\operatorname{sp} A_n = \left\{ \frac{4c_1 c_{-1}}{c_0^2} \cos^2\frac{\pi j}{n+1} : j = 1, \ldots, n \right\}$$

and

$$\operatorname{rad} A_n = \frac{4|c_1|\,|c_{-1}|}{|c_0|^2} \cos^2 \frac{\pi}{n+1}.$$

Anne Greenbaum [141] discussed the example $c(t) = -1.16t^{-1} + 1 + 0.16t$. The spectral radius of A_n is about 0.73. Using (numerically computed) polynomial numerical hulls, she obtained that $\|A_{30}^{29}\|_2 \geq 1.256^{29} \approx 700$ and trying the computer directly, she arrived at the much better result $\|A_{30}^{29}\|_2 \approx 10^4$. Our estimate (8.20) with $c_+(t) = 1+0.16t$ and $c_-(t) = -1.16t^{-1}$ gives

$$\begin{aligned}\liminf_{n\to\infty} \|A_n^{29}\|_2 \geq \|c_+^{-1}c_-\|_\infty^{29} &= \max_{t\in\mathbf{T}} \left|\frac{-1.16t^{-1}}{1+0.16t}\right|^{29} \\ &= \left(\frac{1.16}{1-0.16}\right)^{29} = 1.381^{29} = 10^{4.07}. \end{aligned} \tag{8.21}$$

Thus, although (8.21) is an "$n \to \infty$" result, it is already strikingly good for n about 30. Figure 8.7 shows the norm surface.

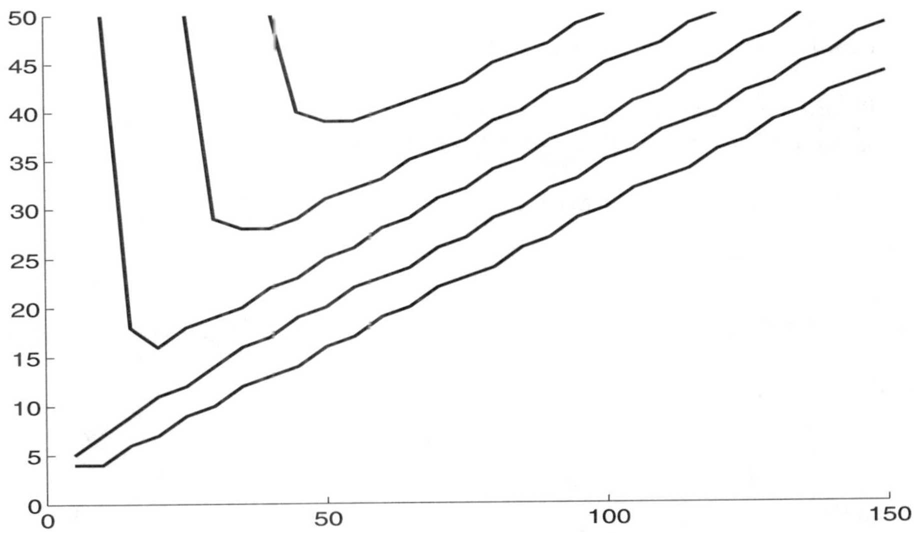

Figure 8.7. *The norm surface for $A_n = T_n(c_+^{-1})T_n(c_-)$ with $c_+(t) = 1 + 0.16t$ and $c_-(t) = -1.16t^{-1}$. The picture shows the level curves $\|A_n^k\|_2 = h$ for $h = 10^{-2}, 1, 10^2, 10^4, 10^6$ (the lower curve corresponds to $h = 10^{-2}$, the upper to 10^6). We took $3 \leq n \leq 50$ and $k = 5, 10, 15, \ldots, 150$.*

Nick Trefethen [270] considered the symbol $c(t) = t^{-1} - 2 + t$. In this case $c_+(t) = -2 + t$ and $c_-(t) = t^{-1}$, so

$$\|c_+^{-1}c_-\|_\infty = \max_{t\in\mathbf{T}} \left|\frac{t^{-1}}{-2+t}\right| = 1,$$

and hence (8.20) does not provide any useful piece of information. However, the brute estimate

$$\|A_n^k\|_2 \le \|T_n(c_+^{-1})\|_2^k \|T_n(c_-)\|_2^k \le \|c_+^{-1}\|_\infty^k \|c_-\|_\infty^k = \max_{t\in\mathbf{T}} \left|\frac{1}{-2+t}\right|^k \cdot 1^k = 1$$

shows that there is no critical behavior. The spectral radius of A_n is $\cos^2 \frac{\pi}{n+1}$, which is approximately 0.9990325 for $n = 100$. MATLAB gives rad $A_{100} = 0.9990$ (fantastic!) and tells us that the norms $\|A_{100}^k\|_2$ are 0.9080, 0.6169, and 0.3804 for $k = 100, 500, 1000$, respectively.

8.6 Genuinely Finite Results

We begin with an observation similar to Proposition 5.11. We denote by $\mathcal{P}_r$ the set of all Laurent polynomials of the form $\sum_{j=-r}^{r} b_j t^j$.

Lemma 8.6. *Let $b \in \mathcal{P}_r$ and $n > 2r(k-1)$. Then*

$$T_n^k(b) = T_n(b^k) + P_n X_k P_n + W_n Y_k W_n$$

with $r(k-1) \times r(k-1)$ matrices X_k and Y_k independent of n.

Proof. For $k = 1$, the assertion is true with $X_1 = Y_1 = 0$. Assume the asserted equality is valid for k. We prove it for $k+1$. Obviously,

$$T_n^{k+1}(b) = T_n(b^k)T_n(b) + P_n X_k P_n T(b) P_n + W_n Y_k W_n P_n T(b) P_n.$$

By Proposition 3.10,

$$T_n(b^k)T_n(b) = T_n(b^{k+1}) - P_n H(b^k) H(\widetilde{b}) P_n - W_n H(\widetilde{b^k}) H(b) W_n.$$

We have

$$H(b^k) = \begin{pmatrix} Z_{11} & \dots & Z_{1k} & 0 & \dots \\ \vdots & & \vdots & \vdots & \\ Z_{k1} & \dots & Z_{kk} & 0 & \dots \\ 0 & \dots & 0 & 0 & \dots \\ \dots & \dots & \dots & \dots & \dots \end{pmatrix}, \quad H(\widetilde{b}) = \begin{pmatrix} U & 0 & \dots \\ 0 & 0 & \dots \\ \dots & \dots & \dots \end{pmatrix}$$

with certain $r \times r$ blocks Z_{ij} and U, which implies that

$$H(b^k)H(\widetilde{b}) = \begin{pmatrix} Z_{11}U & 0 & \dots \\ \vdots & \vdots & \\ Z_{k1}U & 0 & \dots \\ 0 & 0 & \dots \\ \dots & \dots & \dots \end{pmatrix}.$$

Thus, $P_n H(b^k) H(\widetilde{b}) P_n = P_n X'_{k+1} P_n$ with a $kr \times r$ matrix X'_{k+1} independent of n. Analogously, $W_n H(\widetilde{b^k}) H(b) W_n = W_n Y'_{k+1} W_n$ with some $kr \times r$ matrix Y'_{k+1}. Further,

$$X_k = \begin{pmatrix} X_{11} & \dots & X_{1,k-1} & 0 & \dots \\ \vdots & & \vdots & \vdots & \\ X_{k-1,1} & \dots & X_{k-1,k-1} & 0 & \dots \\ 0 & \dots & 0 & 0 & \dots \\ \dots & \dots & \dots & \dots & \dots \end{pmatrix}$$

and

$$T(b) = \begin{pmatrix} A_0 & A_{-1} & & & \\ A_1 & A_0 & A_{-1} & & \\ & A_1 & A_0 & A_{-1} & \\ & & \dots & \dots & \dots \end{pmatrix}$$

with $r \times r$ blocks X_{ij} and A_ℓ independent of n. This shows that at most the first $(k-1)r$ rows of $X_k P_n T(b)$ are nonzero. For $1 \le j \le k-1$ and $\ell \ge k+1$, the j, ℓ block of $X_k P_n T(b)$ is

$$\sum_{m=1}^{k-1} X_{jm} A_{m-\ell} = X_{j,\ell-1} A_{-1} + X_{j,\ell} A_0 + X_{j,\ell+1} A_1,$$

and as $X_{j,\ell-1} = X_{j,\ell} = X_{j,\ell-1} = 0$ for $\ell \ge k+1$, it follows that at most the first kr columns of $X_k P_n T(b)$ are nonzero. Thus, $P_n X_k P_n T(b) P_n = P_n X''_{k+1} P_n$ with some $(k-1)r \times kr$ matrix X''_{k+1} independent of n. Finally,

$$W_n Y_k W_n P_n T(b) P_n = W_n Y_k P_n T(\widetilde{b}) W_n,$$

and the same argument as above shows that this equals $W_n Y''_{k+1} W_n$ with some $(k-1)r \times kr$ matrix Y''_{k+1} that does not depend on n. □

Inequality (8.16) is an asymptotic result. In contrast to this, Corollary 5.19 is genuinely finite. Combining Lemma 8.6 and Corollary 5.19 with $M_2 = M_2^0 = \|b\|_\infty^k$ yields the inequality

$$\|b\|_\infty^k \left(1 - \frac{41r(k-1)}{n}\right) \le \|T_n^k(b)\|_2 \le \|b\|_\infty^k \tag{8.22}$$

for all n and k. Unfortunately, the lower bound of (8.22) is positive for $n > 41r(k-1)$ only, which is not yet of much use for n's below 1000. Better estimates can be derived from Theorem 5.10 on the Fejér means. Here is such an estimate. It is already applicable to $n > 2r(k-1)$.

Theorem 8.7. *If $b \in \mathcal{P}_r$ and $n > 2r(k-1)$, then $\|T_n^k(b)\|_2 \ge \|\sigma_{n-2r(k-1)}(b^k)\|_\infty$.*

Proof. By Lemma 8.6,

$$T_n^k(b) = T_n(b^k) + \begin{pmatrix} X_k & 0 & 0 \\ 0 & 0 & 0 \\ 0 & 0 & W_k Y_k W_k \end{pmatrix} = \begin{pmatrix} * & * & * \\ * & A & * \\ * & * & * \end{pmatrix},$$

where X_k and $W_kY_kW_k$ are $(k-1)r \times (k-1)r$ matrices and $A = T_{n-2r(k-1)}(b^k)$. This implies that

$$\|T_n^k(b)\|_2 \geq \|A\|_2 = \|T_{n-2r(k-1)}(b^k)\|_2.$$

It remains to combine the last inequality and Theorem 5.10. □

Recall that $\|\cdot\|_W$ is the Wiener norm of a Laurent polynomial: $\|b\|_W = \sum_j |b_j|$.

Theorem 8.8. *If $b \in \mathcal{P}_r$ and $n > r(3k-2)$, then*

$$\|T_n^k(b)\|_2 \geq \|b\|_\infty^k - \frac{kr}{n-2r(k-1)}\|b\|_W^k.$$

Proof. For $a \in \mathcal{P}_{kr}$ and $m = n - 2r(k-1) > kr$,

$$(\sigma_m a)(t) = \sum_{|j|\leq kr}\left(1 - \frac{|j|}{m}\right)a_j t^j = \sum_{|j|\leq kr} a_j t^j - \frac{1}{m}\sum_{|j|\leq kr}|j|a_j t^j,$$

whence

$$|(\sigma_m a)(t)| \geq \left|\sum_{|j|\leq kr} a_j t^j\right| - \frac{kr}{m}\sum_{|j|\leq kr}|a_j|$$

and thus $\|\sigma_m a\|_\infty \geq \|a\|_\infty - (kr/m)\|a\|_W$. This inequality in conjunction with Theorem 8.7 gives the assertion. □

Corollary 8.9. *If $b \in \mathcal{P}_r$ and $b_j \geq 0$ for all j, then*

$$\|T_n^k(b)\|_2 \geq \frac{n-3kr}{n-2kr}\left(\sum_j b_j\right)^k \quad \textit{for} \quad n \geq 3kr.$$

Proof. In this case $\|b\|_\infty = \|b\|_W = \sum_j b_j$ and, hence, by Theorem 8.8,

$$\|T_n^k(b)\|_2 \geq \left(1 - \frac{kr}{n-2r(k-1)}\right)\left(\sum b_j\right)^k \geq \frac{n-3kr}{n-2kr}\left(\sum b_j\right)^k. \qquad \square$$

Example 8.10. Let $b(t) = t + \alpha^2 t^{-1}$ with $\alpha \in \mathbf{R}$. From Corollary 8.9 we infer that

$$\frac{1}{2}(1+\alpha^2)^k \leq \|T_{4k}^k(b)\|_2 \leq (1+\alpha^2)^k$$

for all $k \geq 1$. □

Example 8.11. Suppose $T(b)$ is lower triangular. Then $T_n^k(b) = T_n(b^k)$ for all n and k, and hence we can have immediate recourse to Theorem 5.10 to obtain that

$$\|\sigma_n(b^k)\|_\infty \leq \|T_n^k(b)\|_2 \leq \|b\|_\infty^k \text{ for all } n \text{ and } k. \tag{8.23}$$

Let first $b(t) = \lambda + t$. Thus, $T_n(b)$ is the Jordan block $J_n(\lambda)$. For $k \le n$,

$$(\sigma_n(b^k))(t) = \sum_{j=0}^{k} \left(1 - \frac{j}{n}\right) \binom{k}{j} \lambda^{k-j} t^j = (\lambda + t)^k \left(1 - \frac{k}{n} \frac{t}{t+\lambda}\right)$$

and hence (8.23) yields

$$\left(1 - \frac{k}{n} \frac{1}{1+|\lambda|}\right) (1+|\lambda|)^k \le \|J_n^k(\lambda)\|_2 \le (1+|\lambda|)^k \quad (k \le n).$$

In particular,

$$10^{12.62} \le \|J_{100}^{50}(0.8)\| \le 10^{12.77}, \quad 10^{25.17} \le \|J_{100}^{100}(0.8)\| \le 10^{25.53},$$

which is better than the results of Section 8.1.

Now let $a(t) = t + t^2$. Then $T_n(a)$ is a "super Jordan block" [270]. For every natural number k,

$$\|\sigma_{2k}(a^k)\|_\infty = \sum_{j=k}^{2k-1} \left(1 - \frac{j}{2k}\right) \binom{k}{k-j} = \frac{3}{8} 2^k,$$

and thus (8.23) implies that $(3/8)\, 2^k \le \|T_{2k}^k(a)\| \le 2^k$. □

8.7 The Sky Region Contains an Angle

Let $b \in \mathcal{P}_r$ and suppose $\|b\|_\infty > 1$. In Figures 8.1 and 8.3, the sky region looks approximately like an angle: It is bounded by a nearly vertical line on the left and by a curve close to the graph of a linear function $n = ck + d$ from the right and below. The question is whether this remains true beyond the cutouts we see in the pictures and whether this is valid in general. To be more precise, we fix a (large) number B and we call the set

$$S_B = \{(k, n) : \|T_n^k(b)\|_2 > B\}$$

the *sky region* (for our choice of the bound B). The following theorem proves that the lower-right boundary of the sky region is always linear or sublinear.

Theorem 8.12. *If $b \in \mathcal{P}_r$ and $\|b\|_\infty > 1$, then there exist positive constants c and k_0, depending on b and B, such that S_B contains the angle $\{(k, n) : k > k_0, n > ck\}$.*

Proof. This is a simple consequence of inequality (8.22), which shows that

$$\|T_n^k(b)\|_2 \ge \|b\|_\infty^k \left(1 - \frac{41rk}{n}\right) \tag{8.24}$$

for all k and n: if $k > k_0$ where $\|b\|_\infty^{k_0} > 2B$ and $n > 82rk$, then the right-hand side of (8.24) is greater than B. □

Thus, if we walk on the norm surface $(k, n) \mapsto \|T_n^k(b)\|_2$ along a curve whose projection in the k, n plane is given by $n = \varphi(k)$, then we will eventually reach any height B and stay above this height forever provided $\varphi(k)/k \to \infty$. Note that this is satisfied for $\varphi(k) = k \log\log k$. Or in yet other terms, if $\varphi(k)/k \to \infty$, then $\|T_{\varphi(k)}^k(b)\|_2 \to \infty$.

Theorem 8.12 is the deciding argument in support of the statement that $\|T_n^k(b)\|_2$ runs through a critical transient phase if $\lim_{n\to\infty} \operatorname{rad} T_n(b) < 1$ but $\|b\|_\infty > 1$. Suppose, for example, $k = 100$. If the sky region were roughly of the form $S_B = \{(k, n) : k > k_0, n > k^2\}$, then $\|T_n^{100}(b)\|_2$ would be larger than B for $n > 10000$ only. It is the linearity or sublinearity of the lower-right border of the sky region that allows us to conclude that $\|T_n^{100}(b)\|_2$ is already larger than B for n in the hundreds.

We call the number

$$L_B(n) = \#\{k : \|T_n^k(b)\|_2 > B\}$$

the *length of the critical transient phase for the matrix dimension* n. Theorem 8.12 implies that $L_B(n) > n/c - k_0$. In other words, $L_B(n)$ increases at least linearly with the matrix dimension n. Equivalently and a little more elegantly,

$$\liminf_{n\to\infty} \frac{L_B(n)}{n} > 0.$$

May the lower-right boundary of the sky region be strictly sublinear? Or alternatively, may the lowland contain an angle? Let $b \in \mathcal{P}_r$ and $b(t) = \sum_{j=-r}^{r} b_j t^j$. For $\varrho \in (0, \infty)$, we define $b_\varrho \in \mathcal{P}_r$ by

$$b_\varrho(t) = \sum_{j=-r}^{r} b_j \varrho^j t^j.$$

Theorem 8.13. *Let $b \in \mathcal{P}_r$, $\|b\|_\infty > 1$, and $B > 1$. If there exists a number ϱ in $(0, \infty)$ such that $\|b_\varrho\|_\infty < 1$, then S_B is contained in an angle $\{(k, n) : n > ck + d\}$ with $c > 0$ and $d > 0$.*

Proof. The key observation is due to Schmidt and Spitzer [243] and will be extensively exploited in Chapter 11. It consists of the equality $T_n(b) = D_\varrho^{-1} T_n(b_\varrho) D_\varrho$, where $D_\varrho = \operatorname{diag}(1, \varrho, \dots, \varrho^{n-1})$. Letting $M = \max(\varrho, 1/\varrho)$, we get

$$\|T_n^k(b)\|_2 \le \|D_\varrho^{-1} T_n^k(b_\varrho) D_\varrho\|_2 \le \|D_\varrho^{-1}\|_2 \|T_n^k(b)\|_2 \|D_\varrho\|_2 \le M^{n-1} \|b_\varrho\|_\infty^k$$

for all n and k. Thus, if $(k, n) \in S_B$, then $B < M^{n-1} \|b_\varrho\|_\infty^k$. Since $\varrho \ne 1$ and thus $M > 1$, it follows that

$$n > k\frac{\log(1/\|b_\varrho\|_\infty)}{\log M} + \frac{\log B}{\log M} + 1 =: ck + d,$$

and as $\|b_\varrho\|_\infty < 1$ and $B > 1$, we see that $c > 0$ and $d > 0$. □

We will see in Chapter 10 that always

$$\limsup_{n\to\infty} \operatorname{rad} T_n(b) \le \inf_{\varrho\in(0,\infty)} \|b_\varrho\|_\infty$$

and that in certain special cases the equality

$$\lim_{n\to\infty} \operatorname{rad} T_n(b) = \inf_{\varrho\in(0,\infty)} \|b_\varrho\|_\infty \tag{8.25}$$

holds. Equality (8.25) is particularly true if $T(b)$ is Hermitian or tridiagonal or triangular or nonnegative, where nonnegativity means that $b_j \ge 0$ for all j. Hermitian matrices are uninteresting in our context, because for them the value given by (8.25) coincides with $\|a\|_\infty$. However, if $T(b)$ is tridiagonal or triangular or nonnegative and if $\lim \operatorname{rad} T_n(b) < 1$, then (8.25) implies that we can find a $\varrho \in (0,\infty)$ such that $\|b_\varrho\|_\infty < 1$ and hence, by Theorem 8.13, the sky region is contained in an angle. Equivalently, the lowland contains an angle. It also follows that in these cases

$$\limsup_{n\to\infty} \frac{L_B(n)}{n} < \infty.$$

Suppose now that $b \in \mathcal{P}_r$, $\lim \operatorname{rad} T_n(b) < 1$, and $\|b\|_\infty > 1$. If the lower border of the sky region is strictly sublinear, then, by Theorem 8.13, $\inf \|b_\varrho\|_\infty$ must be at least 1. We looked for such symbols in $\mathcal{P}_3$ and observed that they are difficult to identify. The determination of $\inf \|b_\varrho\|_\infty$ for a given $b \in \mathcal{P}_3$ is simple. The problem comes with checking whether $\limsup \operatorname{rad} T_n(b)$ is smaller than 1 for a given candidate $b \in \mathcal{P}_3$. It turns out that $\inf \|b_\varrho\|_\infty$ and $\limsup \operatorname{rad} T_n(b)$ are usually extremely close. We took 3000 random symbols $b \in \mathcal{P}_3$ whose real and imaginary parts of the 7 coefficients $b_{-3}, \dots, b_3$ were drawn from the uniform distribution on $(-1, 1)$. Each time MATLAB computed

$$q = \inf_{\varrho\in(0,\infty)} \|b_\varrho\|_\infty / \operatorname{rad} T_{64}(b).$$

The result was as follows:

$1.00 \le q < 1.02$ in 2793 samples (= 93.1 %)
$1.02 \le q < 1.04$ in 117 samples (= 3.9 %)
$1.04 \le q < 1.06$ in 48 samples (= 1.6 %)
$1.06 \le q < 1.08$ in 23 samples (= 0.77 %)
$1.08 \le q < 1.10$ in 10 samples (= 0.33 %)
$1.10 \le q < 1.20$ in 9 samples (= 0.3 %)

and there was no sample with $q \ge 1.20$. Thus, the dice show that if $\operatorname{rad} T_{64}(b) \le 0.98$, then $\inf \|b_\varrho\|_\infty$ is at most $1.02 \cdot 0.98 = 0.9996 < 1$ with probability about 93 % and if $\operatorname{rad} T_{64}(b) \le 0.83$, then $\inf \|b_\varrho\|_\infty$ does not exceed $1.20 \cdot 0.83 = 0.996 < 1$ nearly surely. This result reveals that if we had a symbol $b \in \mathcal{P}_3$ for which $\limsup \operatorname{rad} T_n(b) < 1$ and $\inf \|b_\varrho\|_\infty \ge 1$, then $\limsup \operatorname{rad} T_n(b)$ would be dramatically close to 1. Ensuring that $\limsup \operatorname{rad} T_n(b)$ is really strictly smaller than 1 and guaranteeing at the same time that $\inf \|b_\varrho\|_\infty \ge 1$ requires subtle tiny adjustments in the higher decimals after the comma of the coefficients. As we were not sure whether these subtleties survive the numerics needed plot the norm surface, that is, to compute $\|T_n^k(b)\|_2$ for n in the 30's and k in the hundreds, we gave up. Thus, we do not know a single symbol with strictly sublinear lower-right border of the sky region or, equivalently, with a lowland that does not contain an angle.

Example 8.14. Figure 8.8 shows the norm surface for the symbol $b(t) = t^{-1} + 0.49^2 t$. The spectral radius of $T_n(b)$ is exactly $0.98\cos\frac{\pi}{n+1}$ and $\|b\|_\infty = 1 + 0.49^2 = 1.2401$. We clearly see the critical transient behavior of $\|T_n^k(b)\|_2$ for n exceeding 20 or 30. We also nicely see the indents in the level curves. The matrix $T(b)$ is nonnegative, and hence, by Theorem 8.13, the sky region must be contained in an angle. The strange thing with Figure 8.8 is that the lower-right pieces of the level curves nevertheless look slightly sublinear. Let $n = \varphi_B(k)$ be the equation of the lower-right piece of the level curve $\|T_n^k(b)\|_2 = B$. If φ_B were sublinear, then $\varphi_B(k)/k$ would approach zero as $k \to \infty$. The left picture of Figure 8.9, showing $10\,\varphi_B(k)/k$ and $10\,\varphi_B(k)\log k/k$ for $100 \le k \le 1000$ does not yet convincingly indicate that $\varphi_B(k)/k$ tends to a positive finite limit. However, the right picture of Figure 8.9, where we extended the range of the k's up to 3000, reveals that there must be a positive finite limit for the lower curve $10\,\varphi_B(k)/k$. □

Theorem 8.12 tells us that the lower-right boundary of the sky region is always sublinear and Theorem 8.13 shows that it is superlinear in many cases. Things are completely different for the $n \times n$ truncations A_n of arbitrary bounded linear operators on ℓ^2.

Let $\psi : \mathbf{N} \to (e, \infty)$ be any function such that $\psi(n) \to \infty$ as $n \to \infty$. We put

$$\xi(n) = \frac{\psi(n)}{\log\psi(n)}, \qquad \lambda_n = e^{-1/\xi(n)}.$$

Clearly, $1/e < \lambda_n < 1$. We define the operator A by

$$\begin{aligned} A &= \operatorname{diag}(J_2(\lambda_1), J_2(\lambda_2), \dots) \\ &:= \operatorname{diag}\left(\begin{pmatrix} \lambda_1 & 0 \\ 1 & \lambda_1 \end{pmatrix}, \begin{pmatrix} \lambda_2 & 0 \\ 1 & \lambda_2 \end{pmatrix}, \dots\right). \end{aligned}$$

Since $\lambda_n < 1$ for all n, the operator A is bounded on ℓ^2. We have

$$A_n^k = \begin{cases} \operatorname{diag}(J_2^k(\lambda_1), \dots, J_2^k(\lambda_m)) & \text{for} \quad n = 2m, \\ \operatorname{diag}(J_2^k(\lambda_1), \dots, J_2^k(\lambda_m), \lambda_{m+1}^k) & \text{for} \quad n = 2m+1. \end{cases}$$

The equality

$$J_2^k(\lambda) = \begin{pmatrix} \lambda^k & 0 \\ k\lambda^{k-1} & \lambda^k \end{pmatrix}$$

and the condition $1/e < \lambda < 1$ imply that

$$k\lambda^k \le k\lambda^{k-1} \le \|J_2^k(\lambda)\|_2 \le \lambda^k + k\lambda^{k-1} \le (1+ek)\lambda^k.$$

Consequently, up to constants independent of n and k, we may replace $\|A_n^k\|_2$ by $k\lambda_m^k$ with $m = [n/2]$. The function $f_m(x) := x\lambda_m^x$ has its maximum at $\xi(m)$ and $f_m(\xi(m)) = (1/e)\xi(m)$. If $k \ge \psi(m)$, then $f_m(k) \le f_m(\psi(m))$ because f_m is monotonically decreasing on the right of $\psi(m)$ (note that $\psi(m) \ge \xi(m)$). As $f_m(\psi(m)) = 1$, it follows that

$$S_B \subset \{(k,n) : k < \psi([n/2])\} \tag{8.26}$$

once B is large enough, say $B > 10$. Thus, on choosing very slowly increasing functions ψ, we obtain very narrow sky regions. In particular, the sky regions need not contain any angles.

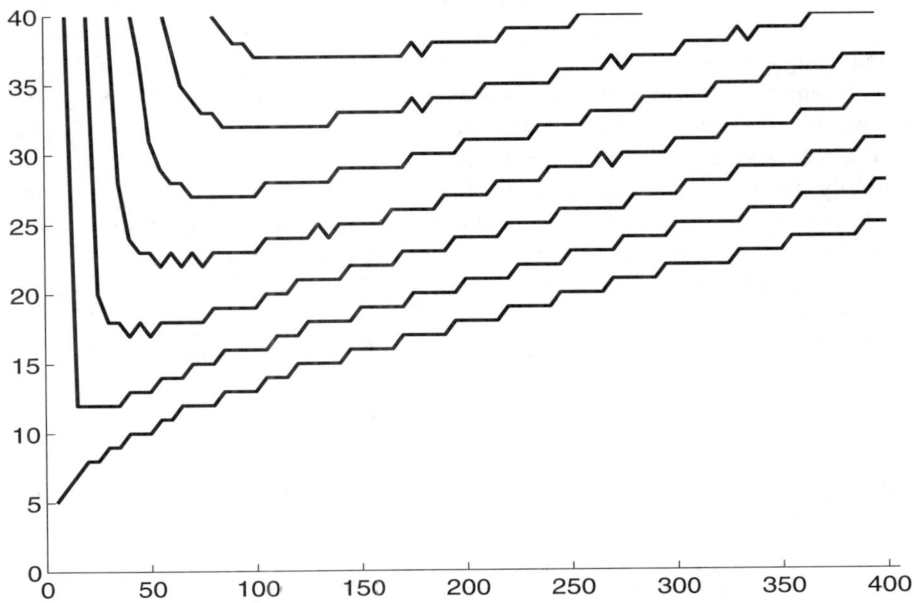

Figure 8.8. *The symbol is* $b(t) = t^{-1} + 0.49^2 t$. *The picture shows the level curves* $\|T_n^k(b)\|_2 = h$ *for* $h = 1, 10, 10^2, 10^3, \ldots, 10^6$ *(the lower curve corresponds to* $h = 1$, *the upper to* 10^6*). We took* $3 \le n \le 40$ *and* $k = 5, 10, 15, \ldots, 400$.

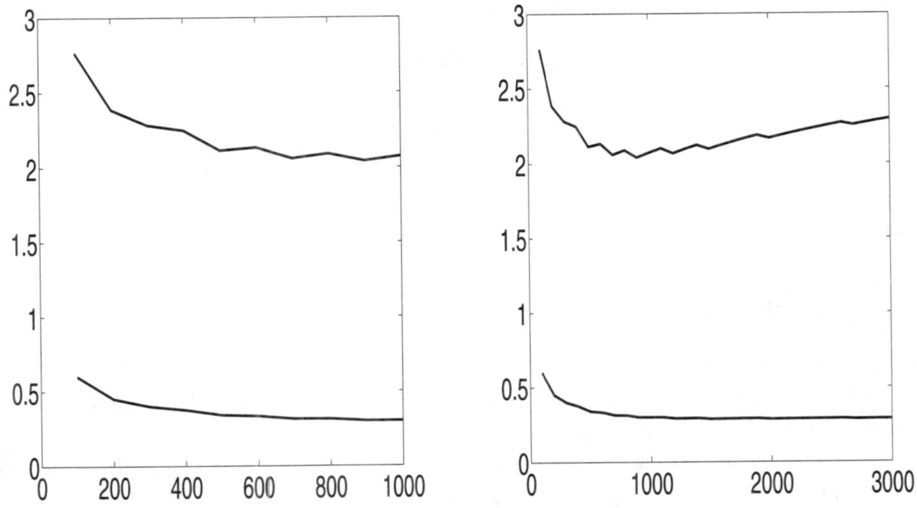

Figure 8.9. *The symbol is* $b(t) = t^{-1} + 0.49^2 t$. *The pictures show* $10\,\varphi_B(k)/k$ *and* $10\,\varphi_B(k) \log k/k$ *for* $B = 10^{-5}$ *over two different ranges of* k.

This is the right moment to return to what was said after Theorem 8.12. For the operator A just constructed, we have $\|A_n^k\|_2 \simeq k\lambda_{[n/2]}^k \to k$ as $n \to \infty$. Thus, when moving along the line $k = 10^4$ on the norm surface, we will eventually be at a height of about 10^4 and may conclude that $\|A_n^k\|_2$ is close to 10^4 for all sufficiently large n. But if, for example, $\psi(n) = \log n$ for large n, then (8.26) shows that we will be in $S_{10^4/2}$ only for the n's satisfying $\log[n/2] > k = 10^4$, that is, for $n > 2\exp(10^4) \approx 2 \cdot 10^{4343}$. We beautifully see that in this case movement along the lines $k =$ constant does practically not provide us with information about the evolution of the norms along the lines $n =$ constant. Moral: Theorem 8.1 is a good reason for expecting critical behavior whenever $\|a\|_\infty > 1$, but *it is Theorem* 8.1 *in conjunction with Theorem* 8.12 *that justifies this expectation within reasonable dimensions.*

In the language of critical transient phase lengths, (8.26) says that

$$\limsup_{n\to\infty} \frac{L_B(n)}{\psi([n/2])} < \infty$$

for the operator constructed above. An estimate in the reverse direction is also easy. Namely, it is readily seen that

$$f_m\left(\frac{1}{2}\,\xi(m)\right) = \frac{1}{2\sqrt{e}}\,\xi(m), \quad f_m(2\,\xi(m)) = \frac{2}{2e^2}\,\xi(m),$$

and since $2/e^2 < 1/(2\sqrt{e})$, this shows that $f_m(k) > \frac{2}{e^2}\,\xi(m)$ for $\frac{1}{2}\,\xi(m) < k < 2\,\xi(m)$. Consequently, $L_B(n) \ge (3/2)\xi([n/2])$ if only $(2/e^2)\xi([n/2]) > B$. As the last inequality is satisfied for all sufficiently large n, we arrive at the conclusion that

$$\liminf_{n\to\infty} \frac{L_B(n)}{\xi([n/2])} \ge \frac{3}{2}.$$

Finally, since $\xi(m) = \psi(m)/\log\psi(m)$, it follows that

$$\liminf_{n\to\infty} \frac{L_B(n)}{\psi([n/2])}\,\log\psi([n/2]) > 0.$$

Rapidly growing functions ψ, such as $\psi(n) = e^n$, therefore yield gigantic critical phase lengths.

8.8 Oscillations

Let $b(t) = t^{-1} + 0.49^2 t$ be as in Example 8.14. When plotting $\|T_{30}^k(b)\|$ for k between 1 and 300, we see Figure 8.10 on the screen, and the purpose of this section is to explain the oscillating behavior in Figure 8.10.

Let A_n be an $n \times n$ matrix and suppose rad $A_n < 1$. For the sake of simplicity, assume that all eigenvalues $\lambda_1, \dots, \lambda_n$ of A_n are simple. We then have $A_n = C\Lambda C^{-1}$, where $\Lambda = \operatorname{diag}(\lambda_1, \dots, \lambda_n)$ and C is an invertible $n \times n$ matrix. It follows that $A_n^k = C\Lambda^k C^{-1}$, and hence

$$A_n^k = C_1\lambda_1^k + \cdots + C_n\lambda_n^k \tag{8.27}$$

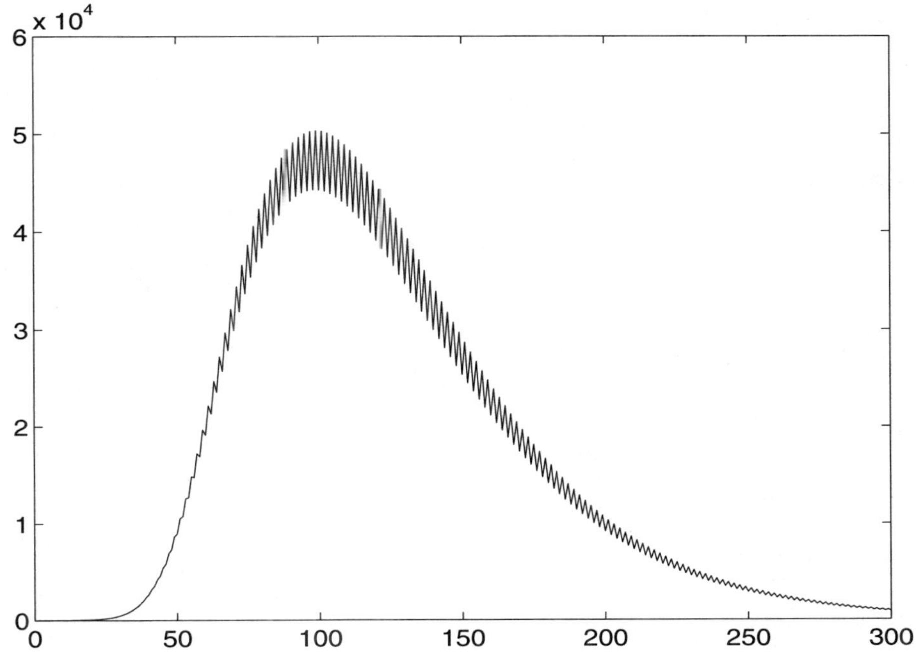

Figure 8.10. *The symbol is $b(t) = t^{-1} + 0.49^2 t$. We see the norms $\|T_{30}^k(b)\|_2$ for $k = 1, 2, 3, \ldots, 300$. A close-up is in the left picture of Figure* 8.11.

with certain $n \times n$ matrices $C_1, \ldots, C_n$ that do not depend on k (note that C_j is simply the product of the jth column of C by the jth row of C^{-1}).

Assume that

$$|\lambda_1| = \cdots = |\lambda_s| > \max_{j \geq s+1} |\lambda_j|.$$

Then (8.27) gives

$$A_n^k = C_1\lambda_1^k + \cdots + C_s\lambda_s^k + O(\sigma^k) \quad \text{as} \quad k \to \infty$$

with $\sigma = \max_{j \geq s+1} |\lambda_j|$. Thus, if k is large, then

$$\|A_n^k\|_2 \approx \|C_1\lambda_1^k + \cdots + C_s\lambda_s^k\|_2, \tag{8.28}$$

which has good chances for oscillatory behavior due to the fact that λ_1 to λ_s have equal moduli.

In the case where A_n is the Toeplitz matrix $T_n(b)$ with $b(t) = t^{-1} + \alpha^2 t$, the eigenvalues are given by

$$2\alpha \cos \frac{\pi j}{n+1} \quad (j = 1, \ldots, n).$$

The two dominating eigenvalues are

$$\lambda_1 = 2\alpha \cos\frac{\pi}{n+1} \quad \text{and} \quad \lambda_2 = 2\alpha\cos\frac{n\pi}{n+1} = -2\alpha\cos\frac{\pi}{n+1}.$$

Consequently,

$$\|T_n^k(b)\|_2 \approx \left(|2\alpha|\,\cos\frac{\pi}{n+1}\right)^k \|C_1 + (-1)^k C_2\|_2 \tag{8.29}$$

for large k. Thus, the damping factor $|2\alpha|^k \cos^k \frac{\pi}{n+1}$ has an amplitude that equals $\|C_1+C_2\|_2$ for even k and $\|C_1 - C_2\|_2$ for odd k. If the damping factor is not too small, then (8.29) (which holds for large k only) should be already valid in the critical transient phase so that we can see it with our eyes. This would be an explanation for the oscillating behavior in Figure 8.10. Let us check our example. Thus, take $\alpha = 0.49$ and $n = 30$. Then

$$\|T_{30}^{100}(b)\|_2 = 10^{4.6457}, \quad \left(0.98\cos\frac{\pi}{31}\right)^{100} \|C_1 + C_2\|_2 = 10^{5.1438},$$

$$\|T_{30}^{101}(b)\|_2 = 10^{4.7014}, \quad \left(0.98\cos\frac{\pi}{31}\right)^{101} \|C_1 - C_2\|_2 = 10^{5.1959},$$

that is, (8.29) cannot be said to be satisfied. The point is that the modulus of the quotient of the dominant eigenvalues and the next eigenvalue is very close to 1, which implies that approximation (8.29) is not yet good enough in the high transient phase. However, consideration of a few more terms does set things right:

$$\left(0.98\cos\frac{\pi}{31}\right)^k \left\| C_1 + (-1)^k C_2 + (C_3 + (-1)^k C_4)\left(\frac{\cos(2\pi/31)}{\cos(\pi/31)}\right)^k \right\|_2$$

equals $10^{4.4597}$ and $10^{4.5201}$ for $k = 100$ and $k = 101$, respectively, and

$$\left(0.98\cos\frac{\pi}{31}\right)^k \Bigg\| C_1 + (-1)^k C_2 + (C_3 + (-1)^k C_4)\left(\frac{\cos(2\pi/31)}{\cos(\pi/31)}\right)^k + (C_5 + (-1)^k C_6)\left(\frac{\cos(3\pi/31)}{\cos(\pi/31)}\right)^k \Bigg\|_2$$

is $10^{4.6512}$ and $10^{4.7067}$ for $k = 100$ and $k = 101$, respectively.

From the paper [96] by Brian Davies, we learned that such oscillation phenomena for semigroup norms are well known. We also learned from [96] that the kind of oscillation may depend on the norm chosen. Figures 8.11 and 8.12 show the different oscillatory behavior of the spectral norms $\|T_n^k(a)\|_2$ and the Frobenius norms $\|T_n^k(a)\|_F$ fairly convincingly. Finally, Figure 8.13 illustrates what happens when walking on the norm surface along the lines $n =$ constant.

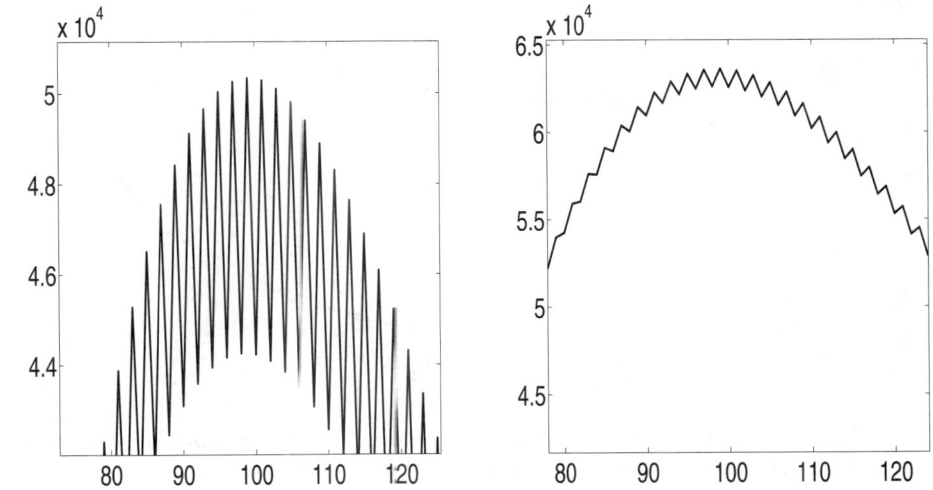

Figure 8.11. *The symbol is given by* $b(t) = t^{-1} + 0.49^2 t$. *The left picture is a close-up of part of Figure* 8.10. *The right picture shows the Frobenius norms* $\|T_{30}^k(b)\|_F$.

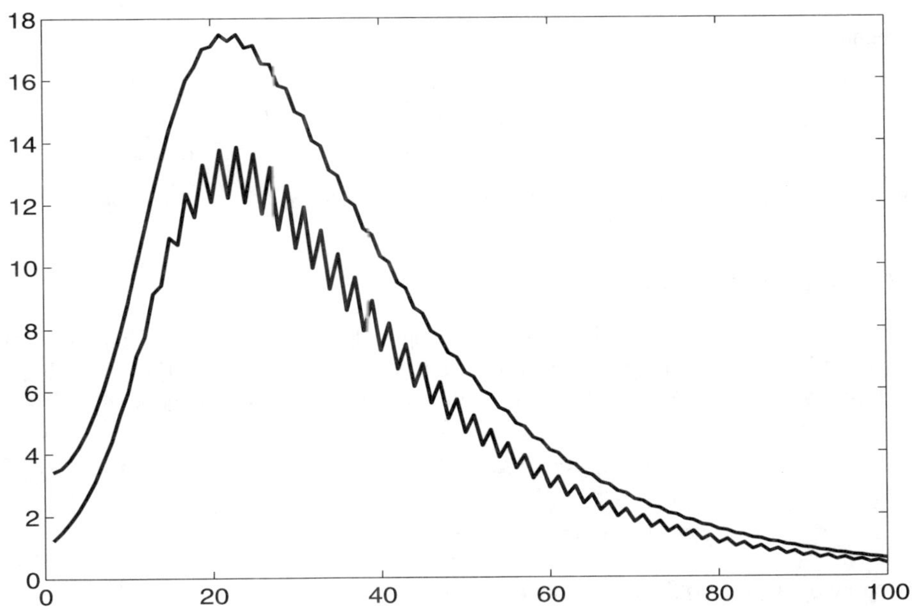

Figure 8.12. *The symbol is* $b(t) = t^{-1} + 0.49^2 t$. *We see the norms* $\|T_{12}^k(b)\|_2$ *(lower curve) and the Frobenius norms* $\|T_{12}^k(b)\|_F$ *(upper curve) for* $k = 1, 2, 3, \dots, 100$.

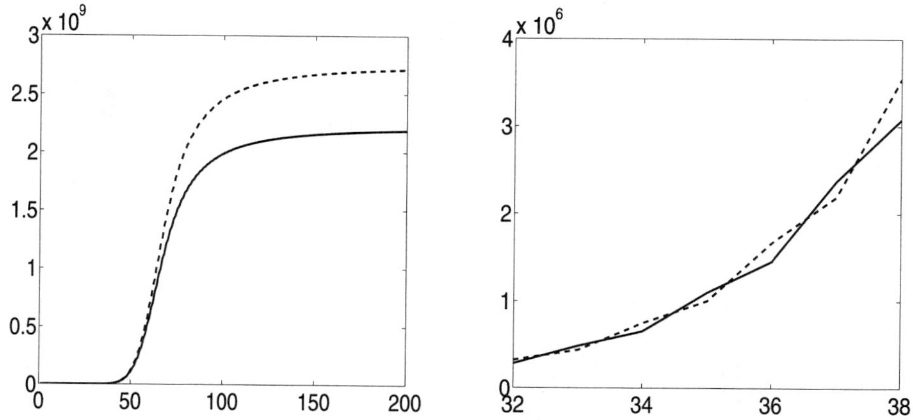

Figure 8.13. *The symbol is again as in Figure* 8.10. *The pictures show the norms* $\|T_n^{100}(a)\|$ *(solid) and* $\|T_n^{101}(a)\|$ *(dashed) for two different ranges of n.*

Another example is considered in Figures 8.14 and 8.15. Now the symbol is $b(t) = \frac{10}{19}(t + t^{-2})$. We have rad $T_{30}(b) = 0.9847$ and $\|b\|_\infty = 1.04$. The dominating eigenvalues of $T_{30}(b)$ are

$$\lambda_1 = \mu, \ \ \lambda_2 = \mu\omega, \ \ \lambda_3 = \mu\omega^2 \quad (\mu = 0.9847, \ \ \omega = e^{2\pi i/3})$$

and hence

$$\|T_{30}^k(b)\|_2 \approx \begin{cases} \|C_1 + C_2 + C_3\|_2 \, |\mu|^k & \text{for} \quad k \equiv 0 \,(\text{mod}\, 3), \\ \|C_1 + C_2\omega + C_3\omega^2\|_2 \, |\mu|^k & \text{for} \quad k \equiv 1 \,(\text{mod}\, 3), \\ \|C_1 + C_2\omega^2 + C_3\omega\|_2 \, |\mu|^k & \text{for} \quad k \equiv 2 \,(\text{mod}\, 3). \end{cases}$$

The period 3 is nicely seen in the right picture of Figure 8.15.

8.9 Exponentials

Let A_n be an $n \times n$ matrix and let $\tau > 0$. Then $\|e^{\tau A_n}\| \to 0$ as $\tau \to \infty$ if and only if the spectrum of A_n is contained in the open left half-plane. Now let $A_n = T_n(b)$ with $b \in W$. The decomposition $b = \text{Re}\, b + i\, \text{Im}\, b$ of b into the real and imaginary parts yields

$$\text{Re}\, A_n := \frac{1}{2}(A_n + A_n^*) = T_n(\text{Re}\, b),$$

$$\text{Im}\, A_n := \frac{1}{2i}(A_n - A_n^*) = T_n(\text{Im}\, b).$$

We denote by max Re b the maximum of Re b on the unit circle $\mathbf{T}$.

Here is the analogue of Theorem 8.1.

Theorem 8.15. *Let* $b \in W$. *Then*

$$\|e^{\tau T_n(b)}\|_2 \le e^{\tau \sup \text{Re}\, b} \ \ \textit{for all} \ \ \tau > 0 \ \ \textit{and all} \ \ n \in \mathbf{N}. \tag{8.30}$$

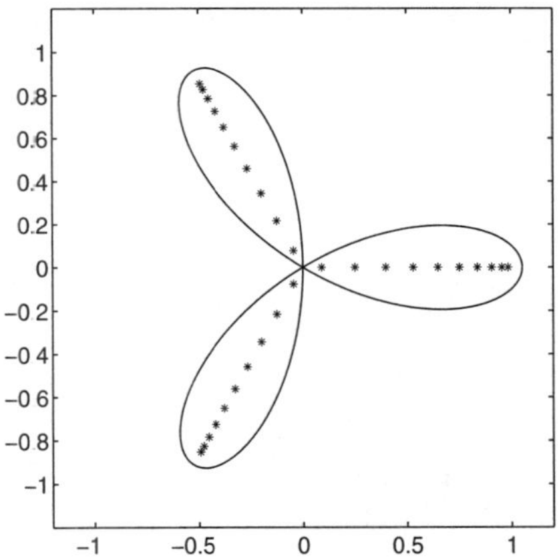

Figure 8.14. *The range $a(\mathbf{T})$ for $a(t) = \frac{10}{19}(t + t^{-2})$ and the eigenvalues of $T_{30}(a)$. The three dominating eigenvalues are 0.9847, 0.9847ω, $0.9847\omega^2$ with $\omega = e^{2\pi i/3}$. The maximum modulus of the remaining 27 eigenvalues is 0.9549.*

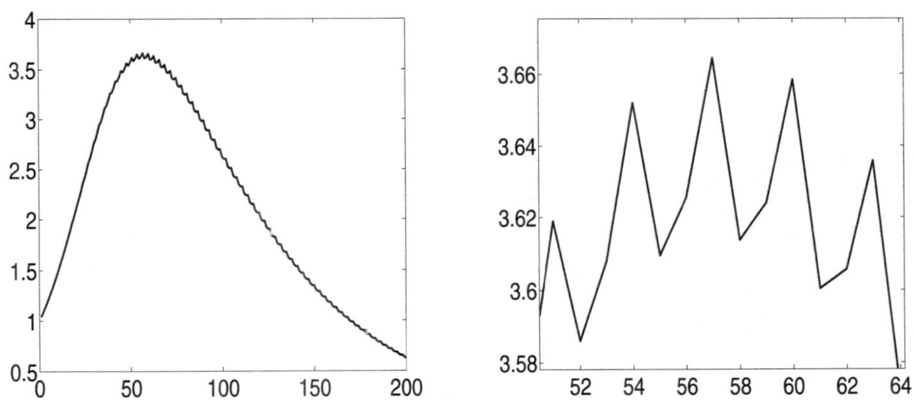

Figure 8.15. *The pictures show the norms $\|T_{30}^k(b)\|_2$ for the symbol defined by $b(t) = \frac{10}{19}(t + t^{-2})$.*

Moreover,

$$\lim_{n\to\infty} \|e^{\tau T_n(b)}\|_2 = e^{\tau \sup \operatorname{Re} b} \textit{ for each } \tau > 0. \tag{8.31}$$

Proof. We have $\|e^{A_n}\|_2 \le \|e^{\operatorname{Re} A_n}\|_2$ for every matrix A_n (see, e.g., [27, p. 258]). Furthermore, $\|e^{\operatorname{Re} A_n}\|_2 = e^{\lambda_{\max}(\operatorname{Re} A_n)}$, where $\lambda_{\max}(\operatorname{Re} A_n)$ is the maximal eigenvalue of the Hermitian matrix $\operatorname{Re} A_n$. Thus,

$$\|e^{\tau T_n(b)}\|_2 \le e^{\tau \lambda_{\max}(T_n(\operatorname{Re} b))}.$$

The number $\lambda_{\max}(T_n(\operatorname{Re} b))$ does not exceed $\max \operatorname{Re} b$. The proof is as follows. Pick

$\mu > \max \operatorname{Re} b$ and assume $T_n(\operatorname{Re} b)x = \mu x$ for some nonzero $x = (x_j)_{j=0}^{n-1} \in \mathbf{C}^n$. Put

$$x(e^{i\theta}) = x_0 + x_1 e^{i\theta} + \cdots + x_{n-1} e^{i(n-1)\theta}.$$

Formula (4.16) shows that

$$(T_n(\mu - \operatorname{Re} b)x, x) = \frac{1}{2\pi} \int_0^{2\pi} (\mu - (\operatorname{Re} b)(e^{i\theta}))|x(e^{i\theta})|^2 d\theta, \tag{8.32}$$

and since $T_n(\mu - \operatorname{Re} b)x = 0$, the right-hand side of (8.32) must be zero. As $x(e^{i\theta})$ vanishes only at finitely many $e^{i\theta} \in \mathbf{T}$, it follows that $\mu - \operatorname{Re} b = 0$ almost everywhere, which is impossible for $\mu > \max \operatorname{Re} b$. Thus, the proof of (8.30) is complete.

Since $e^{\tau T_n(b)} \to e^{\tau T(b)}$ strongly, the Banach-Steinhaus theorem gives

$$\liminf_{n\to\infty} \|e^{\tau T_r(b)}\|_2 \ge \|e^{\tau T(b)}\|_2 \ge \operatorname{rad} e^{\tau T(b)}.$$

From the spectral mapping theorem we deduce that

$$\operatorname{sp} e^{\tau T_n(b)} = e^{\tau \operatorname{sp} T(b)},$$

and hence Corollary 1.12 implies that

$$\liminf_{n\to\infty} \|e^{\tau T_n(b)}\|_2 \ge e^{\tau \max \operatorname{Re} b}.$$

This and (8.30) imply (8.31). □

Thus, Theorem 8.15 tells us that if n is large, then $e^{\tau T_n(b)}$ has critical behavior if and only if $\max \operatorname{Re} b > 0$. To get realistic estimates, one can employ the analogue of (8.6). Trefethen's note [272] and [96], [275] contain an analogue of Theorem 8.3 for exponentials. This result implies that if the pseudospectrum $\operatorname{sp}_\varepsilon A$ contains points in the open right half-plane, then each point $\lambda \in \operatorname{sp}_\varepsilon A$ with $\beta := \operatorname{Re} \lambda > 0$ gives us an estimate of the form

$$\sup_{0<\tau\le\tau_0} \|e^{\tau A}\| \ge e^{\beta\tau_0} \Big/ \left(1 + \varepsilon \, \frac{e^{\beta\tau_0} - 1}{\beta}\right).$$

Things are a little more complicated with the analogue of Theorem 8.12, because we do not know an analogue of (8.22). We can at least prove the following (see [39] for details).

Theorem 8.16. *Let $b \in \mathcal{P}$ and suppose* $\max \operatorname{Re} b > 0$. *For $B > 0$, put*

$$S_B = \{(\tau, n) \in (0, \infty) \times \mathbf{N} : \|e^{\tau T_n(b)}\|_2 > B\}.$$

Then for each $\varepsilon > 0$, there exist positive and finite constants τ_0 and c depending only on B, b, ε such that

$$S_B \supset \left\{(\tau, n) : \ \tau > \tau_0, \ \ n > c\tau(\log \tau)^{1+\varepsilon}\right\}.$$

Clearly, this result is weaker than Theorem 8.12 because of the presence of the factor $(\log \tau)^{1+\varepsilon}$. From the practical point of view we can say that, for large τ, this factor may be ignored in comparison with τ. As for the theoretical side of the problem, we remark that this factor emerges from our techniques and that it can probably be removed by more powerful machinery. We conclude with an example.

Example 8.17. Let $T(b)$ be the tridiagonal matrix generated by $b(t) = t^{-1} + \alpha^2 t - \lambda$ with $\alpha \in [0, \infty)$ and $\lambda \in \mathbf{C}$. The eigenvalues of $T_n(b)$ are densely spread over the interval $(-2\alpha - \lambda, 2\alpha - \lambda)$. Therefore, $\|e^{\tau T_n(b)}\|_2 \to 0$ as $\tau \to \infty$ for each n if and only if $2\alpha < \operatorname{Re}\lambda$. We have that $\max \operatorname{Re} b = -\operatorname{Re}\lambda + 1 + \alpha^2$. Consequently, $\|e^{\tau T_n(b)}\|_2$ has a critical transient phase before decaying to zero for large n if and only if $2\alpha < \operatorname{Re}\lambda < 1+\alpha^2$. Figure 8.16 reveals that the norm surfaces of exponentials look much like their counterparts for powers. □

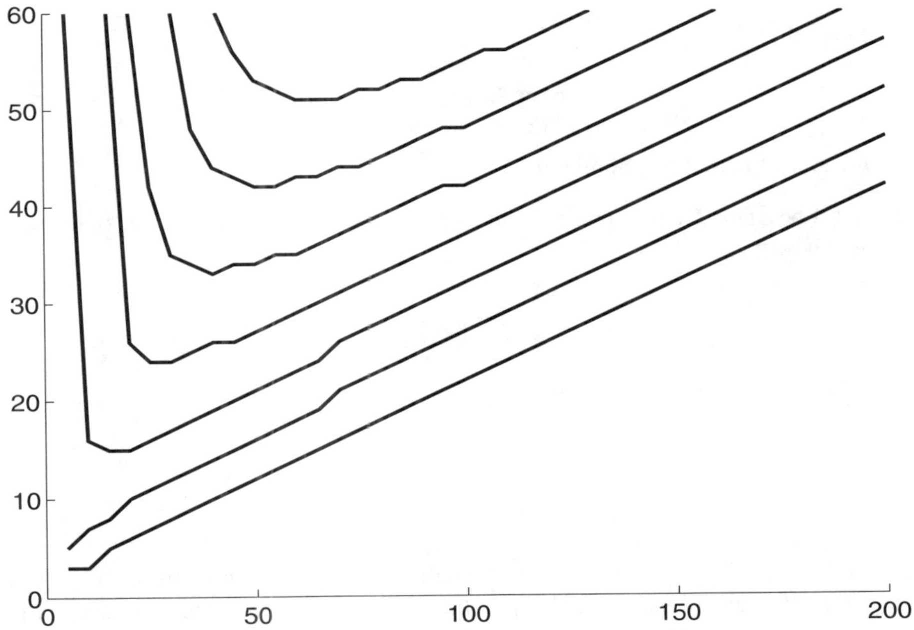

Figure 8.16. *The symbol is* $b(t) = t^{-1} + 3^2 t - 7$. *The picture shows the level curves* $\|e^{\tau T_n(b)}\|_2 = h$ *for* $h = 10^{-2}, 1, 10^2, 10^4, \ldots, 10^{10}$ *(the lower curve corresponds to* $h = 10^{-2}$, *the upper to* 10^{10}*). We took* $3 \le n \le 60$ *and* $\tau = 5, 10, 15, \ldots, 200$.

Exercises

1. Prove that if U_n is unitary, then $G_k(U_n^* A_n U_n) = G_k(A_n)$.
2. Let A_n be an $n \times n$ matrix and let m be the degree of the minimal polynomial of A_n. Prove that

$$\mathcal{H}_2(A_n) = G_1(A_n) \supset G_2(A_n) \supset \cdots \supset G_m(A_n) = G_{m+1}(A_n) = \cdots = \operatorname{sp} A_n.$$

3. Let A_n be a Hermitian matrix. Show that $G_k(A_n)$ equals $\operatorname{conv} \operatorname{sp} A_n$ for $k = 1$ and $\operatorname{sp} A_n$ for $k \ge 2$.

4. The polynomial convex hull of degree k of a set $S \subset \mathbf{C}$ is defined by

$$\operatorname{pco}_k S := \{\zeta \in \mathbf{C} : |p(\zeta)| \le \max_{z \in S} |p(z)| \text{ for all } p \in \mathcal{P}^+_{k+1}\}.$$

Let $b \in \mathcal{P}$. Prove that if n is large enough, then $G_k(C_n(b)) = \operatorname{pco}_k b(\mathbf{T}_n)$, where $\mathbf{T}_n := \{e^{2\pi i \ell/n} : \ell = 0, 1, \dots, n-1\}$. Does $\operatorname{pco}_k b(\mathbf{T}_n)$ converge to $\operatorname{pco}_k b(\mathbf{T})$?

5. Prove that $\operatorname{pco}_k b(\mathbf{T}) \subset G_k(T(b)) \subset \operatorname{conv} \mathcal{R}(b)$ for every $b \in \mathcal{P}$.

6. Let $k \ge 2$. Find a bounded operator A on ℓ^2 such that $G_k(P_n A P_n)$ converges in the Hausdorff metric to a set that contains $G_k(A)$ properly.

7. Prove that

$$\limsup_{n \to \infty} G_k(T_n(b)) \subset G_k(T(b))$$

for every Laurent polynomial b.

8. Let A be a nonzero $n \times n$ matrix and let $\operatorname{rad} A$ denote the spectral radius of the matrix A. Show that

$$\limsup_{k \to \infty} \frac{\|A^k\|_2}{(\operatorname{rad} A)^k}$$

is finite if and only if A is diagonalizable.

Notes

In this chapter we follow [39].

As already said, polynomial numerical hulls were independently invented by Nevanlinna [192], [193] and Greenbaum [141], [142]. It should be mentioned that Faber, Greenbaum, and Marshall pointed out in [112] that the asymptotic formula

$$\varrho_{n-1,n} = 1 - \frac{\log(2n)}{n} + \frac{\log(\log(2n))}{n} + o\left(\frac{1}{n}\right)$$

is in fact a very old result. Namely, the problem of determining $\varrho_{n-1,n}$ is equivalent to a classical problem in complex approximation theory (closely related to the Carathéodory-Fejér interpolation problem) which was explicitly solved by Schur and Szegö [244] and then rediscovered with a different proof by Goluzin [138], [139, Theorem 6, pp. 522–523]. However, the proof in [112] yields more information than the earlier proofs since it allowed the authors to determine $\varrho_{4m-2,4m}$, for example (it is shown that $\varrho_{4m-2,4m} = \sqrt{\varrho_{2m-1,2m}}$).

The exploitation of pseudospectra in connection with the transient behavior of powers of Toeplitz matrices goes back at least to Figure 2 of Reichel and Trefethen's paper [219]. Section 8.3 addresses issues concerning the computation of concrete pseudospectra. Tom

Wright's package EigTool [299] provides us with fantastic software for computing pseudospectra numerically. In 1999, Trefethen [271] wrote: "In 1990, getting a good plot of pseudospectra on a workstation for a 30×30 matrix took me several minutes. Today I would expect the same of a 300×300 matrix, and pseudospectra of matrices with dimensions in the thousands are around the corner." Concerning the computation of pseudospectra of dense matrices, Wright notes in his 2002 thesis [300]: "What was once a very expensive computation has become one that is practically of a similar order of complexity to that of computing eigenvalues." We refer the reader to [110], [271], [275] for more on the subject.

The results of Exercises 1 to 3 are from [141], [192], [193]. In connection with Exercise 7 we remark that James Burke and Anne Greenbaum [80] recently proved that $\liminf G_k(T_n(b))$ contains the closure of the interior of $G_k(T(b))$.

Further results: numerical range of finite Toeplitz matrices. We denote by $\mathcal{P}_r$ the set of the Laurent polynomials of degree r, and for $b(t) = \sum_{j=-r}^{r} b_j t^j$ in $\mathcal{P}_r$, we define the Abel-Poisson mean (= harmonic extension) $h_\varrho b \in \mathcal{P}_r$ by

$$(h_\varrho b)(t) = \sum_{j=-r}^{r} \varrho^{|j|} b_j t^j \quad (t \in \mathbf{T}).$$

It is clear that $\mathcal{H}_2(T_n(b)) \subset \mathcal{H}_2(T(b))$. The closure of the convex set $\mathcal{H}_2(T(b))$ is given by Theorem 7.11. The following lower estimate for $\mathcal{H}_2(T_n(b))$ was communicated to us by Anne Greenbaum [143]: If $b \in \mathcal{P}_r$, then

$$\mathcal{H}_2(T_n(b)) \supset \bigcup_{0 \le \varrho \le \varrho_{r,n}} (h_\varrho b)(\mathbf{T}), \tag{8.33}$$

where $\varrho_{r,n}$ is as in Section 8.2. Anne's proof is as follows. Put

$$b_+(z) = \sum_{j=0}^{r} b_j z^j, \quad b_-(z) = \sum_{j=1}^{r} b_{-j} z^j.$$

Then $T_n(b) = b_+(J^\top) + b_-(J)$, where $J := J_n(0)$ (recall (8.3)). Let $\zeta = (J^\top x, x)$ be a point in the polynomial numerical hull $G_r(J^\top)$. In [141], it is shown that $((J^\top)^\ell x, x) = (J^\top x, x)^\ell$ for $\ell = 1, \dots, r$. This implies that $(J^\ell x, x) = \overline{((J^\top)^\ell x, x)} = \overline{(J^\top x, x)^\ell}$ for $\ell = 1, \dots, r$. Hence

$$(T_n(b)x, x) = b_+((J^\top x, x)) + b_-(\overline{(Jx, x)}),$$

and we arrive at the conclusion that $\mathcal{H}_2(T_n(b))$ contains

$$\begin{aligned} &\{b_+(\zeta) + b_-(\overline{\zeta}) : \zeta \in G_r(J^\top)\} \\ &= \{b_+(\varrho e^{i\theta}) + b_-(\varrho e^{-i\theta}) : 0 \le \varrho \le \varrho_{r,n},\ \theta \in [0, 2\pi)\} \\ &= \{(h_\varrho b)(t) : 0 \le \varrho \le \varrho_{r,n},\ t \in \mathbf{T}\}, \end{aligned}$$

which completes the proof.

The numerical range of a *tridiagonal* $n \times n$ matrix was found by Eiermann [107] (see also Brown and Spitkovsky's paper [78]). If $b(t) = t + \alpha^2 t^{-1}$ with $0 < \alpha < 1$, then $\mathcal{H}_2(T_n(b))$ is the ellipse with the foci $\pm 2\alpha \cos(\pi/(n+1))$, the minor half-axis length $(1-\alpha^2)\cos(\pi/(n+1))$, and the major half-axis length $(1+\alpha^2)\cos(\pi/(n+1))$. In other terms, $\mathcal{H}_2(T_n(b)) = \cos(\pi/(n+1)) \operatorname{clos} \mathcal{H}_2(T(b))$. We know from Section 8.2 that $\cos(\pi/(n+1))$ is just $\varrho_{1,n}$. Thus, in the case at hand estimate (8.33) amounts to the containment

$$\begin{aligned}\mathcal{H}_2(T_n(b)) &\supset \{\varrho(t+\alpha^2 t^{-1}) : 0 \le \varrho \le \varrho_{1,n},\ t \in \mathbf{T}\}\\ &= \left\{\varrho(t+\alpha^2 t^{-1}) : 0 \le \varrho \le \cos\frac{\pi}{n+1},\ t \in \mathbf{T}\right\}\\ &= \cos\frac{\pi}{n+1} \operatorname{clos} \mathcal{H}_2(T(b)).\end{aligned}$$

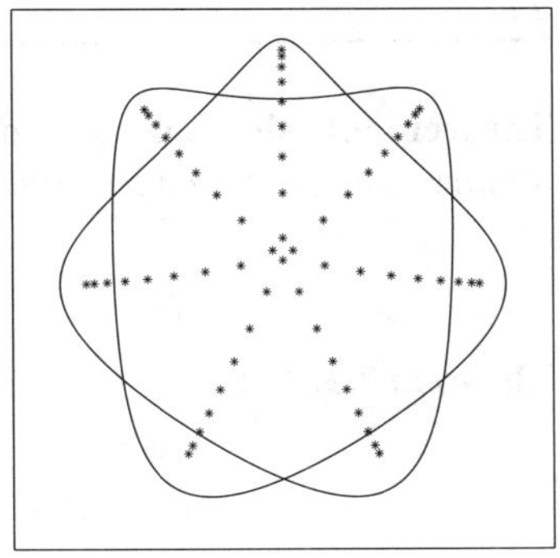

Chapter 9

Singular Values

The asymptotic behavior of the singular values of the matrices $T_n(b)$ is, in a sense, a mirror image of the topological properties of the symbol b: Roch and Silbermann's splitting phenomenon says that if b has no zeros on $\mathbf{T}$ and winding number k about the origin, then the $|k|$ smallest singular values of $T_n(b)$ go exponentially fast to zero while the remaining $n - |k|$ singular values stay away from zero. We also determine the limiting set of the singular values of $T_n(b)$ as $n \to \infty$, and we prove the Avram-Parter theorem, which identifies the corresponding limiting measure. As an application of the Avram-Parter theorem, we show that if x is a randomly chosen vector of length n and n is large, then $\|T_n(b)x\|_2^2$ clusters sharply around a certain value which, moreover, is much smaller than one would predict.

9.1 Approximation Numbers

For $j \in \{0, 1, \ldots, n\}$, let $\mathcal{F}_j^{(n)}$ denote the collection of all $n \times n$ matrices of rank at most j. The jth *approximation number with respect to the ℓ^p norm* of an $n \times n$ matrix A_n is defined by

$$a_j^{(p)}(A_n) = \text{dist}_p(A_n, \mathcal{F}_j^{(n)}) := \min\left\{ \|A_n - F_n\|_p : F_n \in \mathcal{F}_j^{(n)} \right\}.$$

Clearly, $0 = a_n^{(p)}(A_n) \le a_{n-1}^{(p)}(A_n) \le \cdots \le a_1^{(p)}(A_n) \le a_0^{(p)}(A_n) = \|A_n\|_p$. Put

$$\sigma_j^{(p)}(A_n) = a_{n-j}^{(p)}(A_n).$$

Thus, $0 = \sigma_0^{(p)}(A_n) \le \sigma_1^{(p)}(A_n) \le \cdots \le \sigma_{n-1}^{(p)}(A_n) \le \sigma_n^{(p)}(A_n) = \|A_n\|_p$. It is well known that in the case $p = 2$ the numbers $\sigma_j^{(p)}(A_n)$ are the *singular values* of a_n, that is, the nonnegative square roots of the eigenvalues $\lambda_j(A_n^* A_n)$ $(j = 1, \ldots, n)$ of the matrix $A_n^* A_n$:

$$\sigma_j(A_n) := \sigma_j^{(2)}(A_n) = \sqrt{\lambda_j(A_n^* A_n)}. \tag{9.1}$$

The following results are standard.

Theorem 9.1 (singular value decomposition). *If A_n is an $n \times n$ matrix, then there exist unitary matrices U_n and V_n such that*

$$A_n = U_n \operatorname{diag}(\sigma_1(A_n), \ldots, \sigma_n(A_n))\, V_n.$$

Theorem 9.2. *If $1 \le p \le \infty$ and A_n is an $n \times n$ matrix, then*

$$\sigma_1^{(p)}(A_n) = \begin{cases} 1/\|A_n^{-1}\|_p & \text{if } A_n \text{ is invertible,} \\ 0 & \text{if } A_n \text{ is not invertible.} \end{cases}$$

Theorem 9.3. *If A_n, B_n, C_n are $n \times n$ matrices and $1 \le p \le \infty$, then*

$$\sigma_j^{(p)}(A_n B_n C_n) \le \|A_n\|_p\, \sigma_j^{(p)}(B_n)\, \|C_n\|_p$$

for every $j \in \{0, 1, \ldots, n\}$.

Given a Hilbert space H and $A \in \mathcal{B}(H)$, we define

$$\Sigma(A) = \left\{\sigma \in [0, \infty) : \sigma^2 \in \operatorname{sp} A^*A\right\}.$$

In particular, for an $n \times n$ matrix A_n we have

$$\Sigma(A_n) = \{\sigma_1(A_n), \ldots, \sigma_n(A_n)\}.$$

Finally, we set

$$\Sigma_p(A_n) = \left\{\sigma_1^{(p)}(A_n), \ldots, \sigma_n^{(p)}(A_n)\right\}.$$

9.2 The Splitting Phenomenon

The splitting phenomenon is the most striking property of the singular values (approximation numbers) of Toeplitz matrices. It is described by the following theorem. An illustration is in Figure 9.1.

Theorem 9.4 (Roch and Silbermann). *Let b be a Laurent polynomial and suppose $T(b)$ is Fredholm of index $k \in \mathbf{Z}$. Let α be any number satisfying (1.23) and $1 \le p \le \infty$. Then the $|k|$ first approximation numbers $\sigma_1^{(p)}(T_n(b)), \ldots, \sigma_{|k|}^{(p)}(T_n(b))$ of $T_n(b)$ go to zero with exponential speed,*

$$\sigma_{|k|}^{(p)}(T_n(b)) = O(e^{-\alpha n}), \tag{9.2}$$

while the remaining $n - |k|$ approximation numbers $\sigma_{|k|+1}^{(p)}(T_n(b)), \ldots, \sigma_n^{(p)}(T_n(b))$ stay away from zero,

$$\sigma_{|k|+1}^{(p)}(T_n(b)) \ge d > 0 \tag{9.3}$$

for all sufficiently large n, where d is a constant depending only on b.

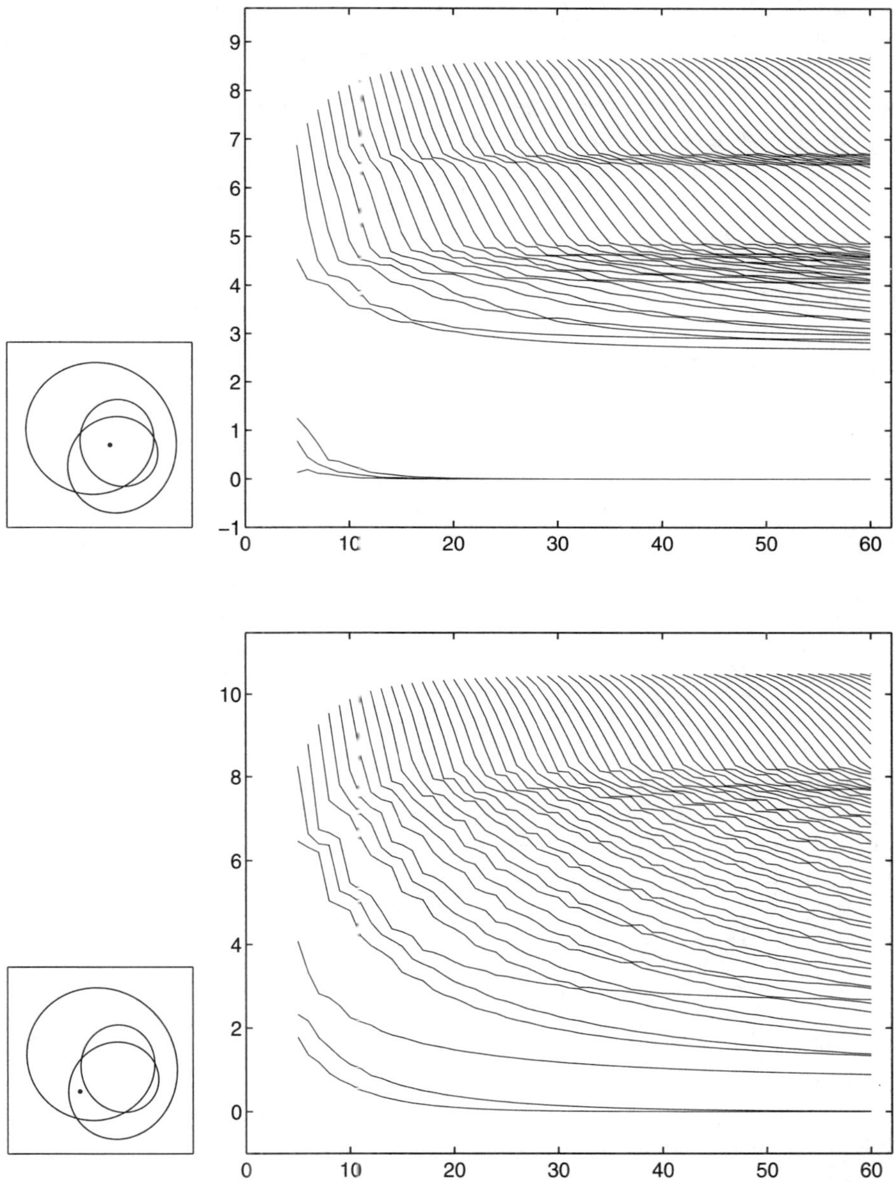

Figure 9.1. *In the small pictures we see the range of the symbol $a(t) = t^{-1} - it + 2it^2 - 5it^3$, which has winding numbers 3 and 2 about the origin and the point $-3 - 2i$, respectively. The singular values of $T_n(a)$ and $T_n(a + 3 + 2i)$ for $5 \le n \le 60$ are shown in the top and bottom pictures. As predicted by Theorem 9.4, three of them go to zero in the top picture and two of them approach zero in the bottom picture. The rest stays away from zero.*

Proof. We first prove (9.2). For the sake of definiteness, let us assume that $k > 0$; there is nothing to be proved for $k = 0$, and the case $k < 0$ can be reduced to the case $k > 0$ by passage to adjoints. Recall that χ_m is defined by $\chi_m(t) := t^m$ ($t \in \mathbf{T}$) and that $\mathcal{P}_n^+$ is the set of all analytic polynomials of degree at most $n-1$. Let $b = b_- \chi_{-k} b_+$ be a Wiener-Hopf factorization of b. If n is sufficiently large, then

$$c_n(t) := \left(\sum_{\ell=0}^{n-k} (b_+^{-1})_\ell t^\ell \right)^{-1}$$

is a function in W. Let $F_n : \mathcal{P}_n^+ \to \mathcal{P}_n^+$ be the linear operator that sends f to $\Pi_n(c_n \chi_{-k} b_- f)$, where Π_n is the orthogonal projection of $L^2(\mathbf{T})$ onto $\mathcal{P}_n^+$. For $j = 0, 1, \dots, k-1$, the function $\chi_j c_n^{-1}$ belongs to $\mathcal{P}_n^+$, and we have

$$F_n(\chi_j c_n^{-1}) = \Pi_n(c_n \chi_{-k} b_- \chi_j c_n^{-1}) = \Pi_n(\chi_{j-k} b_-) = 0.$$

Hence, $\dim \operatorname{Im} F_n = n - \dim \operatorname{Ker} F_n \le n-k$. Let G_n be the matrix representation of F_n with respect to the basis $\{\chi_0, \chi_1, \dots, \chi_{n-1}\}$ of $\mathcal{P}_n^+$. Then $G_n \in \mathcal{F}_{n-k}^{(n)}$ and thus, $\sigma_k^{(p)}(T_n(b)) = a_{n-k}^{(p)}(T_n(b)) \le \|T_n(b) - G_n\|_p$. Since $T_n(b) - G_n$ is the matrix representation of the linear operator $\mathcal{P}_n^+ \to \mathcal{P}_n^+$, $f \mapsto \Pi_n((b_+ - c_n)\chi_{-k} b_- f)$ in the basis $\{\chi_0, \chi_1, \dots, \chi_{n-1}\}$, it follows that

$$\sigma_k^{(p)}(T_n(b)) \le \|b_+ - c_n\|_W \|\chi_{-k}\|_W \|b_-\|_W = O(\|b_+ - c_n\|_W).$$

From Lemma 1.17 we know that $\|b_+^{-1} - c_n^{-1}\|_W = O(e^{-\alpha n})$. This implies that $\|b_+ - c_n\|_W = O(e^{-\alpha n})$ and so gives (9.2).

We now prove inequality (9.3). This time we assume without loss of generality that $k = -j < 0$, since for $k = 0$ the assertion follows from Theorems 3.7 and 9.2 and for $k > 0$ we may pass to adjoints. We have $b = c\chi_j$, where c has no zeros on $\mathbf{T}$ and the winding number of c is zero. As $\|T(\chi_{-j})\|_p = 1$, we deduce from Theorem 9.3 that

$$\begin{aligned}
\sigma_{j+1}^{(p)}(T_n(b)) &= \sigma_{j+1}^{(p)}(T_n(c\chi_j)) = \sigma_{j+1}^{(p)}(T_n(c\chi_j)) \, \|T(\chi_{-j})\|_p \\
&\ge \sigma_{j+1}^{(p)}(T_n(c\chi_j) T_n(\chi_{-j})) = \sigma_{j+1}^{(p)}(T_n(c) - P_n H(c\chi_j) H(\chi_j) P_n)
\end{aligned}$$

(recall Proposition 3.10 for the last equality). Obviously, $\dim \operatorname{Im} H(\chi_j) = j$. Consequently, $H_j := P_n H(c\chi_j) H(\chi_j) P_n \in \mathcal{F}_j^{(n)}$ and hence

$$\begin{aligned}
\sigma_{j+1}^{(p)}(T_n(c) - H_j) &= a_{n-j-1}^{(p)}(T_n(c) - H_j) \\
&= \min\left\{ \|T_n(c) - H_j - K_{n-j-1}\|_p : K_{n-j-1} \in \mathcal{F}_{n-j-1}^{(n)} \right\} \\
&\ge \min\left\{ \|T_n(c) - L_{n-1}\|_p : L_{n-1} \in \mathcal{P}_{n-1}^{(n)} \right\} \\
&= a_{n-1}^{(p)}(T_n(c)) = \sigma_1^{(p)}(T_n(c)).
\end{aligned}$$

Since $T(c)$ is invertible, Theorems 3.7 and 9.2 yield that

$$\liminf_{n\to\infty} \sigma_1^{(p)}(T_n(c)) = \liminf_{n\to\infty} \|T_n^{-1}(c)\|_p^{-1} = d > 0. \qquad \square$$

9.3 Singular Values of Circulant Matrices

Throughout the rest of this chapter we restrict ourselves to the case $p = 2$.

Let b be a Laurent polynomial,

$$b(t) = \sum_{j=-r}^{r} b_j t^j \quad (t \in \mathbf{T}). \tag{9.4}$$

In (9.4) we do not require that both the coefficients b_{-r} and b_r are nonzero. For $n \geq 2r+1$, we define the circulant matrix $C_n(b)$ as in Section 2.1. We have

$$C_n(b) - T_n(b) = \begin{pmatrix} O_{(n-r)\times(n-r)} & D_r \\ E_r & O_{r\times r} \end{pmatrix}, \tag{9.5}$$

where O denotes the zero matrix and

$$D_r = \begin{pmatrix} b_r & b_{r-1} & \dots & b_1 \\ 0 & b_r & \dots & b_2 \\ \vdots & \vdots & \ddots & \vdots \\ 0 & 0 & \dots & b_r \end{pmatrix}, \quad E_r = \begin{pmatrix} b_{-r} & 0 & \dots & 0 \\ b_{-r+1} & b_{-r} & \dots & 0 \\ \vdots & \vdots & \ddots & \vdots \\ b_{-1} & b_{-3} & \dots & b_{-r} \end{pmatrix}.$$

Proposition 9.5. *The singular values of $C_n(b)$ are*

$$|b(1)|, |b(\omega_n)|, \dots, |b(\omega_n^{n-1})|,$$

where $\omega_n := \exp(2\pi i/n)$.

Proof. Clearly, $C_n^*(b) = C_n(\overline{b})$. From (2.6) we infer that $C_n^*(b)C_n(b) = C_n(\overline{b}b) = C_n(|b|^2)$. Thus, by Proposition 2.1, the eigenvalues of $C_n^*(b)C_n(b)$ are just $|b(\omega_n^j)|^2$ $(j = 0, 1, \dots, n-1)$. Formula (9.1) completes the proof. □

Theorem 9.6. *Let b be a Laurent polynomial of the form* (9.4) *and suppose $|b|$ is not constant. Put*

$$m = \min_{t\in\mathbf{T}} |b(t)|, \quad M = \max_{t\in\mathbf{T}} |b(t)|,$$

denote by $\alpha \in \mathbf{N}$ the maximal order of the zeros of $|b| - m$ on $\mathbf{T}$, and let $\beta \in \mathbf{N}$ be the maximal order of the zeros of $M - |b|$ on $\mathbf{T}$. Then for each $k \in \mathbf{N}$ and all sufficiently large n,

$$m \leq \sigma_k(C_n(b)) \leq m + E_k \frac{1}{n^\alpha}, \quad M - D_k \frac{1}{n^\beta} \leq \sigma_{n-k}(C_n(b)) \leq M,$$

where $E_k, D_k \in (0, \infty)$ are constants independent of n.

Proof. Put $f(\theta) = |b(e^{i\theta})| - m$ for $\theta \in [0, 2\pi)$ and let f have a zero of the maximal order α at $\theta_0 \in [0, 2\pi)$. By Proposition 9.5, the singular values of $C_n(b)$ $(n \geq 2r+1)$ are $f(2\pi j/n) + m$ $(j = 0, 1, \dots, n-1)$. Let $U_{k,n}$ be the segment $[\theta_0, \theta_0 + 4\pi k/n]$ and denote by $j_1 < \cdots < j_q$ the numbers j_μ for which $2\pi j_\mu/n$ belongs to $U_{k,n}$. If n is large enough, then f is strictly monotonically increasing on $U_{k,n}$ and q is approximately equal to $2k$. Hence $\sigma_k(C_n(b)) \le f(2\pi j_q/n) + m$. It follows that

$$0 \le \sigma_k(C_n(b)) - m \le f\left(\frac{2\pi j_q}{n}\right) \le f\left(\theta_0 + \frac{4\pi k}{n}\right) \le E\left(\frac{4\pi k}{n}\right)^\alpha = E_k \frac{1}{n^\alpha}.$$

The estimate for $\sigma_{n-k}(C_n(b))$ can be shown analogously. □

9.4 Extreme Singular Values

Theorem 9.4 leaves us with the case where $T(b)$ is not Fredholm. In this section we show that in that case $\sigma_k(T_n(b))$ goes to zero as $n \to \infty$ for each fixed k. This will be done with the help of Theorem 9.6 and the following well-known interlacing result for singular values.

Theorem 9.7. *Let A be a complex $n \times n$ matrix and let $B = P_{n-1}AP_{n-1}$ be the $(n-1)\times(n-1)$ principal submatrix. Then*

$$\begin{array}{ccccc} & & \sigma_1(A) & \le & \sigma_3(B) \\ \sigma_2(B) & \le & \sigma_2(A) & \le & \sigma_4(B) \\ & & \cdots & & \\ \sigma_{n-3}(B) & \le & \sigma_{n-3}(A) & \le & \sigma_{n-1}(B) \\ \sigma_{n-2}(B) & \le & \sigma_{n-2}(A) & \le & \sigma_n(B) \\ \sigma_{n-1}(B) & \le & \sigma_{n-1}(A). & & \end{array}$$

Here is the desired result on the lower singular values.

Theorem 9.8. *Let b be a nonconstant Laurent polynomial and suppose $T(b)$ is not Fredholm. Let $\alpha \in \mathbf{N}$ be the maximal order of the zeros of $|b|$ on $\mathbf{T}$. Then for each natural number $k \ge 1$, $\sigma_k(T_n(b)) = O(1/n^\alpha)$ as $n \to \infty$.*

Proof. First notice that, by Theorem 1.9, $|b|$ does have zeros on $\mathbf{T}$ if $T(b)$ is not Fredholm.

Let b be as in (9.4) and $n \ge 2r+1$. From (9.5) we know that $T_n(b)$ can be successively extended to $C_{n+r}(b)$ by adding one row and one column in each step. We have $\sigma_k(T_n(b)) \le \sigma_{k+1}(T_n(b))$. Since $k+1 \ge 2$, we can r times employ Theorem 9.7 to get $\sigma_k(T_n(b)) \le \sigma_{k+1}(C_{n+r}(b))$ for all sufficiently large n, and Theorem 9.6 with $m = 0$ implies that $\sigma_{k+1}(C_{n+r}(b)) = O(1/n^\alpha)$. □

Note that Theorem 9.8 together with the equality $\|T_n^{-1}(b)\|_2 = 1/\sigma_1(T_n(b))$ yields another proof of Corollary 4.12.

The following result shows that the upper singular values $\sigma_{n-k}(T_n(b))$ always approach $\|T(b)\|_2 = \|b\|_\infty$ as $n \to \infty$, independently of whether $T(b)$ is Fredholm or not.

Theorem 9.9. *Let b be a Laurent polynomial and suppose the modulus $|b|$ is not constant on $\mathbf{T}$. Denote by $\beta \in \mathbf{N}$ the maximal order of the zeros of $\|b\|_\infty - |b|$ on the unit circle $\mathbf{T}$. Then for each $k \geq 0$,*

$$\|b\|_\infty - D_k \frac{1}{n^\beta} \leq \sigma_{n-k}(T_n(b)) \leq \|b\|_\infty$$

with some constant $D_k \in (0, \infty)$ independent of n.

Proof. If n is large enough then, by (9.5) and Theorem 9.7,

$$\sigma_{n-k}(T_n(b)) \geq \sigma_{n-k-2r}(C_{n+r}(b)).$$

The assertion is therefore immediate from Theorem 9.6. □

What happens if $|b|$ is constant? By Proposition 5.6, this occurs if and only if $b(t) = \gamma t^m$ $(t \in \mathbf{T})$ with $\gamma \in \mathbf{C}$ and $m \in \mathbf{Z}$. If $\gamma = 0$, then all singular values of $T_n(b)$ are zero, and if $\gamma \neq 0$, it is easy to see that $|m|$ singular values of $T_n(b)$ are zero and that the $n - |m|$ remaining singular values are equal to $|\gamma|$.

9.5 The Limiting Set

The objective of this section is the determination of the limiting sets $\liminf \Sigma(B_n)$ and $\limsup \Sigma(B_n)$ in the case where B_n is Toeplitz-like. Recall that the set $\Sigma(B)$ of the singular values of a Hilbert space operator is defined as the set of all $\sigma \in [0, \infty)$ for which $\sigma^2 \in \operatorname{sp} B^*B$.

Lemma 9.10. *Let $B_n = T_n(b) + P_n K P_n + W_n L W_n + C_n$, where b is a Laurent polynomial, K and L are matrices with only a finite number of nonzero entries, and $\|C_n\|_2 \to 0$ as $n \to \infty$. Put $B = T(b) + K$ and $\widetilde{B} = T(\widetilde{b}) + L$. Then*

$$\liminf_{n\to\infty} \operatorname{sp}(B_n) \subset \limsup_{n\to\infty} \operatorname{sp}(B_n) \subset \operatorname{sp} B \cup \operatorname{sp} \widetilde{B}, \tag{9.6}$$

and if, in addition, the matrices B_n are all Hermitian, then

$$\liminf_{n\to\infty} \operatorname{sp}(B_n) = \limsup_{n\to\infty} \operatorname{sp}(B_n) = \operatorname{sp} B \cup \operatorname{sp} \widetilde{B}. \tag{9.7}$$

Proof. Let $\lambda \notin \operatorname{sp} B \cup \operatorname{sp} \widetilde{B}$. Then $B - \lambda I$ and $(B - \lambda I)^\sim = \widetilde{B} - \lambda I$ are invertible, and hence Theorem 3.13 implies that there are numbers n_0 and $M \in (0, \infty)$ such that $\|(B_n - \lambda I)^{-1}\|_2 \leq M$ for all $n \geq n_0$. It follows that the spectral radius of $(B_n - \lambda I)^{-1}$ is at most M, which gives $U_{1/M}(0) \cap \operatorname{sp}(B_n - \lambda I) = \emptyset$ for $n \geq n_0$, where $U_\delta(\mu) := \{\lambda \in \mathbf{C} : |\lambda - \mu| < \delta\}$. Hence $U_{1/M}(\lambda) \cap \operatorname{sp} B_n = \emptyset$ for $n \geq n_0$ and thus $\lambda \notin \limsup \operatorname{sp} B_n$. This completes the proof of (9.6).

Now suppose that $B_n = B_n^*$ for all n. Then B and $\widetilde{B}$ are selfadjoint and all spectra occurring in (9.7) are subsets of the real line. We are left to show that if $\lambda \in \mathbf{R}$ and $\lambda \notin \liminf \operatorname{sp} B_n$, then $\lambda \notin \operatorname{sp} B \cup \operatorname{sp} \widetilde{B}$. But if λ is real and not in $\liminf \operatorname{sp} B_n$, then there exists a $\delta > 0$ such that $U_\delta(\lambda) \cap \operatorname{sp} B_{n_k} = \emptyset$ for infinitely many n_k, that is, $U_\delta(0) \cap \operatorname{sp}(B_{n_k} - \lambda I) = \emptyset$

for infinitely many n_k. As $B_{n_k} - \lambda I$ is Hermitian, the spectral radius and the norm of the operator $(B_{n_k} - \lambda I)^{-1}$ coincide, which gives $\|(B_{n_k} - \lambda I)^{-1}\|_2 < 1/\delta$ for infinitely many n_k. It follows that $\{B_{n_k} - \lambda I\}$ and thus also $\{W_{n_k}(B_{n_k} - \lambda I)W_{n_k}\}$ are stable. Lemma 3.4 now shows that $B - \lambda I$ and $\widetilde{B} - \lambda I$ are invertible. This proves (9.7). □

Corollary 9.11. *Let B_n, B, $\widetilde{B}$ be as in Lemma 9.10. Then*

$$\liminf_{n\to\infty} \Sigma(B_n) = \limsup_{n\to\infty} \Sigma(B_n) = \Sigma(B) \cup \Sigma(\widetilde{B}). \tag{9.8}$$

In particular, for every Laurent polynomial b,

$$\liminf_{n\to\infty} \Sigma(T_n(b)) = \limsup_{n\to\infty} \Sigma(T_n(b)) = \Sigma(T(b)) \cup \Sigma(T(\widetilde{b})). \tag{9.9}$$

Proof. We have $B_n^* B_n = T_n(\overline{b}\, b) + P_n X P_n + W_n Y W_n + D_n$, where X and Y have only a finitely many nonzero entries and $\|D_n\|_2 \to 0$ as $n \to \infty$. Equalities (9.8) are therefore straightforward from (9.7). □

Remark 9.12. Let $V : \ell^2 \to \ell^2$ be the map given by $(Vx)_j = \overline{x_j}$. Since

$$\operatorname{sp} VAV = \operatorname{sp} A, \quad \operatorname{sp}(A^*A) \cup \{0\} = \operatorname{sp}(AA^*) \cup \{0\} \tag{9.10}$$

for every $A \in \mathcal{B}(\ell^2)$ and $VT(b)V = T(\widetilde{\overline{b}})$, we obtain

$$\begin{aligned}(\Sigma(T(\overline{b})))^2 &= \operatorname{sp} T(b)T(\overline{b}) = \operatorname{sp} VT(b)T(\overline{b})V \\ &= \operatorname{sp} VT(b)VVT(\overline{b})V = \operatorname{sp} T(\widetilde{\overline{b}})T(\widetilde{b}) = (\Sigma(T(\widetilde{b})))^2,\end{aligned}$$

that is, $\Sigma(T(\widetilde{b})) = \Sigma(T(\overline{b}))$. This and the second equality of (9.10) imply that

$$\Sigma(T(b)) \cup \{0\} = \Sigma(T(\widetilde{b})) \cup \{0\}.$$

However, in general the sets $\Sigma(T(b))$ and $\Sigma(T(\widetilde{b}))$ need not coincide: If $b(t) = t$, then $T^*(b)T(b) = \operatorname{diag}(1, 1, 1, \dots)$, $T^*(\widetilde{b})T(\widetilde{b}) = \operatorname{diag}(0, 1, 1, \dots)$, whence $\Sigma(T(b)) = \{1\}$ and $\Sigma(T(\widetilde{b})) = \{0, 1\}$. □

The set $\Sigma(T(b))$ is available in special cases only. Sometimes the following is useful.

Proposition 9.13. *If b is a Laurent polynomial, then*

$$[\min|b|, \max|b|] \subset \Sigma(T(b)) \subset [0, \max|b|].$$

Proof. From Propositions 1.2 and 1.3 and Corollary 1.10 we see that there is a compact operator K such that

$$\begin{aligned}(\Sigma(T(b)))^2 &= \operatorname{sp} T(\overline{b})T(b) = \operatorname{sp}(T(|b|^2) + K) \\ &\supset \operatorname{sp}_{\text{ess}}(T(|b|^2) + K) = \operatorname{sp}_{\text{ess}} T(|b|^2) = [\min|b|^2, \max|b|^2],\end{aligned}$$

and obviously,

$$(\Sigma(T(b)))^2 = \operatorname{sp} T(\overline{b})T(b) \subset [0, \|T(b)\|_2^2] = [0, \max |b|^2]. \qquad \square$$

Thus, if $T(b)$ is not Fredholm, which is equivalent to the equality $\min |b| = 0$, then

$$\Sigma(T(b)) \cup \Sigma(T(\widetilde{b})) = [0, \max |b|].$$

However, if $T(b)$ is Fredholm, the question of finding

$$\left(\Sigma(T(b)) \cup \Sigma(T(\widetilde{b}))\right) \cap [0, \min |b|]$$

is difficult.

9.6 The Limiting Measure

The purpose of this section is to show that if b is a Laurent polynomial, then

$$\lim_{n\to\infty} \frac{1}{n} \sum_{k=1}^{n} f(\sigma_k(T_n(b))) = \frac{1}{2\pi} \int_0^{2\pi} f(|b(e^{i\theta})|)d\theta \tag{9.11}$$

for every compactly supported function $f : \mathbf{R} \to \mathbf{C}$ of bounded variation. Formula (9.11) is the **Avram-Parter theorem**. The approach of this section is due to Zizler, Zuidwijk, Taylor, and Arimoto [303].

Functions of bounded variation. Let $f : \mathbf{R} \to \mathbf{C}$ be a function with compact support. The function f is said to have *bounded variation on a segment* $[a, b] \subset \mathbf{R}$, $f \in BV[a, b]$, if there exists a constant $V \in [0, \infty)$ such that

$$\sum_{j=1}^{m} |f(x_j) - f(x_{j-1})| \le V \tag{9.12}$$

for every partition $a = x_0 < x_1 < \cdots < x_m = b$ of $[a, b]$. The minimal V for which (9.12) is true for every partition of the segment $[a, b]$ is called the *total variation of f on* $[a, b]$ and is denoted by $V_{[a,b]}(f)$. We let BV stand for the set of all functions $f : \mathbf{R} \to \mathbf{C}$ that have compact support and are of bounded variation on each segment $[a, b] \subset \mathbf{R}$. Such functions are simply referred to as *functions of bounded variation*.

If f is compactly supported and continuously differentiable, then f is clearly BV and $V_{[a,b]}(f) \le \|f'\|_\infty(b - a)$. The characteristic function χ_E of a finite interval E is also of bounded variation and $V_{[a,b]}(\chi_E) = 2$ whenever $[a, b] \supset E$.

If $f \in BV$ and $a \le x \le y \le b$, then

$$|f(y) - f(x)| \le V_{[a,b]}(f); \tag{9.13}$$

indeed, by the definition of $V_{[a,b]}(f)$, we even have

$$|f(x) - f(a)| + |f(y) - f(x)| + |f(b) - f(x)| \le V_{[a,b]}(f).$$

We begin with a result on the singular values of matrices, large blocks of which coincide.

Theorem 9.14. *Let r and n be natural numbers such that $1 \le r < n$, let $K = \{k_1, \dots, k_r\}$ be a subset of $\{1, 2, \dots, n\}$, and put $L = \{1, 2, \dots, n\} \setminus K$. Suppose A and A' are two complex $n \times n$ matrices whose jk entries coincide for all $(j, k) \in L \times L$. If $f \in BV$ and $[a, b]$ is any segment that contains all singular values of A and A', then*

$$\sum_{k=1}^{n} \big| f(\sigma_k(A)) - f(\sigma_k(A')) \big| \le 3r\, V_{[a,b]}(f).$$

Proof. Suppose first that $r = 1$. We can without loss of generality assume that $K = \{n\}$ (the general case can be reduced to this case by permutation similarity). Let $A = (a_{jk})_{j,k=1}^{n}$ and define $B = (a_{jk})_{j,k=1}^{n-1}$. Applying Theorem 9.7 to the pairs (A, B) and (A', B), we get

$$\begin{aligned}
\sigma_1(A), \sigma_1(A') &\in [a, \sigma_3(B)],\\
\sigma_2(A), \sigma_2(A') &\in [\sigma_2(B), \sigma_4(B)],\\
&\dots\\
\sigma_{n-2}(A), \sigma_{n-2}(A') &\in [\sigma_{n-2}(B), b],\\
\sigma_{n-1}(A), \sigma_{n-1}(A') &\in [\sigma_{n-1}(B), b],\\
\sigma_n(A), \sigma_n(A') &\in [\sigma_{n-1}(B), b].
\end{aligned}$$

This in conjunction with (9.13) and the abbreviation $\sigma_j(B) := \sigma_j$ gives

$$\begin{aligned}
&\sum_{k=1}^{n} \big| f(\sigma_k(A)) - f(\sigma_k(A')) \big|\\
&\le V_{[a,\sigma_3]}(f) + V_{[\sigma_2,\sigma_4]}(f) + \dots + V_{[\sigma_{n-2},b]}(f) + V_{[\sigma_{n-1},b]}(f) + V_{[\sigma_{n-1},b]}(f).
\end{aligned}$$

Since each point of $[a, b]$ is covered by at most three of the segments occurring in the last sum, it follows that this sum is at most $3V_{[a,b]}(f)$, which completes the proof for $r = 1$.

Now let $r > 1$. Again we may assume that $K = \{n - r + 1, \dots, n\}$. Define $n \times n$ matrices $A^{(0)}, A^{(1)}, \dots, A^{(r)}$ so that $A^{(0)} = A$, $A^{(r)} = A'$, and the pairs $A^{(\nu-1)}, A^{(\nu)}$ $(\nu = 1, \dots, r)$ are as in the case $r = 1$ considered above. This can be achieved by setting, for $\nu = 0, \dots, r$,

$$a_{jk}^{(\nu)} = \begin{cases} a_{jk} & \text{for} \quad 1 \le j, k \le n - \nu, \\ a'_{jk} & \text{for} \quad n - \nu < j \le n \ \text{ or } \ n - \nu < k \le n. \end{cases}$$

Let $[c, d] \supset [a, b]$ by any segment which contains the singular values of all $A^{(\nu)}$ and define $\widetilde{f} : \mathbf{R} \to \mathbf{C}$ by

$$\widetilde{f} = \begin{cases} 0 & \text{for} \quad x \in (-\infty, a) \cup (b, \infty), \\ f(x) & \text{for} \quad x \in [a, b]. \end{cases}$$

Clearly, $\widetilde{f} \in BV$. From what was proved for $r = 1$ we obtain

$$\sum_{k=1}^{n} |\, f(\sigma_k(A)) - f(\sigma_k(A')) \,| = \sum_{k=1}^{n} |\, \widetilde{f}(\sigma_k(A)) - \widetilde{f}(\sigma_k(A^{(r)})) \,|$$

$$\leq \sum_{\nu=1}^{r} \sum_{k=1}^{n} |\, \widetilde{f}(\sigma_k(A^{(\nu-1)})) - \widetilde{f}(\sigma_k(A^{(\nu)})) \,| \leq \sum_{\nu=1}^{r} 3V_{[c,d]}(\widetilde{f}) = 3r\,V_{[a,b]}(f). \qquad \square$$

Theorem 9.15. *Let $b(t) = \sum_{j=-r}^{r} b_j t^j$ $(t \in \mathbf{T})$ be a Laurent polynomial and let $f \in BV$. If $[c,d]$ is any segment that contains $[0, \|b\|_\infty]$, then for all $n \geq 1$*

$$\left| \sum_{k=1}^{n} f(\sigma_k(T_n(b))) - \frac{n}{2\pi} \int_0^{2\pi} f(|b(e^{i\theta})|)d\theta \right| \leq 7r\,V_{[c,d]}(f).$$

Proof. Suppose first that $|b|$ is constant on $\mathbf{T}$. As observed in the end of Section 9.4, in that case $b(t) = \gamma t^m$, $|m|$ singular values of $T_n(b)$ are zero and $n - |m|$ singular values are equal to $|\gamma|$. Hence

$$\sum_{k=1}^{n} f(\sigma_k(T_n(b))) = |m| f(0) + (n - |m|) f(|\gamma|), \quad \frac{n}{2\pi} \int_0^{2\pi} f(|b(e^{i\theta})|)d\theta = nf(|\gamma|),$$

and the assertion amounts to the inequality $|m|\,|f(0) - f(|\gamma|)| \leq 7|m|\,V_{[c,d]}(f)$, which is certainly true because $|f(0) - f(|\gamma|) \leq V_{[c,d]}(f)$ by virtue of (9.13).

Now suppose that $|b|$ is not constant on $\mathbf{T}$. Define $C_n(b)$ as in Section 2.1 for $n \geq 2r+1$ and put $C_n(b) = T_n(b)$ for $n \leq 2r$. The singular values of $C_n(b)$ and $T_n(b)$ are all contained in $[0, \|b\|_\infty]$. If $n \geq 2r + 1$, then (9.5) implies that $T_n(b)$ and $C_n(b)$ differ only in the last r columns and rows. Consequently, by Theorem 9.14,

$$\left| \sum_{k=1}^{n} f(\sigma_k(T_n(b))) - \sum_{k=1}^{n} f(\sigma_k(C_n(b))) \right| \leq 3r\,V_{[c,d]}(f). \tag{9.14}$$

Put $h(\theta) = f(|b(e^{i\theta})|)$. By Proposition 9.5,

$$\sum_{k=1}^{n} f(\sigma_k(C_n(b))) = \sum_{k=0}^{n-1} h\left(\frac{2\pi k}{n}\right),$$

which gives

$$\left| \sum_{k=1}^{n} f(\sigma_k(C_n(b))) - \frac{n}{2\pi} \int_0^{2\pi} h(\theta)d\theta \right|$$

$$= \left| \frac{n}{2\pi} \sum_{k=0}^{n-1} \int_{2\pi k/n}^{2\pi(k+1)/n} \left(h\left(\frac{2\pi k}{n}\right) - h(\theta) \right) d\theta \right|$$

$$\le \frac{n}{2\pi}\sum_{k=0}^{n-1}\int_{2\pi k/n}^{2\pi(k+1)/n} V_{[2\pi k/n,2\pi(k+1)/n]}(h)d\theta \quad \text{(recall (9.13))}$$

$$= \sum_{k=0}^{n-1} V_{[2\pi k/n,2\pi(k+1)/n]}(h) = V_{[0,2\pi]}(h). \tag{9.15}$$

Now let $u(\theta) = |b(e^{i\theta})|^2$. By assumption, u is a nonconstant and nonnegative Laurent polynomial of degree at most $2r$. Thus, u has at least 2 and at most $4r$ local extrema in $[0, 2\pi)$. Let $\theta_1 < \theta_2 < \cdots < \theta_\ell$ denote the local extrema. As $|b|$ is monotonous on $[\theta_j, \theta_{j+1}]$ ($\theta_{\ell+1} := \theta_1 + 2\pi$), we get

$$\begin{aligned} V_{[0,2\pi]}(h) &= V_{[\theta_1,\theta_1+2\pi]}(f \circ |b|) \\ &= V_{[\theta_1,\theta_2]}(f \circ |b|) + V_{[\theta_2,\theta_3]}(f \circ |b|) + \cdots + V_{[\theta_\ell,\theta_1+2\pi]}(f \circ |b|) \\ &\le V_{[c,d]}(f) + V_{[c,d]}(f) + \cdots + V_{[c,d]}(f) = \ell V_{[c,d]}(f) \le 4r V_{[c,d]}(f). \end{aligned} \tag{9.16}$$

Combining (9.14), (9.15), and (9.16) we arrive at the assertion. □

Corollary 9.16. *Let b be a Laurent polynomial and let $f : \mathbf{R} \to \mathbf{C}$ be a function with compact support. If f is continuous or of bounded variation, then*

$$\lim_{n\to\infty} \frac{1}{n}\sum_{k=1}^{n} f(\sigma_k(T_n(b))) = \frac{1}{2\pi}\int_0^{2\pi} f(|b(e^{i\theta})|)d\theta. \tag{9.17}$$

Proof. For $f \in BV$, the assertion is immediate from Theorem 9.15. So suppose f is continuous. Then f can be uniformly approximated by compactly supported functions of bounded variation (e.g., by continuously differentiable functions) f_m. Given $\varepsilon > 0$, there is an m_0 such that $|f(x) - f_{m_0}(x)| \le \varepsilon$ for $x \in \mathbf{R}$. It follows that

$$\begin{aligned} &\left| \frac{1}{n}\sum_{k=1}^{n} f(\sigma_k(T_n(b))) - \frac{1}{n}\sum_{k=1}^{n} f_{m_0}(\sigma_k(T_n(b))) \right| \\ &\quad \le \frac{1}{n}\sum_{k=1}^{n} \left| f(\sigma_k(T_n(b))) - f_{m_0}(\sigma_k(T_n(b))) \right| \le \frac{1}{n} n\varepsilon = \varepsilon, \end{aligned}$$

$$\begin{aligned} &\left| \frac{1}{2\pi}\int_0^{2\pi} f(|b(e^{i\theta})|)d\theta - \frac{1}{2\pi}\int_0^{2\pi} f_{m_0}(|b(e^{i\theta})|)d\theta \right| \\ &\quad \le \frac{1}{2\pi}\int_0^{2\pi} \left| f(|b(e^{i\theta})|) - f_{m_0}(|b(e^{i\theta})|) \right| d\theta \le \frac{1}{2\pi} 2\pi\varepsilon = \varepsilon, \end{aligned}$$

and as

$$\left| \frac{1}{n}\sum_{k=1}^{n} f_{m_0}(\sigma_k(T_n(b))) - \frac{1}{2\pi}\int_0^{2\pi} f_{m_0}(|b(e^{i\theta})|)d\theta \right| < \varepsilon$$

for all sufficiently large n due to Theorem 9.15, we get

$$\left| \frac{1}{n}\sum_{k=1}^{n} f(\sigma_k(T_n(b))) - \frac{1}{2\pi}\int_0^{2\pi} f(|b(e^{i\theta})|)d\theta \right| < 3\varepsilon$$

whenever n is large enough. This implies (9.17). □

Let E be a (Lebesgue) measurable subset of $\mathbf{R}$. Given $n \in \mathbf{N}$, we denote by $N_n(E)$ the number of singular values of $T_n(b)$ in E (multiplicities taken into account):

$$N_n(E) = \sum_{k=1}^{n} \chi_E(\sigma_k(T_n(b))).$$

We define the measure μ_n by

$$\mu_n(E) = \frac{1}{n} N_n(E)$$

and we let μ denote the measure given by

$$\mu(E) = \frac{1}{2\pi} \int_0^{2\pi} \chi_E(|b(e^{i\theta})|)d\theta = \frac{1}{2\pi} \Big| \{t \in \mathbf{T} : |b(t)| \in E\} \Big|,$$

where $|\cdot|$ stands for the Lebesgue measure on $\mathbf{T}$.

Corollary 9.17. *If b is a Laurent polynomial, then the measures μ_n converge weakly to the measure μ, that is,*

$$\int_{\mathbf{R}} f \, d\mu_n \to \int_{\mathbf{R}} f \, d\mu$$

for every compactly supported continuous function $f : \mathbf{R} \to \mathbf{C}$.

Proof. Since

$$\int_{\mathbf{R}} f \, d\mu_n = \frac{1}{n} \sum_{k=1}^{n} f(\sigma_k(T_n(b))), \qquad \int_{\mathbf{R}} f \, d\mu = \frac{1}{2\pi} \int_0^{2\pi} f(|b(e^{i\theta})|)d\theta,$$

this is a straightforward consequence of Corollary 9.16. □

Obviously, all singular values of $T_n(b)$ lie in $[0, \max|b|]$.

Corollary 9.18. *Let b is a Laurent polynomial of the form $b(t) = \sum_{j=-r}^{r} b_j t^j$ $(t \in \mathbf{T})$. If $E \subset \mathbf{R}$ is any segment, then*

$$|\, N_n(E) - n\mu(E) \,| \le 14r \quad \textit{for all} \quad n \ge 1, \tag{9.18}$$

and if $E = [\min|b|, \max|b|]$, then even

$$N_n(E) - n \,| \le 7r \quad \textit{for all} \quad n \ge 1. \tag{9.19}$$

Proof. Theorem 9.15 with $f = \chi_E$ and $[c, d] = [0, \max|b|]$ gives $|N_n(E) - n\mu(E)| \le 7r V_{[c,d]}(\chi_E)$. Since $V_{[c,d]}(\chi_E) \le 2$, we get (9.18). If $E = [\min|b|, \max|b|]$, then $\mu(E) = 1$ and $V_{[c,d]}(\chi_E) \le 1$. This yields (9.19). □

Our next objective is an improvement of estimate (9.19). For this purpose we need the following analogue of Theorem 9.7.

Theorem 9.19 (Cauchy's interlacing theorem). *Let A be a Hermitian $n \times n$ matrix and let $B = P_{n-1}AP_{n-1}$ be the $(n-1) \times (n-1)$ principal submatrix. Then*

$$\begin{array}{ccccc} & & \lambda_1(A) & \le & \lambda_1(B) \\ \lambda_1(B) & \le & \lambda_2(A) & \le & \lambda_2(B) \\ & & \dots & & \\ \lambda_{n-2}(B) & \le & \lambda_{n-1}(A) & \le & \lambda_{n-1}(B) \\ \lambda_{n-1}(B) & \le & \lambda_n(A). & & \end{array}$$

Theorem 9.20. *Let $b(t) = \sum_{j=-s}^{r} b_j t^j \, (t \in \mathbf{T})$ with $r, s \ge 0$. Then*

$$N_n([0, \min |b|)) \le r + s \text{ for all } n \ge 1. \tag{9.20}$$

Proof. If $n \le r+s$, then (9.20) is trivial. So let $n \ge r+s+1$. We have $T_n(\overline{b})T_n(b) = T_n(|b|^2) - P_n K_s P_n - W_n L_r W_n$, where K_s and L_r are infinite matrices whose entries outside the upper-left $s \times s$ and $r \times r$ blocks, respectively, vanish (see the beginning of the proof of Theorem 5.8). Thus, we may think of $T_n(\overline{b})T_n(b)$ as resulting from $T_{n-r-s}(|b|^2)$ by $r+s$ times adding a row and a column. On $r+s$ times employing Theorem 9.19, we get $\lambda_1(T_{n-r-s}(|b|^2)) \le \lambda_{r+s+1}(T_n(\overline{b})T_n(b))$. As $\lambda_1(T_{n-r-s}(|b|^2) \ge \min |b|^2$ by Corollary 4.28, it follows that $\lambda_{r+s+1}(T_n(\overline{b})T_n(b)) \ge \min |b|^2$. Consequently, at most $r+s$ eigenvalues of $T_n(\overline{b})T_n(b)$ are located in $[0, \min |b|^2)$. This is equivalent to saying that at most $r+s$ singular values of $T_n(b)$ lie in the set $[0, \min |b|)$. □

Let $b(t) = \sum_{j=-s}^{r} b_j t^j$ $(t \in \mathbf{T})$ with $r, s \ge 0$ and suppose $\min |b| > 0$. Denote by k the winding number of b about the origin. Since

$$b(t) = t^{-s}(b_{-s} + b_{-s+1}t + \dots + b_r t^{r+s})$$

and since k is the difference of the number of zeros and the number of poles of $b(z)$ in the unit disk, we see that $|k| \le \max(r, s)$. Form Theorem 9.4 we know that if n is sufficiently large, then at least $|k|$ singular values of $T_n(b)$ lie in $[0, \min |b|)$, and Theorem 9.20 shows that, for every $n \ge 1$, at most $r+s$ singular values of $T_n(b)$ are contained in $[0, \min |b|)$.

9.7 Proper Clusters

Let E be a subset of $\mathbf{R}$ and denote by $\gamma_n(E)$ the number of the singular values of $T_n(b)$ (multiplicities taken into account) that do not belong to E. Thus, with $N_n(E)$ as in Section 9.6, $\gamma_n(E) = n - N_n(E)$. For $\varepsilon > 0$, put $U_\varepsilon(E) = \{\lambda \in \mathbf{R} : \text{dist}(\lambda, E) < \varepsilon\}$. Tyrtyshnikov calls E a *cluster* and a *proper cluster* for $\Sigma(T_n(b))$ if, respectively, $\gamma_n(U_\varepsilon(E)) = o(n)$ and $\gamma_n(U_\varepsilon(E)) = O(1)$ for each $\varepsilon > 0$. Put $\mathcal{R}(|b|) = [\min |b|, \max |b|]$.

Theorem 9.21. *Let b be a Laurent polynomial of degree r. Then $\gamma_n(\mathcal{R}(|b|)) \le 7r$ and hence $\mathcal{R}(|b|)$ is a proper cluster for $\Sigma(T_n(b))$. If E is a subset of $\mathcal{R}(|b|)$ and the closure of E is properly contained in $\mathcal{R}(|b|)$, then E is not a cluster for $\Sigma(T_n(b))$.*

Proof. Formula (9.19) is equivalent to the inequality $\gamma_n(\mathcal{R}(|b|)) \le 7r$. As

$$\gamma_n(U_\varepsilon(\mathcal{R}(|b|))) \le \gamma_n(\mathcal{R}(|b|)),$$

it follows that $\mathcal{R}(|b|)$ is a proper cluster. Now let $E \subset \mathcal{R}(|b|)$ and suppose $\mathcal{R}(|b|) \setminus E$ contains some interval (c, d) with $c < d$. The $\mathcal{R}(|b|) \setminus U_\varepsilon(E)$ also contains some interval $(c_\varepsilon, d_\varepsilon)$ with $c_\varepsilon < d_\varepsilon$ if only $\varepsilon > 0$ is sufficiently small. Clearly,

$$\mu((c_\varepsilon, d_\varepsilon)) = \frac{1}{2\pi} \,|\, \{t \in \mathbf{T} : |b(t)| \in (c_\varepsilon, d_\varepsilon)\} \,|\, =: \delta_\varepsilon > 0.$$

From formula (9.18) we therefore obtain that

$$\begin{aligned} \gamma_n(U_\varepsilon(E)) &= N_n(\mathcal{R}(|b|) \setminus U_\varepsilon(E)) \geq N_n((c_\varepsilon, d_\varepsilon)) \\ &\geq n\mu((c_\varepsilon, d_\varepsilon)) - 14r = n\delta_\varepsilon - 14r, \end{aligned}$$

which shows that $\gamma_n(U_\varepsilon(E))/n$ does not converge to zero. Thus, E cannot be a cluster for $\Sigma(T_n(b))$. □

9.8 Norm of Matrix Times Random Vector

Let A_n be a real $n \times n$ matrix and let $\sigma_1 \leq \sigma_2 \leq \cdots \leq \sigma_n$ be the singular values of A_n. We have $\|A_n x\|_2 \leq \|A_n\|_2$ for every unit vector $x \in \mathbf{R}^n$, and the set $\{\|A_n x\|_2/\|A_n\|_2 : \|x\|_2 = 1\}$ coincides with the segment $[\sigma_1/\sigma_n, 1]$ The purpose of this section is to show that for a randomly chosen unit vector x the value of $\|A_n x\|_2^2/\|A_n\|_2^2$ typically lies near

$$\frac{1}{\sigma_n^2} \frac{\sigma_1^2 + \cdots + \sigma_n^2}{n}. \tag{9.21}$$

Notice that $\sigma_n = \|A_n\|_2$ and that $\sigma_1^2 + \cdots + \sigma_n^2 = \|A_n\|_{\mathrm{F}}^2$, where $\|A_n\|_{\mathrm{F}}$ is the Frobenius (or Hilbert-Schmidt norm). Thus, if $\|A_n\|_2 = 1$, then for a typical unit vector x the value of $\|A_n x\|_2^2$ is close to $\|A_n\|_{\mathrm{F}}^2/n$.

Obviously, in the case where A_n is a large Toeplitz matrix, the expression (9.21) can be tackled by the Avram-Parter formula (9.11).

Let $\mathbf{B}_n = \{x \in \mathbf{R}^n : \|x\|_2 \leq 1\}$ and $\mathbf{S}_{n-1} = \{x \in \mathbf{R}^n : \|x\|_2 = 1\}$. For a given real $n \times n$ matrix A_n, we consider the random variable

$$X_n(x) = \frac{\|A_n x\|_2}{\|A_n\|_2},$$

where x is uniformly distributed on $\mathbf{S}_{n-1}$. For $k \in \mathbf{N}$, the expectation of X_n^k is

$$EX_n^k = \frac{1}{|\mathbf{S}_{n-1}|} \int_{\mathbf{S}_{n-1}} \frac{\|A_n x\|_2^k}{\|A_n\|_2^k} \, d\sigma(x),$$

where $d\sigma$ is the surface measure on $\mathbf{S}_{n-1}$. The variance of X_n^k is

$$\sigma^2 X_n^k = E\left(X_n^k - EX_n^k\right)^2 = EX_n^{2k} - \left(EX_n^k\right)^2.$$

Lemma 9.22. *For every natural number k,*

$$\frac{1}{|\mathbf{S}_{n-1}|} \int_{\mathbf{S}_{n-1}} \frac{\|A_n x\|_2^k}{\|A_n\|_2^k} \, d\sigma(x) = \frac{1}{|\mathbf{B}_n|} \int_{\mathbf{B}_n} \frac{\|A_n x\|_2^k}{\|A_n\|_2^k \|x\|_2^k} \, dx.$$

Proof. Using spherical coordinates, $x = rx'$ with $x' \in \mathbf{S}_{n-1}$, we get

$$\int_{\mathbf{B}_n} \frac{\|A_n x\|_2^k}{\|x\|_2^k} dx = \int_0^1 \int_{\mathbf{S}_{n-1}} \frac{r^k \|A_n x'\|_2^k}{r^k} r^{n-1}\, d\sigma(x') dr = \frac{1}{n} \int_{\mathbf{S}_{n-1}} \|A_n x'\|_2^k\, d\sigma(x'),$$

and since

$$|\mathbf{S}_{n-1}| = \frac{2\pi^{n/2}}{\Gamma(n/2)} \quad \text{and} \quad |\mathbf{B}_n| = \frac{\pi^{n/2}}{\Gamma(n/2+1)} \tag{9.22}$$

and thus $|\mathbf{S}_{n-1}|/n = |\mathbf{B}_n|$, the assertion follows. □

Theorem 9.23. *If $A_n \neq 0$, then*

$$EX_n^2 = \frac{1}{\sigma_n^2} \frac{\sigma_1^2 + \cdots + \sigma_n^2}{n}, \tag{9.23}$$

$$\sigma^2 X_n^2 = \frac{2}{n+2} \frac{1}{\sigma_n^4} \left(\frac{\sigma_1^4 + \cdots + \sigma_n^4}{n} - \left(\frac{\sigma_1^2 + \cdots + \sigma_n^2}{n} \right)^2 \right). \tag{9.24}$$

Proof. Let $A_n = U_n D_n V_n$ be the singular value decomposition. Thus, U_n and V_n are orthogonal matrices and $D_n = \operatorname{diag}(\sigma_1, \ldots, \sigma_n)$. By Lemma 9.22,

$$\begin{aligned} EX_n^2 &= \frac{1}{|\mathbf{B}_n|} \int_{\mathbf{B}_n} \frac{\|U_n D_n V_n x\|_2^2}{\|U_n D_n V_n\|_2^2\, \|x\|_2^2} dx \\ &= \frac{1}{|\mathbf{B}_n|} \int_{\mathbf{B}_n} \frac{\|D_n V_n x\|_2^2}{\|D_n\|_2^2\, \|V_n x\|_2^2} dx = \frac{1}{|\mathbf{B}_n|} \int_{\mathbf{B}_n} \frac{\|D_n x\|_2^2}{\|D_n\|_2^2\, \|x\|_2^2} dx \\ &= \frac{1}{|\mathbf{B}_n|} \int_{\mathbf{B}_n} \frac{\sigma_1^2 x_1^2 + \cdots + \sigma_n^2 x_n^2}{\sigma_n^2 (x_1^2 + \cdots + x_n^2)} dx_1 \ldots dx_n. \end{aligned} \tag{9.25}$$

A formula by Liouville states that if $\lambda < (p_1 + \cdots + p_n)/2$, then

$$\begin{aligned} &\underset{\substack{x_1, \ldots, x_n \geq 0 \\ x_1^2 + \cdots + x_n^2 \leq 1}}{\int \cdots \int} \frac{x_1^{p_1 - 1} \ldots x_n^{p_n - 1}}{(x_1^2 + \cdots + x_n^2)^\lambda} dx_1 \ldots dx_n \\ &= \frac{1}{2^n \left(\frac{p_1 + \cdots + p_n}{2} - \lambda \right)} \frac{\Gamma\left(\frac{p_1}{2}\right) \ldots \Gamma\left(\frac{p_n}{2}\right)}{\Gamma\left(\frac{p_1 + \cdots + p_n}{2}\right)} \end{aligned} \tag{9.26}$$

(see, e.g., [120, No. 676.14]). From (9.22) and (9.26) we infer that

$$\begin{aligned} &\frac{1}{|\mathbf{B}_n|} \int_{\mathbf{B}_n} \frac{x_j^2}{x_1^2 + \cdots + x_n^2} dx \\ &= \frac{\Gamma\left(\frac{n}{2} + 1\right)}{\pi^{n/2}} \frac{2^n}{2^n \left(\frac{n-1}{2} + \frac{3}{2} - 1 \right)} \frac{\Gamma\left(\frac{1}{2}\right)^{n-1} \Gamma\left(\frac{3}{2}\right)}{\Gamma\left(\frac{n-1}{2} + \frac{3}{2}\right)} = \frac{1}{n}. \end{aligned}$$

This together with (9.25) gives (9.23). In analogy to (9.25),

$$EX_n^4 = \frac{1}{|\mathbf{B}_n|} \int_{\mathbf{B}_n} \frac{(\sigma_1^2 x_1^2 + \cdots + \sigma_n^2 x_n^2)^2}{\sigma_n^4 (x_1^2 + \cdots + x_n^2)^2} \, dx_1 \ldots dx_n. \tag{9.27}$$

From (9.26) we obtain

$$\frac{1}{|\mathbf{B}_n|} \int_{\mathbf{B}_n} \frac{x_j^4}{(x_1^2 + \cdots + x_n^2)^2} \, dx$$

$$= \frac{\Gamma\left(\frac{n}{2}+1\right)}{\pi^{n/2}} \frac{2^n}{2^n \left(\frac{n-1}{2} + \frac{5}{2} - 2\right)} \frac{\Gamma\left(\frac{1}{2}\right)^{n-1} \Gamma\left(\frac{5}{2}\right)}{\Gamma\left(\frac{n-1}{2} + \frac{5}{2}\right)} = \frac{3}{n(n+2)},$$

$$\frac{1}{|\mathbf{B}_n|} \int_{\mathbf{B}_n} \frac{x_j^2 x_k^2}{(x_1^2 + \cdots + x_n^2)^2} \, dx$$

$$= \frac{\Gamma\left(\frac{n}{2}+1\right)}{\pi^{n/2}} \frac{2^n}{2^n \left(\frac{n-2}{2} + \frac{3}{2} + \frac{3}{2} - 2\right)} \frac{\Gamma\left(\frac{1}{2}\right)^{n-2} \Gamma\left(\frac{3}{2}\right)^2}{\Gamma\left(\frac{n-2}{2} + + \frac{3}{2} + \frac{3}{2}\right)} = \frac{1}{n(n+2)},$$

whence, by (9.27),

$$EX_n^4 = \sum_{j=1}^{n} \frac{\sigma_j^4}{\sigma_n^4} \frac{3}{n(n+2)} + 2 \sum_{j<k} \frac{\sigma_j^2 \sigma_k^2}{\sigma_n^4} \frac{1}{n(n+2)}$$

$$= \frac{1}{n(n+2)} \frac{1}{\sigma_n^4} \left(2(\sigma_1^4 + \cdots + \sigma_n^4) + (\sigma_1^2 + \cdots + \sigma_n^2)^2 \right). \tag{9.28}$$

Since $\sigma^2 X_n^2 = EX_n^4 - (EX_n^2)^2$, formula (9.24) follows from (9.23) and (9.28). □

From (9.24) we see that always $\sigma^2 X_n^2 \le 2/(n+2)$. Thus, by Chebyshev's inequality,

$$P\left(\left| X_n^2 - \frac{1}{\sigma_n^2} \frac{\sigma_1^2 + \cdots + \sigma_n^2}{n} \right| \ge \varepsilon \right) \le \frac{2}{(n+2)\varepsilon^2}$$

for each $\varepsilon > 0$. This reveals that for large n the values of $\|A_n x\|_2^2/(\|A_n\|_2^2 \|x\|_2^2)$ cluster around

$$\frac{1}{\sigma_n^2} \frac{\sigma_1^2 + \cdots + \sigma_n^2}{n}.$$

Notice also that $\sigma^2 X_n^2$ can be written as

$$\sigma^2 X_n^2 = \frac{2}{n-2} \frac{1}{\sigma_n^4} \sum_{i<j} \left(\frac{\sigma_j^2 - \sigma_i^2}{n} \right)^2.$$

Figures 9.2 to 9.4 illustrate the phenomenon by two examples.

Obvious modifications of the proof of Theorem 9.23 show that Theorem 9.23 remains true for complex matrices on $\mathbf{C}^n$ with the ℓ^2 norm.

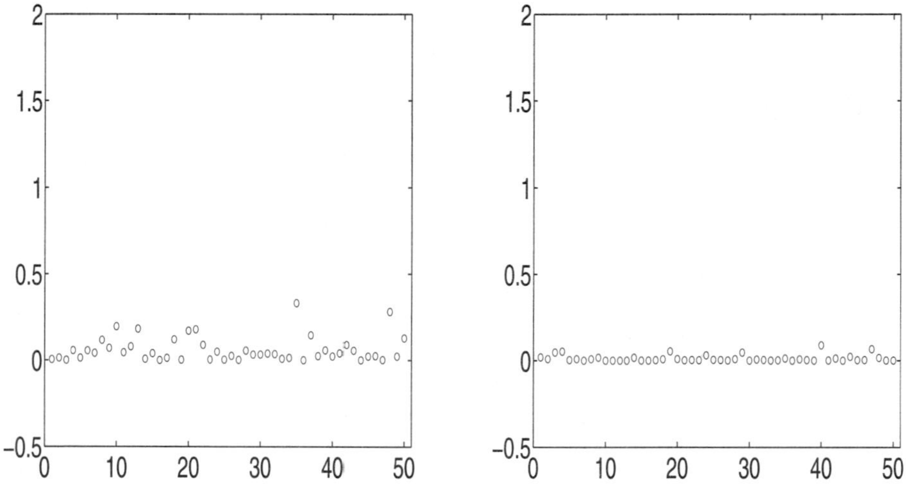

Figure 9.2. *Let A_n be the $n \times n$ matrix all entries of which are 1. We see the values $\|A_n x\|_2^2/\|A_n\|_2^2$ for 50 vectors x that were randomly drawn from the unit sphere of $\mathbf{R}^n$ with the uniform distribution. Note that the expected value of $\|A_n x\|_2^2/\|A_n\|_2^2$ is $1/n$ and that the variance is less than $2/n^2$. The n is 20 in the left picture and 100 in the right.*

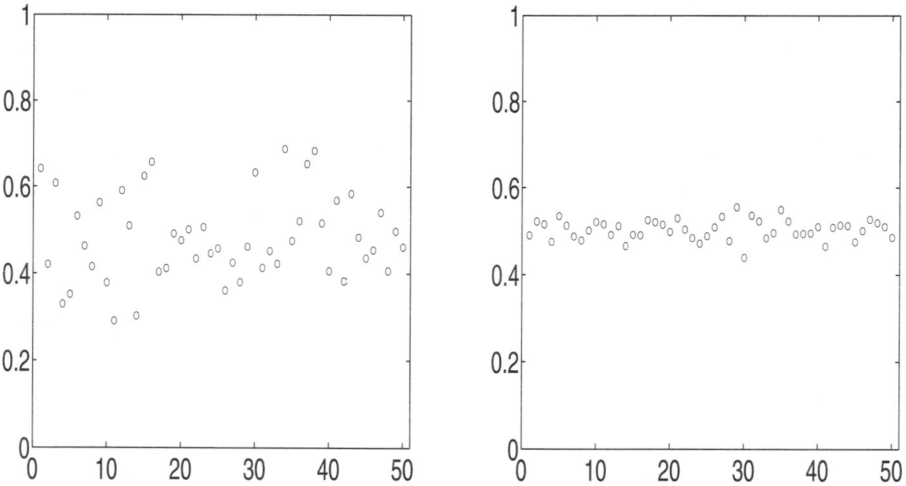

Figure 9.3. *Let $b(t) = t + t^{-1}$. The pictures show $\|T_n(b)x\|_2^2/\|T_n(b)\|_2^2$ for 50 vectors x that were randomly drawn from the unit sphere of $\mathbf{R}^n$ with the uniform distribution. We have $n = 30$ in the left picture and $n = 600$ in the right. The expected value for $\|T_n(b)x\|_2^2/\|T_n(b)\|_2^2$ converges to 0.5 as $n \to \infty$.*

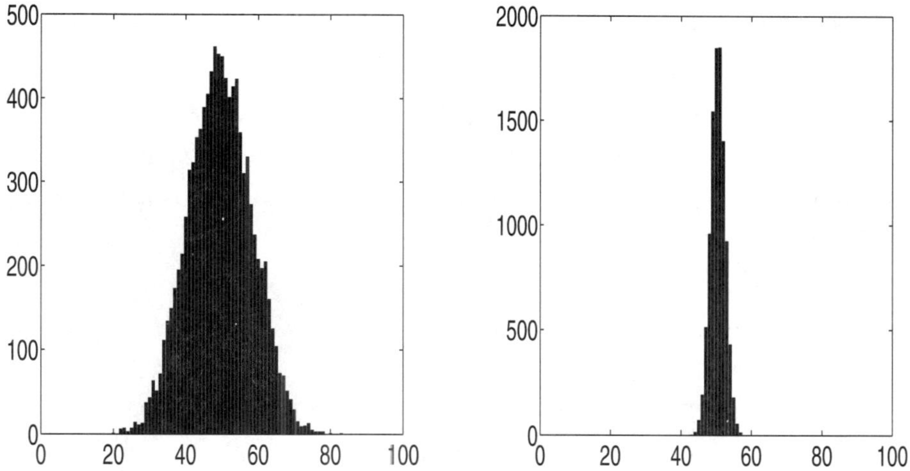

Figure 9.4. *The symbol is again* $b(t) = t + t^{-1}$. *The pictures show the distribution of* $100\,\|T_n(b)x\|_2^2/\|T_n(b)\|_2^2$ *for* 10000 *vectors* x *that were randomly drawn from the unit sphere of* $\mathbf{R}^n$ *with the uniform distribution. In the left picture we took* $n = 30$ *and in the right* $n = 600$. *Notice the different scales of the vertical axes.*

9.9 The Case of Toeplitz and Circulant Matrices

We need one more simple auxilary result.

Lemma 9.24. *Let* $EX_n^2 = \mu_n^2$ *and suppose* $\mu_n \to \mu$ *as* $n \to \infty$. *If* $\varepsilon > 0$ *and* $|\mu_n - \mu| < \varepsilon$, *then*

$$P(|X_n - \mu| \geq \varepsilon) \leq \frac{\sigma^2 X_n^2}{\mu_n^2(\varepsilon - |\mu_n - \mu|)^2}.$$

Proof. We have

$$\begin{aligned} P(|X_n - \mu| \geq \varepsilon) &\leq P\Big(|X_n - \mu_n| \geq \varepsilon - |\mu_n - \mu|\Big) \\ &\leq P\Big(|X_n - \mu_n|(X_n + \mu_n) \geq \mu_n(\varepsilon - |\mu_n - \mu|)\Big) \\ &= P\Big(|X_n^2 - \mu_n^2| \geq \mu_n(\varepsilon - |\mu_n - \mu|)\Big), \end{aligned}$$

and the assertion is now immediate from Chebyshev's inequality. □

Now let b be a Laurent polynomial and let $\sigma_1(T_n(b)) \leq \cdots \leq \sigma_n(T_n(b))$ be the singular values of $T_n(b)$. We abbreviate $\sigma_j(T_n(b))$ to σ_j. The Avram-Parter formula (9.11) tells us that

$$\lim_{n\to\infty} \frac{f(\sigma_1) + \cdots + f(\sigma_n)}{n} = \frac{1}{2\pi}\int_0^{2\pi} f(|b(e^{i\theta})|)d\theta \tag{9.29}$$

for every compactly supported function $f : \mathbf{R} \to \mathbf{C}$ with bounded variation. In particular,

$$\lim_{n\to\infty} \frac{\sigma_1^k + \cdots + \sigma_n^k}{n} = \|b\|_k^k := \frac{1}{2\pi}\int_0^{2\pi} |b(e^{i\theta})|^k d\theta \tag{9.30}$$

for every natural number k. Moreover, if $T(b)$ is invertible, then $\sigma_1 = \sigma_1(T_n(b))$ stays away from zero as $n \to \infty$ (Theorem 3.7), and hence (9.29) with $f(s)$ equal to a negative integral power of s times the characteristic function of $[m, M]$ for appropriate $0 < m < M$ shows that (9.30) is true for every integer k.

Theorem 9.25. *If $|b|$ is not constant on the unit circle $\mathbf{T}$, then for each $\varepsilon > 0$ there is an n_0 such that*

$$P\left(\left|\frac{\|T_n(b)x\|_2}{\|T_n(b)\|_2\,\|x\|_2} - \frac{\|b\|_2}{\|b\|_\infty}\right| \ge \varepsilon\right) \le \frac{3}{n+2}\,\frac{1}{\varepsilon^2}\,\frac{\|b\|_4^4 - \|b\|_2^2}{\|b\|_2^2\,\|b\|_\infty^2} \tag{9.31}$$

for all $n \ge n_0$. If $|b|$ is constant throughout $\mathbf{T}$, then

$$P\left(\frac{\|T_n(b)x\|_2}{\|T_n(b)\|_2\,\|x\|_2} \le 1 - \varepsilon\right) = o\left(\frac{1}{n}\right) \tag{9.32}$$

for each $\varepsilon > 0$.

Proof. Put

$$\mu_n = \frac{1}{\sigma_n}\sqrt{\frac{\sigma_1^2 + \cdots + \sigma_n^2}{n}}, \qquad \mu = \frac{\|b\|_2}{\|b\|_\infty}.$$

Suppose first that $|b|$ is not constant. Then $\|b\|_4 > \|b\|_2$. From (9.30) we know that $\mu_n \to \mu$. Moreover, (9.30) and Theorem 9.23 imply that

$$\frac{n+2}{2}\sigma^2 X_n^2 \to \frac{1}{\|b\|_\infty^4}\left(\|b\|_4^4 - \|b\|_2^4\right).$$

Thus, Lemma 9.24 shows that

$$P(|X_n - \mu| \ge \varepsilon) \le \frac{3}{n+2}\,\frac{1}{\|b\|_\infty^4}\left(\|b\|_4^4 - \|b\|_2^4\right)\frac{1}{\mu^2\varepsilon^2}$$

for all sufficiently large n, which is (9.31). On the other hand, if $|b|$ is constant, we infer from (9.30) and Theorem 9.23 that

$$\mu_n \to 1 \quad \text{and} \quad \frac{n+2}{2}\sigma^2 X_n^2 = o(1),$$

whence, by Lemma 9.24,

$$P(X_n \le 1 - \varepsilon) \le \frac{3}{n+2}\,o(1)\,\frac{1}{\varepsilon^2} = o\left(\frac{1}{n}\right),$$

which is (9.32). □

We now turn to inverses of Toeplitz matrices.

Theorem 9.26. *If b has no zeros on $\mathbf{T}$ and* $\operatorname{wind} b = 0$, *then*

$$P\left(\left|\frac{\|T_n^{-1}(b)x\|_2}{\|T_n^{-1}(b)\|_2\,\|x\|_2} - \frac{\|b^{-1}\|_2}{\|T^{-1}(b)\|_2}\right| \geq \varepsilon\right) = O\left(\frac{1}{n}\right)$$

for each $\varepsilon > 0$.

Proof. The singular values of $T_n^{-1}(b)$ are $1/\sigma_j$ $(j = 1, \ldots, n)$. Thus, by Theorem 9.23,

$$EX_n^2 = \frac{1}{n}\sigma_1^2\left(\frac{1}{\sigma_1^2} + \cdots + \frac{1}{\sigma_n^2}\right),$$

$$\sigma^2 X_n^2 = \frac{2}{n(n+2)}\sigma_1^4\left(\frac{1}{\sigma_1^4} + \cdots + \frac{1}{\sigma_n^4} - \frac{1}{n}\left(\frac{1}{\sigma_1^2} + \cdots + \frac{1}{\sigma_n^2}\right)^2\right).$$

Since $T(b)$ is invertible, we can invoke formula (9.30) with $k = -2$ and the first of formulas (6.16) to deduce that

$$\mu_n^2 := EX_n^2 \to \|T^{-1}(b)\|^{-2}\|b^{-1}\|_2^2 =: \mu^2.$$

As always $\sigma^2 X_n^2 \leq 2/(n+2)$, we obtain from Lemma 9.24 that

$$P(|X_n - \mu| \geq \varepsilon) \leq \frac{3}{n+2}\frac{1}{\mu^2}\frac{1}{\varepsilon^2} = O\left(\frac{1}{n}\right). \qquad \square$$

If $|\operatorname{wind} b| = k \geq 1$, then $T_n(b)$ need not be invertible for all sufficiently large dimensions n. We therefore consider the Moore-Penrose inverse $T_n^+(b)$, which coincides with $T_n^{-1}(b)$ in the case of invertibility.

Theorem 9.27. *If b has no zeros on $\mathbf{T}$ and* $|\operatorname{wind} b| \geq 1$, *then for each* $\varepsilon > 0$,

$$P\left(\frac{\|T_n^+(b)x\|_2}{\|T_n^+(b)\|_2\,\|x\|_2} \geq \varepsilon\right) = O\left(\frac{1}{n^2}\right).$$

Proof. Theorem 9.4 tells us that if $|\operatorname{wind} b| = k \geq 1$, then k singular values of $T_n(b)$ converge to zero with exponential speed, $\sigma_\ell \leq Ce^{-\gamma n}$ $(\gamma > 0)$ for $\ell \leq k$, while the remaining singular values stay away from zero, $\sigma_\ell \geq \lambda > 0$ for $\ell \geq k+1$. Thus, the singular values of $T_n^+(b)$ are

$$\underbrace{0, \ldots, 0}_{j}, \frac{1}{\sigma_n}, \ldots, \frac{1}{\sigma_{j+1}}, \frac{1}{\sigma_j}$$

with $0 < \sigma_j \le \sigma_{j+1} \le \cdots \le \sigma_n$ and $j \le k$, and from Theorem 9.23 we infer that

$$\begin{aligned} EX_n^2 &= \frac{1}{n}\sigma_j^2\left(\frac{1}{\sigma_j^2}+\cdots+\frac{1}{\sigma_n^2}\right) \\ &= \frac{1}{n}\left(\frac{\sigma_j^2}{\sigma_j^2}+\cdots+\frac{\sigma_j^2}{\sigma_k^2}\right)+\frac{\sigma_j^2}{n}\left(\frac{1}{\sigma_{k+1}^2}+\cdots+\frac{1}{\sigma_n^2}\right) \\ &\le \frac{1}{n}(k-j+1)+\frac{C^2e^{-2\gamma n}}{n}\frac{n-k}{\lambda^2} \\ &\le \frac{k}{n}+\frac{C^2e^{-2\gamma n}}{\lambda^2} \le \frac{k+1}{n} \end{aligned}$$

for all sufficiently large n. Also by Theorem 9.23,

$$\sigma^2 X_n^2 = \frac{2}{n(n+2)}\sigma_j^4\left(\frac{1}{\sigma_j^4}+\cdots+\frac{1}{\sigma_n^4}-\frac{1}{n}\left(\frac{1}{\sigma_j^2}+\cdots+\frac{1}{\sigma_n^2}\right)^2\right),$$

and, analogously,

$$\sigma_j^4\left(\frac{1}{\sigma_j^4}+\cdots+\frac{1}{\sigma_n^4}\right) \le k+\frac{C^4e^{-4\gamma n}(n-k)}{\lambda^4} \le k+1,$$

$$\sigma_j^4\left(\frac{1}{\sigma_j^2}+\cdots+\frac{1}{\sigma_n^2}\right)^2 \le \left(k+\frac{C^2e^{-2\gamma n}(n-k)}{\lambda^2}\right)^2 \le (k+1)^2,$$

which gives $\sigma^2 X_n^2 = O(1/n^2)$. If $\varepsilon > 0$, then

$$P(X_n \ge \varepsilon) = P(X_n^2 \ge \varepsilon^2) \le P\left(X_n^2 \ge \frac{k+1}{n}+\frac{\varepsilon^2}{2}\right)$$

for all sufficiently large n, and thus,

$$\begin{aligned} P(X_n \ge \varepsilon) &\le P\left(X_n^2 \ge EX_n^2+\frac{\varepsilon^2}{2}\right) \\ &\le P\left(|X_n^2 - EX_n^2| \ge \frac{\varepsilon^2}{2}\right) \le \frac{4}{\varepsilon^4}\sigma^2 X_n^2 = O\left(\frac{1}{n^2}\right). \qquad \square \end{aligned}$$

Define the circulant matrices $C_n(b)$ as in Section 2.1. The singular values of $C_n(b)$ are $|b(\omega_n^j)|$ $(j = 0, \ldots, n-1)$, where $\omega_n = e^{2\pi i/n}$. The only Laurent polynomials b of constant modulus are $b(t) = \alpha t^k$ $(t \in \mathbf{T})$ with $\alpha \in \mathbf{C}$; in this case $\|C_n(b)x\|_2 = |\alpha|\,\|x\|_2$ for all x.

Theorem 9.28. *If $|b|$ is not constant, then for each $\varepsilon > 0$ there exists an n_0 such that*

$$P\left(\left|\frac{\|C_n(b)x\|_2}{\|C_n(b)\|_2\,\|x\|_2}-\frac{\|b\|_2}{\|b\|_\infty}\right| \ge \varepsilon\right) \le \frac{3}{n+2}\frac{1}{\varepsilon^2}\frac{\|b\|_4^4-\|b\|_2^2}{\|b\|_2^2\,\|b\|_\infty^2} \quad \textit{for all} \quad n \ge n_0.$$

Proof. The proof is analogous to the proof of (9.31). Note that now (9.30) amounts to the fact that the integral sum

$$\frac{\sigma_1^k + \cdots + \sigma_n^k}{n} = \sum_{j=0}^{n-1} |b(e^{2\pi ij/n})|^k \frac{1}{n}$$

converges to the Riemann integral

$$\int_0^1 |b(e^{2\pi i\theta})|^k d\theta = \int_0^{2\pi} |b(e^{i\theta})|^k \frac{d\theta}{2\pi} =: \|b\|_k^k.$$

Furthermore, it is obvious that $\sigma_n = \max |b(\omega_n^j)| \to \|b\|_\infty$. □

If b has no zeros on $\mathbf{T}$, then $C_n^{-1}(b) = C_n(b^{-1})$, and hence Theorem 9.28 delivers

$$P\left(\left|\frac{\|C_n^{-1}(b)x\|_2}{\|C_n^{-1}(b)\|_2 \, \|x\|_2} - \frac{\|b^{-1}\|_2}{\|b^{-1}\|_\infty}\right| \geq \varepsilon\right) \leq \frac{3}{n+2} \frac{1}{\varepsilon^2} \frac{\|b^{-1}\|_4^4 - \|b^{-1}\|_2^2}{\|b^{-1}\|_2^2 \, \|b^{-1}\|_\infty^2} \tag{9.33}$$

for all sufficiently large n.

9.10 The Nearest Structured Matrix

We denote by $M_n(\mathbf{R})$ the linear space of all $n \times n$ matrices with real entries. Let $A_n \in M_n(\mathbf{R})$ and let $0 \leq \sigma_1 \leq \cdots \leq \sigma_n$ be the singular values of A_n. Suppose $\sigma_n > 0$. The random variable $X_n^2 = \|A_n x\|_2^2 / \|A_n\|_2^2$ assumes its values in $[0, 1]$. In this section we establish a few results on the distribution function of this random variable and we give a nice application to the problem of describing in probabilistic terms the distance of a matrix to the nearest matrix of a given structure. With notation as in the proof of Theorem 9.23,

$$E_\xi := \left\{x \in S_{n-1} : \frac{\|A_n x\|_2^2}{\|A_n\|_2^2} < \xi\right\} = \left\{x \in S_{n-1} : \frac{\|D_n V_n x\|_2^2}{\sigma_n^2} < \xi\right\}.$$

Put $G_\xi = \{x \in S_{n-1} : \|D_n x\|_2^2/\sigma_n^2 < \xi\}$. Clearly, $G_\xi = V_n(E_\xi)$. Since V_n is an orthogonal matrix, it leaves the surface measure on S_{n-1} invariant. It follows that $|G_\xi| = |V_n(E_\xi)|$ and hence

$$F_n(\xi) := P(X_n^2 < \xi) = P\left(\frac{\|D_n x\|_2^2}{\sigma_n^2} < \xi\right) = P\left(\frac{\sigma_1^2 x_1^2 + \cdots + \sigma_n^2 x_n^2}{\sigma_n^2} < \xi\right). \tag{9.34}$$

This reveals first of all that the distribution function $F_n(\xi)$ depends only on the singular values of A_n. We let $f_n(\xi)$ stand for the density function corresponding to the distribution function $F_n(\xi)$.

A real-valued random variable X is said to be $B(\alpha, \beta)$ distributed on (a, b) if

$$P(c \leq X < d) = \int_c^d f(\xi)\, d\xi,$$

where the density function $f(\xi)$ is zero on $(-\infty, a]$ and $[b, \infty)$ and equals

$$\frac{(b-a)^{1-\alpha-\beta}}{B(\alpha, \beta)} (\xi - a)^{\alpha-1} (b - \xi)^{\beta-1}$$

on the interval (a, b). Here

$$B(\alpha, \beta) = \frac{\Gamma(\alpha)\Gamma(\beta)}{\Gamma(\alpha + \beta)}$$

is the common beta function and it is assumed that $\alpha > 0$ and $\beta > 0$.

We first consider 2×2 matrices, that is, we let $n = 2$. From (9.34) we infer that

$$F_2(\xi) = P\left(\frac{\sigma_1^2}{\sigma_2^2} x_1^2 + x_2^2 < \xi\right). \tag{9.35}$$

The constellation $\sigma_1 = \sigma_2$ is uninteresting, because $F_2(\xi) = 0$ for $\xi < 1$ and $F_2(\xi) = 1$ for $\xi \geq 1$ in this case.

Theorem 9.29. *If $\sigma_1 < \sigma_2$, then the random variable X_2^2 is subject to the $B(\frac{1}{2}, \frac{1}{2})$ distribution on $(\sigma_1^2/\sigma_2^2, 1)$.*

Proof. Put $\tau = \sigma_1/\sigma_2$. By (9.35), $F_2(\xi)$ is $\frac{1}{2\pi}$ times the length of the piece of the unit circle $x_1^2 + x_2^2 = 1$ that is contained in the interior of the ellipse $\tau^2 x_1^2 + x_2^2 = \xi$. This gives $F_2(\xi) = 0$ for $\xi \leq \tau^2$ and $F_2(\xi) = 1$ for $\xi \geq 1$. Thus, let $\xi \in (\tau^2, 1)$. Then the circle and the ellipse intersect at the four points

$$\left(\pm\sqrt{\frac{1-\xi}{1-\tau^2}}, \pm\sqrt{\frac{\xi - \tau^2}{1-\tau^2}}\right),$$

and consequently,

$$F_2(\xi) = \frac{2}{\pi} \arctan \sqrt{\frac{\xi - \tau^2}{1 - \xi}},$$

which implies that $F_2'(\xi)$ equals

$$\frac{1}{\pi}(\xi - \tau^2)^{-1/2}(1-\xi)^{-1/2} = \frac{1}{B(1/2, 1/2)}(\xi - \tau^2)^{-1/2}(1-\xi)^{-1/2}$$

and proves that X_2^2 has the $B(\frac{1}{2}, \frac{1}{2})$ distribution on $(\tau^2, 1)$. □

In the case $n \geq 3$, things are more involved. An idea of the variety of possible distribution functions is provided by the class of matrices whose singular values satisfy

$$0 = \sigma_1 = \cdots = \sigma_{n-2} < \sigma_{n-1} < \sigma_n.$$

For $y \in (0, 1)$, the complete elliptic integrals $\mathrm{K}(y)$ and $\mathrm{E}(y)$ are defined by

$$\mathrm{K}(y) = \int_0^{\pi/2} \frac{d\varphi}{\sqrt{1 - y^2 \sin^2 \varphi}}, \qquad \mathrm{E}(y) = \int_0^{\pi/2} \sqrt{1 - y^2 \sin^2 \varphi}\, d\varphi.$$

Put $\mu = \sigma_n/\sigma_{n-1}$. In [52], we showed that on $(0, 1/\mu^2)$ one has the following densities:

$$f_3(\xi) = \frac{\mu}{\pi\sqrt{1-\xi}}\,\mathrm{K}\left(\sqrt{\frac{\xi(\mu^2-1)}{1-\xi}}\right),$$

$$f_4(\xi) = \mu \quad \text{(uniform distribution)},$$

$$f_5(\xi) = \frac{3\mu\sqrt{1-\xi}}{\pi}\,\mathrm{E}\left(\sqrt{\frac{\xi(\mu^2-1)}{1-\xi}}\right),$$

$$f_6(\xi) = 2\mu - \mu(\mu^2+1)\xi,$$

and $f_7(\xi)$ equals

$$\frac{5\mu}{3\pi}\sqrt{1-\xi}\left((4-2\xi-2\xi\mu^2)\,\mathrm{E}\left(\sqrt{\frac{\xi(\mu^2-1)}{1-\xi}}\right) - (1-\xi\mu^2)\,\mathrm{K}\left(\sqrt{\frac{\xi(\mu^2-1)}{1-\xi}}\right)\right).$$

In some particular cases, one gets a complete answer. Here is an example.

Theorem 9.30. *Let $n \geq 3$. If $\sigma_1 = \cdots = \sigma_{n-m} = 0$ and $\sigma_{n-m+1} = \cdots = \sigma_n > 0$, then the random variable X_n^2 is $B(\frac{m}{2}, \frac{n-m}{2})$ distributed on $(0, 1)$.*

This can be proved by the argument of the proof of Theorem 9.29, the only difference being that now one has to compute some multidimensional integrals. A full proof is in [52].

Orthogonal projections have just the singular value pattern of Theorem 9.30. This leads to some pretty nice conclusions. Let E be an N-dimensional Euclidean space and let U be an m-dimensional linear subspace of E. We denote by $\mathcal{P}_U$ the orthogonal projection of E onto U. Then for $y \in E$, the element $\mathcal{P}_U y$ is the best approximation of y in U and we have $\|y\|^2 = \|\mathcal{P}_U y\|^2 + \|y - \mathcal{P}_U y\|^2$. The singular values of $\mathcal{P}_U$ are $N - m$ zeros and m units. Thus, Theorem 9.30 implies that if y is uniformly distributed on the unit sphere of E, then $\|\mathcal{P}_U y\|^2$ has the $B(\frac{m}{2}, \frac{N-m}{2})$ distribution on $(0, 1)$. In particular, if N is large, then $\mathcal{P}_U y$ lies with high probability close to the sphere of radius $\sqrt{\frac{m}{N}}$ and the squared distance $\|y - \mathcal{P}_U y\|^2$ clusters sharply around $1 - \frac{m}{N}$.

Now take $E = M_n(\mathbf{R})$. With the Frobenius norm $\|\cdot\|_F$, E is an n^2-dimensional Euclidean space. Let $U = \mathrm{Str}_n(\mathbf{R})$ denote any class of structured matrices that form an m-dimensional linear subspace of $M_n(\mathbf{R})$. Examples include

the Toeplitz matrices, $\mathrm{Toep}_n(\mathbf{R})$
the Hankel matrices, $\mathrm{Hank}_n(\mathbf{R})$
the tridiagonal matrices, $\mathrm{Tridiag}_n(\mathbf{R})$
the tridiagonal Toeplitz matrices, $\mathrm{TridiagToep}_n(\mathbf{R})$
the symmetric matrices, $\mathrm{Symm}_n(\mathbf{R})$
the lower-triangular matrices, $\mathrm{Lowtriang}_n(\mathbf{R})$
the matrices with zero main diagonal, $\mathrm{Zerodiag}_n(\mathbf{R})$
the matrices with zero trace, $\mathrm{Zerotrace}_n(\mathbf{R})$.

The dimensions of these linear spaces are

$$\dim \mathrm{Toep}_n(\mathbf{R}) = 2n - 1, \quad \dim \mathrm{Hank}_n(\mathbf{R}) = 2n - 1,$$
$$\dim \mathrm{Tridiag}_n(\mathbf{R}) = 3n - 2, \quad \dim \mathrm{TridiagToep}_n(\mathbf{R}) = 3,$$
$$\dim \mathrm{Symm}_n(\mathbf{R}) = \frac{n^2+n}{2}, \quad \dim \mathrm{Lowtriang}_n(\mathbf{R}) = \frac{n^2+n}{2},$$
$$\dim \mathrm{Zerodiag}_n(\mathbf{R}) = n^2 - n, \quad \dim \mathrm{Zerotrace}_n(\mathbf{R}) = n^2 - 1.$$

Suppose n is large and $Y_n \in M_n(\mathbf{R})$ is uniformly distributed on the unit sphere on $M_n(\mathbf{R})$, $\|Y_n\|_F^2 = 1$. Let $\mathcal{P}_{\mathrm{Str}} Y_n$ be the best approximation of Y_n by a matrix in $\mathrm{Str}_n(\mathbf{R})$. Notice that the determination of $\mathcal{P}_{\mathrm{Str}} Y_n$ is a least squares problem that can be easily solved. For instance, $\mathcal{P}_{\mathrm{Toep}} Y_n$ is the Toeplitz matrix whose kth diagonal, $k = -(n-1), \ldots, n-1$, is formed by the arithmetic mean of the numbers in the kth diagonal of Y_n. Recall that $\dim \mathrm{Str}_n(\mathbf{R}) = m$. From what was said in the preceding paragraph, we conclude that $\|\mathcal{P}_{\mathrm{Str}} Y_n\|_F^2$ is $B(\frac{m}{2}, \frac{n^2-m}{2})$ distributed on $(0, 1)$. For example, $\|\mathcal{P}_{\mathrm{Toep}} Y_n\|^2$ has the $B(\frac{2n-1}{2}, \frac{n^2-2n+1}{2})$ distribution on $(0, 1)$. The expected value of the variable $\|Y_n - \mathcal{P}_{\mathrm{Toep}} Y_n\|^2$ is $1 - \frac{2}{n} + \frac{1}{n^2}$ and the variance does not exceed $\frac{4}{n^3}$. Hence, Chebyshev's inequality gives

$$P\left(1 - \frac{2}{n} + \frac{1}{n^2} - \frac{\varepsilon}{n} < \|Y_n - \mathcal{P}_{\mathrm{Toep}} Y_n\|^2 < 1 - \frac{2}{n} + \frac{1}{n^2} + \frac{\varepsilon}{n}\right) \geq 1 - \frac{4}{n\varepsilon^2}. \tag{9.36}$$

Consequently, $\mathcal{P}_{\mathrm{Toep}} Y_n$ is with high probability found near the sphere with the radius $\sqrt{\frac{2}{n} - \frac{1}{n^2}}$ and $\|Y_n - \mathcal{P}_{\mathrm{Toep}} Y_n\|_F^2$ is tightly concentrated around $1 - \frac{2}{n} + \frac{1}{n^2}$.

We arrive at the conclusion that nearly all $n \times n$ matrices of Frobenius norm 1 are at nearly the same distance to the set of all $n \times n$ Toeplitz matrices!

This does not imply that the Toeplitz matrices are at the center of the universe. In fact, the conclusion is true for each of the classes $\mathrm{Str}_n(\mathbf{R})$ listed above. For instance, from Chebyshev's inequality we obtain

$$P\left(\frac{1}{2} - \frac{1}{2n} - \varepsilon < \|Y_n - \mathcal{P}_{\mathrm{Symm}} Y_n\|^2 < \frac{1}{2} - \frac{1}{2n} + \varepsilon\right) \geq 1 - \frac{1}{2n^2\varepsilon^2} \tag{9.37}$$

and

$$P\left(\frac{1}{n^2} - \frac{\varepsilon}{n^2} < \|Y_n - \mathcal{P}_{\mathrm{Zerotrace}} Y_n\|^2 < \frac{1}{n^2} + \frac{\varepsilon}{n^2}\right) \geq 1 - \frac{2}{n^2\varepsilon^2}.$$

If the expected value of $\|Y_n - \mathcal{P}_{\mathrm{Struct}} Y_n\|^2$ stays away from 0 and 1 as $n \to \infty$, we have much sharper estimates. Namely, Lemma 2.2 of [93] in conjunction with Theorem 9.30 implies that if X_n^2 has the $B(\frac{m}{2}, \frac{N-m}{2})$ distribution on $(0, 1)$, then

$$P\left(X_n^2 \leq \sigma \frac{m}{N}\right) \leq \left(\sigma e^{1-\sigma}\right)^{m/2}, \quad P\left(X_n^2 \geq \tau \frac{m}{N}\right) \leq \left(\tau e^{1-\tau}\right)^{m/2} \tag{9.38}$$

for $0 < \sigma < 1 < \tau$. This yields, for example,

$$P\left(\sigma\left(\frac{1}{2} - \frac{1}{2n}\right) < \|Y_n - \mathcal{P}_{\mathrm{Symm}} Y_n\|_F^2 < \tau\left(\frac{1}{2} - \frac{1}{2n}\right)\right)$$
$$\geq 1 - \left(\sigma e^{1-\sigma}\right)^{(n^2+n)/4} - \left(\tau e^{1-\tau}\right)^{(n^2+n)/4} \tag{9.39}$$

whenever $0 < \sigma < 1 < \tau$. Clearly, (9.39) is better than (9.37). On the other hand, let $\varepsilon > 0$ be small and choose τ such that $\tau\,(1 - \frac{2}{n} + \frac{1}{n^2}) = 1 - \frac{2}{n} + \frac{1}{n^2} + \frac{\varepsilon}{n}$. Then $(\tau e^{1-\tau})^{n-1/2} = 1 - \frac{\varepsilon^2}{2n} + O\,(\frac{1}{n^2})$, the O depending on ε, and hence (9.38) amounts to

$$P\left(\|Y_n - \mathcal{P}_{\text{Toep}} Y_n\|^2 \geq 1 - \frac{2}{n} + \frac{1}{n^2} + \frac{\varepsilon}{n} \right) \leq 1 - \frac{\varepsilon^2}{2n} + O\left(\frac{1}{n^2}\right),$$

which is worse than the Chebyshev estimate (9.36).

Exercises

1. Let ℓ_n^p be the space $\mathbf{C}^n$ with the ℓ^p norm. Does every $n \times n$ matrix A_n have a representation $A_n = U_n S_n V_n$ where U_n and V_n induce invertible isometries on ℓ_n^p and S_n is a diagonal matrix?

2. Let $\{A_n\}$, $\{B_n\}$, $\{E_n\}$, $\{R_n\}$ be sequences of $n \times n$ matrices and suppose

$$A_n = B_n + E_n + R_n, \quad \|E_n\|_{\text{F}}^2 = o(n), \quad \text{rank}\, R_n = o(n).$$

Let $\lambda_j(C_n)$ and $\sigma_j(C_n)$ $(j = 1, \ldots, n)$ denote the eigenvalues and singular values of an $n \times n$ matrix C_n.

(a) Prove that if, in addition, A_n, B_n, E_n, R_n are all Hermitian, then the eigenvalues of A_n and B_n are tied by the relation

$$\lim_{n\to\infty} \frac{1}{n} \sum_{j=1}^{n} \left[\varphi(\lambda_j(A_n)) - \varphi(\lambda_j(B_n))\right] = 0$$

for every continuous function $\varphi : \mathbf{R} \to \mathbf{C}$ with compact support.

(b) Show that if C_n is an arbitrary $n \times n$ matrix and H_n denotes the Hermitian matrix $\left(\begin{smallmatrix} 0 & C_n \\ C_n^* & 0 \end{smallmatrix}\right)$, then

$$\{\lambda_1(H_n), \ldots, \lambda_{2n}(H_n)\} = \{\sigma_1(C_n), \ldots, \sigma_n(C_n), -\sigma_1(C_n), \ldots, -\sigma_n(C_n)\}.$$

(c) Deduce from (a) and (b) that the singular values of A_n and B_n satisfy

$$\lim_{n\to\infty} \frac{1}{n} \sum_{j=1}^{n} \left[\varphi(\sigma_j(A_n)) - \varphi(\sigma_j(B_n))\right] = 0$$

for every compactly supported continuous function $\varphi : \mathbf{R} \to \mathbf{C}$.

3. Let $X_1, \ldots, X_n$ be independent random variables subject to the Gaussian normal distribution with mean 0 and variance 1. Show that

$$\frac{(X_1, \ldots, X_n)}{\sqrt{X_1^2 + \cdots + X_n^2}}$$

is uniformly distributed on the unit sphere $\mathbf{S}_{n-1}$.

4. Let A_n be the $n \times n$ matrix

$$A_n = \begin{pmatrix} 1 & 1 & \dots & 1 \\ 1 & 1 & \dots & 1 \\ \dots & \dots & \dots & \dots \\ 1 & 1 & \dots & 1 \end{pmatrix},$$

let $x = (x_1, \dots, x_n)$ be uniformly distributed on $\mathbf{S}_{n-1}$, and consider the random variable $X_n^2 = \|A_n x\|_2^2 / \|A_n\|_2^2$.

(a) Show that the inequality $|A_n x\|_2^2 \le \|A_n\|_2^2 \|x\|_2^2$ is the inequality

$$(x_1 + \dots + x_n)^2 \le n\,(x_1^2 + \dots + x_n^2).$$

(b) Compute the singular values of A_n.

(c) Show that

$$EX_n^2 = \frac{1}{n}, \quad \sigma^2 X_n^2 = \frac{2}{n+2}\,\frac{1}{n}\left(1 - \frac{1}{n}\right).$$

Use Chebyshev's inequality to deduce that the inequality

$$(x_1 + \dots + x_n)^2 \le \frac{n}{2}\,(x_1^2 + \dots + x_n^2) \tag{9.40}$$

is true with probability of at least 90 % for $n \ge 18$ and with probability of at least 99 % for $n \ge 57$ and that the inequality

$$(x_1 + \dots + x_n)^2 \le \frac{n}{100}\,(x_1^2 + \dots + x_n^2) \tag{9.41}$$

is true with probability of at least 90 % for $n \ge 895$ and with probability of at least 99 % for $n \ge 2829$.

(d) Prove that X_n^2 is subject to the $B\,(\frac{1}{2}, \frac{n-1}{2})$ distribution on $(0, 1)$ and use this insight to show that (9.40) is true with probability of at least 90 % for $n \ge 6$ and with probability of at least 99 % for $n \ge 12$ and that (9.41) is true with probability of at least 90 % for $n \ge 271$ and with probability of at least 99 % for $n \ge 662$.

5. Every Hilbert space operator A with closed range has a well-defined Moore-Penrose inverse A^+. Let $A = T(a) : \ell^2 \to \ell^2$ with $a \in W$ and suppose $T(a)$ has closed range. Heinig and Hellinger [154] (also see [71]) showed that $T_n^+(a)$ converges strongly to $T^+(a)$ if and only if $a(t) \ne 0$ for $t \in \mathbf{T}$ and one of the following conditions is satisfied:

$$\operatorname{wind} a = 0, \tag{9.42}$$

$$\operatorname{wind} a > 0 \text{ and } (a^{-1})_{-m} = 0 \text{ for all sufficiently large } m, \tag{9.43}$$

$$\operatorname{wind} a < 0 \text{ and } (a^{-1})_{m} = 0 \text{ for all sufficiently large } m, \tag{9.44}$$

where $(a^{-1})_j$ denotes the jth Fourier coefficient of a^{-1}. Show that if $a \in \mathcal{P}$, then (9.43) or (9.44) are only possible if a is of the form $a(t) = t^k p_+(t)$ with $k \ge 1$ and a polynomial $p_+ \in \mathcal{P}^+$ such that $p_+(z) \ne 0$ for $|z| \le 1$ or $a(t) = t^{-k} p_-(t)$ with $k \ge 1$ and a polynomial $p_- \in \mathcal{P}^-$ such that $p_-(1/z) \ne 0$ for $|z| \le 1$.

6. To compute the numerical range $\mathcal{H}_2(A_n) = \{(A_n x, x) : \|x\|_2 = 1\}$ of an $n \times n$ matrix A_n one could try drawing a large number N of random vectors x_j from the uniform distribution on the unit sphere of $\mathbf{C}^n$ and plotting the superposition of the values $(A_n x_j, x_j)$ $(j = 1, \ldots, N)$. In the right picture of Figure 9.5 we see the result for $A_n = T_n(b)$ with b as in Figure 7.1, $n = 50$ and $N = 500$. As the numerical range contains at least the convex hull of the eigenvalues, which are shown in the left picture, we conclude that our experiment failed dramatically. Why did it fail?

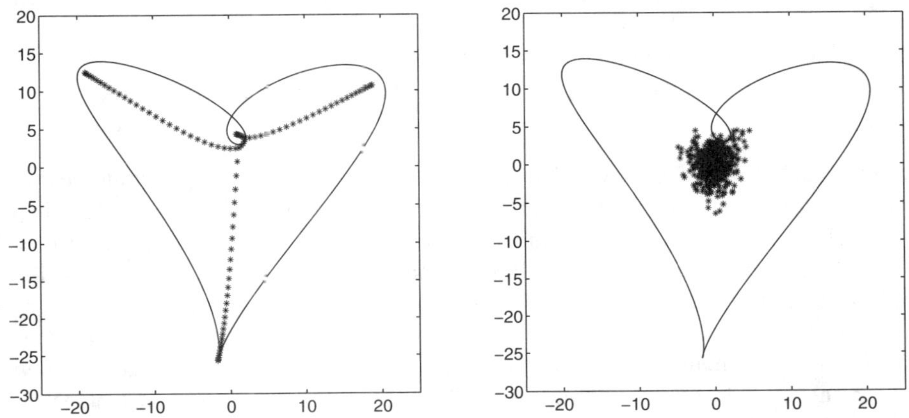

Figure 9.5. *We see the range $b(\mathbf{T})$ and the 50 eigenvalues of $T_{50}(b)$ (left) and the values $(T_{50}(b)x, x)$ for 500 vectors x drawn randomly from the uniform distribution on the unit sphere of $\mathbf{C}^{50}$ (right).*

Notes

Proofs of the theorems of Section 9.1 are in [27], [133], [166], for example.

The splitting phenomenon, Theorem 9.4, was discovered by Roch and Silbermann [224], [225]. They proved Theorem 9.4 for $p = 2$ and their proof is based on C^*-algebra techniques. In [35], another proof was given and the result was extended to $1 \le p \le \infty$. Estimate (9.2) was established in our paper [47]. The proof of Theorem 9.4 presented here is a combination of arguments of [35] and [48].

The idea of deriving results on Toeplitz matrices by comparing Toeplitz matrices with their circulant cousins has been developed by Shigeru Arimoto and coauthors in a series of papers since about 1985 which deal with problems of theoretical chemistry and in which sequences of banded circulant matrices are called alpha matrices. Independently, the same idea has emerged in papers by Beam and Warming [20], Tyrtyshnikov [284], and Serra Capizzano and Tilli [251]. Theorems 9.6 to 9.9 are from our book [48], but we are sure that versions of these theorems had been known earlier. We noticed in [48] that we found these theorems in our manuscripts but that we cannot remember whether we obtained them ourselves some time ago or whether we took them from somewhere.

Theorems 9.7 and 9.19 can be found in [27] and [167]. We thank Estelle Basor for pointing out that Theorem 9.7 was incorrectly stated in our previous book [48].

Formula (9.9) is Widom's [295] and its generalization (9.8) is due to Roch and Silbermann [222].

Formula (9.11), the so-called Avram-Parter theorem, was established by Parter [200] for symbols $b \in L^\infty$ which are locally normal, that is, which can be written as the product of a continuous and a real-valued function. Avram [10] proved (9.11) for general $b \in L^\infty$. We refer the reader to Sections 5.6 and 5.8 of [71] and to [301] for further generalizations of the Avram-Parter theorem. The elegant approach and the results of Section 9.6 are due to Zizler, Zuidwijk, Taylor, and Arimoto [303]. The notions of the cluster and the proper cluster were introduced by Tyrtyshnikov in [284].

Sections 9.8 to 9.10 are based on our paper [52].

A solution to Exercise 1 is in [71] (see also [250]). Exercise 2 is a result of Tyrtyshnikov [282], [284].

Further results: C^*-algebras I. The purpose of the following is to illustrate how a few simple C^*-algebra arguments yield part of the results established in the previous chapters very quickly. Of course, the C^* machinery forces us to limit ourselves to the case $p = 2$.

A *Banach algebra* is a complex Banach space $\mathcal{A}$ with an associative and distributive multiplication such that $\|ab\| \le \|a\| \, \|b\|$ for all $a, b \in \mathcal{A}$. If a Banach algebra has a unit element, which is usually denoted by e, 1, or I, it is referred to as a *unital Banach algebra*. A conjugate-linear map $a \mapsto a^*$ of a Banach algebra into itself is called an *involution* if $a^{**} = a$ and $(ab)^* = b^*a^*$ for all $a, b \in \mathcal{A}$. Finally, a C^*-algebra is a Banach algebra with an involution such that $\|a^*a\| = \|a\|^2$ for all $a \in \mathcal{A}$. Notice that C^*-algebras are especially nice Banach algebras, although the terminology suggests the contrary.

If H is a Hilbert space, then the sets $\mathcal{B}(H)$ and $\mathcal{K}(H)$ of all bounded and compact linear operators on H are C^*-algebras under the operator norm and passage to the Hermitian adjoint as involution. This is in particular the case for $\mathcal{B} := \mathcal{B}(\ell^2)$ and $\mathcal{K} := \mathcal{K}(\ell^2)$. Clearly, $\mathcal{B}$ is unital, but $\mathcal{K}$ has no unit element. The set $C := C(\mathbf{T})$ is a C^*-algebra under the norm $\|\cdot\|_\infty$ and the involution $a \mapsto \overline{a}$ (passage to the complex conjugate). The Wiener algebra W is a Banach algebra with an involution, $a \mapsto \overline{a}$, but it is not a C^*-algebra, because the equality $\|\overline{a}a\|_W = \|a\|_W^2$ is not satisfied for all $a \in W$. In the case $p = 2$, the Banach algebras $\mathbf{F}$ and $\mathbf{S}$ introduced in Section 3.4 are C^*-algebras. The involution is defined by $\{A_n\}^* := \{A_n^*\}$.

A subset of a C^*-algebra $\mathcal{A}$ that is itself a C^*-algebra is called a *C^*-subalgebra*. Let $\mathcal{A}$ be a C^*-algebra and let E be a subset of $\mathcal{A}$. The *C^*-algebra generated by E* is the smallest C^*-subalgebra of $\mathcal{A}$ that contains E. (In other words, the C^*-algebra generated by E is the intersection of all C^*-subalgebras of $\mathcal{A}$ that contain E.) If this C^*-algebra is $\mathcal{A}$ itself, one says that *$\mathcal{A}$ is generated by E*.

One of the many excellent properties of C^*-algebras is their *inverse closedness*. This means the following. If $\mathcal{A}_1$ is a unital C^*-algebra with the unit element e and $\mathcal{A}_2$ is a C^*-subalgebra of $\mathcal{A}_1$ which contains e, then every element $a \in \mathcal{A}_2$ that is invertible in $\mathcal{A}_1$ is automatically invertible in $\mathcal{A}_2$.

So far we have considered the Toeplitz operator $T(a)$ for $a \in W$ only. For $a \in C$ (or even $a \in L^\infty(\mathbf{T})$), this operator is also defined via the matrix $(a_{j-k})_{j,k=1}^\infty$ formed of the Fourier coefficients. It is not difficult to prove that $T(a)$ is bounded on ℓ^2 for $a \in C$ (or even $a \in L^\infty(\mathbf{T})$). If $b \in \mathcal{P}$, then the sequence $\{T_n(b)\}$ is an element of the C^*-algebra $\mathbf{F}$. Let $\mathbf{A}$ denote the C^*-algebra generated by $E = \{\{T_n(b)\} : b \in \mathcal{P}\}$ in $\mathbf{F}$. Let finally $\mathbf{G}$ be the set of all $\{A_n\} \in \mathbf{F}$ for which $\|A_n\|_2 \to 0$ as $n \to \infty$.

Theorem on A. *The C^*-algebra* **A** *is the set of all sequences* $\{A_n\}$ *of the form*

$$A_n = T_n(a) + P_n K P_n + W_n L W_n + C_n \tag{9.45}$$

with $a \in C$, $K \in \mathcal{K}$, $L \in \mathcal{K}$, $\{C_n\} \in \mathbf{G}$.

This theorem was established in [66] (proofs are also in [70, Proposition 7.27] and [62, Proposition 2.2]).

A *C^*-algebra homomorphism* is a map $f : \mathcal{A}_1 \to \mathcal{A}_2$ of a C^*-algebra $\mathcal{A}_1$ into a C^*-algebra $\mathcal{A}_2$ satisfying

$$f(\alpha a) = \alpha f(a), \quad f(a+b) = f(a) + f(b), \quad f(ab) = f(a)f(b), \quad f(a^*) = f(a)^*$$

for all $\alpha \in \mathbf{C}$, $a \in \mathcal{A}_1$, $b \in \mathcal{A}_1$. The set **G** is obviously a closed two-sided ideal of the C^*-algebra **A**. Therefore the quotient algebra **A**/**G** is a C^*-algebra with the usual quotient operations and the usual quotient norm. The sum $\mathcal{B} \oplus \mathcal{B}$ is the C^*-algebra of all ordered pairs $(A, B) \in \mathcal{B}^2$ with the natural operations and the norm $\|(A, B)\| := \max(\|A\|_2, \|B\|_2)$. Let $\{A_n\}$ be given by (9.45). Then

$$A_n \to A := T(a) + K \quad \text{and} \quad W_n A_n W_n \to \widetilde{A} := T(\widetilde{a}) + L$$

strongly as $n \to \infty$.

Theorem on A/G. *The map* Sym *defined by*

$$\text{Sym} : \mathbf{A}/\mathbf{G} \to \mathcal{B} \oplus \mathcal{B}, \quad \{A_n\} + \mathbf{G} \mapsto (A, \widetilde{A})$$

is a C^*-algebra homomorphism that preserves spectra and norms.

It is easily verified that Sym is a C^*-algebra homomorphism. Since the only compact Toeplitz operator is the zero operator (this is Corollary 1.13 for $a \in C$), it follows that Sym is injective. As injective homomorphisms of unital C^*-algebras automatically preserve spectra and norms (which is another exquisite property of C^*-algebras that is not shared by general Banach algebras), we arrive at the conclusion of the theorem. This simple reasoning goes back to [66], [70, Theorem 7.11] and is explicit in [34].

Here are some immediate consequences of the theorem on **A**/**G**.

Consequence 1. A sequence $\{A_n\} \in \mathbf{A}$ is stable on ℓ^2 if and only if $A = T(a) + K$ and $\widetilde{A} = T(\widetilde{a}) + L$ are invertible on ℓ^2 (Theorem 3.13 for $p = 2$). Indeed, due to the inverse closedness of **A**/**G** in **F**/**G**, the stability of $\{A_n\}$ is equivalent to the condition that 0 does not belong to the spectrum of $\{A_n\} + \mathbf{G}$ in **A**/**G**.

Consequence 2. If $\{A_n\} \in \mathbf{A}$ and both A and $\widetilde{A}$ are invertible on ℓ^2, then

$$A_n^{-1} = T_n(a^{-1}) + P_n X P_n + W_n Y W_n + D_n \tag{9.46}$$

with $X \in \mathcal{K}$, $Y \in \mathcal{K}$, $\{D_n\} \in \mathbf{G}$ for all sufficiently large n (recall Section 3.5). Indeed, the assumption implies that $(\{A_n\} + \mathbf{G})^{-1} \in \mathbf{A}/\mathbf{G}$ and the theorem on **A** therefore yields (9.46). Passing to the strong limit $n \to \infty$ in (9.46) we get

$$T^{-1}(a) = T(a^{-1}) + X \quad \text{and} \quad T^{-1}(\widetilde{a}) = T(\widetilde{a}^{-1}) + Y,$$

that is, we recover (3.15) and (3.16).

Consequence 3. If $\{A_n\} \in \mathbf{A}$, then

$$\lim_{n\to\infty} \|A_n\|_2 = \max(\|A\|_2, \|\widetilde{A}\|_2)$$

(Corollary 5.14 for $p = 2$). This follows from Theorem 3.1 and the equalities

$$\limsup_{n\to\infty} \|A_n\|_2 = \|\{A_n\} + \mathbf{G}\|_{\mathbf{A}/\mathbf{G}} = \|\mathrm{Sym}\,(\{A_n\} + \mathbf{G})\|_{\mathcal{B}\oplus\mathcal{B}} = \max(\|A\|_2, \|\widetilde{A}\|_2).$$

Consequence 4. If $\{A_n\} \in \mathbf{A}$ and A and $\widetilde{A}$ are invertible on ℓ^2, then

$$\lim_{n\to\infty} \|A_n^{-1}\|_2 = \max(\|A^{-1}\|_2, \|\widetilde{A}^{-1}\|_2)$$

(Theorem 6.3 for $p = 2$). To see this, combine Consequences 2 and 3.

Consequence 5. If $\{A_n\} \in \mathbf{A}$ and at least one of the operators A and $\widetilde{A}$ is nonzero, then

$$\lim_{n\to\infty} \kappa_2(A_n) = \max(\|A\|_2, \|\widetilde{A}\|_2)\,\max(\|A^{-1}\|_2, \|\widetilde{A}^{-1}\|_2)$$

(Corollary 6.4 for $p = 2$). This is straightforward from Consequences 3 and 4.

Consequence 6. If $\{A_n\} \in \mathbf{A}$ and $\varepsilon > 0$, then

$$\liminf_{n\to\infty} \mathrm{sp}_\varepsilon^{(2)} A_n = \limsup_{n\to\infty} \mathrm{sp}_\varepsilon^{(2)} A_n = \mathrm{sp}_\varepsilon^{(2)} A \cup \mathrm{sp}_\varepsilon^{(2)} \widetilde{A}$$

(generalization of Theorem 7.7 in the case $p = 2$). Once Consequence 4 is available, this can be proved by the argument of the proof of Theorem 7.7.

Consequence 7. If $\{A_n\} \in \mathbf{A}$ and $A_n^* = A_n$ for all n, then

$$\liminf_{n\to\infty} \mathrm{sp}\, A_n = \limsup_{n\to\infty} \mathrm{sp}\, A_n = \mathrm{sp}\, A \cup \mathrm{sp}\, \widetilde{A}$$

(Lemma 9.10). To see this, let $\lambda \notin \mathrm{sp}\, A \cup \mathrm{sp}\, \widetilde{A}$. Then $\{A_n - \lambda I\}$ is stable (Consequence 1) and hence the spectral radius of $(A_n - \lambda I)^{-1}$ remains bounded as $n \to \infty$. This implies that $\lambda \notin \limsup \mathrm{sp}\, A_n$. Conversely, let $\lambda \in \mathbf{R}$ and $\lambda \notin \liminf \mathrm{sp}\, A_n$. Then there exists a $\delta > 0$ such that $U_\delta(\lambda) \cap \mathrm{sp}\, A_n = \emptyset$ for infinitely many n, that is, $U_\delta(0) \cap \mathrm{sp}\,(A_n - \lambda I) = \emptyset$ for infinitely many n. As $A_n - \lambda I$ is Hermitian, the spectral radius and the norm of $(A_n - \lambda I)^{-1}$ coincide, which gives that $\|(A_n - \lambda I)^{-1}\|_2 \le 1/\delta$ for infinitely many n. Consequently, we arrive at a subsequence $\{n_k\}$ such that $\{A_{n_k} - \lambda I\}$ and thus also $\{W_{n_k}(A_{n_k} - \lambda I)W_{n_k}\}$ is stable. Lemma 3.4 now yields the invertibility of $A - \lambda I$ and $\widetilde{A} - \lambda I$.

Consequence 8. If $\{A_n\} \in \mathbf{A}$, then

$$\liminf_{n\to\infty} \Sigma(A_n) = \limsup_{n\to\infty} \Sigma(A_n) = \Sigma(A) \cup \Sigma(\widetilde{A})$$

(Corollary 9.11). Since $\{A_n^* A_n\} \in \mathbf{A}$, this follows from Consequence 7.

Summary. Thus, we have demonstrated that many sharp convergence results can be obtained very comfortably by working with appropriate C^*-algebras. For more details and for further developments of this idea we refer the reader to [36], [48], [223] and especially to Hagen, Roch, and Silbermann's monograph [149]. Another approach to questions of numerical analysis via C^*-algebras was worked out by Arveson [7], [8], [9]. Fragments of this approach will be outlined in the notes to Chapter 14.

We nevertheless want to emphasize that the C^*-algebra approach has its limitations. For example, it is restricted to Hilbert space operators. Moreover, refinements of the consequences cited above, such as estimates of the convergence speed, require hard analysis and hence the tools presented in the preceding chapters.

Chapter 10

Extreme Eigenvalues

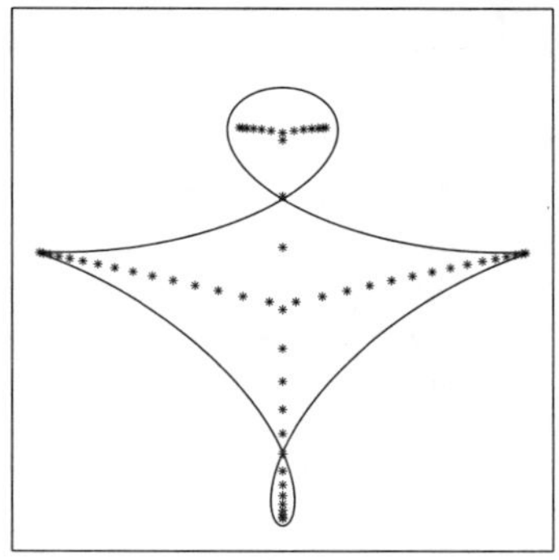

In this chapter we embark on the extreme eigenvalues of Hermitian Toeplitz matrices and on estimates for the spectral radius of not necessarily Hermitian Toeplitz matrices.

10.1 Hermitian Matrices

Let b be a nonconstant Laurent polynomial. The matrix $T_n(b)$ is Hermitian if and only if $\overline{b}_j = b_{-j}$, that is, if and only if b is real valued. So suppose b is real valued and let

$$m = \min_{t \in \mathbf{T}} b(t), \quad M = \max_{t \in \mathbf{T}} b(t).$$

By Lemma 4.7,

$$(T_n(b)x, x) = \frac{1}{2\pi} \int_0^{2\pi} b(e^{i\theta}) |f(e^{i\theta})|^2 d\theta, \tag{10.1}$$

where $f(e^{i\theta}) = x_0 + x_1 e^{i\theta} + \cdots + x_{n-1} e^{i(n-1)\theta}$. This implies that all eigenvalues of $T_n(b)$ are contained in the open interval (m, M),

$$m < \lambda_1(T_n(b)) \le \lambda_2(T_n(b)) \le \cdots \le \lambda_n(T_n(b)) < M. \tag{10.2}$$

Theorem 10.1. *Let b be a nonconstant real-valued Laurent polynomial, let $\mathcal{R}(b) = [m, M]$, and denote by 2α and 2β the maximal order of the zeros of $b - m$ and $M - b$, respectively. Then for each fixed k,*

$$\lambda_k(T_n(b)) - m \simeq \frac{1}{n^{2\alpha}}, \quad M - \lambda_{n-k}(T_n(b)) \simeq \frac{1}{n^{2\beta}},$$

where the notation $x_n \simeq y_n$ means that there are constants $C_1, C_2 \in (0, \infty)$ such that $C_1 y_n \le x_n \le C_2 y_n$ for all sufficiently large n.

Proof. Put $a = b - m$. Then, by (10.1), $(T_n(a)x, x) \ge 0$ for all $x \in \ell_n^2$. This shows that $(T_n^*(a)T_n(a))^{1/2} = (T_n(a)T_n(a))^{1/2} = T_n(a)$. Consequently, the eigenvalues of $T_n(a)$

coincide with its singular values. Theorem 9.8 therefore gives $\lambda_k(T_n(a)) \le E_k n^{-2\alpha}$ for all n with some $E_k \in (0, \infty)$ independent of n. On the other hand, from Theorem 4.32 we infer that $\lambda_k(T_n(a)) \ge \lambda_1(T_n(a)) = \|T_n^{-1}(a)\|^{-1} \ge D_k n^{-2\alpha}$ for all n with some constant $D_k \in (0, \infty)$ which does not depend on n. Thus, $\lambda_k(T_n(b)) - m = \lambda_k(T_n(a)) \simeq n^{-2\alpha}$. Repeating the above reasoning with $a = M - b$, we obtain that $M - \lambda_{n-k}(T_n(b)) = \lambda_{k+1}(T_n(a)) \simeq n^{-2\beta}$. □

In concrete cases, the estimates provided by Theorem 10.1 can be made more precise by comparing Toeplitz matrices with appropriate circulants. For $n \ge 2r + 1$, define the circulant $C_n(b)$ as in Sections 2.1 and 9.3. On using Theorem 9.19 r times, we obtain

$$\lambda_k(C_{n+r}(b)) \le \lambda_k(T_n(b)) \le \lambda_{k+r}(C_{n+r}(b)), \tag{10.3}$$

$$\lambda_{n-k}(C_{n+r}(b)) \le \lambda_{n-k}(T_n(b)) \le \lambda_{n+r-k}(C_{n+r}(b)). \tag{10.4}$$

From Proposition 2.1 we know that if $n \ge r + 1$, then the eigenvalues of $C_{n+r}(b)$ are

$$b(e^{2\pi ij/(n+r)}) \quad (j = 0, 1, \dots, n + r - 1).$$

This in conjunction with (10.3) and (10.4) often gives good bounds for the extreme eigenvalues of $T_n(b)$.

Example 10.2. Let b be a nonnegative and nonconstant Laurent polynomial of degree $r \ge 1$. Suppose b has exactly one zero on $\mathbf{T}$. Without loss of generality assume that this is a zero of order $2\alpha \ge 2$ at the point 1. Thus, if we write $h(\theta) = b(e^{i\theta})$, then

$$h(\theta) = \frac{h^{(2\alpha)}(0)}{(2\alpha)!}\,\theta^{2\alpha}\left(1 + O(\theta^{2\alpha+1})\right).$$

Put $\mu_\alpha = h^{(2\alpha)}(0)/(2\alpha)!$. It is clear that, for sufficiently large n,

$$\begin{aligned}
&\lambda_1(C_{n+r}(b)) = h(0) = 0,\\
&\lambda_2(C_{n+r}(b)) \sim h\left(\frac{2\pi}{n+r}\right) \sim \mu_\alpha\left(\frac{2\pi}{n}\right)^{2\alpha},\\
&\lambda_3(C_{n+r}(b)) \sim h\left(\frac{2\pi}{n+r}\right) \sim \mu_\alpha\left(\frac{2\pi}{n}\right)^{2\alpha},\\
&\lambda_4(C_{n+r}(b)) \sim h\left(\frac{4\pi}{n+r}\right) \sim \mu_\alpha\left(\frac{4\pi}{n}\right)^{2\alpha},\\
&\dots\\
&\lambda_j(C_{n+r}(b)) \sim h\left(\frac{2\pi[j/2]}{n+r}\right) \sim \mu_\alpha\left(\frac{2\pi[j/2]}{n}\right)^{2\alpha}.
\end{aligned} \tag{10.5}$$

Combining (10.3) and (10.5) we see that, for fixed $k \ge 1$,

$$\frac{9}{10}\,\mu_\alpha\left(\frac{2\pi[k/2]}{n}\right)^{2\alpha} \le \lambda_k(T_n(b)) \le \frac{10}{9}\,\mu_\alpha\left(\frac{2\pi[(k+r)/2]}{n}\right)^{2\alpha}$$

for all sufficiently large n. □

Although the eigenvalue distribution of Toeplitz band matrices will be the subject of the next chapter, we already now have everything at our disposal in order to treat the Hermitian case.

Since $\operatorname{sp} T_n(b) = m + \operatorname{sp} T_n(b-m)$ and since the eigenvalues of $T_n(b-m)$ coincide with the singular values of $T_n(b-m)$, we can have immediate recourse to the results of Section 9.4. However, since Theorem 9.19 is a little bit sharper than Theorem 9.7, repetition of the reasoning of Section 9.6 with Theorem 9.7 replaced by Theorem 9.19 yields slightly better results. We begin with the analogue of Theorem 9.14.

Theorem 10.3. *Let r and n be natural numbers such that $1 \le r < n$, let $K = \{k_1, \dots, k_r\}$ be a subset of $\{1, 2, \dots, n\}$ consisting of r distinct elements, and put $L = \{1, 2, \dots, n\} \setminus K$. Suppose A and A' are two Hermitian $n \times n$ matrices whose jk entries coincide for all $(j,k) \in L \times L$. If $f \in BV$ and $[a,b]$ is any segment which contains the eigenvalues of both A and A', then*

$$\sum_{k=1}^{n} \left| f(\lambda_k(A)) - f(\lambda_k(A')) \right| \le r V_{[a,b]}(f).$$

Proof. We proceed exactly as in the proof of Theorem 9.14. Define B as in that proof. Now Theorem 9.19 gives

$$\begin{aligned}
\lambda_1(A), \lambda_1(A') &\in [a, \lambda_1(B)], \\
\lambda_2(A), \lambda_2(A') &\in [\lambda_1(B), \lambda_2(B)], \\
&\dots \\
\lambda_{n-1}(A), \lambda_{n-1}(A') &\in [\lambda_{n-2}(B), \lambda_{n-1}(B)], \\
\lambda_n(A), \lambda_n(A') &\in [\lambda_{n-1}(B), b],
\end{aligned}$$

whence, as in the proof of Theorem 9.14,

$$\sum_{k=1}^{n} \left| f(\lambda_k(A)) - f(\lambda_k(A')) \right| \le V_{[a,\lambda_1]}(f) + V_{[\lambda_1,\lambda_2]}(f) + \dots + V_{[\lambda_{n-1},b]}(f) = V_{[a,b]}(f)$$

with $\lambda_j := \lambda_j(B)$. Thus, the factor 3 that we encountered in the proof of Theorem 9.14 disappears. □

Theorem 10.4. *Let $b(t) = \sum_{j=-r}^{r} b_j t^j$ $(t \in \mathbf{T})$ be a nonconstant real-valued Laurent polynomial and let $f \in BV$. If $[c,d]$ is any segment which contains $\mathcal{R}(b) = [\min b, \max b]$, then for all $n \ge 1$,*

$$\left| \sum_{k=1}^{n} f(\lambda_k(T_n(b))) - \frac{n}{2\pi} \int_0^{2\pi} f(b(e^{i\theta}))d\theta \right| \le 3r V_{[c,d]}(f).$$

Proof. This follows from the reasoning of the proof of Theorem 9.15. In (9.14) we can drop the factor 3, and considering $u(\theta) = b(e^{i\theta})$ instead of $u(\theta) = |b(e^{i\theta})|^2$, we can replace the factor 4 in (9.16) by the factor 2. This gives the assertion. □

Corollary 10.5. *Let b be a real-valued Laurent polynomial and let $f : \mathbf{R} \to \mathbf{C}$ be a function with compact support. If f is continuous or of bounded variation, then*

$$\lim_{n\to\infty} \frac{1}{n} \sum_{k=1}^{n} f(\lambda_k(T_n(b))) = \frac{1}{2\pi} \int_0^{2\pi} f(b(e^{i\theta}))d\theta.$$

Proof. Use the same arguments as in the proof of Corollary 9.16. □

Given a (Lebesgue) measurable subset E of $\mathbf{R}$, we put

$$N_n(E) = \sum_{k=1}^{n} \chi_E(\lambda_k(T_n(b))), \quad \mu_n(E) = \frac{1}{n} N_n(E),$$

$$\mu(E) = \frac{1}{2\pi} \int_0^{2\pi} \chi_E(b(e^{i\theta}))d\theta = \frac{1}{2\pi} |\{t \in \mathbf{T} : b(t) \in E\}|,$$

where $|\cdot|$ is Lebesgue measure on $\mathbf{T}$. In the same way we proved Corollaries 9.17 and 9.18 we now obtain the following two results.

Corollary 10.6. *Let b be a real-valued Laurent polynomial. Then*

$$\int_{\mathbf{R}} f d\mu_n \to \int_{\mathbf{R}} f d\mu$$

for every compactly supported continuous function $f : \mathbf{R} \to \mathbf{C}$. □

Corollary 10.7. *If b is a real-valued Laurent polynomial of degree r and $E \subset \mathbf{R}$ is a segment, then $|N_n(E) - n\mu(E)| \le 6r$ for every $n \ge 1$.* □

Corollary 10.8. *Let b be a real-valued Laurent polynomial. Then*

$$\mathrm{sp}\, T_n(b) \subset [\min b, \max b] \tag{10.6}$$

for every $n \ge 1$ and

$$\liminf_{n\to\infty} \mathrm{sp}\, T_n(b) = \limsup_{n\to\infty} \mathrm{sp}\, T_n(b) = [\min b, \max b] \; (= \mathrm{sp}\, T(b)).$$

Proof. The inclusion (10.6) follows from (10.2). Let $\lambda \in [\min b, \max b]$ and put $E = [\lambda - \varepsilon, \lambda + \varepsilon]$, where $\varepsilon > 0$ can be chosen as small as desired. Then $\mu(E) > 0$, and Corollary 10.7 therefore implies that $N_n(E) \ge 1$ for all sufficiently large n. This shows that $[\min b, \max b] \subset \liminf \mathrm{sp}\, T_n(b)$. □

10.2 First-Order Trace Formulas

In this section we establish some simple results which are of interest on their own and will be needed in the next section.

The *trace* $\mathrm{tr}\, A$ of an $n \times n$ matrix $A = (a_{jk})_{j,k=1}^n$ is defined as usual:

$$\mathrm{tr}\, A = a_{11} + a_{22} + \cdots + a_{nn}.$$

Denoting by $\lambda_1(A), \ldots, \lambda_n(A)$ the eigenvalues of A, we have

$$\operatorname{tr} A^k = \lambda_1^k(A) + \cdots + \lambda_n^k(A)$$

for every natural number k. The *trace norm* of A is defined by

$$\|A\|_{\mathrm{tr}} = \sigma_1(A) + \cdots + \sigma_n(A),$$

where $\sigma_1(A), \ldots, \sigma_n(A)$ are the singular values of A. From Theorem 9.3 we deduce that

$$\|ABC\|_{\mathrm{tr}} \le \|A\|_2 \|B\|_{\mathrm{tr}} \|C\|_2. \tag{10.7}$$

It is also well known that

$$|\operatorname{tr} A| \le \|A\|_{\mathrm{tr}}. \tag{10.8}$$

Finally, we denote by $\mathcal{O}$ the collection of all sequences $\{K_n\}_{n=1}^{\infty}$ of complex $n \times n$ matrices K_n such that

$$\frac{1}{n} \|K\|_{\mathrm{tr}} \to 0.$$

Lemma 10.9. *If a and b are Laurent polynomials, then $\{T_n(a)T_n(b) - T_n(ab)\} \in \mathcal{O}$.*

Proof. By Proposition 3.10,

$$T_n(a)T_n(b) - T_n(ab) = -P_n H(a) H(\widetilde{b}) P_n - W_n H(\widetilde{a}) H(b) W_n.$$

The matrices $H(a)H(\widetilde{b})$ and $H(\widetilde{a})H(b)$ have only finitely many nonzero entries. Thus, since $\|P_n\|_2 = \|W_n\|_2 = 1$, inequality (10.7) yields

$$\frac{1}{n} \|P_n H(a) H(\widetilde{b}) P_n\|_{\mathrm{tr}} \le \frac{1}{n} \|P_n\|_2 \|H(a)H(\widetilde{b})\|_{\mathrm{tr}} \|P_n\|_2 = o(1),$$

$$\frac{1}{n} \|W_n H(\widetilde{a}) H(b) W_n\|_{\mathrm{tr}} \le \frac{1}{n} \|W_n\|_2 \|H(\widetilde{a})H(b)\|_{\mathrm{tr}} \|W_n\|_2 = o(1). \qquad \square$$

Lemma 10.10. *If b is a Laurent polynomial and $k \in \mathbf{N}$, then $\{T_n^k(b) - T_n(b^k)\} \in \mathcal{O}$.*

Proof. The assertion is trivial for $k = 1$. Now suppose that the assertion is true for some $k \in \mathbf{N}$. Then $T_n^{k+1}(b) = T_n^k(b)T_n(b) = T_n(b^k)T_n(b) + K_n T_n(b)$ with some $\{K_n\} \in \mathcal{O}$. Since $\|K_n T_n(b)\|_{\mathrm{tr}} \le \|K_n\|_{\mathrm{tr}} \|T_n(b)\|_2 \le \|K_n\|_{\mathrm{tr}} \|b\|_\infty$, it is clear that $\{K_n T_n(b)\} \in \mathcal{O}$. Lemma 10.9 implies that $\{T_n(b^k)T_n(b) - T_n(b^{k+1})\} \in \mathcal{O}$. This shows that $\{T_n^{k+1}(b) - T_n(b^{k+1})\}$ is a sequence in $\mathcal{O}$. $\square$

Theorem 10.11. *Let b be a Laurent polynomial and $k \in \mathbf{N}$. Then*

$$\lim_{n\to\infty} \frac{1}{n} \sum_{j=1}^{n} \lambda_j^k(T_n(b)) = \frac{1}{2\pi} \int_0^{2\pi} (b(e^{i\theta}))^k \, d\theta. \tag{10.9}$$

Proof. First notice that

$$\frac{1}{n}\sum_{j=1}^{n}\lambda_j^k(T_n(b)) = \frac{1}{n}\mathrm{tr}\; T_n^k(b).$$

By Lemma 10.10,

$$\frac{1}{n}\mathrm{tr}\; T_n^k(b) = \frac{1}{n}\mathrm{tr}\; T_n(b^k) + \frac{1}{n}\mathrm{tr}\; K_n$$

with $\{K_n\} \in \mathcal{O}$. Since

$$\frac{1}{n}\mathrm{tr}\; T_n(b^k) = \frac{1}{n}\left((b^k)_0 + \cdots + (b^k)_0\right) = \frac{1}{n}\, n(b^k)_0 = (b^k)_0 = \frac{1}{2\pi}\int_0^{2\pi}(b(e^{i\theta}))^k d\theta$$

and, by (10.8), $|\mathrm{tr}\; K_n|/n \le \|K_n\|_{\mathrm{tr}}/n = o(1)$, we arrive at (10.9). □

10.3 The Spectral Radius

Given a matrix or an operator A, we denote by rad A its *spectral radius*,

$$\mathrm{rad}\, A = \max\{|\lambda| : \lambda \in \mathrm{sp}\, A\}.$$

In general, rad $T_n(b)$ does not converge to rad $T(b)$; indeed, if $b(t) = t$, then rad $T_n(b) = 0$ and rad $T(b) = 1$.

Suppose $b(t) = \sum_{j=-r}^{r} b_j t^j$ $(t \in \mathbf{T})$. For $\varrho \in (0, \infty)$, we define the Laurent polynomial b_ϱ by

$$b_\varrho(e^{i\theta}) = b(\varrho e^{i\theta}) = \sum_{j=-r}^{r} b_j \varrho^j e^{ij\theta}.$$

Theorem 10.12 (Schmidt and Spitzer). *If b is a Laurent polynomial, then*

$$\limsup_{k\to\infty}\left|\frac{1}{2\pi}\int_0^{2\pi}(b(e^{i\theta}))^k d\theta\right|^{1/k} \le \liminf_{n\to\infty}\mathrm{rad}\; T_n(b)$$

$$\le \limsup_{n\to\infty}\mathrm{rad}\; T_n(b) \le \inf_{\varrho\in(0,\infty)}\|b_\varrho\|_\infty. \tag{10.10}$$

Proof. Let $\lambda_j(T_n(b))$ be the eigenvalues of $T_n(b)$. Obviously,

$$\left|\frac{1}{n}\sum_{j=1}^{n}\lambda_j^k(T_n(b))\right|^{1/k} \le \max_{1\le j\le n}|\lambda_j(T_n(b))| = \mathrm{rad}\; T_n(b).$$

From Theorem 10.11 we therefore get

$$\left|\frac{1}{2\pi}\int_0^{2\pi}(b(e^{i\theta}))^k d\theta\right|^{1/k} \le \liminf_{n\to\infty}\mathrm{rad}\; T_n(b),$$

which is stronger than the first inequality of (10.10). Since

$$T_n(b_\varrho) = \operatorname{diag}(1, \varrho, \dots, \varrho^{n-1})\, T_n(b)\, \operatorname{diag}(1, \varrho^{-1}, \dots, \varrho^{-(n-1)}),$$

we have $\operatorname{sp} T_n(b) = \operatorname{sp} T_n(b_\varrho)$ and thus

$$\operatorname{rad} T_n(b) = \operatorname{rad} T_n(b_\varrho) \le \|T_n(b_\varrho)\|_2 \le \|b_\varrho\|_\infty.$$

This implies that

$$\operatorname{rad} T_n(b) \le \inf_{\varrho\in(0,\infty)} \|b_\varrho\|_\infty$$

and gives the last inequality of (10.10). □

By Hadamard's three circles theorem,

$$\log M(\varrho) := \log \|b_\varrho\|_\infty = \log \max_{\theta\in[0,2\pi)} |b(\varrho e^{i\theta})|$$

is a convex function of $\log \varrho$. Since $\log M(\varrho) \to +\infty$ as $\varrho \to 0$ and $\varrho \to +\infty$, it follows that there is a unique $\varrho_0 \in (0, \infty)$ such that

$$\inf_{\varrho\in(0,\infty)} \|b_\varrho\|_\infty = \|b_{\varrho_0}\|_\infty.$$

We exhibit two cases in which all inequalities of (10.10) become equalities.

Theorem 10.13. *Let b be a Laurent polynomial and suppose $T(b)$ is Hermitian. Then*

$$\lim_{n\to\infty} \operatorname{rad} T_n(b) = \operatorname{rad} T(b) = \|b\|_\infty$$
$$= \inf_{\varrho\in(0,\infty)} \|b_\varrho\|_\infty = \limsup_{k\to\infty} \left| \frac{1}{2\pi} \int_0^{2\pi} (b(e^{i\theta}))^k d\theta \right|^{1/k}.$$

Proof. The function b is real valued. Let $\mathcal{R}(b) = [m, M]$. We know from (10.2) and Theorem 10.1 that

$$\operatorname{sp} T_n(b) \subset [m, M], \quad \lambda_1(T_n(b)) \to m, \quad \lambda_n(T_n(b)) \to M.$$

This implies that

$$\operatorname{rad} T_n(b) = \max\Big(|\lambda_1(T_n(b))|, |\lambda_n(T_n(b))|\Big) \to \max(m, M) = \|b\|_\infty = \operatorname{rad} T(b).$$

If f is a nonnegative continuous function on $\mathbf{T}$, then

$$\lim_{k\to\infty} \left(\frac{1}{2\pi} \int_0^{2\pi} (f(e^{i\theta}))^k d\theta \right)^{1/k} = \|f\|_\infty.$$

Since b^2 is nonnegative, we therefore get

$$\limsup_{k\to\infty} \left| \frac{1}{2\pi} \int_0^{2\pi} (b(e^{i\theta}))^k d\theta \right|^{1/k} \ge \limsup_{k\to\infty} \left| \frac{1}{2\pi} \int_0^{2\pi} (b(e^{i\theta}))^{2k} d\theta \right|^{1/(2k)}$$
$$= \left(\limsup_{k\to\infty} \left| \frac{1}{2\pi} \int_0^{2\pi} (b(e^{i\theta}))^{2k} d\theta \right|^{1/k} \right)^{1/2} = \|b^2\|_\infty^{1/2} = \|b\|_\infty.$$

On the other hand, it is obvious that

$$\limsup_{k\to\infty}\left|\frac{1}{2\pi}\int_0^{2\pi}(b(e^{i\theta}))^k d\theta\right|^{1/k}\le \|b\|_\infty.$$

Consequently,

$$\limsup_{k\to\infty}\left|\frac{1}{2\pi}\int_0^{2\pi}(b(e^{i\theta}))^k d\theta\right|^{1/k}= \|b\|_\infty.$$

We are left with showing that

$$\inf_{\varrho\in(0,\infty)}\|b_\varrho\|_\infty=\|b\|_\infty.$$

Assume the contrary, that is,

$$\inf_{\varrho\in(0,\infty)}\|b_\varrho\|_\infty<\|b\|_\infty.$$

Because $b_{-j}=\overline{b}_j$, we have $b(\varrho e^{i\theta})=\overline{b(\varrho^{-1}e^{i\theta})}$. Thus, we can assume that

$$\|b_\varrho\|_\infty=\|b_{\varrho^{-1}}\|_\infty<\|b\|_\infty.$$

This means that the analytic function $b(z)$ does not attain its maximum modulus on the boundary of the annulus $\{z\in\mathbf{C}:\varrho\le|z|\le\varrho^{-1}\}$, which contradicts the maximum modulus principle. □

Theorem 10.14. *Let b be a Laurent polynomial and suppose $T(b)$ is triangular. Let b_0 denote the* 0*th Fourier coefficient of b. Then*

$$\operatorname{rad} T_n(b)=|b_0|=\inf_{\varrho\in(0,\infty)}\|b_\varrho\|_\infty=\left|\frac{1}{2\pi}\int_0^{2\pi}(b(e^{i\theta}))^k d\theta\right|^{1/k}$$

for every $n\ge 1$ and every $k\ge 1$.

Proof. For the sake of definiteness, assume that

$$b(z)=b_0+b_1z+\cdots+b_sz^s.$$

Clearly, $\operatorname{sp} T_n(b)=b_0$ and hence $\operatorname{rad} T_n(b)=|b_0|$. The maximum modulus principle implies that $M(\varrho):=\|b_\varrho\|_\infty$ decreases monotonically as $\varrho\to 0$. Therefore

$$\inf_{\varrho\in(0,\infty)}M(\varrho)=\lim_{\varrho\to 0}M(\varrho)=|b(0)|=|b_0|.$$

Finally,

$$\frac{1}{2\pi}\int_0^{2\pi}(b(e^{i\theta}))^k d\theta$$

is the 0th Fourier coefficient of b^k and thus equal to b_0^k. Consequently

$$\left|\frac{1}{2\pi}\int_0^{2\pi}(b(e^{i\theta}))^k d\theta\right|^{1/k}=|b_0^k|^{1/k}=|b_0|. \qquad \square$$

10.4 Matrices with Nonnegative Entries

We now consider Toeplitz band matrices generated by Laurent polynomials

$$b(t) = \sum_{j=-r}^{r} b_j t^j$$

for which $b_j \ge 0$ for all j. Schmidt and Spitzer showed that in this case the inequalities in (10.10) also become equalities. We here follow Elsner and Friedland [109], who developed an alternative approach to the problem.

A nonempty subset $K \subset \mathbf{R}^n$ is called a *cone* if $x+y \in K$ and $\alpha x \in K$ for all $x, y \in K$ and all $\alpha \ge 0$. A cone K is said to be *proper with respect to a linear subspace* $L \subset \mathbf{R}^n$ if $K \subset L$, K is a closed and convex subset of $\mathbf{R}^n$, $K \cap (-K) = \{0\}$ (property of being a pointed cone), and $K - K = L$ (property of being solid in L).

Theorem 10.15 (Krein and Rutman). *Let L be a linear subspace of $\mathbf{R}^n$ and let $K \subset L$ be a cone that is proper with respect to L. If a linear operator $A : \mathbf{R}^n \to \mathbf{R}^n$ leaves L invariant, then* rad $(A|L)$ *is an eigenvalue of A with an associated eigenvector in K.*

For a proof see, e.g., [26, p. 6].

Let A be a real $n \times n$ matrix and let $x \in \mathbf{R}^n$. In this section we write $A \ge 0$ if all entries of A are nonnegative and $x \ge 0$ in case all components of x are nonnegative.

Corollary 10.16. *Suppose A is a real $n \times n$ matrix and $y \in \mathbf{R}^n$ is a vector such that*

$$y \ge 0, \quad Ay \ge 0, \quad A^2 y \ge 0, \quad \ldots .$$

Then A has a nonnegative real eigenvalue with an associated eigenvector $z \ge 0$.

Proof. Let K be the smallest closed and convex cone which contains $y, Ay, A^2y, \ldots$ and put $L = K - K$. Since K is a subset of the standard cone of the nonnegative vectors of $\mathbf{R}^n$, it is clear that $K \cap (-K) = \{0\}$. Thus, K is proper with respect to L. Obviously, A leaves K and therefore L invariant. By Theorem 10.15, rad $(A|L)$ is a nonnegative eigenvalue of the matrix A with an associated eigenvector $z \in K$. Hence $z \ge 0$. □

Theorem 10.17. *Let b be a Laurent polynomial and suppose $T_n(b) \ge 0$ for all n. Then*

$$\lim_{n\to\infty} \operatorname{rad} T_n(b) = \inf_{\varrho \in (0,\infty)} \|b_\varrho\|_\infty .$$

Proof. To make the proof transparent, we only consider the special case where

$$b(t) = b_{-2}t^{-2} + b_{-1}t^{-1} + b_0 + b_1 t;$$

it is easily seen that the argument used in the following works also in the general case.

Let μ_n be the spectral radius of $T_n(b)$ and, for $n \ge 3$, consider the equation

$$T_n(b)x^{(n)} = \mu_n x^{(n)}. \tag{10.11}$$

Theorem 10.15 (applied to the case where K is the standard cone of the nonnegative vectors of $\mathbf{R}^n$) implies that (10.11) has a solution

$$x^{(n)} = (x_0^{(n)}, x_1^{(n)}, \dots, x_{n-1}^{(n)})^\top \geq 0, \quad x_{n-1}^{(n)} \neq 0.$$

The first $n-2$ equations of (10.11) yield

$$\begin{aligned}
b_{-2}x_2^{(n)} &= (\mu_n - b_0)x_0^{(n)} - b_{-1}x_1^{(n)} \\
b_{-2}x_3^{(n)} &= -b_1x_0^{(n)} + (\mu_n - b_0)x_1^{(n)} - b_{-1}x_2^{(n)} \\
b_{-2}x_4^{(n)} &= -b_1x_1^{(n)} + (\mu_n - b_0)x_2^{(n)} - b_{-1}x_3^{(n)} \\
&\dots \\
b_{-2}x_{n-1}^{(n)} &= -b_1x_{n-4}^{(n)} + (\mu_n - b_0)x_{n-3}^{(n)} - b_{-1}x_{n-2}^{(n)}.
\end{aligned}$$

Put

$$A_n = \begin{pmatrix} 0 & 1 & 0 \\ 0 & 0 & 1 \\ -b_{-2}^{-1}b_1 & -b_{-2}^{-1}(\mu_n - b_0) & -b_{-2}^{-1}b_{-1} \end{pmatrix},$$

$$y^{(0,n)} = \begin{pmatrix} x_0^{(n)} \\ x_1^{(n)} \\ x_2^{(n)} \end{pmatrix}, \quad y^{(1,n)} = \begin{pmatrix} x_1^{(n)} \\ x_2^{(n)} \\ x_3^{(n)} \end{pmatrix}, \quad \dots, \quad y^{(n-3,n)} = \begin{pmatrix} x_{n-3}^{(n)} \\ x_{n-2}^{(n)} \\ x_{n-1}^{(n)} \end{pmatrix}.$$

The last $n-3$ equations of the above $n-2$ equations can then be written in the form

$$A_n y^{(0,n)} = y^{(1,n)}, \quad A_n y^{(1,n)} = y^{(2,n)}, \quad \dots, \quad A_n y^{(n-2,n)} = y^{(n-3,n)}. \tag{10.12}$$

We can without loss of generality assume that $\|y^{(0,n)}\|_2 = 1$. Then there is a sequence $n_k \to \infty$ such that $y^{(0,n_k)}$ converges in ℓ^2 to some $y \in \ell^2$ with $\|y\|_2 = 1$. Since $T_n(b)$ is a principal submatrix of $T_{n+1}(b)$, we deduce that $\mu_n \leq \mu_{n+1}$ (see, e.g., [26, p. 28]). As $\mu_n \leq \|b\|_\infty$ for all n, it follows that the sequence $\{\mu_n\}$ has a limit $\mu \in (0, \infty)$. Consequently, the matrices A_n converge to the matrix

$$A = \begin{pmatrix} 0 & 1 & 0 \\ 0 & 0 & 1 \\ -b_{-2}^{-1}b_1 & -b_{-2}^{-1}(\mu - b_0) & -b_{-2}^{-1}b_{-1} \end{pmatrix}.$$

For every integer $j \geq 0$,

$$A^j y = \lim_{k\to\infty} A_{n_k}^j y^{(0,n_k)} = \lim_{k\to\infty} y^{(j,n_k)} \geq 0.$$

Thus, by Corollary 10.16, the matrix A has a nonnegative eigenvalue λ with an associated eigenvector $z = (z_1, z_2, z_3)^\top \geq 0$. The companion structure of A implies that z can be assumed to be $z = (1, \lambda, \lambda^2)$.

Considering the last row of the equality $Az = \lambda z$ we obtain

$$-b_{-2}(b_1 + (\mu - b_0)\lambda + b_{-1}\lambda^2) = \lambda \cdot \lambda^2,$$

whence

$$\mu = b_1(1/\lambda) + b_0 + b_{-1}(1/\lambda)^{-1} + b_{-2}(1/\lambda)^{-2}.$$

Thus, $\mu = b(1/\lambda)$. Since $\|b_{1/\lambda}\|_\infty \le b(1/\lambda)$, it results that

$$\mu \ge \inf_{\varrho\in(0,\infty)} \|b_\varrho\|_\infty.$$

The reverse inequality follows from Theorem 10.12. □

Exercises

1. Let $A \in \mathcal{B}(\ell^2)$ be selfadjoint and put

$$m = \inf_{\|x\|_2=1} (Ax, x), \quad M = \sup_{\|x\|_2=1} (Ax, x).$$

(a) Prove that $A_n = P_n A P_n|\mathrm{Im}\, P_n$ has the following properties:

$$m \le \lambda_{\min}(A_n) \le \lambda_{\max}(A_n) \le M,$$

$$\lim_{n\to\infty} \lambda_{\min}(A_n) = m, \quad \lim_{n\to\infty} \lambda_{\max}(A_n) = M,$$

$$\{m, M\} \subset \mathrm{sp}\, A \subset \liminf_{n\to\infty} \mathrm{sp}\, A_n \subset \limsup_{n\to\infty} \mathrm{sp}\, A_n \subset [m, M].$$

(b) Find an A for which $m = 0$, $M = 1$, $\mathrm{sp}\, A = \{0, 1\}$, and $\mathrm{sp}\, A_n = \{0, 1\}$ for all $n \ge 2$.

(c) Find an A such that $m = 0$, $M = 1$, $\mathrm{sp}\, A = \{0, 1\}$, and

$$\liminf_{n\to\infty} \mathrm{sp}\, A_n = \limsup_{n\to\infty} \mathrm{sp}\, A_n = [0, 1].$$

2. Let $a, b \in \mathcal{P}$ be real-valued Laurent polynomials such that $a \le b$ on $\mathbf{T}$ and denote by $\lambda_j(T_n(a))$ and $\lambda_j(T_n(b))$ the eigenvalues of $T_n(a)$ and $T_n(b)$ in nondecreasing order. Show that $\lambda_j(T_n(a)) \le \lambda_j(T_n(b))$ for all j.

3. Let $a, b \in \mathcal{P}$ be nonnegative on $\mathbf{T}$. Prove that the eigenvalues of $T_n^{-1}(a)T_n(b)$ are all located in (r, R) where

$$r = \inf_{t\in\mathbf{T},\, a(t)>0} \frac{b(t)}{a(t)}, \quad R = \sup_{t\in\mathbf{T},\, a(t)>0} \frac{b(t)}{a(t)}.$$

4. Let $c_k = a\varrho^{|k|} + b\varrho^{-|k|}$ with real numbers a, b, ϱ such that $b \ne 0$ and $0 < \varrho < 1$. Denote by $\lambda_1(A_n) \le \cdots \le \lambda_n(A_n)$ the eigenvalues of the Toeplitz matrix $A_n = (c_{j-k})_{j,k=1}^n$. Prove that there are finite numbers m and M such that

$$m \le \lambda_2(A_n) \le \cdots \le \lambda_{n-1}(A_n) \le M \tag{10.13}$$

for all n and that

$$\lim_{n\to\infty} \lambda_1(A_n) = -\infty, \quad \lim_{n\to\infty} \lambda_n(A_n) = +\infty.$$

5. Let $\mathcal{H}$ denote the set of all Hermitian Toeplitz matrices $(a_{j-k})_{j,k=1}^n$ with $a_0 \in [\alpha_0, \beta_0]$ and $a_k \in [\alpha_k, \beta_k] + i[\gamma_k, \delta_k]$ for $1 \le k \le n-1$, and let $\mathcal{H}_0$ stand for the subset of $\mathcal{H}$ constituted by the matrices with $a_0 = 0$ and $a_k \in \{\alpha_k, \beta_k\} + i\{\gamma_k, \delta_k\}$. The set $\mathcal{H}$ is called a Hermitian Toeplitz interval matrix and the 4^{n-1} matrices in $\mathcal{H}_0$ are referred to as the vertex matrices of $\mathcal{H}$. Prove that the maximum (minimum) of the eigenvalues of the matrices in $\mathcal{H}$ equals β_0 (α_0) plus the maximum (minimum) of the eigenvalues of the matrices in $\mathcal{H}_0$.

Notes

The asymptotic behavior of the extreme eigenvalues of Hermitian Toeplitz matrices has been extensively studied for a long time and by many authors, including Kac, Murdock, and Szegö [176], Grenander and Szegö [145], Widom [290], [291], [292], Parter [197], [198], [199], and Serra Capizzano [247], [248]. We will say more on the contributions by Parter and Widom below. In the case of a single zero, Theorem 10.1 is classical. For symbols with several zeros, this theorem is due to Serra Capizzano [247]. The proof given here is from our book [48] and the same comments as on Theorem 9.8 apply to Theorem 10.1. Beginning with Theorem 10.3, we follow [303] in Section 10.1.

Theorem 10.11 is a special case of what is called Szegö's first limit theorem. For symbols $b \in L^\infty$, this theorem was established by Szegö [262]. Versions of the theorem are now known to be true even for symbols $b \in L^1$ or for Toeplitz matrices generated by the Fourier coefficients of Radon measures [285], [301]. We refer the reader to [71] for some more details. The set $\mathcal{O}$ was introduced by SeLegue [245], and the simple approach of Section 10.2 is based on ideas of SeLegue [245] and Fasino and Tilli [116].

The results of Sections 10.3 and 10.4 are due to Schmidt and Spitzer [243]. The arguments of Section 10.4 are from Elsner and Friedland's paper [109].

Exercise 2 is from [290]. For Exercise 3 see [23]. Exercise 4 is due to Trench [279] (see also [281]). He even proved the following. If we put

$$g(\theta) = \frac{(a-b)(1-\varrho^2)}{1-2\varrho\cos\theta+\varrho^2},$$

then (10.13) is true with $m = \min g(\theta)$ and $M = \max g(\theta)$ for all sufficiently large n and

$$\lim_{n\to\infty} \frac{1}{n}\sum_{j=2}^{n-1} \varphi(\lambda_j(A_n)) = \frac{1}{2\pi}\int_0^{2\pi} \varphi(g(\theta))\,d\theta$$

for every $\varphi \in C[m, M]$. Exercise 5 is from [159].

Further results: Parter and Widom. These two outstanding mathematicians have no joint published work, but their parallel efforts made the late 1950s and early 1960s the heyday of extreme eigenvalues of Hermitian Toeplitz matrices. Citing their results with proofs would go beyond the scope of this book. We think the following does nevertheless provide an idea of the exciting territory that was explored by Seymour Parter and Harold Widom more than 40 years ago.

Let α be a natural number and let $c \in \mathcal{P}$ be a real-valued and positive Laurent polynomial, $c(\mathbf{T}) \subset (0, \infty)$. Put $b(t) = |t-1|^{2\alpha} c(t)$. The matrices $T_n(b)$ are all Hermitian and positive definite. We denote by

$$0 < \lambda_1(T_n(b)) \le \lambda_2(T_n(b)) \le \cdots \le \lambda_n(T_n(b))$$

their eigenvalues. As in the notes to Chapter 4, we consider the functions

$$[T_n^{-1}(b)]_{[nx],[ny]}, \quad (x, y) \in [0, 1]^2.$$

Let $H_\alpha^{(n)}$ be the integral operator on $L^2(0, 1)$ with the kernel $n\,[T_n^{-1}(b)]_{[nx],[ny]}$. Widom [291], [292] observed that $\operatorname{sp} H_\alpha^{(n)} = \operatorname{sp} T_n^{-1}(b)$ and showed that $n^{-2\alpha} H_\alpha^{(n)}$ converges in the operator norm on $L^2(0, 1)$ to an integral operator H_α with a certain kernel $F_\alpha(x, y)$. The expression for $F_\alpha(x, y)$ was quite complicated, but it resembled $G_\alpha(x, y)/c(1)$ with $G_\alpha(x, y)$ given by (4.62). To see that this should not come as a surprise, let us shortly jump to the year 2004. Formula (4.60) says that

$$n^{-2\alpha}\, n\,[T_n^{-1}(b)]_{[nx],[ny]} \to \frac{1}{c(1)}\, G_\alpha(x, y) \quad \text{as} \quad n \to \infty$$

uniformly on $[0, 1]^2$. This clearly implies that $H_\alpha^{(n)}$ converges in the operator norm on $L^2(0, 1)$ to the integral operator with the kernel $G_\alpha(x, y)/c(1)$. Thus, Widom's kernel $F_\alpha(x, y)$ must coincide with $G_\alpha(x, y)/c(1)$.

Let us denote by K_α the integral operator on $L^2(0, 1)$ whose kernel is $G_\alpha(x, y)$. Obviously, $K_\alpha = c(1)\, H_\alpha$. The operator K_α is compact, selfadjoint, and positive definite. Let

$$\|K_\alpha\| = \mu_1(K_\alpha) \ge \mu_2(K_\alpha) \ge \cdots > 0$$

be its eigenvalues (repeated according to their multiplicities). The largest eigenvalues of $n^{-2\alpha} H_\alpha^{(n)}$ are

$$\frac{1}{n^{2\alpha}\lambda_1(T_n(b))} \ge \frac{1}{n^{2\alpha}\lambda_2(T_n(b))} \ge \cdots.$$

Since $n^{-2\alpha} H_\alpha^{(n)}$ converges in the operator norm to $(1/c(1))\, K_\alpha$, we arrive at the conclusion that

$$\lim_{n\to\infty} n^{2\alpha}\lambda_j(T_n(b)) = \frac{c(1)}{\mu_j(K_\alpha)}$$

for each j. We write $q_n \sim r_n$ if $r_n \ne 0$ for all n and $q_n/r_n \to 1$ as $n \to \infty$. Thus, we culminate with

$$\lambda_j(T_n(b)) \sim \frac{c(1)}{\mu_j(K_\alpha)}\, \frac{1}{n^{2\alpha}}.$$

This formula was proved by Kac, Murdock, and Szegö [176] for $\alpha = 1$ and by Parter in [198] for $\alpha = 2$ and then in [197] for general α. The above derivation is Widom's [291], [292].

What can be said about the eigenvalues $\mu_j(K_\alpha)$? Parter [197] and Widom [291] showed that $G_\alpha(x, y)$ is the Green kernel of the boundary value problem

$$(-1)^\alpha u^{(2\alpha)}(x) = v(x), \quad x \in [0, 1], \tag{10.14}$$
$$u^{(k)}(0) = u^{(k)}(1) = 0, \quad k = 0, 1, \dots, \alpha - 1. \tag{10.15}$$

For $\alpha = 1$, this is a classical result [90]. For general α, this was independently rediscovered by Rambour and Seghier [215], [216]. A very short and self-contained proof is also in [40]. Thus, the solution of (10.14), (10.15) is given by

$$u(x) = \int_0^1 G_\alpha(x, y) v(y)\, dy.$$

Consequently, if we denote by

$$0 < \gamma_1(\alpha) \le \gamma_2(\alpha) \le \cdots$$

the eigenvalues of the boundary value problem (10.14), (10.15), then $\gamma_j(\alpha) = 1/\mu_j(K_\alpha)$. It follows that

$$\lambda_j(T_n(b)) \sim \frac{c(1)\, \gamma_j(\alpha)}{n^{2\alpha}}$$

for each j. In particular,

$$\lambda_{\min}(T_n(b)) \sim \frac{c(1)\, \gamma_{\min}(\alpha)}{n^{2\alpha}},$$

where $\lambda_{\min} = \lambda_1$ and $\gamma_{\min} = \gamma_1$.

We have $\gamma_{\min}(\alpha) = \min(Au, u)/(u, u)$, where A is the operator given by (10.14), (10.15). Due to the boundary conditions, we can partially integrate α times to get

$$\frac{(Au, u)}{(u, u)} = \frac{\int_0^1 [u^{(\alpha)}(x)]^2 dx}{\int_0^1 [u(x)]^2 dx}.$$

Hence, $\gamma_{\min}(\alpha)$ can also be characterized as $1/C_\alpha$, where C_α is the best constant for which the Wirtinger-Sobolev type inequality

$$\int_0^1 [u(x)]^2 dx \le C_\alpha \int_0^1 [u^{(\alpha)}]^2 dx$$

is true for all smooth functions u satisfying (10.15).

It is well known (and easily seen) that $\gamma_{\min}(1) = \pi^2$. Computation of an appropriate 4×4 determinant yields that $\gamma_{\min}(2) = \delta^4$, where δ is the smallest positive solution of the equation $\cosh \delta \cos \delta = 1$. This yields $\gamma_{\min}(2) = 500.5467$. The constant $\gamma_{\min}(3)$ can be found as the smallest zero of a function that is given by a 6×6 determinant. We arrived numerically at $\gamma_{\min}(3) \approx 61529$. In [74] it is shown that $\gamma_{\min}(\alpha)$ has the asymptotics

$$\gamma_{\min}(\alpha) = \sqrt{8\pi\alpha} \left(\frac{4\alpha}{e}\right)^{2\alpha} \left[1 + O\left(\frac{1}{\sqrt{\alpha}}\right)\right] \quad \text{as} \quad \alpha \to \infty.$$

In the case $\alpha = 1$ we also have $\gamma_j(1) = j^2\pi^2$. Parter [198] considered the case $\alpha = 2$ and proved the formula

$$\gamma_j(2) = 16\left(\frac{(2j+1)\pi + E_j}{4}\right)^4,$$

where E_j is determined by the equation

$$\tan\left(\frac{(2j+1)\pi + E_j}{4}\right) = (-1)^j \tanh\left(\frac{(2j+1)\pi + E_j}{4}\right).$$

He also computed the first four values:

$$\frac{3\pi + E_1}{4} = 2.3650, \qquad \frac{5\pi + E_2}{4} = 3.9266,$$
$$\frac{7\pi + E_3}{4} = 5.4978, \qquad \frac{9\pi + E_4}{4} = 7.0686.$$

The above results are first order asymptotics. Parter and Widom also established second-order asymptotics. Here is a sample result. Put $h(\theta) = b(e^{i\theta})$. Widom [290] showed that if c is even, that is, $c(t) = c(1/t)$ for all $t \in \mathbf{T}$, then

$$\lambda_j(T_n(|t-1|^2 c(t))) = \frac{\sigma^2\pi^2 j^2}{2(n+1)^2}\left(1 + \frac{\varrho}{n+1}\right) + o\left(\frac{1}{n^3}\right),$$

where $\sigma^2 = h''(0)\ (> 0)$ and

$$\varrho = \int_{-\pi}^{\pi} \frac{1}{\sin^2(\theta/2)} \log\left(\frac{h(\theta)}{2\sigma^2}\cot^2\frac{\theta}{2}\right) d\theta.$$

Further results: preconditioning. Let b be a nonzero Laurent polynomial and suppose $b(t) \geq 0$ for all $t \in \mathbf{T}$. Then the matrices $T_n(b)$ are positive definite. To solve the system $T_n(b)x = y$ numerically, one can use preconditioning, that is, one can pass to the equivalent system

$$A_n^{-1} T_n(b) x = A_n^{-1} y \tag{10.16}$$

with an appropriate positive definite matrix A_n. System (10.16) can be solved by conjugate gradient iteration (see, e.g., [11], [85], [92], [101], [121], [211], [269], [273]). One starts with an initial vector x_0, and at the kth iteration step one obtains an approximate solution x_k satisfying the error estimate

$$\frac{(A_n^{-1/2} T_n(b) A_n^{-1/2}(x - x_k), x - x_k)}{(A_n^{-1/2} T_n(b) A_n^{-1/2}(x - x_0), x - x_0)} \leq 4\left(\frac{\sqrt{\kappa_n} - 1}{\sqrt{\kappa_n} + 1}\right)^{2k} \tag{10.17}$$

with

$$\kappa_n = \frac{\lambda_{\max}(A_n^{-1/2} T_n(b) A_n^{-1/2})}{\lambda_{\min}(A_n^{-1/2} T_n(b) A_n^{-1/2})} = \frac{\lambda_{\max}(A_n^{-1} T_n(b))}{\lambda_{\min}(A_n^{-1} T_n(b))}. \tag{10.18}$$

In view of (10.17), the task of designing a good preconditioner A_n^{-1} includes the problem of keeping κ_n (≥ 1) as close to 1 as possible, which, by (10.18), means that the extreme eigenvalues of $A_n^{-1}T_n(b)$ should be as close to one another as possible.

Gilbert Strang [258] and Tony Chan [89] proposed using circulant matrices A_n (see also [87]). This works if b has no zeros on $\mathbf{T}$, but Tyrtyshnikov [283] proved that such preconditioners may fail if b has zeros on $\mathbf{T}$. Potts and Steidl [212] showed that so-called ω-circulant matrices A_n lead to better results.

Raymond Chan [83] proposed the use of banded Toeplitz matrices as preconditioners. Suppose, for example, b has a single zero on $\mathbf{T}$, say at t_0, and let 2α be the order of this zero. Theorem 10.1 implies that $\kappa(T_n(b)) \simeq n^{2\alpha}$. Taking $A_n = T_n(a)$ with $a(t) = |t - t_0|^{2\alpha}$ one gets

$$\kappa_n = \frac{\lambda_{\max}(A_n^{-1}T_n(b))}{\lambda_{\min}(A_n^{-1}T_n(b))} = O(1)$$

(see [83], [86]). This idea was elaborated also in [23], [246].

Chan, Ng, and Yip's survey [86] and Potts' dissertation [211] are excellent and up-to-date introductions to the advanced precoditioning business for positive definite Toeplitz matrices. For additional issues addressing banded Toeplitz matrices we also refer to [151], [190], [260].

Chapter 11

Eigenvalue Distribution

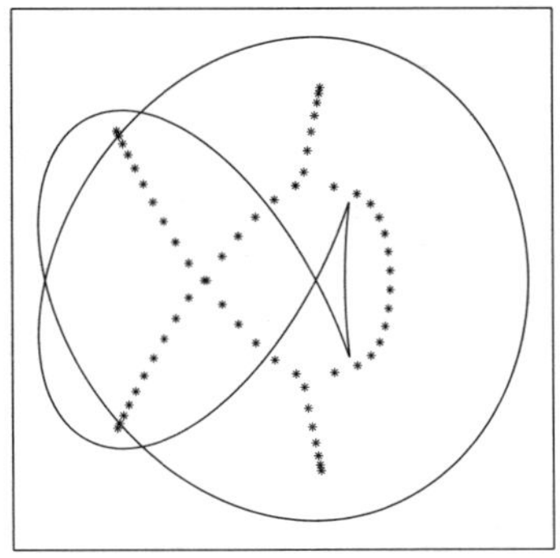

This chapter is devoted to the results of Schmidt, Spitzer, and Hirschman on the asymptotic eigenvalue distribution of $T_n(b)$ as $n \to \infty$. We show that the spectra $\mathrm{sp}\, T_n(b)$ converge in the Hausdorff metric to a certain limiting set $\Lambda(b)$, which is either a singleton or the union of finitely many analytic arcs. We also determine the corresponding limiting measure, which characterizes the density of the asymptotic distribution of the eigenvalues along the limiting set.

11.1 Toward the Limiting Set

Because things are trivial in the case where $T(b)$ is triangular, we will throughout this chapter assume that

$$b(t) = \sum_{j=-r}^{s} b_j t^j, \quad r \geq 1, \quad s \geq 1, \quad b_{-r} \neq 0, \quad b_s \neq 0.$$

As first observed by Schmidt and Spitzer, it turns out that the eigenvalue distribution of Toeplitz band matrices is in no obvious way related to the spectrum of the corresponding infinite matrices. To see this, choose $\varrho \in (0, \infty)$ and put

$$b_\varrho(t) = \sum_{j=-r}^{s} b_j \varrho^j t^j.$$

Clearly, $b_\varrho(\mathbf{T}) = b(\varrho\mathbf{T})$. We have

$$T_n(b_\varrho) = \mathrm{diag}\,(\varrho, \varrho^2, \ldots, \varrho^n)\, T_n(b)\, \mathrm{diag}\,(\varrho^{-1}, \varrho^{-2}, \ldots, \varrho^{-n}), \tag{11.1}$$

and hence

$$\mathrm{sp}\, T_n(b_\varrho) = \mathrm{sp}\, T_n(b). \tag{11.2}$$

Thus, if there were any reason for sp $T_n(b)$ to mimic $b(\mathbf{T})$ or sp $T(b)$, this reason would also force sp $T_n(b)$ to mimic $b(\varrho\mathbf{T})$ or sp $T(b_\varrho)$.

There are at least two possible definitions of the limiting set of the spectra sp $T_n(b)$. We define

$$\Lambda_s(b) := \liminf_{n\to\infty} \operatorname{sp} T_n(b)$$

as the set of all $\lambda \in \mathbf{C}$ for which there exist $\lambda_n \in \operatorname{sp} T_n(b)$ such that $\lambda_n \to \lambda$, and we let

$$\Lambda_w(b) := \limsup_{n\to\infty} \operatorname{sp} T_n(b)$$

stand for the set of all $\lambda \in \mathbf{C}$ for which there are $n_1 < n_2 < n_3 < \cdots$ and $\lambda_{n_k} \in \operatorname{sp} T_{n_k}(b)$ such that $\lambda_{n_k} \to \lambda$. Obviously, $\Lambda_s(b) \subset \Lambda_w(b)$.

Lemma 11.1. *We have* $\Lambda_s(b) \subset \Lambda_w(b) \subset \operatorname{sp} T(b)$.

Proof. Let $\lambda_0 \notin \operatorname{sp} T(b)$. Then, by Theorem 3.7, $\{T_n(b-\lambda_0)\}$ is stable, that is, $\|T_n^{-1}(b-\lambda_0)\|_2 \le M < \infty$ for all $n \ge n_0$. It follows that if $|\lambda - \lambda_0| < 1/(2M)$, then $\|T_n^{-1}(b-\lambda)\|_2 \le 2M$ for all $n \ge n_0$, which shows that λ_0 has a neighborhood $U(\lambda_0)$ such that $U(\lambda_0) \cap \operatorname{sp} T_n(a) = \emptyset$ for all $n \ge n_0$. Consequently, $\lambda_0 \notin \Lambda_w(b)$. □

Corollary 11.2. *We even have*

$$\Lambda_s(b) \subset \Lambda_w(b) \subset \bigcap_{\varrho\in(0,\infty)} \operatorname{sp} T(b_\varrho). \tag{11.3}$$

Proof. From (11.2) we know that $\Lambda_s(b) = \Lambda_s(b_\varrho)$ and $\Lambda_w(b) = \Lambda_w(b_\varrho)$. The assertion is therefore an immediate consequence of Lemma 11.1. □

We will show that all inclusions of (11.3) are actually equalities. At the present moment, we restrict ourselves to giving another description of the intersection occurring in (11.3). For $\lambda \in \mathbf{C}$, put

$$Q(\lambda, z) = z^r(b(z) - \lambda) = b_{-r} + \cdots + (b_0 - \lambda)z^r + \cdots + b_s z^{r+s}$$

and denote by $z_1(\lambda), \ldots, z_{r+s}(\lambda)$ the zeros of $Q(\lambda, z)$ for fixed λ:

$$Q(\lambda, z) = b_s \prod_{j=1}^{r+s} (z - z_j(\lambda)).$$

Label the zeros so that

$$|z_1(\lambda)| \le |z_2(\lambda)| \le \cdots \le |z_{r+s}(\lambda)|$$

and define

$$\Lambda(b) = \{\lambda \in \mathbf{C} : |z_r(\lambda)| = |z_{r+1}(\lambda)|\}. \tag{11.4}$$

Theorem 11.3. *The following equality holds:*

$$\bigcap_{\varrho\in(0,\infty)} \operatorname{sp} T(b_\varrho) = \Lambda(b).$$

Proof. By Corollary 1.11, $T(b)-\lambda$ is invertible if and only if $b(z)-\lambda$ has no zeros on $\mathbf{T}$ and wind $(b-\lambda)=0$. As wind $(b-\lambda)$ equals the difference of the zeros and the poles of $b(z)-\lambda$ in $\mathbf{D} := \{z \in \mathbf{C} : |z| < 1\}$ and as the only pole of

$$b(z)-\lambda = b_{-r}z^{-r} + \cdots + (b_0-\lambda) + \cdots + b_s z^s$$

is a pole of the multiplicity r at $z=0$, it results that $T(b)-\lambda$ is invertible if and only if $b(z)-\lambda$ has no zeros on $\mathbf{T}$ and exactly r zeros in $\mathbf{D}$. Equivalently, $T(b)-\lambda$ is invertible exactly if $Q(\lambda, z)$ has no zeros on $\mathbf{T}$ and precisely r zeros in $\mathbf{D}$.

Analogously, $T(b_\varrho)-\lambda$ is invertible if and only if $Q(\lambda, z)$ has no zero on $\varrho^{-1}\mathbf{T}$ and exactly r zeros in $\varrho^{-1}\mathbf{D}$.

Now suppose $\lambda \notin \Lambda(b)$. Then $|z_r(\lambda)| < |z_{r+1}(\lambda)|$. Consequently, there is a ϱ such that $|z_r(\lambda)| < \varrho < |z_{r+1}(\lambda)|$. It follows that $Q(\lambda, z)$ has no zero on $\varrho\mathbf{T}$ and exactly r zeros in $\varrho\mathbf{D}$. Thus, $T(b_{1/\varrho}-\lambda)$ is invertible and therefore λ is not in $\bigcap_{\varrho\in(0,\infty)} \operatorname{sp} T(b_\varrho)$.

Conversely, suppose there is a $\varrho \in (0,\infty)$ such that $\lambda \notin \operatorname{sp} T(b_\varrho)$. Then, by what was said above, $Q(\lambda, z)$ has no zeros on $\varrho^{-1}\mathbf{T}$ and precisely r zeros in $\varrho^{-1}\mathbf{D}$. This implies that $|z_r(\lambda)| < \varrho^{-1} < |z_{r+1}(\lambda)|$, whence $\lambda \notin \Lambda(b)$. □

We will henceforth in this chapter always suppose that the greatest common divisor of the indices k with $b_k \neq 0$ is 1:

$$\text{g.c.d.}\,\{k : b_k \neq 0\} = 1. \tag{11.5}$$

This is no loss of generality. Indeed, consider, for example, $T_5(b)$ with

$$b(t) = \sum_k b_{2k} t^{2k}.$$

It is easily seen that

$$T_5(b) = \begin{pmatrix} b_0 & 0 & b_{-2} & 0 & b_{-4} \\ 0 & b_0 & 0 & b_{-2} & 0 \\ b_2 & 0 & b_0 & 0 & b_{-2} \\ 0 & b_2 & 0 & b_0 & 0 \\ b_4 & 0 & b_2 & 0 & b_0 \end{pmatrix} \tag{11.6}$$

is unitarily equivalent (via a permutation matrix) to

$$\begin{pmatrix} b_0 & b_{-2} & b_{-4} & 0 & 0 \\ b_2 & b_0 & b_{-2} & 0 & 0 \\ b_4 & b_2 & b_0 & 0 & 0 \\ 0 & 0 & 0 & b_0 & b_{-2} \\ 0 & 0 & 0 & b_2 & b_0 \end{pmatrix}, \tag{11.7}$$

which shows that (11.6) and (11.7) have the same eigenvalues. It follows that $\Lambda_s(b) = \Lambda_s(b^\#)$ and $\Lambda_w(b) = \Lambda_w(b^\#)$, where

$$b^\#(t) := \sum_k b_{2k} t^k.$$

11.2 Structure of the Limiting Set

A point $\lambda_0 \in \mathbf{C}$ is called a *branch point* if $Q(\lambda_0, z)$ has multiple zeros.

Lemma 11.4. *There are at most* $2(r+s)-1$ *branch points.*

Proof. A point λ is a branch point if and only if the polynomials

$$Q(\lambda, z) = b_s z^{r+s} + \cdots + (b_0 - \lambda)z^r + \cdots + b_{-r}$$

and

$$\frac{\partial}{\partial z} Q(\lambda, z) = (r+s)b_s z^{r+s-1} + \cdots + r(b_0 - \lambda)z^{r-1} + \cdots + b_{-r+1}$$

have a common zero. This happens exactly if the resultant of Q and $\partial Q/\partial z$ is zero. The resultant of Q and $\partial Q/\partial z$ is the determinant

$$\begin{vmatrix} b_s & \cdots & b_0 - \lambda & \cdots & b_{-r} & \\ & \cdots & \cdots & \cdots & \cdots & \\ & b_s & \cdots & b_0 - \lambda & \cdots & b_{-r} \\ (r+s)b_{r+s} & \cdots & r(b_0 - \lambda) & \cdots & b_{-r+1} & \\ & \cdots & \cdots & \cdots & \cdots & \\ & (r+s)b_s & \cdots & r(b_0-\lambda) & \cdots & b_{-r+1} \end{vmatrix}, \tag{11.8}$$

which has $r+s-1$ rows starting with b_s and $r+s$ rows beginning with $(r+s)b_s$. The determinant (11.8) equals

$$\begin{aligned} &\pm(b_0 - \lambda)^{r+s-1}\Big(r(b_0 - \lambda)\Big)^{r+s} + O\left(\lambda^{2(r+s)-2}\right) \\ &= \pm r^{r+s}\lambda^{2(r+s)-1} + O\left(\lambda^{2(r+s)-2}\right), \end{aligned}$$

which is a polynomial of degree $2(r+s)-1$ and therefore has at most $2(r+s)-1$ distinct zeros. □

Since $b_{-r} \neq 0$, none of the zeros $z_j(\lambda)$ of $Q(\lambda, z)$ is zero.

Lemma 11.5. *If* λ_0 *is not a branch point, then* $z_j(\lambda)/z_k(\lambda)$ $(j \neq k)$ *is not constant in some open neighborhood of* λ_0.

Proof. Assume $z_j(\lambda) = \gamma z_k(\lambda)$ with some $\gamma \in \mathbf{C} \setminus \{0\}$ for all λ in some open neighborhood of λ_0. Then

$$Q(\lambda, z_k(\lambda)) = 0, \quad Q(\lambda, \gamma z_k(\lambda)) = 0.$$

Consequently,

$$\begin{aligned} 0 &= \gamma^{-r} Q(\lambda, \gamma z_k(\lambda)) - Q(\lambda, z_k(\lambda)) \\ &= \gamma^{-r}\left(b_{-r} + \cdots + (b_0 - \lambda)(\gamma z_k(\lambda))^r + \cdots + b_s(\gamma z_k(\lambda))^{r+s}\right) \\ &\quad - \left(b_{-r} + \cdots + (b_0 - \lambda)(z_k(\lambda))^r + \cdots + b_s(z_k(\lambda))^{r+s}\right) \\ &= (\gamma^{-r} - 1)b_{-r} + \cdots + 0 + \cdots + (\gamma^s - 1)b_s(z_k(\lambda))^{r+s}. \end{aligned} \tag{11.9}$$

The function $z_k(\lambda)$ is not constant in a neighborhood of λ_0: If $Q(\lambda, z_0) = 0$ for all λ sufficiently close to λ_0, then

$$\lambda z_0^r = b_{-r} + \cdots + b_0 z_0^r + \cdots + b_s z_0^{r+s},$$

and since $b_{-r} \neq 0$, it follows that $z_0 \neq 0$, whence

$$\lambda = b_{-r} z_0^{-r} + \cdots + b_0 + \cdots + b_s z_0^s = \text{const},$$

which is impossible. Thus, from (11.9) we infer that $\gamma^k = 1$ whenever $b_k \neq 0$. By (11.5), this shows that $\gamma = 1$. Thus, $z_j(\lambda) = z_k(\lambda)$ and, in particular, $z_j(\lambda_0) = z_k(\lambda_0)$. This, however, contradicts our hypothesis that $Q(\lambda_0, z)$ have no multiple zeros. □

Recall that we labelled the zercs $z_j(\lambda)$ of $Q(\lambda, z)$ so that

$$|z_1(\lambda)| \leq |z_2(\lambda)| \leq \cdots \leq |z_{r+s}(\lambda)|$$

and that we defined

$$\Lambda(b) = \{\lambda \in \mathbf{C} : |z_r(\lambda)| = |z_{r+1}(\lambda)|\}.$$

Proposition 11.6. *Let $\lambda_0 \in \Lambda(b)$ and suppose λ_0 is not a branch point. Then*

(a) *λ_0 is not an isolated point cf $\Lambda(b)$,*

(b) *there is an open neighborl:ood U of λ_0 such that $\Lambda(b) \cap U$ is a finite union of analytic arcs.*

Proof. Since $\lambda_0 \in \Lambda(b)$, there are $p \geq 1$ and $q \geq 1$ such that

$$\begin{aligned} |z_1(\lambda_0)| \leq \cdots \leq |z_{r-p}(\lambda_0)| \\ < |z_{r-p+1}(\lambda_0)| = \cdots = |z_{r+q}(\lambda_0)| \\ < |z_{r+q+1}(\lambda_0)| \leq \cdots \leq |z_{r+s}(\lambda_0)|. \end{aligned}$$

There is an open neighborhood U of λ_0 such that

$$\begin{aligned} |z_1(\lambda)|, \ldots, |z_{r-p}(\lambda)| \\ < |z_{r-p+1}(\lambda)|, \ldots, |z_{r+q}(\lambda)| \\ < |z_{r+q+1}(\lambda)|, \ldots, |z_{r+s}(\lambda_0)| \quad \text{for} \quad \lambda \in U. \end{aligned}$$

Pick $j, k \in \{r - p + 1, \ldots, r + q\}$ and let $\gamma_{jk} = \{\lambda \in U : z_j(\lambda)/z_k(\lambda) \in \mathbf{T}\}$. Clearly, $\lambda_0 \in \gamma_{jk}$. By Lemma 11.5, the function $\varphi(z) := z_j(\lambda)/z_k(\lambda)$ is not constant. Hence, there is a smallest $m \in \mathbf{N}$ such that $\varphi^{(m)}(\lambda_0) \neq 0$. We have

$$\varphi(\lambda) = \varphi(\lambda_0) + \frac{\varphi^{(m)}(\lambda_0)}{m!}(\lambda - \lambda_0)^m (1 + \psi(\lambda)),$$

where ψ is analytic in U and $\psi(\lambda_0) = 0$. This implies that γ_{jk} is the union of $2m$ analytic arcs starting at λ_0 and terminating on ∂U. Put $\Gamma = \bigcup_{j \neq k} \gamma_{jk}$. Of course, Γ is also a finite union of analytic arcs beginning at λ_0 and ending on ∂U. Each arc of Γ is the carrier of a set of relations

$$|z_{j_1}(\lambda)| = |z_{j_2}(\lambda)| = \cdots = |z_{j_n}(\lambda)| \quad (n \geq 2). \tag{11.10}$$

An arc of Γ is contained in $\Lambda(b)$ if and only if

$$|z_r(\lambda)| = |z_{r+1}(\lambda)| \tag{11.11}$$

is a part of the relations (11.10). If (11.11) is not contained in (11.10), then λ_0 is an isolated point of $\Lambda(b)$. Thus, we show that λ_0 cannot be an isolated point of $\Lambda(b)$. This will complete the proof of (a) and (b).

Assume λ_0 is isolated. Then, after suitably relabelling the zeros,

$$|z_j(\lambda)| < |z_k(\lambda)| \text{ for } j \in \{1,\ldots,r\},\ k \in \{r+1,\ldots,r+s\},\ \lambda \in U \setminus \{\lambda_0\}.$$

Let $\chi(\lambda) = z_r(\lambda)/z_{r+1}(\lambda)$. It follows that $|\chi(\lambda)| < 1$ for $\lambda \in U \setminus \{\lambda_0\}$ and that $|\chi(\lambda_0)| = 1$. As χ is analytic in U, this contradicts the maximum modulus principle. Hence λ_0 cannot be isolated. □

If λ_0 is a branch point, we can introduce a uniformization parameter t such that

$$\lambda = \lambda_0 + t^m, \quad Q(\lambda_0 + t^m, z) = b_s \prod_{j=1}^{r+s} (z - \tau_j(t)),$$

and the functions τ_j are analytic in some open neighborhood $V(0)$ of the origin.

Lemma 11.7. *If λ_0 is a branch point and $j \neq k$, then $\tau_j(t)/\tau_k(t)$ is not constant in $V(0)$.*

Proof. We proceed as in the proof of Lemma 11.5. Assume $\tau_j(t) = \gamma\tau_k(t)$ for some $\gamma \in \mathbf{C} \setminus \{0\}$. Then $Q(\lambda_0 + t^m, \tau_k(t)) = 0$ and $Q(\lambda_0 + t^m, \gamma\tau_k(t)) = 0$, whence

$$\begin{aligned} 0 &= \gamma^{-r} Q(\lambda_0 + t^m, \gamma\tau_k(t)) - Q(\lambda_0 + t^m, \tau_k(t)) \\ &= (\gamma^{-r} - 1)b_{-r} + \cdots + 0 + \cdots + (\gamma^s - 1)b_s(\tau_k(t))^{r+s}. \end{aligned} \tag{11.12}$$

If $\tau_k(t) = z_0$ is constant, then $Q(\lambda_0 + t^m, z_0) = 0$ for all $t \in V(0)$, so

$$(\lambda_0 + t^m)z_0^r = b_{-r} + \cdots + b_0 z_0^r + \cdots + b_s z_0^{r+s}$$

for all $t \in V(0)$, and as $z_0 \neq 0$, it results that

$$\lambda_0 + t^m = b_{-r} z_0^{-r} + \cdots + b_0 + \cdots + b_s z_0^s = \text{const}$$

for all $t \in V(0)$, which is impossible. Thus, $\tau_k(t)$ cannot be constant. From (11.12) and (11.5) we therefore obtain that $\gamma = 1$ and, consequently, $\tau_j(t) = \tau_k(t)$. By Lemma 11.4, $Q(\lambda_0 + t^m, z)$ has only simple zeros for $t \in V(0) \setminus \{0\}$. Hence $\tau_j(t) \neq \tau_k(t)$ for $t \in V(0) \setminus \{0\}$. This contradiction completes the proof. □

Proposition 11.8. *If $\lambda_0 \in \Lambda(b)$ is a branch point, then*

(a) λ_0 *is not an isolated point of* $\Lambda(b)$,

(b) *there exists an open neighborhood U of λ_0 such that $\Lambda(b) \cap U$ is a finite union of analytic arcs.*

Proof. Using Lemma 11.7, this can be proved by the argument of the proof of Proposition 11.6, the only difference being that the γ_{jk} appearing there must be replaced by $\gamma_{jk} = \{t \in V(0) : \tau_j(t)/\tau_k(t) \in \mathbf{T}\}$. □

A point $\lambda \in \Lambda(b)$ will be called an *exceptional point* if λ is a branch point or if there is no open neighborhood U of λ such that $\Lambda(b) \cap U$ is an analytic arc starting and terminating on ∂U.

Theorem 11.9. *The set $\Lambda(b)$ is the union of a finite number of pairwise disjoint (open) analytic arcs and a finite number of exceptional points, and the set $\Lambda(b)$ has no isolated points.*

Proof. By Lemma 11.4, $\Lambda(b)$ has at most finitely many branch points. Theorem 11.3 implies that $\Lambda(b)$ is compact, which in conjunction with Propositions 11.6 and 11.8 shows that $\Lambda(b)$ contains at most a finite number of exceptional points. The assertion is now immediate from Propositions 11.6 and 11.8. □

11.3 Toward the Limiting Measure

If $\lambda \notin \Lambda(b)$, then, by definition (11.4), there is a real number ϱ satisfying

$$|z_r(\lambda)| < \varrho < |z_{r+1}(\lambda)|. \tag{11.13}$$

As usual, let $D_n(a) = \det T_n(a)$.

Lemma 11.10. *There is a continuous function*

$$g : \mathbf{C} \setminus \Lambda(b) \to (0, \infty)$$

such that

$$\lim_{n\to\infty} |D_n(b-\lambda)|^{1/n} = g(\lambda)$$

uniformly on compact subsets of $\mathbf{C} \setminus \Lambda(b)$. If ϱ is given by (11.13), *then*

$$g(\lambda) = \exp \int_0^{2\pi} \log |b_\varrho(e^{i\theta}) - \lambda| \frac{d\theta}{2\pi}. \tag{11.14}$$

Proof. Let $\lambda \notin \Lambda(b)$ and pick any ϱ such that (11.13) is satisfied. By (11.1),

$$D_n(b-\lambda) = D_n(b_\varrho - \lambda).$$

Clearly, $b_\varrho(t)-\lambda \neq 0$ for $t \in \mathbf{T}$. The function $b_\varrho(z)-\lambda$ has exactly one pole in $\mathbf{D}$, namely, a pole of the order r at $z=0$. Due to (11.13), $b_\varrho(z)-\lambda$ has exactly r zeros in $\mathbf{D}$. Hence $\operatorname{wind}(b_\varrho-\lambda)=0$. From Theorem 2.11 we deduce that

$$D_n(b_\varrho-\lambda)=G(b_\varrho-\lambda)^n E(b_\varrho-\lambda)\left(1+o(q_\lambda^n)\right), \tag{11.15}$$

where, by (2.30), $q_\lambda \in (0,1)$ can be taken as

$$q_\lambda=\frac{1}{2}\left(\frac{|z_r(\lambda)|}{|z_{r+1}(\lambda)|}+1\right).$$

Since q_λ depends continuously on $\lambda \in \mathbf{C}\setminus\Lambda(b)$, it follows that (11.15) holds uniformly with respect to compact subsets of $\mathbf{C}\setminus\Lambda(b)$. From (11.15) we obtain that

$$|D_n(b_\varrho-\lambda)|^{1/n} \to |G(b_\varrho-\lambda)| =: g(\lambda),$$

and (2.25) implies that

$$|G(b_\varrho-\lambda)|=\exp\int_0^{2\pi}\log|b_\varrho(e^{i\theta})-\lambda|\,\frac{d\theta}{2\pi}. \qquad \square$$

Now let $\lambda_0 \in \Lambda(b)$ and suppose λ_0 is not an exceptional point. Let U be a sufficiently small open neighborhood of λ_0. Then $U\setminus\Lambda(b)$ has exactly two connected components. We denote them by D_1 and D_2. For $\lambda \in \Lambda(b)\cap U$, there are $p \geq 1$ and $q \geq 1$ such that

$$\begin{aligned}|z_1(\lambda)| &\leq \cdots \leq |z_{r-p}(\lambda)|\\ &< |z_{r-p+1}(\lambda)| = \cdots = |z_{r+q}(\lambda)|\\ &< |z_{r+q+1}(\lambda)| \leq \cdots \leq |z_{r+s}(\lambda)|.\end{aligned} \tag{11.16}$$

We can label the numbers $|z_j(\lambda)|$ so that

$$\max\left(|z_1(\lambda)|,\ldots,|z_r(\lambda)|\right) < \min\left(|z_{r+1}(\lambda)|,\ldots,|z_{r+s}(\lambda)|\right)$$

for $\lambda \in D_1$. Put $N_1=\{r+1,\ldots,r+s\}$. If $\lambda \in D_2$, there is a (unique) set N_2 of integers drawn from $\{1,2,\ldots,r+s\}$ such that

$$\max_{j\notin N_2}|z_j(\lambda)| < \min_{k\in N_2}|z_k(\lambda)|.$$

Clearly, N_2 is the union of $\{r+q+1,\ldots,r+s\}$ and a set formed by $s-q$ natural numbers from $\{r-p+1,\ldots,r+q\}$. Consequently, if $\lambda \in \Lambda(b)\cap U$, then

$$\prod_{k\in N_1}|z_k(\lambda)|=\prod_{k=r+1}^{r+s}|z_k(\lambda)|=\prod_{k\in N_2}|z_k(\lambda)|. \tag{11.17}$$

Lemma 11.11. *Let $i \in \{1,2\}$. Then*

$$g(\lambda)=|b_s|\prod_{k\in N_i}|z_k(\lambda)| \text{ for } \lambda \in D_i.$$

Proof. Fix $\lambda \in D_1$ and choose a ϱ satisfying (11.13). We have

$$Q(\lambda, \varrho e^{i\theta}) = b_s \prod_{j=1}^{r+s} \left(\varrho e^{i\theta} - z_j(\lambda)\right) = \varrho^r e^{ir\theta} \left(b(\varrho e^{i\theta}) - \lambda\right)$$

and thus

$$b(\varrho e^{i\theta}) - \lambda = b_s \varrho^{-r} e^{-ir\theta} \prod_{j=1}^{r+s} \left(\varrho e^{i\theta} - z_j(\lambda)\right)$$

$$= b_s \prod_{j=1}^{r} \left(1 - \varrho^{-1} e^{-i\theta} z_j(\lambda)\right) \prod_{k=r+1}^{r+s} \left(\varrho e^{i\theta} - z_k(\lambda)\right).$$

It follows that

$$\int_0^{2\pi} \log\left(b(\varrho e^{i\theta}) - \lambda\right) \frac{d\theta}{2\pi} = \int_0^{2\pi} \log b_s \frac{d\theta}{2\pi}$$

$$+ \sum_{j=1}^{r} \int_0^{2\pi} \log\left(1 - \varrho^{-1} e^{-i\theta} z_j(\lambda)\right) \frac{d\theta}{2\pi}$$

$$+ \sum_{k=r+1}^{r+s} \int_0^{2\pi} \log\left(\varrho e^{i\theta} - z_k(\lambda)\right) \frac{d\theta}{2\pi} + 2m\pi i$$

with some $m \in \mathbf{Z}$. Because

$$\log\left(1 - \varrho^{-1} e^{-i\theta} z_j(\lambda)\right) = -\sum_{\ell=1}^{\infty} \frac{1}{\ell} \left(\varrho^{-1} e^{-i\theta} z_j(\lambda)\right)^{\ell},$$

$$\log\left(\varrho e^{i\theta} - z_k(\lambda)\right) = \log(-z_k(\lambda)) + \log\left(1 - \varrho e^{i\theta}/z_k(\lambda)\right)$$

$$= \log(-z_k(\lambda)) - \sum_{\ell=1}^{\infty} \frac{1}{\ell} \left(\varrho e^{i\theta}/z_k(\lambda)\right)^{\ell},$$

we get

$$\int_0^{2\pi} \log\left(b(\varrho e^{i\theta}) - \lambda\right) \frac{d\theta}{2\pi} = \log b_s + \sum_{k=r+1}^{r+s} \log(-z_k(\lambda)) + 2m\pi i$$

and hence

$$\exp \int_0^{2\pi} \log\left(b(\varrho e^{i\theta}) - \lambda\right) \frac{d\theta}{2\pi} = b_s \prod_{k=r+1}^{r+s} (-z_k(\lambda)).$$

Taking moduli and using Lemma 11.10 we arrive at the equality

$$g(\lambda) = |b_s| \prod_{k=r+1}^{r+s} |z_k(\lambda)| = |b_s| \prod_{k \in N_1} |z_k(\lambda)|.$$

This proves the assertion for $i = 1$. The proof is analogous for $i = 2$. □

Put

$$G_i(\lambda) = b_s \prod_{k \in N_i} (-z_k(\lambda)) \quad (i = 1, 2).$$

Obviously, G_i is analytic in U. Lemma 11.11 shows that

$$g(\lambda) = |G_i(\lambda)| \text{ for } \lambda \in D_i. \tag{11.18}$$

This implies that g extends to a continuous function on the closure $\overline{D}_i$ of D_i.

Lemma 11.12. *We have*

$$|G_1(\lambda)| = |G_2(\lambda)| \textit{ for } \lambda \in \Lambda(b) \cap U, \tag{11.19}$$

$$|G_1(\lambda)| > |G_2(\lambda)| \textit{ for } \lambda \in D_1, \tag{11.20}$$

$$|G_2(\lambda)| > |G_1(\lambda)| \textit{ for } \lambda \in D_2. \tag{11.21}$$

Figure 11.1 illustrates the situation.

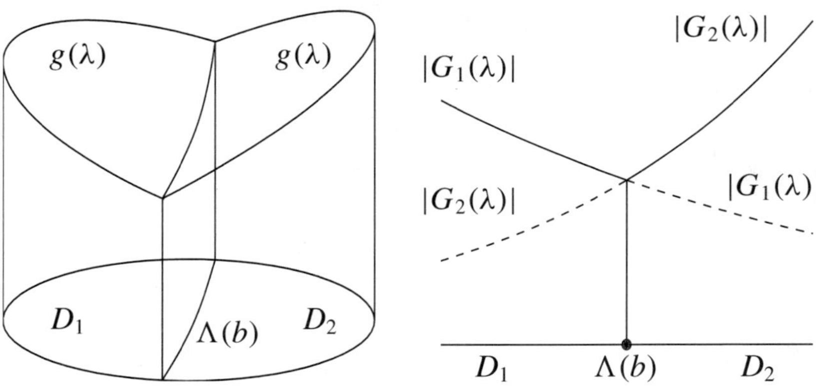

Figure 11.1. *An illustration to Lemma* 11.12.

Proof. Equality (11.19) is obvious from (11.17). Let us prove inequality (11.20). For $j \in \{r - p + 1, \ldots, r\}$ and $k \in \{r + 1, \ldots, r + q\}$, consider

$$\varphi_{jk}(\lambda) := z_j(\lambda)/z_k(\lambda).$$

By Lemma 11.5, φ_{jk} is not constant on U. Consequently, there is a $\lambda_0 \in \Lambda(b) \cap U$ such that $\varphi'_{jk}(\lambda_0) \neq 0$ for all j and all k. By the labelling of the numbers $|z_\ell(\lambda)|$,

$$\varphi_{jk}(\lambda_0) \in \mathbf{T} \text{ and } |\varphi_{jk}(\lambda)| < 1 \text{ for } \lambda \in D_1. \tag{11.22}$$

As φ_{jk} maps a small neighborhood of λ_0 univalently onto a region of $\mathbf{C}$, we can assume that φ_{jk} is univalent on U (simply choose U small enough). Thus, by (11.22),

$$|\varphi_{jk}(\lambda)| > 1 \text{ for } \lambda \in D_2,$$

which implies that $|z_j(\lambda)| > |z_k(\lambda)|$ for $\lambda \in D_2$. Taking into account the definition of N_2, we therefore see that N_2 contains at least $\min(p, q) \geq 1$ numbers from $\{r - p + 1, \ldots, r\}$. Consequently, for $\lambda \in D_2$ we have

$$|G_1(\lambda)| = |b_s| \prod_{k=r+1}^{r+s} |z_k(\lambda)| > |b_s| \prod_{k \in N_2} |z_k(\lambda)| = |G_2(\lambda)|.$$

This completes the proof of (11.20).

The inequality (11.21) follows from (11.20) by symmetry. □

For $\lambda \in \Lambda(b) \cap U$, let $n_1 = n_1(\lambda)$ and $n_2 = n_2(\lambda)$ denote the outer normal vector of D_1 and D_2 at λ, respectively.

Corollary 11.13. (a) *The function g extends to a function that is continuous and positive on $\mathbf{C} \setminus \Lambda(b)$ and at the nonexceptional points of $\Lambda(b)$.*

(b) *If $\lambda \in \Lambda(b)$ is a nonexceptional point, then the normal derivatives $\partial g/\partial n_1$ and $\partial g/\partial n_2$ exist at λ and*

$$\frac{\partial g}{\partial n_1}(\lambda) + \frac{\partial g}{\partial n_2}(\lambda) \neq 0. \tag{11.23}$$

Proof. (a) We already observed that g admits the asserted continuous extension (see the paragraph before Lemma 11.12). Since all zeros of $Q(\lambda, z)$ are nonzero, Lemma 11.11 implies that $g(\lambda) > 0$.

(b) Since G_1 and G_2 are analytic in U, we see from (11.18) that the normal derivatives of g exist. Lemma 11.12 in conjunction with (11.18) implies (11.23). □

11.4 Limiting Set and Limiting Measure

We denote by $\mu_n = \mu_{n,b}$ the measure that assigns each eigenvalue $\lambda_j(T_n(b))$ $(j = 1, \ldots, n)$ measure $1/n$. Thus, if E is a subset of $\mathbf{C}$, then $\mu_n(E)$ is $1/n$ times the number of eigenvalues of $T_n(b)$ (counted up in accordance with their multiplicities) which lie in E. Let $C(\mathbf{C})$ be the set of all continuous functions on $\mathbf{C}$. Obviously,

$$\int_{\mathbf{C}} \varphi(\lambda)\, d\mu_n(\lambda) = \frac{1}{n} \sum_{j=1}^{n} \varphi(\lambda_j(T_n(b))) \text{ for all } \varphi \in C(\mathbf{C}). \tag{11.24}$$

Let g be as in Section 11.3. Then $\log g$ is locally integrable in $\mathbf{C} \setminus \Lambda(b)$. As the two-dimensional Lebesgue measure of $\Lambda(b)$ is zero (Theorem 11.9), we see that $\log g$ is defined almost everywhere in $\mathbf{C}$. Let $\Delta \log g$ stand for the distributional Laplacian of $\log g$.

We let $C_0^\infty(\mathbf{C})$ denote the set of all infinitely differentiable and compactly supported functions of $\mathbf{C} = \mathbf{R}^2$ into $\mathbf{C}$. Furthermore, dA and ds will denote area and length measures.

Lemma 11.14. *The measures μ_n converge in the distributional sense to the measure $\frac{1}{2\pi}\Delta \log g\, dA$, that is,*

$$\int_{\mathbf{C}} \varphi(\lambda)\, d\mu_n(\lambda) \to \frac{1}{2\pi} \int_{\mathbf{C}} \log g(\lambda)\, \Delta\varphi(\lambda)\, dA(\lambda) \tag{11.25}$$

for every $\varphi \in C_0^\infty$.

Proof. Evidently,

$$\int_{\mathbf{C}} \log|z-\lambda|\, d\mu_n(z) = \frac{1}{n}\sum_{j=1}^{n} \log|\lambda_j(T_n(b)) - \lambda| = \log|D_n(b-\lambda)|^{1/n}.$$

Thus, by Lemma 11.10,

$$\int_{\mathbf{C}} \log|z-\lambda|\, d\mu_n(z) \to \log g(\lambda) \tag{11.26}$$

uniformly on compact subsets of $\mathbf{C} \setminus \Lambda(b)$. It follows that (11.26) is also true in the distributional sense, whence

$$\frac{1}{2\pi}\Delta_\lambda \int_{\mathbf{C}} \log|z-\lambda|\, d\mu_n(z) \to \frac{1}{2\pi}\Delta_\lambda \log g(\lambda) \tag{11.27}$$

in the distributional sense. It is well known from potential theory that

$$\frac{1}{2\pi}\Delta_\lambda \int_{\mathbf{C}} \log|z-\lambda|\, d\nu(z) = \nu(\lambda)$$

for every compactly supported finite Borel measure ν. Consequently, (11.27) implies that

$$\mu_n(\lambda) \to \frac{1}{2\pi}\Delta_\lambda \log g(\lambda)$$

in the distributional sense, which says that (11.25) holds for all $\varphi \in C_0^\infty$. □

By Corollary 11.13, g and the normal derivatives $\partial g/\partial n_1$ and $\partial g/\partial n_2$ are well defined at the nonexceptional points of $\Lambda(b)$. From Theorem 11.9 we know that $\Lambda(b)$ has at most finitely many exceptional points.

Lemma 11.15. *If $\varphi \in C_0^\infty(\mathbf{C})$, then*

$$\int_{\mathbf{C}} \log g(\lambda)(\Delta\varphi)(\lambda)\, dA(\lambda) = \int_{\Lambda(b)} \varphi(\lambda)\frac{1}{g(\lambda)}\left|\frac{\partial g}{\partial n_1}(\lambda) + \frac{\partial g}{\partial n_2}(\lambda)\right| ds(\lambda).$$

Proof. Let $\lambda_0 \in \Lambda(b)$ be a nonexceptional point and let U be a sufficiently small open neighborhood of λ_0. We denote the boundaries of the two connected components D_1 and D_2 of $U \setminus \Lambda(b)$ by ∂D_1 and ∂D_2, and we put $\Lambda_1 = \partial D_1 \cap \Lambda(b)$, $\Lambda_2 = \partial D_2 \cap \Lambda(b)$. Clearly, Λ_1 and Λ_2 coincide as sets. However, the outer normal n is n_1 at the points of Λ_1 and it is n_2 at the points of Λ_2 (see Figure 11.2).

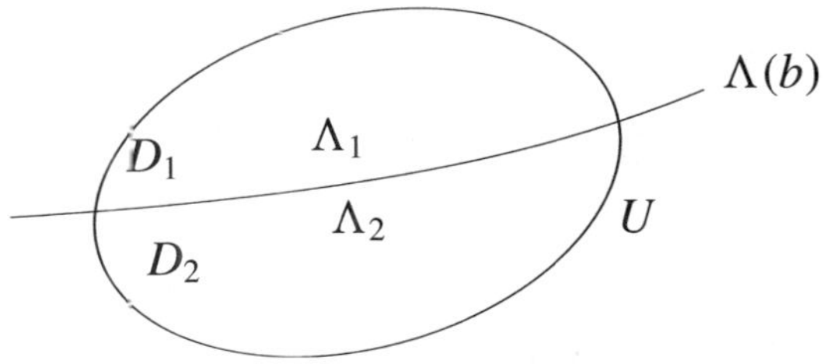

Figure 11.2. *An illustration to the proof.*

From Green's formula,

$$\int_{\Omega}(u\,\Delta v - v\,\Delta u)\,dA = \int_{\partial\Omega}\left(u\,\frac{\partial v}{\partial n} - v\,\frac{\partial u}{\partial n}\right)ds, \tag{11.28}$$

we deduce that

$$\int_{D_1}(\log g\,\Delta\varphi - \varphi\,\Delta\,\log g)\,dA = \int_{\partial D_1}\left(\log g\,\frac{\partial\varphi}{\partial n_1} - \varphi\,\frac{\partial}{\partial n_1}\log g\right)ds.$$

By (11.14), $\log g$ is harmonic in D_1. Thus, $\Delta \log g = 0$, whence

$$\int_{D_1}\log g\,\Delta\varphi\,dA = \int_{\partial D_1}\log g\,\frac{\partial\varphi}{\partial n_1}\,ds - \int_{\partial D_1}\varphi\,\frac{\partial}{\partial n_1}\log g\,ds. \tag{11.29}$$

Analogously,

$$\int_{D_2}\log g\,\Delta\varphi\,dA = \int_{\partial D_2}\log g\,\frac{\partial\varphi}{\partial n_2}\,ds - \int_{\partial D_2}\varphi\,\frac{\partial}{\partial n_2}\log g\,ds. \tag{11.30}$$

Adding (11.29) and (11.30) we obtain

$$\begin{aligned}\int_{U}\log g\,\Delta\varphi\,dA = &\int_{\partial U}\log g\,\frac{\partial\varphi}{\partial n}\,ds + \int_{\Lambda_1}\log g\,\frac{\partial\varphi}{\partial n_1}\,ds + \int_{\Lambda_2}\log g\,\frac{\partial\varphi}{\partial n_2}\,ds\\ &- \int_{\partial U}\varphi\,\frac{\partial}{\partial n}\log g\,ds - \int_{\Lambda_1}\varphi\,\frac{\partial}{\partial n_1}\log g\,ds - \int_{\Lambda_2}\varphi\,\frac{\partial}{\partial n_2}\log g\,ds,\end{aligned}$$

and since

$$\frac{\partial\varphi}{\partial n_1} + \frac{\partial\varphi}{\partial n_2} = 0 \quad\text{and}\quad \frac{\partial}{\partial n}\log g = \frac{1}{g}\,\frac{\partial g}{\partial n},$$

it follows that

$$\begin{aligned}\int_{U}\log g\,\Delta\varphi\,dA = &\int_{\partial U}\log g\,\frac{\partial\varphi}{\partial n}\,ds - \int_{\partial U}\varphi\,\frac{1}{g}\,\frac{\partial g}{\partial n}\,ds\\ &- \int_{\Lambda(b)}\varphi\,\frac{1}{g}\left(\frac{\partial g}{\partial n_1} + \frac{\partial g}{\partial n_2}\right)ds.\end{aligned} \tag{11.31}$$

From (11.18) and Lemma 11.12 we infer that

$$\frac{\partial g}{\partial n_1} < 0 \quad \text{and} \quad \frac{\partial g}{\partial n_2} < 0 \quad \text{on} \quad \Lambda(b).$$

Hence

$$-\int_{\Lambda(b)} \varphi \frac{1}{g} \left(\frac{\partial g}{\partial n_1} + \frac{\partial g}{\partial n_2} \right) ds = \int_{\Lambda(b)} \varphi \frac{1}{g} \left| \frac{\partial g}{\partial n_1} + \frac{\partial g}{\partial n_2} \right| ds. \tag{11.32}$$

If $\operatorname{supp} \varphi \subset U$, then (11.31) and (11.32) give

$$\int_U \log g \, \Delta\varphi \, dA = \int_{\Lambda(b)} \varphi \frac{1}{g} \left| \frac{\partial g}{\partial n_1} + \frac{\partial g}{\partial n_2} \right| ds.$$

Using an appropriate partition of unity it is easily seen that the last equality holds for all $\varphi \in C_0^\infty(\mathbf{C})$. □

We are now in a position to establish the main results of this chapter.

Theorem 11.16 (Hirschman). *The measures $d\mu_n$ converge weakly to the measure that is supported on $\Lambda(b)$ and equals*

$$\frac{1}{2\pi} \frac{1}{g} \left| \frac{\partial g}{\partial n_1} + \frac{\partial g}{\partial n_2} \right| ds \quad \textit{on} \quad \Lambda(b). \tag{11.33}$$

In other terms,

$$\frac{1}{n} \sum_{j=1}^{n} \varphi(\lambda_j(T_n(b))) \to \frac{1}{2\pi} \int_{\Lambda(b)} \varphi(\lambda) \frac{1}{g(\lambda)} \left| \frac{\partial g}{\partial n_1}(\lambda) + \frac{\partial g}{\partial n_2}(\lambda) \right| ds(\lambda) \tag{11.34}$$

for every $\varphi \in C(\mathbf{C})$ with compact support.

Proof. Combining (11.24) and Lemmas 11.14 and 11.15, we see that (11.34) is valid for every $\varphi \in C_0^\infty(\mathbf{C})$. As $C_0^\infty(\mathbf{C})$ is dense in the space of all continuous functions with compact support, we arrive at the assertion. □

Theorem 11.17 (Schmidt and Spitzer). *We have $\Lambda_s(b) = \Lambda_w(b) = \Lambda(b)$.*

Proof. To prove that $\Lambda(b) \subset \Lambda_s(b)$, pick $\lambda_0 \in \Lambda(b)$ and let U be any open neighborhood of λ_0. Choose a function $\varphi \in C_0^\infty(\mathbf{C})$ so that $\operatorname{supp} \varphi \subset U$ and $\varphi(\lambda_0) = 1$. From Theorem 11.16 we get

$$\frac{1}{n} \sum_{j=1}^{n} \varphi(\lambda_j(T_n(b))) \to \frac{1}{2\pi} \int_{\Lambda(b)} \varphi(\lambda) \frac{1}{g(\lambda)} \left| \frac{\partial g}{\partial n_1}(\lambda) + \frac{\partial g}{\partial n_2}(\lambda) \right| ds(\lambda) > 0.$$

As

$$\frac{\#\{j : \lambda_j(T_n(b)) \in U\}}{n} = \frac{1}{n}\sum_{j=1}^{n} \chi_U(\lambda_j(T_n(b))) \geq \frac{1}{n}\sum_{j=1}^{n} \varphi(\lambda_j(T_n(b))),$$

it follows that $\#\{j : \lambda_j(T_n(b)) \in U\} > 0$ for all sufficiently large n, which implies that $\lambda_0 \in \Lambda_s(b)$.

We are left to prove that $\Lambda_w(b) \subset \Lambda(b)$. This follows from Corollary 11.2 and Theorem 11.3. We can also argue as follows. Let $\lambda_0 \notin \Lambda(b)$. Then $|z_r(\lambda_0)| < |z_{r+1}(\lambda_0)|$. Since $|z_r(\lambda)|$ and $|z_{r+1}(\lambda)|$ depend continuously on λ, there is an open neighborhood U of λ_0 such that $|z_r(\lambda)| < |z_{r+1}(\lambda)|$ for all $\lambda \in U$. By Lemma 11.10, $|D_n(b-\lambda)|^{1/n} \to g(\lambda) > 0$ for all $\lambda \in U$. Hence, $D_n(b-\lambda) \neq 0$ for all sufficiently large n and all $\lambda \in U$. This shows that $\operatorname{sp} T_n(b) \cap U = \emptyset$ for all sufficiently large n and implies that $\lambda_0 \notin \Lambda_w(b)$. □

Example 11.18. We already know $\Lambda(b)$ in the case where b is a trinomial of the form $b(t) = b_{-1}t^{-1} + b_0 + b_1 t$. One can also describe $\Lambda(b)$ explicitly if b is a general trinomial, that is,

$$b(t) = b_{-r}t^{-r} + b_0 + b_s t^s.$$

By translation of the plane, we may assume that $b_0 = 0$, and by further rotation and change of the scale, we may confine ourselves to the case where

$$b(t) = t^{-r} + t^s, \quad r \geq 1, \quad s \geq 1, \quad \text{g.c.d.}(r, s) = 1.$$

Schmidt and Spitzer showed that in this case $\Lambda(b)$ is the star

$$\Lambda(b) = \left\{\omega d : \omega = e^{2\pi i/(r+s)}, 0 \leq d \leq R\right\},$$

where $R = (r+s)s^{-s/(r+s)}r^{-r/(r+s)}$. □

The pictures at the top right corners of the beginning pages of the chapters are examples of symbol curves $b(\mathbf{T})$ and the eigenvalues of $T_{60}(b)$. Figures 11.3 to 11.5 show some more sophisticated examples. These figures were scanned from printouts left by Olga Grudskaya.

11.5 Connectedness of the Limiting Set

Theorem 11.19 (Ullman). *If b is a Laurent polynomial, then $\Lambda(b)$ is a connected set.*

Proof. Assume $\Lambda(b)$ is not connected. Then we can find a subset $K \subset \Lambda(b)$ and a function $\varphi \in C_0^\infty(\mathbf{C})$ such that $\varphi|K = 1$ and $\varphi|\Lambda(b) \setminus K = 0$. Moreover, φ can be chosen so that $\varphi|\Omega_1 = 1$ and $\varphi|\Omega_2 = 0$ where Ω_1 and Ω_2 are open sets with smooth Jordan boundaries which contain K and $\Lambda(b) \setminus K$, respectively. Let Ω_0 and Ω_3 be open subsets of $\mathbf{C}$ which contain K and $\Lambda(b) \setminus K$ and are contained in the interior of Ω_1 and Ω_2, respectively. We may assume that Ω_0 and Ω_3 have smooth Jordan boundaries. Finally, put $\Omega = \mathbf{C} \setminus (\Omega_0 \cup \Omega_3)$. Abbreviate the limiting measure (11.33) to $d\mu$.

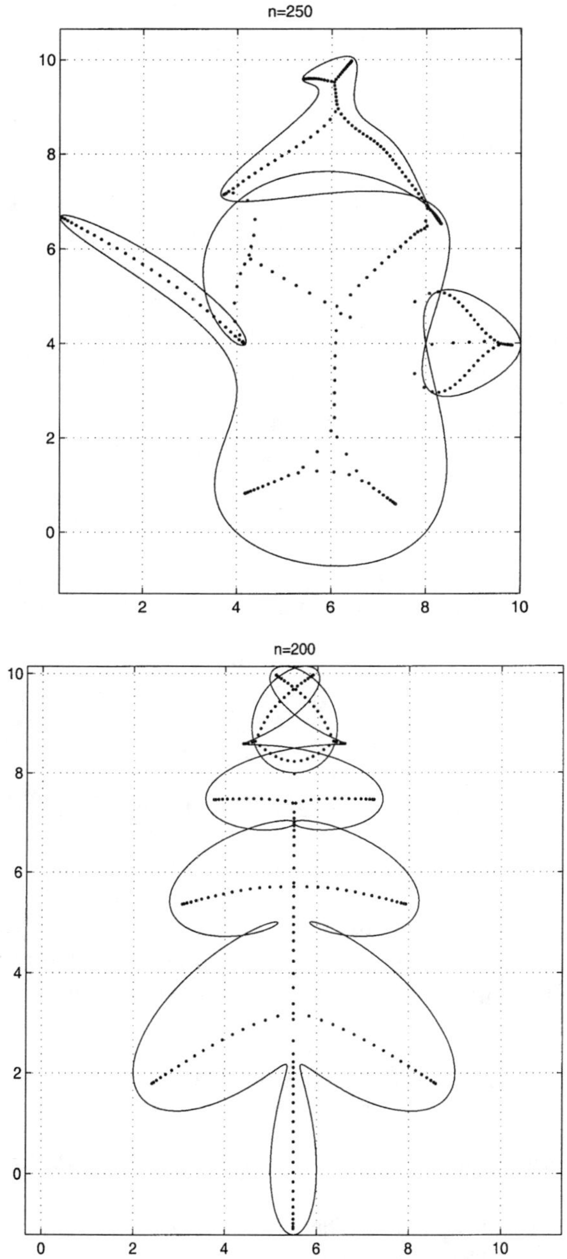

Figure 11.3. *Legacy of Olga Grudskaya: The ranges $b(\mathbf{T})$ for two Laurent polynomials and the eigenvalues of the matrices $T_n(b)$.*

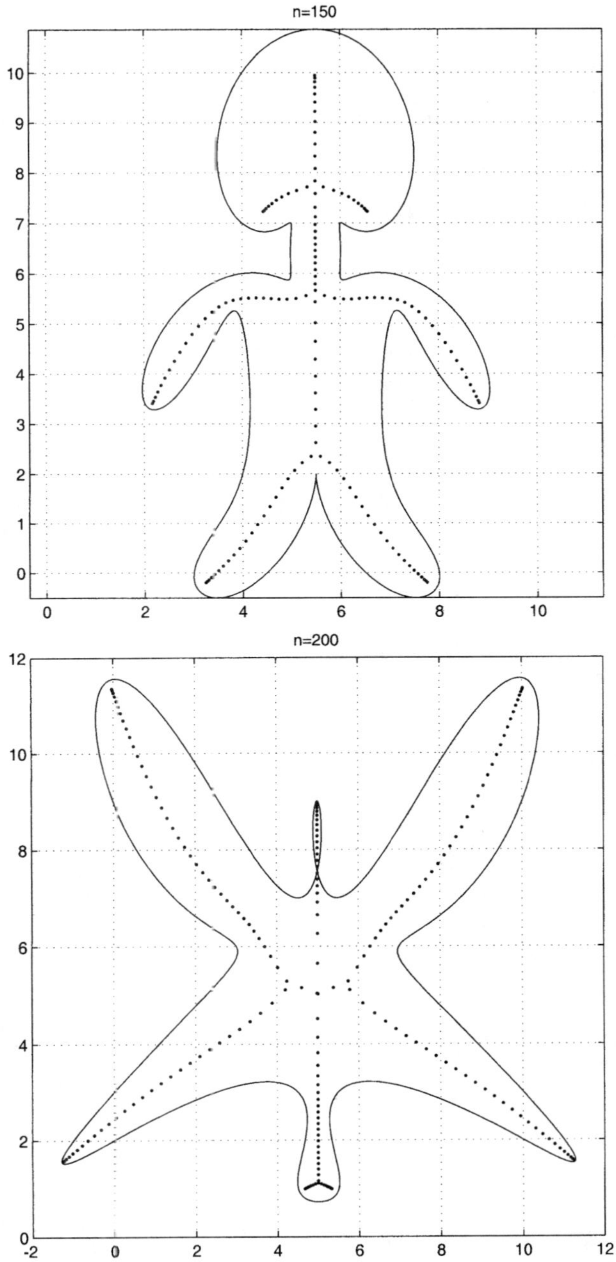

Figure 11.4. *Legacy of Olga Grudskaya: The ranges $b(\mathbf{T})$ for two Laurent polynomials and the eigenvalues of the matrices $T_n(b)$.*

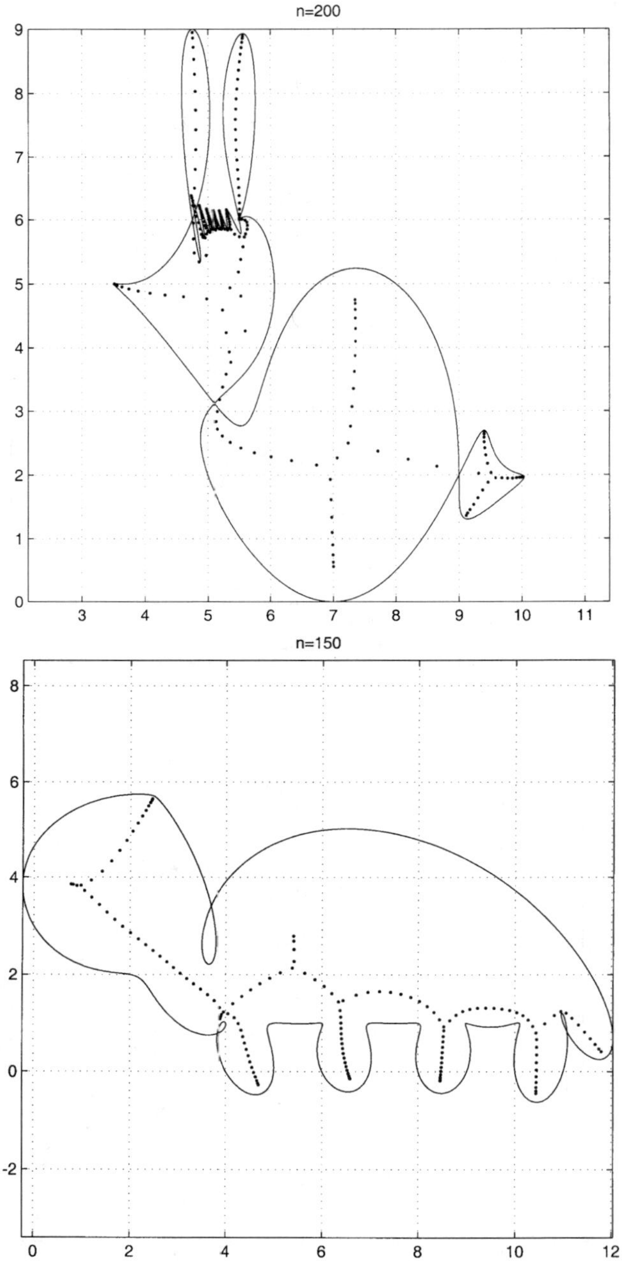

Figure 11.5. *Legacy of Olga Grudskaya: The ranges $b(\mathbf{T})$ for two Laurent polynomials and the eigenvalues of the matrices $T_n(b)$.*

Since $\mu(\Lambda(b)) = 1$ and neither K nor $\Lambda(b) \setminus K$ can degenerate to a point (Theorem 11.9), it follows that

$$0 < \mu(K) < 1. \tag{11.35}$$

From Lemma 11.15 we obtain

$$\mu(K) = \int_K d\mu = \int_{\Lambda(b)} \varphi\, d\mu = \int_{\mathbf{C}} \log g\, \Delta\varphi\, dA,$$

and since $\Delta\varphi = 0$ on Ω_0 and Ω_2, we get

$$\mu(K) = \int_{\Omega} \log g\, \Delta\varphi\, dA.$$

As $\Delta \log g = 0$ on Ω, Green's formula (11.28) implies that

$$\mu(K) = \int_{\partial\Omega} \left(\log g \frac{\partial\varphi}{\partial n} - \varphi \frac{\partial \log g}{\partial n} \right) ds,$$

where n is now the outer normal to Ω. Taking into account that

$$\varphi \,|\, \partial\Omega_3 = 0, \quad \frac{\partial\varphi}{\partial n} \Big| \,\partial\Omega_3 = 0, \quad \varphi \,|\, \partial\Omega_0 = 1, \quad \frac{\partial\varphi}{\partial n} \Big| \,\partial\Omega_0 = 0,$$

we get

$$\mu(K) = \int_{\partial\Omega_0} \frac{\partial \log g}{\partial n}\, ds, \tag{11.36}$$

where n is now the outer normal to Ω_0. From (11.15) we infer that if λ_0 is a point in $\mathbf{C} \setminus \Lambda(b)$, then the analytic functions $D_{n+1}(b-\lambda)/D_n(b-\lambda)$ converge uniformly in some open neighborhood of λ_0 to some analytic function. Thus, there is an analytic function $G : \mathbf{C} \setminus \Lambda(b) \to \mathbf{C}$ such that

$$\lim_{n\to\infty} \frac{D_{n+1}(b-\lambda)}{D_n(b-\lambda)} = G(\lambda) \neq 0$$

for all $\lambda \in \mathbf{C} \setminus \Lambda(b)$. Lemma 11.10 implies that $g(\lambda) = |G(\lambda)|$. Fix a point $\lambda_0 \in \partial\Omega_0$ and choose an argument $v(\lambda) = \arg G(\lambda)$ of $G(\lambda)$ which is continuous on $\partial\Omega_0 \setminus \{\lambda_0\}$. Put $u = \log g$. Since

$$\log G = \log|G| + i \arg G = u + iv$$

is analytic in a neighborhood of $\partial\Omega_0 \setminus \{\lambda_0\}$, the Cauchy-Riemann equations give $u_x = v_y$ and $u_y = -v_x$. Let $x = x(t)$ and $y = y(t)$ $(t \in (0, 2\pi))$ be a parametrization of of $\partial\Omega_0 \setminus \{\lambda_0\}$. From (11.36) we now obtain that

$$\begin{aligned} \mu(K) &= \int_{\partial\Omega_0} \frac{\partial \log g}{\partial n}\, ds = \int_0^{2\pi} (u_x \dot{y} - u_y \dot{x})\, dt \\ &= \int_0^{2\pi} (v_x \dot{x} + v_y \dot{y})\, dt = \int_0^{2\pi} \frac{\partial v}{\partial s}\, ds = \frac{1}{2\pi} v \Big|_{\partial\Omega_0 \setminus \{\lambda_0\}}. \end{aligned} \tag{11.37}$$

But the number (11.37) is always an integer, which contradicts (11.35). □

As the following result shows, $\Lambda(b)$ may separate the plane.

Theorem 11.20. *Let $b(t) = \mu + t^{-r}(t-\alpha)^r(t-\beta)^r$, where μ, α, β are complex numbers and $\alpha\beta \neq 0$. If $r = 1$ or $r = 2$, then $\mathbf{C} \setminus \Lambda(b)$ is connected. If $r \geq 3$, then $\mathbf{C} \setminus \Lambda(b)$ has at most $[(r+1)/2]$ components (including the unbounded component), and for each natural number j between 1 and $[(r+1)/2]$ there exist α and β such that $\mathbf{C} \setminus \Lambda(b)$ has exactly j components.*

Proof. Put $a(t) = t^{-1}(t-\alpha)(t-\beta)$. The curve $a(\mathbf{T})$ is an ellipse with the foci $-(\alpha+\beta) \pm 2\sqrt{\alpha\beta}$ and $\Lambda(a)$ is the line segment between the foci. Obviously,

$$b(t) = \mu + (a(t))^r.$$

We claim that

$$\Lambda(b) = \mu + (\Lambda(a))^r. \tag{11.38}$$

To prove our claim, we assume for the sake of definiteness that $a(t)$ traces out $a(\mathbf{T})$ counterclockwise as t moves around $\mathbf{T}$ counterclockwise. Notice that $b_\varrho = \mu + a_\varrho^r$ for all $\varrho \in (0, \infty)$.

Pick $\lambda \in \Lambda(a)$. Then $\lambda \in \operatorname{sp} T(a_\varrho)$ for all $\varrho \in (0, \infty)$ due to Theorem 11.3. Put $\omega = e^{2\pi i/r}$. If $\omega^k\lambda \in a_\varrho(\mathbf{T})$ for some $k \in \{1, \dots, r\}$, then

$$\mu + \lambda^r = \mu + (\omega^k\lambda)^r \in b_\varrho(\mathbf{T})$$

and hence $\mu + \lambda^r \in \operatorname{sp} T(b_\varrho)$. Now assume that $\omega^k\lambda \notin a_\varrho(\mathbf{T})$ for all $k \in \{1, \dots, r\}$. Since $\lambda \in \operatorname{sp} T(a_\varrho)$, it follows that $\operatorname{wind}(a_\varrho - \lambda) = 1$. As $b_\varrho - \mu - \lambda^r = a_\varrho^r - \lambda^r$, we have

$$\operatorname{wind}(b_\varrho - \mu - \lambda^r) = \operatorname{wind} \prod_{k=1}^{r}(a_\varrho - \omega^k\lambda) = \sum_{k=1}^{r} \operatorname{wind}(a_\varrho - \omega^k\lambda),$$

and as $\operatorname{wind}(a_\varrho - \omega^k\lambda)$ is either 0 or 1, we conclude that $\operatorname{wind}(b_\varrho - \mu - \lambda^r) \geq 1$. Consequently, $\mu + \lambda^r \in \operatorname{sp} T(b_\varrho)$. From Theorem 11.3 we now obtain the inclusion "$\supset$" in (11.38). To verify the reverse inclusion, let $\varrho \in \mathbf{C} \setminus \{0\}$ be any number such that $\varrho^2 = \alpha\beta$. Then $a_\varrho(\mathbf{T}) = \Lambda(a)$. It results that $b_\varrho(\mathbf{T}) = \mu + (\Lambda(a))^r$, and as $\operatorname{wind}(b_\varrho - \zeta) = 0$ for all $\zeta \notin b_\varrho(\mathbf{T})$, we see that $\operatorname{sp} T(b_\varrho) = \mu + (\Lambda(a))^r$. Since

$$\bigcap_{\varrho \in \mathbf{C}\setminus\{0\}} \operatorname{sp} T(b_\varrho) = \bigcap_{\varrho \in (0,\infty)} \operatorname{sp} T(b_\varrho),$$

we infer from Theorem 11.3 that $\Lambda(b) \subset \operatorname{sp} T(b_\varrho) = \mu + (\Lambda(a))^r$, which is the inclusion "$\subset$" of (11.38).

Since neither a line segment nor the square of a line segment does separate the plane, we obtain from equality (11.38) the assertion for $r = 1$ and $r = 2$. Combining (11.38) with the fact that $\Lambda(a)$ is a line segment, one can easily see that the set $\mathbf{C} \setminus \Lambda(b)$ has at most $[(r+1)/2]$ components and that each number of components between 1 and $[(r+1)/2]$ can indeed be realized (for example, we get exactly $[(r+1)/2]$ components if $|\mu - (\alpha+\beta)| > 0$ is sufficiently small and $|\alpha\beta|$ is sufficiently large. □

Exercises

1. Gershgorin's theorem states that if $A_n = (a_{ij})$ is an $n \times n$ matrix and

$$R_i'(A_n) := \left(\sum_{j=1}^{n} |a_{ij}| \right) - |a_{ii}|,$$

then

$$\mathrm{sp}\, A_n \subset \bigcup_{i=1}^{n} \{\lambda \in \mathbf{C} : |\lambda - a_{ii}| \le R_i'(A_n)\}.$$

Show that in the case of large Toeplitz band matrices this theorem amounts to the trivial estimate $\mathrm{rad}\, T_n(b) \le \|b\|_W$.

2. Let $b \in \mathcal{P}$ and suppose $\mathcal{R}(b) \subset \mathbf{R}$. Let further $\{a_1, a_2, a_3, \ldots\}$ be a convergent sequence of real numbers. Put

$$A_n = T_n(b) + \mathrm{diag}\,(a_1, \ldots, a_n), \quad A = T(b) + \mathrm{diag}\,(a_1, a_2, \ldots).$$

Show that

$$\liminf_{n\to\infty} \mathrm{sp}\, A_n = \limsup_{n\to\infty} \mathrm{sp}\, A_n = \mathrm{sp}\, A.$$

3. Let $a \in \mathcal{P}$ and let $H_n(a) = P_n H(a) P_n | \mathrm{Im}\, P_n$.

(a) Show that $\Sigma(H_n(a))$ and $\mathrm{sp}\, H_n(a)$ converge to $\Sigma(H(a))$ and $\mathrm{sp}\, H(a)$, respectively, in the Hausdorff metric.

(b) Show that $\Sigma(H(a))$ and $\mathrm{sp}\, H(a)$ are finite sets containing the origin.

(c) Prove that

$$\lim_{n\to\infty} \frac{1}{n} \sum_{j=1}^{n} \varphi(\sigma_j(H_n(a))) = \lim_{n\to\infty} \frac{1}{n} \sum_{j=1}^{n} \varphi(\lambda_j(H_n(a))) = \varphi(0)$$

for every measurable function φ that is continuous at the origin.

4. Let b be a Laurent polynomial.

(a) Show that $\mathrm{sp}\, T_n(b) \subset \mathrm{conv}\, \mathcal{R}(b)$ for all $n \ge 1$.

(b) Prove that

$$\lim_{n\to\infty} \frac{1}{n} \sum_{j=1}^{n} \varphi(\lambda_j(T_n(b))) = \frac{1}{2\pi} \int_0^{2\pi} \varphi(e^{i\theta})\, d\theta \tag{11.39}$$

for every function φ that is analytic in a disk containing $\mathrm{conv}\, \mathcal{R}(b)$.

(c) Let $\Omega \subset \mathbf{C}$ be an open set such that conv $\mathcal{R}(b) \subset \Omega$. Prove that (11.39) is true for every harmonic function φ on Ω.

(d) What does (11.39) say for $b(t) = b_0 + b_1(t)$?

(e) Let Ω be an open set in the plane and $\varphi \in C(\Omega)$. Prove that if (11.39) is true for every b with conv $\mathcal{R}(b) \subset \Omega$, then φ is harmonic on Ω.

5. Let $a, b \in \mathcal{P}$ and consider $A_n = T_n(b) + H_n(a)$. Fix an open subset Ω of $\mathbf{C}$ that contains sp $T(b) \cup$ sp $(T(b) + H(a))$.

(a) Show that sp $A_n \subset \Omega$ for all sufficiently large n.

(b) Show that

$$\lim_{n\to\infty} \frac{1}{n} \left[\sum_{j=1}^{n} p(\lambda_j(A_n)) - \sum_{j=1}^{n} p(\lambda_j(T_n(b))) \right]$$
$$= \lim_{n\to\infty} \frac{1}{n} \operatorname{tr} [p(A_n) - p(T_n(b))] = 0$$

for every polynomial $p \in \mathcal{P}^+$.

(c) Prove that if φ is harmonic in Ω, then

$$\lim_{n\to\infty} \frac{1}{n} \sum_{j=1}^{n} \varphi(\lambda_j(A_n)) = \frac{1}{2\pi} \int_{\Lambda(b)} \varphi(\lambda) d\mu(\lambda),$$

where $d\mu$ is the measure (11.33).

6. Do the mass centers of the sets $\Lambda(b)$ and sp $T(b)$ always coincide?

7. Compute the determinant and the eigenvalues of the Toeplitz matrix

$$A_n = \begin{pmatrix} c+1 & a & \dots & a^{n-1} \\ a^{-1} & c+1 & \dots & a^{n-2} \\ \vdots & \vdots & \ddots & \vdots \\ a^{-(n-1)} & a^{-(n-2)} & \dots & c+1 \end{pmatrix}.$$

Notes

Sections 11.1 to 11.4 are based on the pioneering papers on this topic, Schmidt and Spitzer [243] and Hirschman [165], and on some significant improvements of the original proofs that were introduced by Widom [296], [297]. Theorem 11.19 was established in [286].

Exercise 3 is of course easy. For the spectral distribution of more general Hankel matrices we refer to Fasino and Tilli's papers [115] and [116]. Exercise 4 is a result of Tilli [264]. The analogue of Exercise 5 for the singular values is in [116], for example. The result of Exercise 5 can probably be generalized to a broader class of test functions φ. Exercise 7 is from Sakhnovich's paper [241].

Further results: limiting sets separating the plane. For a long time it had been an open question whether the set $\Lambda(b)$ may separate the plane. In 1992, Anselone and Sloan [6] studied the truncated Wiener-Hopf operator

$$(A_\tau f)(x) = 2\int_0^x e^{-(x-t)} f(t)dt + \int_x^\tau e^{x-t} f(t)dt \quad (0 < x < \tau)$$

on $L^2(0,\tau)$. They were interested in the set Λ of all $\lambda \in \mathbf{C}$ such that $\lambda = \lim \lambda_n$ with $\lambda_n \in \operatorname{sp} A_{\tau_n}$ and $\tau_n \to \infty$, and on the basis of numerical experiments, they conjectured that Λ is the union of the circle $\{\lambda \in \mathbf{C} : |\lambda - 1/12| = 1/12\}$ and the line segment $[3/2-\sqrt{2}, 3/2+\sqrt{2}\,]$. Note that this set separates the plane. The symbol of the corresponding "infinite" Wiener-Hopf operator

$$(Af)(x) = 2\int_0^x e^{-(x-t)} f(t)dt + \int_x^\infty e^{x-t} f(t)dt \quad (0 < x < \infty)$$

on $L^2(0,\infty)$ is the Fourier transform of the kernel,

$$a(x) = \int_{-\infty}^0 e^{ixt}e^t dt + \int_0^\infty e^{xt} \cdot 2e^{-t} dt = \frac{3+ix}{1+x^2} \quad (x \in \dot{\mathbf{R}} := \mathbf{R} \cup \{\infty\}).$$

This is a rational function and $a(\dot{\mathbf{R}})$ is an ellipse. In a sense, Anselone and Sloan's conjecture answered the continuous analogue on the question whether $\Lambda(b)$ may separate the plane in the negative. In [72], the continuous analogue of Theorem 11.17 was established, and this result implied that the conjecture of Anselone and Sloan was true. Thus, it was Wiener-Hopf operators with rational symbols that showed us for the first time a limiting set Λ for which $\mathbf{C} \setminus \Lambda$ is disconnected (see Figure 11.6).

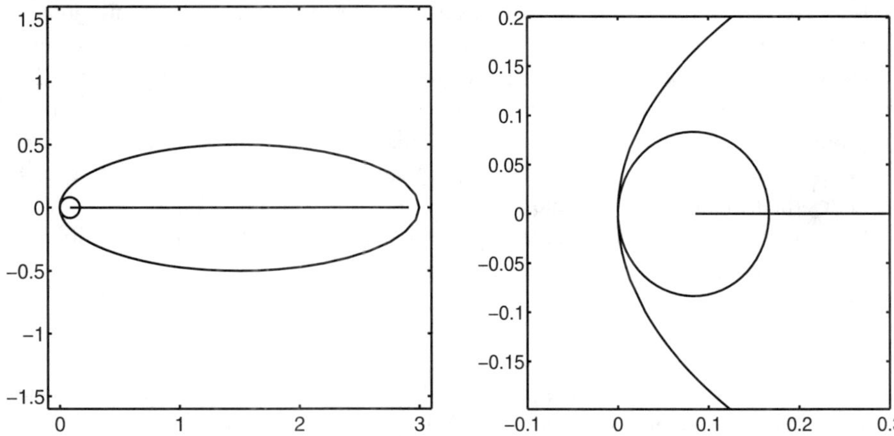

Figure 11.6. *The ellipse $a(\dot{\mathbf{R}})$ and the limiting set Λ on the left and a zoom in on the right.*

In the context of Toeplitz band matrices, symbols b for which $\Lambda(b)$ separates the plane were detected numerically in [20] (Figure 5.3(c)) and [71] (lower left picture of

Figure 38, the "smiling shark"). The picture on the first page of this chapter also shows a disconnected set $\mathbf{C} \setminus \Lambda(b)$; the corresponding symbol is $b(t) = t^{-3} + 0.99\,t^{-2} + 0.1\,t^2 - 0.44\,t^3$. Theorem 11.20, which is a first "analytic" result in this direction, is from our paper [50].

Further results: dense Toeplitz matrices. The sets $\Lambda_s(b)$ and $\Lambda_w(b)$ can be defined for Toeplitz matrices $T(b)$ with arbitrary symbols b, but for general symbols b things remain mysterious. In the case of rational symbols, the limiting sets were characterized by K. M. Day [97], [98] (the result is also cited in [71, Section 5.9], and for the underlying determinant formula see also [32]).

For certain classes of continuous or piecewise continuous symbols b, the asymptotic eigenvalue distribution of $T_n(b)$ can be described by formulas of the Szegö type. These say that if $\lambda_1^{(n)}, \ldots, \lambda_n^{(n)}$ are the eigenvalues of $T_n(b)$, then

$$\lim_{n\to\infty} \frac{1}{n} \sum_{j=1}^{n} \varphi(\lambda_j^{(n)}) = \frac{1}{2\pi} \int_0^{2\pi} \varphi(b(e^{i\theta}))d\theta \tag{11.40}$$

for every continuous function $\varphi : \mathbf{C} \to \mathbf{R}$ with compact support (see [15], [71], [265], [296], [297] for details). Notice that (11.40) implies that, up to $o(n)$ possible outliers, the eigenvalues cluster along the (essential) range $b(\mathbf{T})$ of the symbol but that (11.40) does not tell us whether the possible $o(n)$ outliers produce additional pieces of $\Lambda_w(b)$.

In order to find $\Lambda_w(b)$, one can try computing sp $T_n(b)$ for some large values of n numerically, having hopes that the result is a more or less good approximation to $\Lambda_w(b)$. This works for rational symbols and, in general, also for the symbols for which (11.40) holds. Another possibility of determining $\Lambda_w(b)$ is to approximate b by a Laurent polynomial b_n, surmising that $\Lambda_w(b_n)$ is close to $\Lambda_w(b)$. This approach fails for piecewise continuous symbols, which should not come as a surprise, since a properly piecewise continuous function can never be approximated uniformly by Laurent polynomials as closely as desired. Unexpectedly, this approach does in general also not work for continuous symbols. This is a consequence of the main result of [51], which shows that the asymptotic spectrum $\Lambda_w(\cdot)$ is discontinuous on the space of continuous symbols. The result of [51] is as follows: There exist $b \in W$ such that $\Lambda_w(S_n b)$ does not converge to $\Lambda_w(b)$ in the Hausdorff metric. Here $S_n b$ denotes the nth partial sum of the Fourier series (see Section 5.3).

Figures 11.7 and 11.8, which are from [51] and were produced by Olga Grudskaya, convincingly illustrate what happens. The symbol b is

$$b(t) = t^{-1}(33 - (t + t^2)(1 - t^2)^{3/4}).$$

This is a continuous function which is piecewise C^∞ but not C^∞. The matrix $T(b)$ is a lower Hessenberg matrix. The results of Widom [296] imply that (11.40) is valid, that is, we expect that the eigenvalues of $T_n(b)$ cluster along the range $b(\mathbf{T})$ of the symbol. The two pictures of Figure 11.7 show the range $b(\mathbf{T})$ and the eigenvalues of $T_m(b)$ for $m = 128$ (left) and $m = 512$ (right). In Figure 11.8, we plot the eigenvalues of $T_{128}(S_n b)$ for $n = 4, 6, 8, 12$. These eigenvalue distributions mimic the sets $\Lambda_w(S_n b)$ sufficiently well, and it is clearly seen that $\Lambda_w(S_n a)$ grows like a rampant tree that in the $n \to \infty$ limit has nothing in common with $\Lambda_w(b)$.

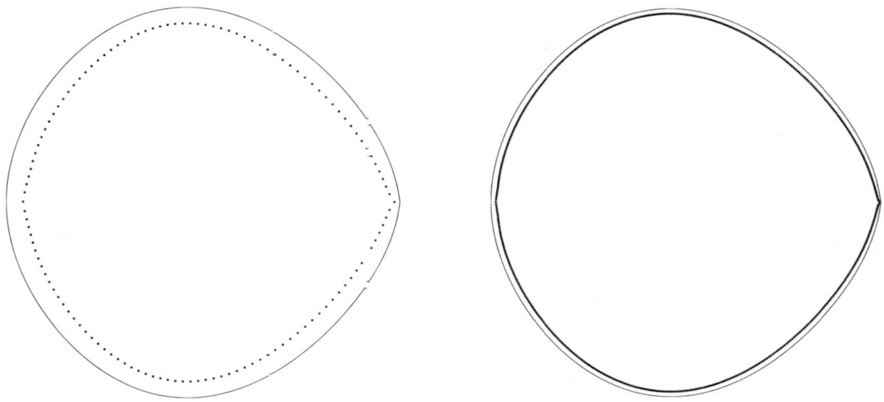

Figure 11.7. *The range $b(\mathbf{T})$ and the eigenvalues of $T_{128}(b)$ (left) and $T_{512}(b)$ (right).*

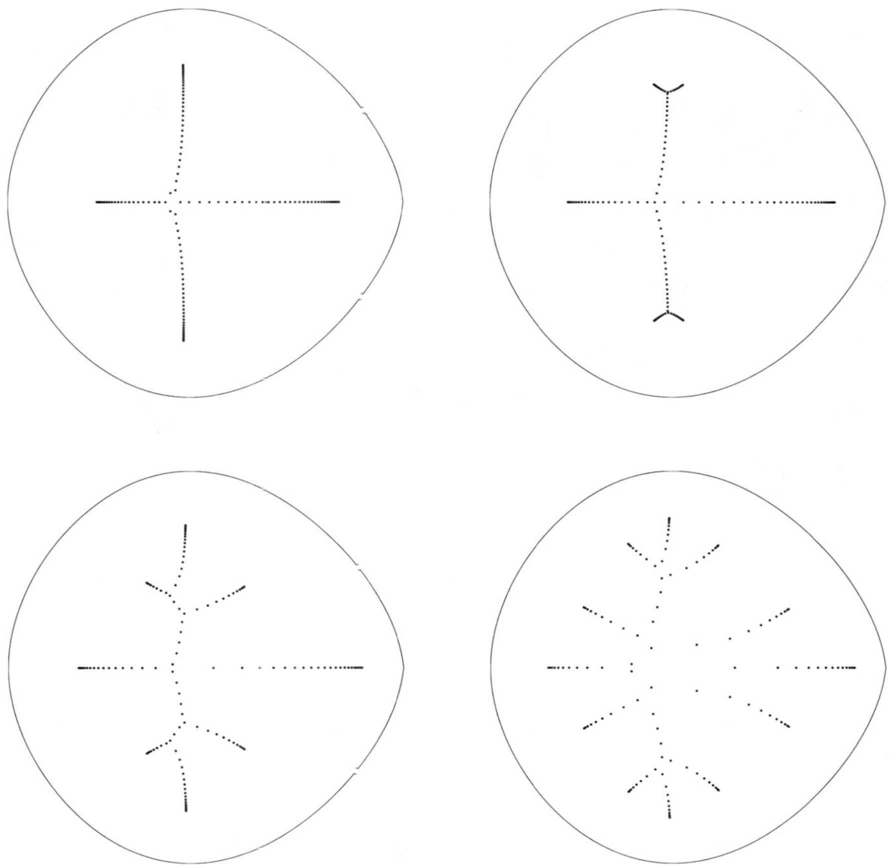

Figure 11.8. *The range $b(\mathbf{T})$ and the eigenvalues of $T_{128}(S_n b)$ for the values* $n = 4, 6, 8, 12$.

Chapter 12

Eigenvectors and Pseudomodes

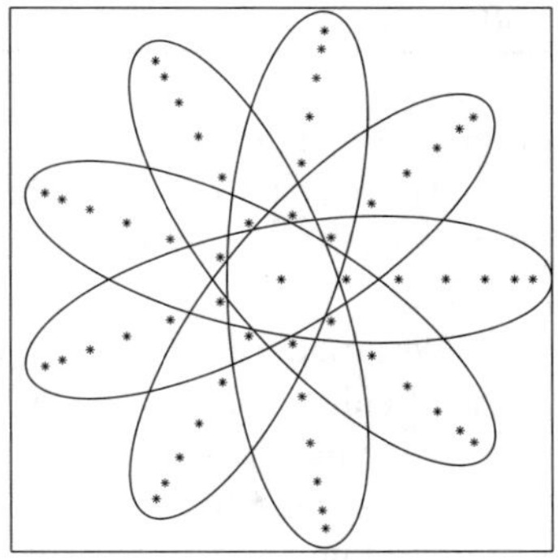

In previous chapters, we studied the asymptotic behavior of eigenvalues and pseudoeigenvalues (= points in the pseudospectrum) of large Toeplitz matrices. This chapter concerns asymptotic results on eigenvectors and pseudomodes (= pseudoeigenvectors). We will in particular point out that there are striking differences between the Toeplitz and circulant cases.

12.1 Tridiagonal Circulant and Toeplitz Matrices

Let $b(t) = t + \alpha^2 t^{-1}$. In Sections 2.1 and 2.2 we found explicit formulas for the eigenvectors of $C_n(b)$ and $T_n(b)$. The purpose of this section is to give the reader some delight with a few pictures that can be easily produced with MATLAB.

Let $n \geq 3$. We know from Proposition 2.1 that the eigenvectors of $C_n(b)$ normalized to ℓ^2 norm 1 are $x_1, \dots, x_n$ with

$$x_j = \frac{1}{\sqrt{n}} \left(1,\ \overline{\omega}_n^{\,j-1},\ \overline{\omega}_n^{\,2(j-1)},\ \dots,\ \overline{\omega}_n^{\,(n-1)(j-1)}\right), \tag{12.1}$$

where $\omega_n = e^{2\pi i/n}$. Theorem 2.4 tells us that after normalization to the ℓ^2 norm 1 the eigenvectors of $T_n(b)$ are $x_1, \dots, x_n$ with

$$x_j = c_n(\alpha) \left(\frac{1}{\alpha} \sin \frac{\pi j}{n+1},\ \left(\frac{1}{\alpha}\right)^2 \sin \frac{2\pi j}{n+1},\ \dots,\ \left(\frac{1}{\alpha}\right)^n \sin \frac{n\pi j}{n+1}\right), \tag{12.2}$$

where $c_n(\alpha)$ is the normalization constant. Note first of all that (12.1) is independent of b, while (12.2) depends on b. A second immediate observation is that the absolute values of the components of the eigenvectors (12.1) are constant, whereas those of (12.2) are localized in the left for $|\alpha| > 1$ and in the right for $|\alpha| < 1$.

The eigenvectors $x_1, \dots, x_n$ form a basis in $\mathbf{C}^n$. When identifying $\mathbf{C}^n$ with $\mathbf{R}^{2n}$, these eigenvectors deliver the basis

$$(\operatorname{Re} x_1, \operatorname{Im} x_1), \dots, (\operatorname{Re} x_n, \operatorname{Im} x_n), (-\operatorname{Im} x_1, \operatorname{Re} x_1), \dots, (-\operatorname{Im} x_n, \operatorname{Re} x_n) \tag{12.3}$$

in $\mathbf{R}^{2n}$. Notice that $\operatorname{Im} x_j = 0$ for all j in the Toeplitz case provided α is real.

The second half of the basis (12.3) clearly mimics the first half. In Figure 12.1 we plotted the 10 vectors of the first halves of the bases (12.3) for $C_{10}(b)$ and $T_{10}(b)$ with $\alpha = 3/2$.

We arrange the $2n$ vectors (12.3) to a $(2n) \times (2n)$ matrix and denote this matrix by EC_n in the circulant case and by $ET_n(\alpha)$ in the Toeplitz case. Figure 12.2 shows pseudocolor plots of the matrices EC_{30} and $ET_{30}(3/2)$. A gray point indicates a matrix entry near zero, dark points stand for positive entries, and light points represent negative ones. The right picture of Figure 12.1 and especially the bottom picture of Figure 12.2 show that (12.3) is a very ill-conditioned basis in the Toeplitz case. Finally, Figures 12.3 and 12.4 depict surface plots of the matrices EC_{30} and $ET_{30}(3/2)$ from two different viewpoints. Whatever insights these two figures might provide, they show us at least the bizarre difference between two worlds.

12.2 Eigenvectors of Triangular and Tridiagonal Matrices

We now turn to the eigenvectors of general banded Toeplitz matrices. Let $b(t) = \sum_{j=-r}^{s} b_j t^j$ be a Laurent polynomial. As in Chapter 11, we write

$$b(t) - \lambda = t^{-r} b_s (t - z_1(\lambda)) \dots (t - z_{r+s}(\lambda)), \quad |z_1(\lambda)| \le \dots \le |z_{r+s}(\lambda)|. \tag{12.4}$$

The system $T_n(b - \lambda)x_n = 0$ ($x_n \in \mathbf{C}^n$) is a difference equation with constant coefficients and $r+s$ boundary conditions. Analogously, the equation $T(b-\lambda)x_0 = 0$ ($x_0 \in \ell^2$) is a difference equation that has constant coefficients, and the constraints are s boundary conditions and the requirement that x_0 be in ℓ^2. Thus, once λ is an eigenvalue, the eigenvectors can be expressed in terms of the zeros $z_1(\lambda), \dots, z_{r+s}(\lambda)$ of (12.4) and certain coefficients that can be determined from the boundary constraints. The resulting formulas are nevertheless very complicated and inappropriate for answering the question we are interested in here in a straightforward fashion. Trench [278] derived simpler equations for the eigenvalues of $T_n(b)$ and constructed simpler formulas for the eigenvectors (see Exercise 2). Again, these formulas cannot be immediately employed in order to see the asymptotic behavior of the eigenvectors.

An interesting result on the eigenvectors was established by Zamarashkin and Tyrtyshnikov [301], [302]. They proved that the eigenvectors of $T_n(b)$ are asymptotically distributed as the columns of the Hermitian adjoint of the Fourier matrix,

$$F_n^* = \frac{1}{\sqrt{n}} (e^{-i\frac{2\pi}{n}jk})_{j,k=0}^{n-1} =: (\mathrm{col}_1^{(n)}, \dots, \mathrm{col}_n^{(n)}),$$

in the following sense: For each $\varepsilon > 0$, the number of integers $k \in \{1, \dots, n\}$ for which $\min_\lambda \|T_n(b - \lambda)\, \mathrm{col}_k^{(n)}\|_2 > \varepsilon$ is $o(n)$ as $n \to \infty$.

We here consider an entirely different question. Take a λ_n in $\operatorname{sp} T_n(b)$ for each $n \ge 1$ and suppose the points λ_n converge to some point λ. Clearly, λ belongs to the limiting set $\Lambda(b)$ studied in Chapter 11. Conversely, given any $\lambda \in \Lambda(b)$, we can find $\lambda_n \in \operatorname{sp} T_n(b)$ such that $\lambda_n \to \lambda$. If λ is an eigenvalue of $T(b)$, are the eigenvectors of $T_n(b)$ corresponding to the eigenvalues λ_n related to the eigenvectors of $T(b)$ corresponding to λ? What can be said about the eigenvectors of $T_n(b)$ for λ_n in the case where λ is not an eigenvalue of $T(b)$?

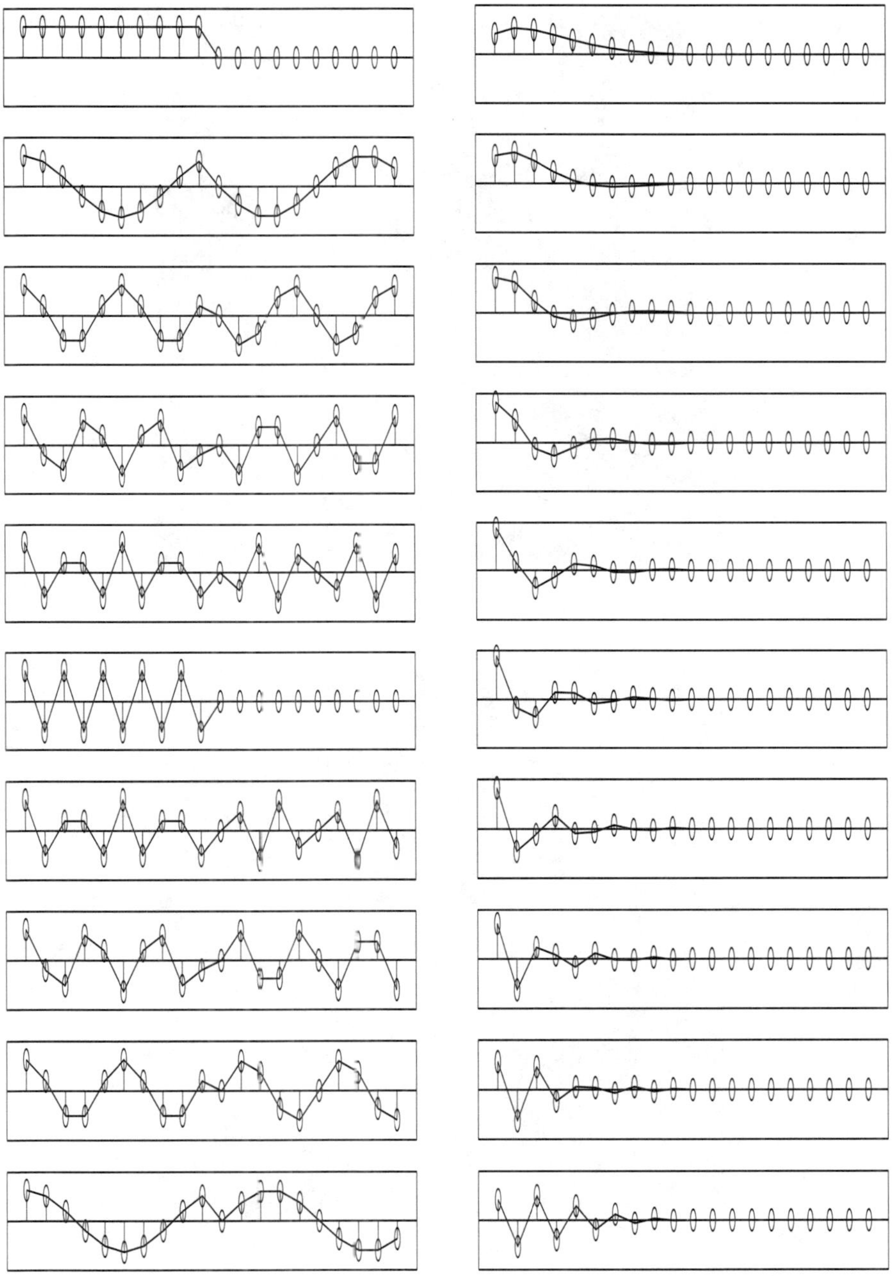

Figure 12.1. *Eigenvectors of a tridiagonal circulant matrix (left) and a tridiagonal Toeplitz matrix (right).*

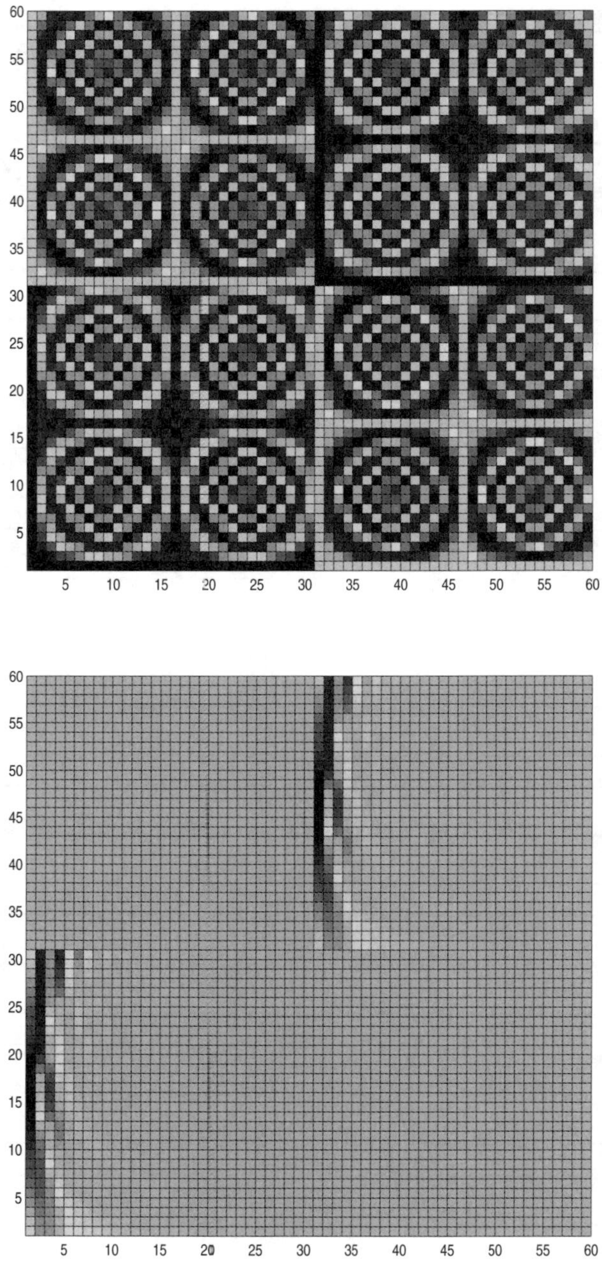

Figure 12.2. *A pseudocolor plot of the real eigenvector basis of a tridiagonal circulant matrix (top) and a tridiagonal Toeplitz matrix (bottom).*

Eigenvectors of infinite Toeplitz matrices were investigated in Section 1.8. We repeat Proposition 1.20 in slightly modified form.

Proposition 12.1. *Let* $\lambda \in \operatorname{sp} T(b)$ *and suppose that* $\lambda \notin b(\mathbf{T})$*. The point* λ *is an eigenvalue of* $T(b)$ *as an operator on* ℓ^2 *if and only if* wind $(b, \lambda) = -m < 0$*. In that case* $\dim \operatorname{Ker} T(b-\lambda) = m$ *and* $\operatorname{Ker} T(b-\lambda) = \operatorname{lin}\{x_0, Vx_0, \ldots, V^{m-1}x_0\}$*, where* V *is the shift operator*

$$V : \ell^2 \to \ell^2, \{w_1, w_2, \ldots\} \mapsto \{0, w_1, w_2, \ldots\}$$

and $x_0 = \{x_k^{(0)}\}_{k=1}^{\infty}$ *is an exponentially decaying sequence with* $x_1^{(0)} = 1$.

Proof. The proof of Proposition 1.20 gives

$$\operatorname{Ker} T(b-\lambda) = \operatorname{lin}\{T(b_+^{-1})e_0, \ldots, T(b_+^{-1})e_{m-1}\}.$$

We may think of $T(b_+^{-1})e_j$ as the jth column of the lower-triangular Toeplitz matrix $T(b_+^{-1})$. As the 0th Fourier coefficient of b_+^{-1} may be assumed to be 1 and the Fourier coefficients of b_+^{-1} decay exponentially, we get all assertions. □

Triangular Toeplitz matrices. Suppose $b(t) = b_0 + b_{-1}t^{-1} + \cdots + b_{-r}t^{-r}$ with $r \geq 1$ and $b_{-r} \neq 0$. The matrix $T(b)$ is upper triangular. It is clear that $\operatorname{sp} T_n(b) = \{b_0\}$ for all n and hence $\Lambda(b) = \{b_0\}$. We assume that $b_0 \notin b(\mathbf{T})$. Let $b_{-1} = \cdots = b_{-m_0+1} = 0$ and $b_{-m_0} \neq 0$. Then

$$b(t) - b_0 = b_{-m_0}t^{-r}(t-\delta_1)\ldots(t-\delta_p)(t-\mu_1)\ldots(t-\mu_q),$$

where $|\delta_k| < 1$ and $|\mu_k| > 1$ for all k. Since $p+q = r-m_0$, we get

$$\operatorname{wind}(b, b_0) = -r + p = -m_0 - q \leq -m_0 \leq -1. \tag{12.5}$$

Thus, by Proposition 12.1, $\operatorname{Ker} T(b-b_0)$ has the dimension m_0+q and from the proofs of Propositions 1.20 and 12.1 we infer that

$$\operatorname{Ker} T(b-b_0) = \operatorname{lin}\{T(b_+^{-1})e_1, \ldots, T(b_+^{-1})e_{m_0+q}\}$$

with

$$b_+^{-1}(t) = \prod_{k=1}^{q}\left(1 + \frac{t}{\mu_1} + \frac{t^2}{\mu_1^2} + \cdots\right).$$

On the other hand, it is easily seen that $\operatorname{Ker} T_n(b-b_0) = \operatorname{lin}\{e_1, \ldots, e_{m_0}\}$. Consequently, in general $\operatorname{Ker} T_n(b-b_0)$ and $\operatorname{Ker} T(b-b_0)$ are in no way related. If $b_{-1} \neq 0$ ($\Leftrightarrow m_0 = 1$), which is the generic case, then $\operatorname{Ker} T_n(b-b_0)$ is spanned by e_1. If, in addition, wind $(b, b_0) = -1$, then $q = 0$ and hence $\operatorname{Ker} T(b-b_0)$ is also spanned by e_1. This is in accordance with Theorem 12.3, which will be proved below. However, in the case where $b_{-1} \neq 0$ and wind $(b, b_0) \leq -2$, no eigenvector of $T(b)$ that is outside $\operatorname{lin}\{e_1\}$ can be approximated by eigenvectors of $T_n(b)$.

If $b(t) = b_0 + b_1t + \cdots + b_st^s$ ($s \geq 1$, $b_s \neq 0$), then $T(b)$ is lower triangular. Again $\operatorname{sp} T_n(b) = \{b_0\}$ for all values of n and $\Lambda(b) = \{b_0\}$. From (12.5) we obtain that

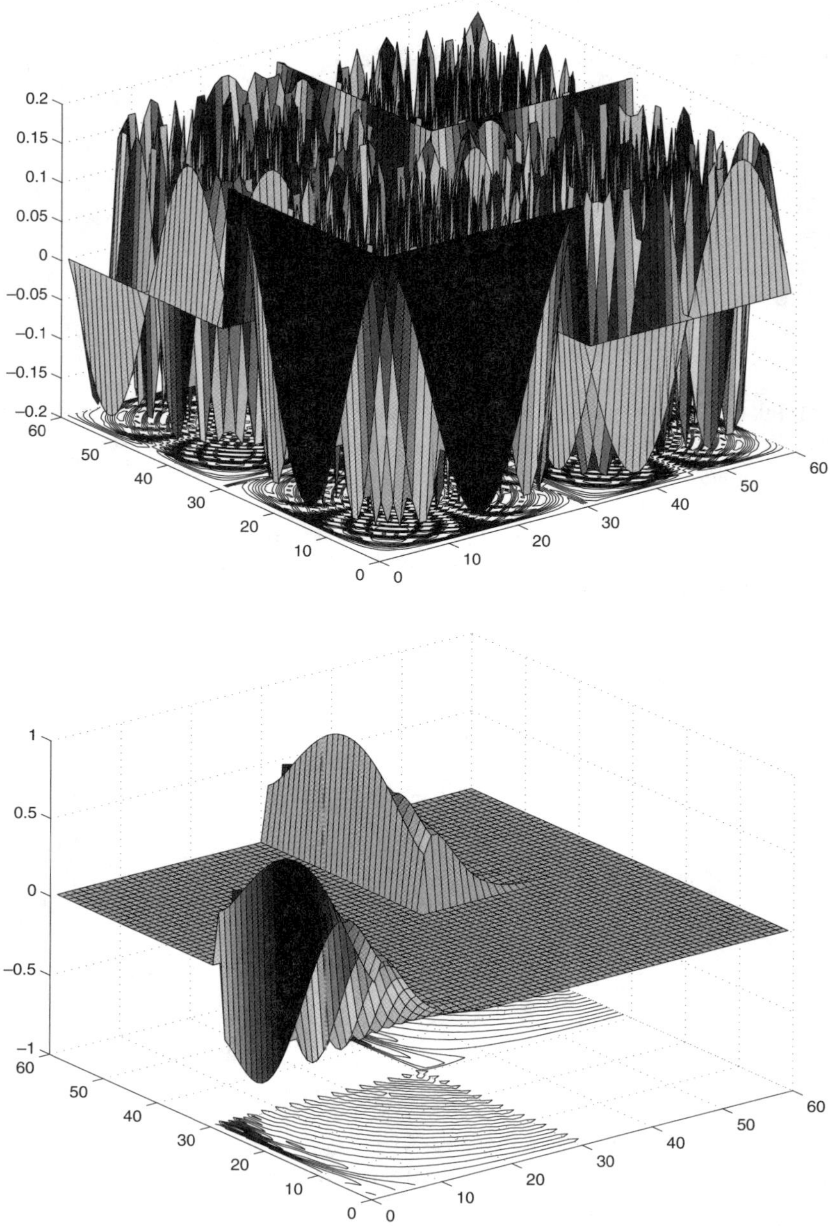

Figure 12.3. *A surface plot of the real eigenvector basis of a tridiagonal circulant matrix (top) and a tridiagonal Toeplitz matrix (bottom).*

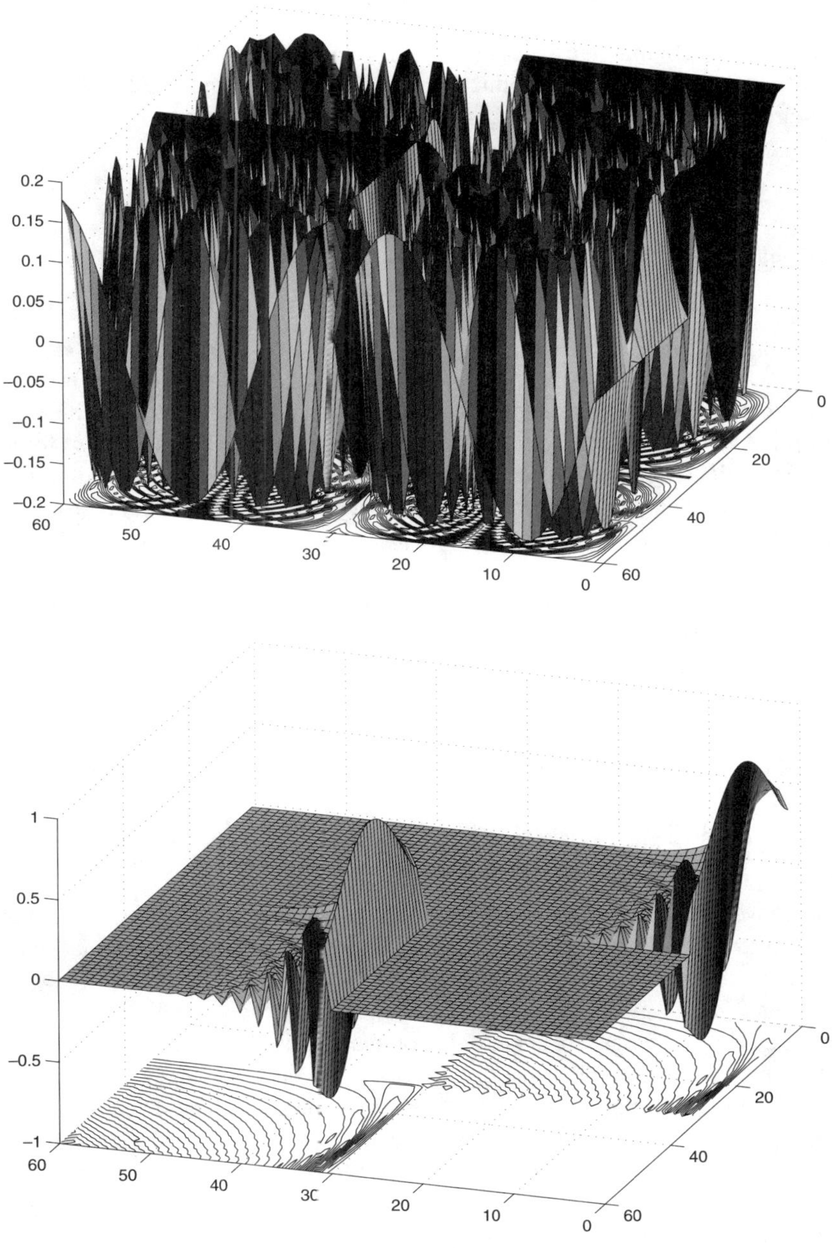

Figure 12.4. *Another view at the surface plots of Figure* 12.3.

$\operatorname{wind}(b, b_0) = -\operatorname{wind}(\widetilde{b}, b_0) \geq 1$, and hence $\operatorname{Ker} T(b - b_0) = \{0\}$ by virtue of Proposition 12.1. If $b_1 = \cdots = b_{m_0-1} = 0$ and $b_{m_0} \neq 0$, then $\operatorname{Ker} T_n(b - b_0) = \operatorname{lin}\{e_n, \ldots, e_{n-m_0+1}\}$.

Tridiagonal Toeplitz matrices. Let $b(t) = b_{-1}t^{-1} + b_0 + b_1 t$. Fix one of the two values of $\sqrt{b_{-1}/b_1}$ and denote it by α. Define $\sqrt{b_1 b_{-1}}$ as $b_1\alpha$. By Theorem 2.4, the eigenvalues of $T_n(b)$ are

$$\lambda_j = b_0 + 2\sqrt{b_1 b_{-1}} \cos \frac{\pi j}{n+1} \quad (j = 1, \ldots, n),$$

and an eigenvector for λ_j is $x_{j,n} = (x_{j,k}^{(n)})_{k=1}^n$ with

$$x_{j,k}^{(n)} = \left(\frac{1}{\alpha}\right)^{k-1} \sin \frac{k\pi j}{n+1} \Big/ \sin \frac{\pi j}{n+1} \quad (k = 1, \ldots, n).$$

It follows that $\Lambda(b)$ is the line segment $[\, b_0 - 2\sqrt{b_1 b_{-1}}, b_0 + 2\sqrt{b_1 b_{-1}}\,]$. The range $b(\mathbf{T})$ is an ellipse with the foci $b_0 \pm 2\sqrt{b_1 b_{-1}}$ for $|\alpha| \neq 1$ and the line segment $\Lambda(b)$ for $|\alpha| = 1$. Consequently, the spectrum of $T(b)$ consists of the points on the ellipse and in its interior for $|\alpha| \neq 1$ and coincides with the line segment $\Lambda(b)$ for $|\alpha| = 1$. If $|\alpha| < 1$, then $\operatorname{wind}(b, \lambda) = 1$ for all $\lambda \in \Lambda(b)$, while if $|\alpha| > 1$, we have $\operatorname{wind}(b, \lambda) = -1$ for all $\lambda \in \Lambda(b)$.

Now pick a point $\lambda = b_0 + 2\sqrt{b_1 b_{-1}} \cos\theta \in \Lambda(b)$ $(\theta \in [0, \pi])$ and chose any $j_n \in \{1, \ldots, n\}$ such that $\pi j_n/(n+1) \to \theta$ as $n \to \infty$. Then

$$\lambda_n := b_0 + 2\sqrt{b_1 b_{-1}} \cos \frac{\pi j_n}{n+1} \in \operatorname{sp} T_n(b)$$

and $\lambda_n \to \lambda$. For the kth component of the eigenvector $x_n := x_{j_n,n}$ we obtain

$$x_k^{(n)} = \left(\frac{1}{\alpha}\right)^{k-1} \sin \frac{k\pi j_n}{n+1} \Big/ \sin \frac{\pi j_n}{n+1} \to \left(\frac{1}{\alpha}\right)^{k-1} \frac{\sin k\theta}{\sin\theta} =: x_k^{(0)} \quad \text{as } n \to \infty,$$

with the convention that $\sin(k\theta) / \sin\theta = k$ in the cases $\theta = 0$ and $\theta = \pi$. Put $x_0 = \{x_k^{(0)}\}_{k=1}^{\infty}$. If $|\alpha| \leq 1$, then $x_0 \notin \ell^2$, and $T(b)$ is known to have no eigenvalues in this case (Proposition 12.1 for $|\alpha| < 1$ and Theorem 1.31 for $|\alpha| = 1$). However, if $|\alpha| > 1$, then $x_0 \in \ell^2$, $\operatorname{Ker} T(b - \lambda)$ is one dimensional and x_0 is an eigenvector of $T(b)$ for λ. Clearly, if $|\alpha| > 1$, then x_n converges to x_0 not only componentwise but even in ℓ^2.

The conclusion is as follows. Let $\lambda \in \Lambda(b)$, $\lambda_n \in \operatorname{sp} T_n(b)$, and $\lambda_n \to \lambda$. The matrix $T_n(b)$ has exactly one eigenvector x_n for λ_n satisfying the normalization condition $x_1^{(n)} = 1$. If λ is an eigenvalue of $T(b)$ and x_0 is an eigenvector for λ, then x_0 can be normalized so that $x_1^{(0)} = 1$. In the case where λ is an eigenvalue of $T(b)$, the vectors x_n converge to x_0 in ℓ^2, but in the case where λ is not an eigenvalue of $T(b)$, the vectors x_n converge componentwise to some vector x_0 that does not belong to the space ℓ^2.

12.3 Asymptotics of Eigenvectors

Now take $\lambda \in \mathbf{C} \setminus b(\mathbf{T})$ and suppose $\operatorname{wind}(b, \lambda) = -m < 0$. We then can write $b - \lambda = \chi_{-m}c$, where $\chi_k(t) := t^k$ $(t \in \mathbf{T})$ and c is a Laurent polynomial without zeros on $\mathbf{T}$ and with wind $(c, 0) = 0$. Let $c(t) = \sum_j c_j t^j$ $(t \in \mathbf{T})$. Since $0 \notin c(\mathbf{T})$ and wind $(c, 0) = 0$, it follows that the operator $T(c)$ is invertible on ℓ^2, that the matrices $T_n(c)$ are invertible for all sufficiently large n, and that $T_n^{-1}(c)P_n$ converges strongly to $T^{-1}(c)$ on ℓ^2 (Corollaries 1.11 and 3.8).

Lemma 12.2. *Let $\lambda \in \operatorname{sp} T_n(b)$. Suppose $\lambda \notin b(\mathbf{T})$ and* wind $(b, \lambda) = -m < 0$. *A vector $x = (x_j)_{j=1}^n$ belongs to* Ker $T_n(b - \lambda)$ *if and only if*

$$-T_n(c)\begin{pmatrix} x_{m+1} \\ x_{m+2} \\ \vdots \\ x_n \\ 0 \\ \vdots \\ 0 \end{pmatrix} = x_m \begin{pmatrix} c_1 \\ c_2 \\ \vdots \\ c_n \end{pmatrix} + \cdots + x_1 \begin{pmatrix} c_m \\ c_{m+1} \\ \vdots \\ c_{m+n-1} \end{pmatrix}. \tag{12.6}$$

Proof. We have $b - \lambda = c\chi_{-m}$ and hence, by Proposition 3.10,

$$T_n(b - \lambda) = T_n(c\chi_{-m}) = T_n(c)T_n(\chi_{-m}) + P_n H(c) H(\chi_m) P_n.$$

It follows that $T_n(b - \lambda)x = 0$ if and only if

$$-T_n(c)T_n(\chi_{-m})x = P_n H(c) H(\chi_m) P_n x,$$

which is the same as (12.6). □

The following result shows that things are remarkably nice in the case $m = 1$.

Theorem 12.3. *Let $\lambda \in \Lambda(b)$, $\lambda_n \in \operatorname{sp} T_n(b)$, $\lambda_n \to \lambda$. Suppose that $\lambda \notin b(\mathbf{T})$ and that* wind $(b, \lambda) = -1$. *Then there exist $n_0 \in \mathbf{N}$, $x_n \in \mathbf{C}^n$ $(n \geq n_0)$, and $x_0 \in \ell^2$ such that $x_1^{(n)} = 1$ and*

$$\operatorname{Ker} T_n(b - \lambda_n) = \operatorname{lin}\{x_n\}, \quad \operatorname{Ker} T(b - \lambda) = \operatorname{lin}\{x_0\}, \quad x_n \to x_0 \ \text{in} \ \ell^2.$$

Proof. We have $b - \lambda = \chi_{-1}c$ with c as above and

$$b - \lambda_n = b - \lambda + \lambda - \lambda_n = \chi_{-1}(c + (\lambda - \lambda_n)\chi_1) =: \chi_{-1}(c + \delta_n \chi_1).$$

Since $T_n(c + \delta_n \chi_1) = T_n(c)\,[I + \delta_n T_n^{-1}(c)T_n(\chi_1)]$ and $\|\delta_n T_n^{-1}(c)T_n(\chi_1)\| \leq 1/2$ for all n large enough, we see that $T_n(c + \delta_n \chi_1)$ is invertible for all sufficiently large n and that $T_n^{-1}(c + \delta_n \chi_1)P_n$ converges strongly to $T^{-1}(c)$ on ℓ^2.

Lemma 12.2 shows that $x_n = (x_j^{(n)})_{j=1}^n$ is in Ker $T_n(b - \lambda_n)$ if and only if

$$\begin{pmatrix} x_2^{(n)} \\ x_3^{(n)} \\ \vdots \\ x_n^{(n)} \\ 0 \end{pmatrix} = -T_n^{-1}(c + \delta_n \chi_1)\, x_1^{(n)} \begin{pmatrix} c_1 + \delta_n \\ c_2 \\ \vdots \\ c_n \end{pmatrix}. \tag{12.7}$$

Since $\dim \operatorname{Ker} T_n(b-\lambda_n) \ge 1$, the system (12.7) must have a solution. As equation (12.7) determines $x_2^{(n)}, \dots, x_n^{(n)}$ uniquely once $x_1^{(n)}$ is given, it follows that $\dim \operatorname{Ker} T_n(b-\lambda_n) = 1$ and that $\operatorname{Ker} T_n(b-\lambda_n) = \operatorname{lin}\{x_n\}$, where $x_n = (x_j^{(n)})_{j=1}^n$ is specified by $x_1^{(n)} = 1$ and

$$\begin{pmatrix} x_2^{(n)} \\ x_3^{(n)} \\ \vdots \\ x_n^{(n)} \\ 0 \end{pmatrix} = -T_n^{-1}(c+\delta_n\chi_1) \begin{pmatrix} c_1+\delta_n \\ c_2 \\ \vdots \\ c_n \end{pmatrix}. \tag{12.8}$$

The right-hand side of (12.8) converges to

$$-T^{-1}(c) \begin{pmatrix} c_1 \\ c_2 \\ \vdots \end{pmatrix}.$$

Consequently, x_n converges in ℓ^2 to $x_0 = \{x_j^{(0)}\}_{j=1}^\infty$ with $x_1^{(0)} = 1$ and

$$\begin{pmatrix} x_2^{(0)} \\ x_3^{(0)} \\ \vdots \end{pmatrix} = -T^{-1}(c) \begin{pmatrix} c_1 \\ c_2 \\ \vdots \end{pmatrix}.$$

This can be written as $T(c)T(\chi_{-1})x_0 + H(c)H(\chi_1)x_0 = 0$, which is equivalent to the equality $0 = T(c\chi_{-1})x_0 = T(b-\lambda)x_0$. Thus, x_0 is a nonzero element of $\operatorname{Ker} T(b-\lambda)$. Because $\dim \operatorname{Ker} T(b-\lambda) = 1$ by Proposition 12.1, it follows that $\operatorname{Ker} T(b-\lambda) = \operatorname{lin}\{x_0\}$. □

Example 12.4. This example illustrates Theorem 12.3. Let

$$b(t) = i\,t^2 + 2 + t^{-1} + \frac{1}{2}t^{-2} - 2t^{-3}.$$

The range $b(\mathbf{T})$ and $\operatorname{sp} T_{50}(b)$ are shown in Figure 12.5. We consider the following points $\lambda_n \in \operatorname{sp} T_n(b)$:

$$\begin{aligned}
\lambda_{10} &= 4.4190 + 0.2617i \in \operatorname{sp} T_{10}(b), \\
\lambda_{20} &= 4.7423 + 0.1834i \in \operatorname{sp} T_{20}(b), \\
\lambda_{30} &= 4.8177 + 0.1639i \in \operatorname{sp} T_{30}(b), \\
\lambda_{40} &= 4.8463 + 0.1563i \in \operatorname{sp} T_{40}(b), \\
\lambda_{50} &= 4.8601 + 0.1526i \in \operatorname{sp} T_{50}(b).
\end{aligned}$$

These points are the beginning of a sequence of points $\lambda_n \in \operatorname{sp} T_n(b)$ that converge to some point $\lambda \in \Lambda(b)$, which is indicated by the arrow in Figure 12.5. For each of these λ_n, we compute an eigenvector $x_n \in \mathbf{C}^n$ of $T_n(b)$ and normalize it so that its first component is 1. By Theorem 12.3, the vectors x_n must converge to some $x_0 \in \ell^2$. This is convincingly seen

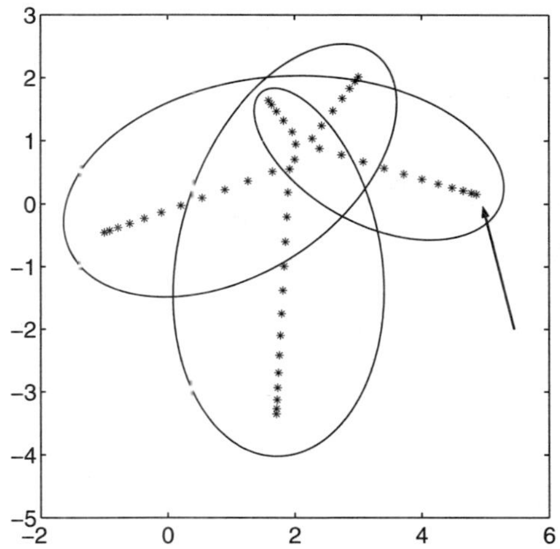

Figure 12.5. *The range* $b(\mathbf{T})$ *and the* 50 *eigenvalues of* $T_{50}(b)$ *for the symbol of Example* 12.4. *The arrow points to a point in* $\Lambda(b)$ *that is approximated by points* $\lambda_n \in \operatorname{sp} T_n(b)$.

in Figure 12.6, which shows the real part of x_n in the left column and the imaginary part of x_n in the right. The top row corresponds to $n = 10$, the bottom row to $n = 50$. □

We now turn to the case where wind $(b, \lambda) = -m \le -2$. We will then prove that generically the kernels of $T_n(b - \lambda_n)$ are one dimensional, $T_n(b - \lambda_n) = \operatorname{lin}\{x_n\}$, that the vectors $x_n \in \mathbf{C}^n$ converge to some $x_0 \in \operatorname{Ker} T(b - \lambda)$, and that the limits x_0 corresponding to different choices of the sequence $\lambda_n \to \lambda$ all belong to a single one-dimensional subspace of $\operatorname{Ker} T(b - \lambda)$.

Given $\lambda \in \Lambda(b)$, we write the factorization (12.4) now in the form

$$b(t) - \lambda = b_s t^{-r}(t - z_1^{(0)}) \dots (t - z_{r+s}^{(0)}), \quad |z_1^{(0)}| \le \dots \le |z_{r+s}^{(0)}|.$$

Since $\Lambda(b) = \{\lambda \in \mathbf{C} : |z_r(\lambda)| = |z_{r+1}(\lambda)|\}$ by virtue of Theorem 11.17, the point λ belongs to $\Lambda(b)$ if and only if $|z_r^{(0)}| = |z_{r+1}^{(0)}|$.

Theorem 12.5. *Let* $r \ge 1$, $s \ge 1$, $\lambda \in \Lambda(b)$, $\lambda \notin b(\mathbf{T})$, $\lambda_n \in \operatorname{sp} T_n(b)$, *and* $\lambda_n \to \lambda$. *Suppose* $|z_{r-1}^{(0)}| < |z_r^{(0)}|$. *Then, for all sufficiently large n,*

$$\operatorname{Ker} T_n(b - \lambda_n) = \operatorname{lin}\{x_n\}$$

with certain $x_n = (x_j^{(n)})_{j=1}^n \in \mathbf{C}^n$ *satisfying* $x_1^{(n)} = 1$. *The limits*

$$x_k^{(0)} = \lim_{n \to \infty} x_k^{(n)}$$

exist for all $k \ge 2$. *If* wind $(b, \lambda) \le -1$, *then*

$$\{1, x_2^{(0)}, x_3^{(0)}, \dots\} \in \operatorname{Ker} T(b - \lambda),$$

while if wind $(b, \lambda) \ge 0$, *we have* $\operatorname{Ker} T(b - \lambda) = \{0\}$ *and*

$$\{1, x_2^{(0)}, x_3^{(0)}, \dots\} \notin \ell^2.$$

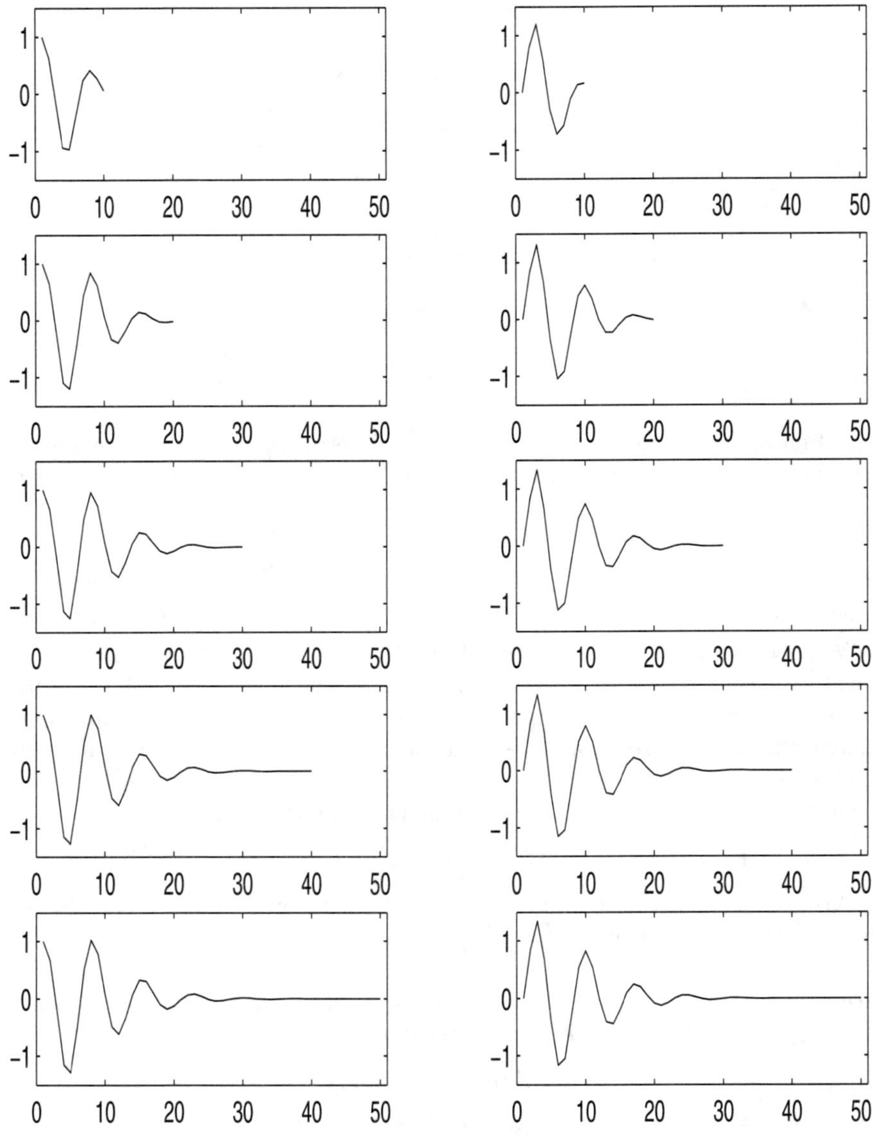

Figure 12.6. *The figure shows the real parts (left) and imaginary parts (right) of eigenvectors of $T_n(b)$ corresponding to the eigenvalues λ_n of Example* 12.4 *for* $n = 10, 20, 30, 40, 50$ *(from the top to the bottom). As predicted by Theorem* 12.3, *these eigenvectors converge to some (infinite) vector in* ℓ^2.

Proof. Recall the definition of b_ϱ given in Section 11.1. The key observation, already used in Chapter 11, is the identity $T_n(b_\varrho - \lambda_n) = D_\varrho T_n(b - \lambda_n) D_\varrho^{-1}$, where $D_\varrho = \operatorname{diag}(1, \varrho, \dots, \varrho^{n-1})$. This identity implies that

$$\operatorname{sp} T_n(b) = \operatorname{sp} T_n(b_\varrho), \quad \operatorname{Ker} T_n(b - \lambda_n) = D_\varrho^{-1} \operatorname{Ker} T_n(b_\varrho - \lambda_n). \tag{12.9}$$

We have

$$\begin{aligned} b_\varrho(t) - \lambda &= b_s \varrho^{-r} t^{-r} (\varrho t - z_1^{(0)}) \dots (\varrho t - z_{r+s}^{(0)}) \\ &= b_s(-z_1^{(0)}) \dots (-z_{r+s}^{(0)}) \varrho^{-r} t^{-r} \left(1 - \frac{\varrho t}{z_1^{(0)}}\right) \dots \left(1 - \frac{\varrho t}{z_{r+s}^{(0)}}\right). \end{aligned}$$

By assumption, we can choose a $\varrho \in (0, \infty)$ such that

$$\frac{\varrho}{|z_1^{(0)}|} \ge \dots \ge \frac{\varrho}{|z_{r-1}^{(0)}|} > 1 > \frac{\varrho}{|z_r^{(0)}|} \ge \dots \ge \frac{\varrho}{|z_{r+s}^{(0)}|}.$$

This gives $\operatorname{wind}(1 - \varrho t / z_j^{(0)}) = 1$ for $j = 1, \dots, r-1$ and $\operatorname{wind}(1 - \varrho t / z_j^{(0)}) = 0$ for $j = r, \dots, r+s$, and hence $\operatorname{wind}(b_\varrho, \lambda) = -r + r - 1 = -1$. Thus, we can use Theorem 12.3 to conclude that there are $y_n \in \mathbf{C}^n$ and $y_0 \in \ell^2$ such that $y_1^{(n)} = 1$ and

$$\operatorname{Ker} T_n(b_\varrho - \lambda_n) = \operatorname{lin}\{y_n\}, \quad \operatorname{Ker} T(b_\varrho - \lambda) = \operatorname{lin}\{y_0\}, \quad y_n \to y_0 \text{ in } \ell^2.$$

From (12.9) we obtain that $x_n = (x_j^{(n)})_{j=1}^n \in \operatorname{Ker} T_n(b - \lambda_n)$ if and only if $x_j^{(n)} = \varrho^{-j} y_j^{(n)}$ with $(y_j^{(n)})_{j=1}^n$ in $\operatorname{Ker} T_n(b_\varrho - \lambda_n)$. As

$$\operatorname{Ker} T_n(b_\varrho - \lambda_n) = \operatorname{lin}\{(1, y_2^{(n)}, \dots, y_n^{(n)})\},$$

where $y_k^{(n)} \to y_k$ as $n \to \infty$ for each k, it follows that

$$\operatorname{Ker} T_n(b - \lambda_n) = \operatorname{lin}\{(1, x_2^{(n)}, \dots, x_n^{(n)})\} \tag{12.10}$$

with $x_k^{(n)} = \varrho^{-k} y_k^{(n)} \to \varrho^{-k} y_k =: x_k^{(0)}$ as $n \to \infty$ for each k.

Now suppose $\operatorname{wind}(b, \lambda) = -m \le -1$. By Lemma 12.2 and (12.10),

$$\begin{pmatrix} x_{m+1}^{(n)} \\ \vdots \\ x_n^{(n)} \\ 0 \\ \vdots \\ 0 \end{pmatrix} = -T_n^{-1}(c + \delta_n \chi_m) \left[x_m^{(n)} \begin{pmatrix} c_1 \\ \vdots \\ c_m + \delta_n \\ \vdots \\ c_n \end{pmatrix} + \dots + x_1^{(n)} \begin{pmatrix} c_m + \delta_n \\ \vdots \\ \vdots \\ \vdots \\ c_{m+n-1} \end{pmatrix} \right]$$

with $x_1^{(n)} = 1$. As

$$x_2^{(n)} \to x_2^{(0)}, \dots, x_m^{(n)} \to x_m^{(0)}, \quad \delta_n \to 0, \quad T_n^{-1}(c + \delta_n \chi_m) \to T^{-1}(c) \text{ (strongly)},$$

we obtain that

$$\{x_{m+1}^{(n)}, \ldots, x_n^{(n)}, 0, \ldots\} \to \{x_{m+1}^{(0)}, x_{m+2}^{(0)}, \ldots\}$$

in ℓ^2. Thus, $x_n \to x_0$ in ℓ^2. Because $T_n(b - \lambda_n)x_n = 0$, it follows that $T(b - \lambda)x_0 = 0$, that is, x_0 belongs to $\operatorname{Ker} T(b - \lambda)$.

Finally, let $\operatorname{wind}(b, \lambda) \geq 0$. Then $\operatorname{Ker} T(b - \lambda) = \{0\}$ by Proposition 12.1. Put $x_j^{(n)} = 0$ and $x_j^{(0)} = 0$ for $j < 0$. The kth equation of the system $T_n(b - \lambda_n)x_n = 0$ reads

$$b_s x_{k-s}^{(n)} + \cdots + b_1 x_{k-1}^{(n)} + (b_0 - \lambda_n)x_k^{(n)} + b_{-1}x_{k+1}^{(n)} + \cdots + b_{-r}x_{k+r}^{(n)} = 0.$$

Passage to the limit $n \to \infty$ gives

$$b_s x_{k-s}^{(0)} + \cdots + b_1 x_{k-1}^{(0)} + (b_0 - \lambda)x_k^{(0)} + b_{-1}x_{k+1}^{(0)} + \cdots + b_{-r}x_{k+r}^{(0)} = 0,$$

which is the kth equation of the infinite system $T(b - \lambda)x_0 = 0$. Thus, if the sequence $\{1, x_2^{(0)}, x_3^{(0)}, \ldots\}$ would belong to ℓ^2, then it would be in $\operatorname{Ker} T(b - \lambda)$, which is impossible. □

Example 12.6. We let b be as in Example 12.4, but we now consider the transpose matrices $T_n(\widetilde{b})$. Choose $\lambda_n \in \operatorname{sp} T_n(\widetilde{b}) = \operatorname{sp} T_n(b)$ and the limiting point $\lambda \in \Lambda(\widetilde{b}) = \Lambda(b)$ exactly as in Example 12.4. While $\operatorname{wind}(b, \lambda) = -1$ in Example 12.4, we now have $\operatorname{wind}(\widetilde{b}, \lambda) = 1$, and hence Theorem 12.5 implies that the corresponding eigenvectors (with first component equal to 1) do not converge in ℓ^2. Figure 12.7 indicates that this is indeed the case. This figure also shows that nevertheless, as predicted by Theorem 12.5, for each fixed k the kth components of the eigenvectors converge. □

We finally remark that the assumption of Theorem 12.5 (that is, the requirement that $|z_{r-1}^{(0)}| < |z_r^{(0)}|$) is generically satisfied in the following sense. Let $\mathcal{P}$ denote the set of all Laurent polynomials on $\mathbf{T}$ and equip $\mathcal{P}$ with the L^∞ metric. Then the set of all $b \in \mathcal{P}$ satisfying the condition of Theorem 12.5 is open and dense in $\mathcal{P}$.

12.4 Pseudomodes of Circulant Matrices

Let A be a bounded linear operator on a complex Hilbert space $\mathcal{H}$. A point λ in $\mathbf{C}$ is said to be an *ε-pseudoeigenvalue* of A if $\|(A - \lambda I)^{-1}\| \geq 1/\varepsilon$ (with the convention that $\|(A - \lambda I)^{-1}\| := \infty$ in case $A - \lambda I$ is not invertible). In the language of Section 7.1, the ε-pseudoeigenvalues are just the points of the ε-pseudospectrum $\operatorname{sp}_\varepsilon A$. If λ is an ε-pseudoeigenvalue, then there exists a nonzero $x \in \mathcal{H}$ such that $\|(A - \lambda I)x\| \leq \varepsilon\|x\|$. Each such x is called an *ε-pseudomode* (or ε-pseudoeigenvector) for A at λ.

Now suppose we are given a sequence $\{A_n\}_{n=1}^{\infty}$ of matrices $A_n \in \mathbf{C}^{n\times n}$. We think of A_n as an operator on $\mathbf{C}^n$ with the ℓ^2 norm. We call a point $\lambda \in \mathbf{C}$ an *asymptotically good pseudoeigenvalue* for $\{A_n\}$ if $\|(A_n - \lambda I)^{-1}\|_2 \to \infty$ as $n \to \infty$. In that case we can find nonzero vectors $x_n \in \mathbf{C}^n$ satisfying

$$\|(A_n - \lambda I)x_n\|_2 / \|x_n\|_2 \to 0 \text{ as } n \to \infty,$$

and each sequence $\{x_n\}$ with this property will be called an *asymptotically good pseudomode* for $\{A_n\}$ at λ.

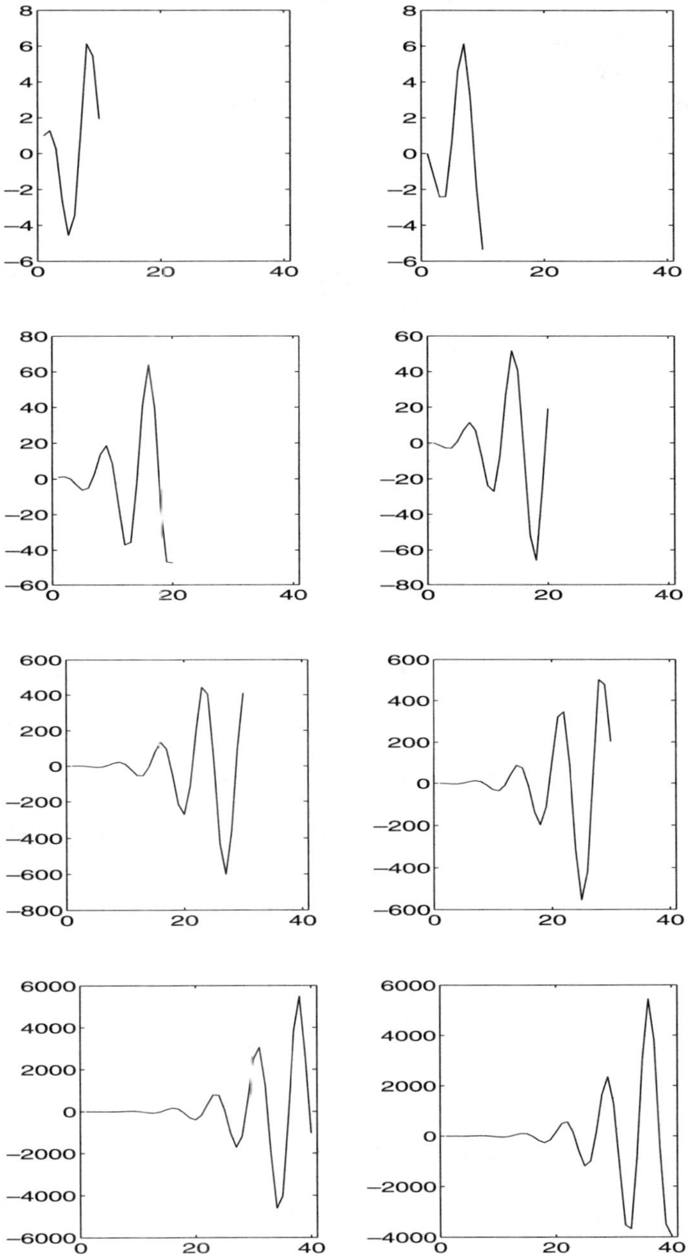

Figure 12.7. *This figure is the analogue of the upper eight pictures of Figure* 12.6 *and shows the real parts (left) and imaginary parts (right) of eigenvectors of* $T_n(\widetilde{b})$ *corresponding to the eigenvalues* λ_n *of Example* 12.6 *for* $n = 10, 20, 30, 40$ *(from the top to the bottom). Notice the different scales of the vertical axes.*

This and the following section are devoted to the structure of asymptotically good pseudomodes for sequences constituted by the circulant cousins of Toeplitz band matrices (called α-matrices in theoretical chemistry [303]) and by Toeplitz matrices themselves.

Given a subset J_n of $\{1, 2, \dots, n\}$, we denote by P_{J_n} the projection on $\mathbf{C}^n$ defined by

$$(P_{J_n} y)_j = \begin{cases} y_j & \text{for} \quad j \in J_n, \\ 0 & \text{for} \quad j \notin J_n. \end{cases} \tag{12.11}$$

The number of elements in J_n will be denoted by $|J_n|$. Let $\{y_n\}_{n=1}^\infty$ be a sequence of nonzero vectors $y_n \in \mathbf{C}^n$. We say that $\{y_n\}$ is *asymptotically localized* if there exists a sequence $\{J_n\}_{n=1}^\infty$ of sets $J_n \subset \{1, \dots, n\}$ such that

$$\lim_{n\to\infty} \frac{|J_n|}{n} = 0 \quad \text{and} \quad \lim_{n\to\infty} \frac{\|P_{J_n} y_n\|_2}{\|y_n\|_2} = 1.$$

We denote by $F_n \in \mathbf{C}^{n\times n}$ the Fourier matrix:

$$F_n = \frac{1}{\sqrt{n}} \left(\omega_n^{jk}\right)_{j,k=0}^{n-1}, \quad \omega_n := e^{2\pi i/n}.$$

A sequence $\{y_n\}_{n=1}^\infty$ of nonzero vectors $y_n \in \mathbf{C}^n$ will be called *asymptotically extended* if $\{F_n y_n\}$ is asymptotically localized.

We define the circulant matrix $C_n(b)$ as in Section 2.1.

Theorem 12.7. *Let b be a Laurent polynomial. A point $\lambda \in \mathbf{C}$ is an asymptotically good pseudoeigenvalue for $\{C_n(b)\}$ if and only if $\lambda \in b(\mathbf{T})$, in which case every asymptotically good pseudomode for $\{C_n(b)\}$ is asymptotically extended.*

Proof. Clearly, $C_n(b) - \lambda I = C_n(b - \lambda)$. We know from Proposition 2.1 that

$$C_n(b-\lambda) = F_n^* \operatorname{diag}\left(b(\omega_n^j) - \lambda\right)_{j=0}^{n-1} F_n =: F_n^* D_n F_n.$$

Since F_n is unitary, it follows that

$$\|C_n^{-1}(b-\lambda)\|_2 = \frac{1}{\min\limits_{0\le j\le n-1} |b(\omega_n^j) - \lambda|},$$

which shows that $\|C_n^{-1}(b-\lambda)\|_2 \to \infty$ if and only if $\lambda \in b(\mathbf{T})$.

Now pick $\lambda \in b(\mathbf{T})$ and suppose $\{x_n\}$ is an asymptotically good pseudomode for $\{C_n(b)\}$ at λ. We may without loss of generality assume that $\|x_n\|_2 = 1$. Put $y_n = (y_j^{(n)})_{j=1}^n = F_n x_n$. We have

$$\|C_n(b-\lambda)x_n\|_2 = \|F_n^* D_n F_n x_n\|_2 = \|D_n y_n\|_2. \tag{12.12}$$

Fix an $\varepsilon > 0$. For $\delta > 0$, we put

$$G_n(\delta) = \{j \in \{1, \dots, n\} : |b(\omega_n^{j-1}) - \lambda| \le \delta\},$$
$$E(\delta) = \{\theta \in [0, 2\pi) : |b(e^{i\theta}) - \lambda| < \delta\}.$$

Since b is analytic in $\mathbf{C} \setminus \{0\}$, the set $E(\delta)$ is a finite union of intervals. Hence $|G_n(\delta)|/n \to |E(\delta)|/(2\pi)$ as $n \to \infty$, where $|E(\delta)|$ denotes the (length) measure of $E(\delta)$. Because $|E(\delta)| \to 0$ as $\delta \to 0$, there exist $\delta(\varepsilon) > 0$ and $N_1(\varepsilon) \geq 1$ such that $|G_n(\delta(\varepsilon))|/n < \varepsilon$ for all $n \geq N_1(\varepsilon)$. From (12.12) we infer that $\|D_n y_n\|_2^2 \to 0$ as $n \to \infty$. Consequently, $\|D_n y_n\|_2^2 < \varepsilon\delta(\varepsilon)^2$ for all $n \geq N_2(\varepsilon)$. Since

$$\|D_n y_n\|_2^2 = \sum_{j=1}^{n} |b(\omega_n^{j-1}) - \lambda|^2 |y_j^{(n)}|^2 \geq \delta(\varepsilon)^2 \sum_{j \notin G_n(\delta(\varepsilon))} |y_j^{(n)}|^2,$$

it follows that $\sum_{j \notin G_n(\delta(\varepsilon))} |y_j^{(n)}|^2 < \varepsilon$ for $n \geq N_2(\varepsilon)$. Thus, $\|P_{G_n(\delta(\varepsilon))} y_n\|_2^2 > 1 - \varepsilon$ for all $n \geq N_2(\varepsilon)$. Put $n(\varepsilon) = \max(N_1(\varepsilon), N_2(\varepsilon))$.

Now let $\varepsilon_k = 1/k$ ($k \geq 2$). With $\delta_k := \delta(\varepsilon_k)$ and $n_k := n(\varepsilon_k)$ we then have

$$\frac{|G_n(\delta_k)|}{n} < \frac{1}{k} \quad \text{and} \quad \|P_{G_n(\delta_k)} y_n\|_2^2 > 1 - \frac{1}{k} \quad \text{for } n \geq n_k. \tag{12.13}$$

We may without loss of generality assume that $1 < n_2 < n_3 < \cdots$. For $1 \leq n < n_2$, we let J_n denote an arbitrary subset of $\{1, \ldots, n\}$. For $n \geq n_2$, we define the sets $J_n \subset \{1, \ldots, n\}$ by

$$\begin{aligned} &J_{n_2} = G_{n_2}(\delta_2),\ J_{n_2+1} = G_{n_2+1}(\delta_2), \ldots, J_{n_3-1} = G_{n_3-1}(\delta_2), \\ &J_{n_3} = G_{n_3}(\delta_3),\ J_{n_3+1} = G_{n_3+1}(\delta_3), \ldots, J_{n_4-1} = G_{n_4-1}(\delta_3), \quad \ldots . \end{aligned}$$

From (12.13) we see that

$$\begin{aligned} &\frac{|J_{n_2}|}{n_2} < \frac{1}{2}, \ldots, \frac{|J_{n_3-1}|}{n_3-1} < \frac{1}{2}, \\ &\frac{|J_{n_3}|}{n_3} < \frac{1}{3}, \ldots, \frac{|J_{n_4-1}|}{n_4-1} < \frac{1}{3}, \quad \ldots, \end{aligned}$$

which shows that $|J_n|/n \to 0$ as $n \to \infty$. Also by (12.13),

$$\begin{aligned} &\|P_{J_{n_2}} y_{n_2}\|_2^2 > 1 - \frac{1}{2}, \ldots, \|P_{J_{n_3-1}} y_{n_3-1}\|_2^2 > 1 - \frac{1}{2}, \\ &\|P_{J_{n_3}} y_{n_3}\|_2^2 > 1 - \frac{1}{3}, \ldots, \|P_{J_{n_4-1}} y_{n_4-1}\|_2^2 > 1 - \frac{1}{3}, \quad \ldots, \end{aligned}$$

and hence $\|P_{J_n} y_n\|_2 \to 1$ as $n \to \infty$. Since $\|y_n\|_2 = 1$ for all n, it results that $\{y_n\}$ is asymptotically localized. Consequently, $\{x_n\}$ is asymptotically extended. □

12.5 Pseudomodes of Toeplitz Matrices

Let b be a Laurent polynomial. Suppose that $\lambda \notin b(\mathbf{T})$ and wind $(b, \lambda) = -m < 0$. We then can write $b - \lambda = c\chi_{-m}$, where $0 \notin c(\mathbf{T})$, wind $(c, 0) = 0$, and χ_k is defined by $\chi_k(t) = t^k$ ($t \in \mathbf{T}$). The operator $T(c)$ is invertible on ℓ^2 and, moreover, the matrices $T_n(c)$ are invertible for all sufficiently large n,

$$\lim_{n\to\infty} \|T_n^{-1}(c)\|_2 = \|T^{-1}(c)\|_2 \quad \text{and} \quad T_n^{-1}(c) P_n \to T^{-1}(c) \text{ strongly} \tag{12.14}$$

(see Corollaries 3.8 and 6.5). We know from the proof of Proposition 1.20 that the m elements

$$u_j := T^{-1}(c)e_j \quad (j = 1, \dots, m) \tag{12.15}$$

form a basis in Ker $T(b - \lambda)$, where $e_j \in \ell^2$ is the sequence whose jth term is 1 and the remaining terms of which are zero.

By Theorem 3.7, each point $\lambda \in \mathbf{C} \setminus b(\mathbf{T})$ with wind $(b, \lambda) \neq 0$ is an asymptotically good pseudoeigenvalue for $\{T_n(b)\}$. The following theorem provides us with a complete description of the structure of asymptotically good pseudomodes.

Theorem 12.8. *Suppose $\lambda \notin b(\mathbf{T})$ and* wind $(b, \lambda) = -m < 0$. *Let $x_n \in \mathbf{C}^n$ be unit vectors. The sequence $\{x_n\}$ is an asymptotically good pseudomode for $\{T_n(b)\}$ at λ if and only if there exist $\gamma_1^{(n)}, \dots, \gamma_m^{(n)} \in \mathbf{C}$ and $z_n \in \mathbf{C}^n$ such that*

$$x_n = \gamma_1^{(n)} P_n u_1 + \cdots + \gamma_m^{(n)} P_n u_m + z_n, \tag{12.16}$$

$$\sup_{n \geq 1,\, 1 \leq j \leq m} |\gamma_j^{(n)}| < \infty, \quad \lim_{n \to \infty} \|z_n\|_2 = 0, \tag{12.17}$$

where $u_1, \dots, u_m$ are given by (12.15).

Proof. Assume that (12.16) and (12.17) hold. Since $\|T_n(b - \lambda)P_n\|_2 \leq \|b - \lambda\|_\infty$, we see that $T_n(b - \lambda)z_n \to 0$. As the numbers $|\gamma_j^{(n)}|$ are bounded by a constant independent of n and as $P_n \to I$ strongly and $T(b - \lambda)u_j = 0$, we obtain that

$$\lim_{n \to \infty} T_n(b - \lambda)x_n = \sum_{j=1}^{m} \lim_{n \to \infty} \gamma_j^{(n)} T_n(b - \lambda)u_j = 0.$$

Thus, $\{x_n\}$ is an asymptotically good pseudomode.

Conversely, suppose $\|T_n(b - \lambda)x_n\|_2 \to 0$. Put $y_n = T_n(b - \lambda)x_n$. With $Q_n = I - P_n$, we have

$$\begin{aligned} T_n(b - \lambda) &= T_n(\chi_{-m}c) = P_n T(\chi_{-m}c)P_n = P_n T(\chi_{-m})T(c)P_n \\ &= P_n T(\chi_{-m})P_n T(c)P_n + P_n T(\chi_{-m})Q_n T(c)P_n =: A_n + B_n. \end{aligned}$$

Since $T(\chi_{-m})$ is nothing but the shift operator $\{\xi_1, \xi_2, \dots\} \mapsto \{\xi_{m+1}, \xi_{m+2}, \dots\}$, it follows that

$$\operatorname{Im} A_n \subset \operatorname{Im} P_{n-m}, \quad \operatorname{Im} B_n \subset \operatorname{Im} P_{\{n-m+1,\dots,n\}}, \tag{12.18}$$

where Im C refers to the image (= range) of the operator C. Also recall (12.11). This implies that

$$\|y_n\|_2^2 = \|A_n x_n + B_n x_n\|_2^2 = \|A_n x_n\|_2^2 + \|B_n x_n\|_2^2,$$

and hence $\|A_n x_n\|_2 \to 0$ because $\|y_n\|_2 \to 0$. The equality $T_n(\chi_{-m})T_n(c)x_n = A_n x_n$ gives

$$T_n(c)x_n = \gamma_1^{(n)} e_1 + \cdots + \gamma_m^{(n)} e_m + T_n(\chi_m)A_n x_n$$

with certain complex numbers $\gamma_1^{(n)}, \ldots, \gamma_m^{(n)}$. Since

$$\left(\sum_{j=1}^{m} |\gamma_j^{(n)}|^2\right)^{1/2} \le \|T_n(c)x_n\|_2 + \|T(\chi_m)\|_2 \, \|A_n x_n\|_2 \le \|b\|_\infty + \|A_n x_n\|_2,$$

we conclude that there is an $M < \infty$ such that $|\gamma_j^{(n)}| \le M$ for all n and j. Finally, from (12.14), (12.15) and the equality

$$x_n = \gamma_1^{(n)} T_n^{-1}(c)e_1 + \cdots + \gamma_m^{(n)} T_n^{-1}(c)e_m + T_n^{-1}(c)T_n(\chi_m)A_n x_n$$

we get (12.16) and (12.17) with

$$z_n = T_n^{-1}(c)T_n(\chi_m)A_n x_n + \sum_{j=1}^{m} \gamma_j^{(n)} (T_n^{-1}(c)e_j - P_n T^{-1}(c)e_j).$$

This completes the proof. □

We now sharpen the definition of an asymptotically localized sequence. We say that a sequence $\{y_n\}$ of vectors $y_n \in \mathbf{C}^n$ is *asymptotically strongly localized in the beginning part* if

$$\lim_{n\to\infty} \frac{\|P_{\{1,\ldots,j_n\}} y_n\|_2}{\|y_n\|_2} = 1 \qquad (12.19)$$

for every sequence $\{j_n\}_{n=1}^{\infty}$ such that $j_n \to \infty$ and $1 \le j_n \le n$. Asymptotic strong localization in the beginning part implies, for example, that (12.19) is true with $j_n = \log\log n$ for sufficiently large n.

Theorem 12.9. *Suppose* $\lambda \notin b(\mathbf{T})$ *and* wind $(b, \lambda) = -m < 0$. *Then every asymptotically good pseudomode for* $\{T_n(b)\}$ *at* λ *is asymptotically strongly localized in the beginning part.*

Proof. Let $\{x_n\}$ be an asymptotically good pseudomode for $\{T_n(b)\}$ at λ. We may without loss of generality assume that $\|x_n\|_2 = 1$ for all n. By Theorem 12.8,

$$x_n = \gamma_1^{(n)} P_n u_1 + \cdots + \gamma_m^{(n)} P_n u_m + z_n =: w_n + z_n,$$

where $u_1, \ldots, u_m$ are given by (12.15) and $\gamma_1^{(n)}, \ldots, \gamma_m^{(n)}, z_n$ satisfy (12.17). Choose $M < \infty$ so that $|\gamma_i^{(n)}| \le M$ for all i and n. Let $\{j_n\}$ be any sequence such that $j_n \to \infty$ and $1 \le j_n \le n$. Put $J_n = \{1, \ldots, j_n\}$ and $J_n^c = \{j_n + 1, \ldots, n\}$. From (12.15) we infer that $u_1, \ldots, u_m \in \ell^2$. We have $\|P_{J_n^c} w_n\|_2 \le M \sum_{i=1}^{m} \|P_{J_n^c} u_i\|_2$. Since $u_i = \{u_k^{(i)}\}_{k=1}^{\infty}$ is in ℓ^2 and hence

$$\|P_{J_n^c} u_i\|_2^2 = \sum_{k=j_n+1}^{n} |u_k^{(i)}|^2 \le \sum_{k=j_n+1}^{\infty} |u_k^{(i)}|^2 = o(1) \quad \text{as} \quad j_n \to \infty,$$

it follows that $\|P_{J_n^c} w_n\|_2 \to 0$ as $n \to \infty$. Finally,

$$\begin{aligned} 1 &\ge \|P_{J_n} x_n\|_2^2 = 1 - \|P_{J_n^c} x_n\|_2^2 = 1 - \|P_{J_n^c}(w_n + z_n)\|_2^2 \\ &\ge 1 - \left(\|P_{J_n^c} w_n\|_2 + \|P_{J_n^c} z_n\|_2\right)^2, \end{aligned}$$

and because $\| P_{J_n^c} w_n \|_2 \to 0$ and $\| P_{J_n^c} z_n \|_2 \to 0$ as $n \to \infty$, we arrive at the conclusion that $\| P_{J_n} x_n \|_2 \to 1$. $\square$

To conclude this section, suppose that $\lambda \in \mathbf{C} \setminus b(\mathbf{T})$ and $\operatorname{wind}(b, \lambda) = m > 0$. We then have $\lambda \notin \widetilde{b}(\mathbf{T})$ and $\operatorname{wind}(\widetilde{b}, \lambda) = -m < 0$. Moreover, with W_n defined by (3.9), $W_n T_n(b - \lambda) W_n = T_n(\widetilde{b} - \lambda)$ and hence $\|T_n(b - \lambda)x_n\|_2 = \|T_n(\widetilde{b} - \lambda) W_n x_n\|_2$. Consequently, by Theorem 12.8, a sequence $\{x_n\}$ of unit vectors is an asymptotically good pseudomode of $\{T_n(b)\}$ at λ if and only if

$$W_n x_n = \gamma_1^{(n)} P_n \widetilde{u}_1 + \cdots + \gamma_m^{(n)} P_n \widetilde{u}_m + z_n, \tag{12.20}$$

where $|\gamma_j^{(n)}| \le M < \infty$ for all j and n, $\|z_n\|_2 \to 0$ as $n \to \infty$, and $\widetilde{u}_1, \ldots, \widetilde{u}_m$ are given by $\widetilde{u}_j = T^{-1}(\widetilde{c}) e_j$. Clearly, (12.20) can be rewritten in the form

$$x_n = \gamma_1^{(n)} W_n \widetilde{u}_1 + \cdots + \gamma_m^{(n)} W_n \widetilde{u}_m + \widetilde{z}_n$$

with $\widetilde{z}_n = W_n z_n$. The analogue of Theorem 12.9 says that every asymptotically good pseudomode $\{x_n\}$ for $\{T_n(b)\}$ at λ is asymptotically strongly localized in the terminating part, that is, the sequence $\{W_n x_n\}$ is asymptotically strongly localized in the beginning part.

Exercises

1. Let $T_{2n}(b)$ be a real symmetric Toeplitz matrix of order $2n$. An eigenvalue λ of $T_{2n}(b)$ is said to be even (odd) if there exists an eigenvector x for λ such that $W_n x = x$ ($W_n x = -x$). Show that the even and odd eigenvalues of $T_{2n}(b)$ are the eigenvalues of $T_n(b) + H_n(b)$ and $T_n(b) - H_n(b)$, respectively, where $H_n(b)$ is the principal $n \times n$ submatrix of the infinite Hankel matrix $H(b)$.

2. Let $b(t) = \sum_{j=-r}^{s} b_j t^j$ with $r s b_{-r} b_s \neq 0$. Suppose λ is an eigenvalue of $T_n(b)$. Denote by $\xi_1, \ldots, \xi_m$ the distinct zeros of the polynomial $z^r(b(z) - \lambda)$ and let α_i be the multiplicity of ξ_i. Define the matrix A_{r+s} as in Section 2.5. Finally, put $d_n(\lambda) = \dim \operatorname{Ker}(T_n(b) - \lambda I)$.

   (a) Prove that $x = (\, x_0 \; x_1 \; \ldots \; x_{n-1} \,)^\top$ is an eigenvector of $T_n(b)$ for λ if and only if

   $$x_k = \sum_{i=1}^{m} \sum_{j=0}^{\alpha_i - 1} C_{ji} \, j! \binom{n-k}{j} \xi_i^{n-k-j} \quad (k = 0, 1, \ldots, n-1),$$

   where

   $$(\, C_{01} \; \ldots \; C_{\alpha_1 - 1, 1} \; \ldots \; C_{0m} \; \ldots \; C_{\alpha_m - 1, m} \,) \, A_{r+s} = (\, 0 \; \ldots \; 0 \,).$$

   (b) Show that $d_n(\lambda) = r + s - \operatorname{rank} A_{r+s}$.

   (c) Show that $d_n(\lambda) \le \min(r, s)$.

   (d) Show that if $d_n(\lambda) = m \ge 2$, then λ is also an eigenvalue of $T_{n+1}(b)$ and $T_{n-1}(b)$ and $d_{n+1}(\lambda) \ge m - 1$ and $d_{n-1}(\lambda) \ge m - 1$.

3. Let k be a function in $L^1(\mathbf{R})$ and suppose $k(x) = 0$ for $|x| \geq r$. For $\tau > 2r$, there is a unique continuation of k to a τ-periodic function k_τ on all of $\mathbf{R}$. A continuous analogue of the circulant matrix $C_n(b)$ is the operator on $L^2(0, \tau)$ that is defined by

$$(A_\tau f)(x) = \gamma f(x) + \int_0^\tau k_\tau(x-t) f(t) dt, \quad x \in (0, \tau),$$

where γ is a fixed complex number. Prove that a point $\lambda \in \mathbf{C}$ is an asymptotically good eigenvalue, which means that $\|(A_\tau - \lambda I)^{-1}\|_2 \to \infty$ as $\tau \to \infty$, if and only if $\lambda = \gamma$ or

$$\lambda = \gamma + \int_{-r}^{r} k(x) e^{i\xi x} dx$$

for some $\xi \in \mathbf{R}$.

Notes

The results of Sections 12.2 and 12.3 are from our paper [61] with Ramírez de Arellano. Sections 12.4 and 12.5 are based on our article [55].

Exercise 1 is a special case of results of [5] and [82]; there one can also find a corresponding result for $T_{2n+1}(b)$. Exercise 2 summarizes the basic results of Trench's paper [278]. A solution to Exercise 3 is in [55].

Further results: real symmetric Toeplitz matrices. Let $b \in \mathcal{P}$. The matrix $T_n(b)$ is real and symmetric if and only if $b(e^{i\theta})$ $(\theta \in (-\pi, \pi))$ is real and even. Assume this condition is satisfied. We then have

$$b_n = \frac{1}{2\pi} \int_{-\pi}^{\pi} b(e^{i\theta}) e^{-in\theta} d\theta = \frac{1}{\pi} \int_0^\pi b(e^{i\theta}) \cos n\theta \, d\theta.$$

If λ is an eigenvalue of $T_n(b)$, then there exists an eigenvector x for λ such that $W_n x = x$ or $W_n x = -x$. In the former case λ is called an even eigenvalue and in the latter case λ is said to be odd (recall Exercise 1). In [5], [82] it is shown that $T_n(b)$ has exactly $[\frac{n+1}{2}]$ even and exactly $[\frac{n}{2}]$ odd eigenvalues (a repeated eigenvalue is necessarily both even and odd [99]). For $\alpha < \beta$, we denote by $N_{\text{even}}(\alpha, \beta; n)$ and $N_{\text{odd}}(\alpha, \beta; n)$ the number of even and odd eigenvalues of $T_n(b)$ in $[\alpha, \beta]$. Trench [280] proved that

$$\lim_{n\to\infty} \frac{N_{\text{even}}(\alpha, \beta; n)}{n} = \lim_{n\to\infty} \frac{N_{\text{odd}}(\alpha, \beta; n)}{n} = \frac{1}{\pi} |\{\theta \in (0, \pi) : \alpha \leq b(e^{i\theta}) \leq \beta\}|,$$

where $|\cdot|$ denotes Lebesgue measure.

Chapter 13

Structured Perturbations

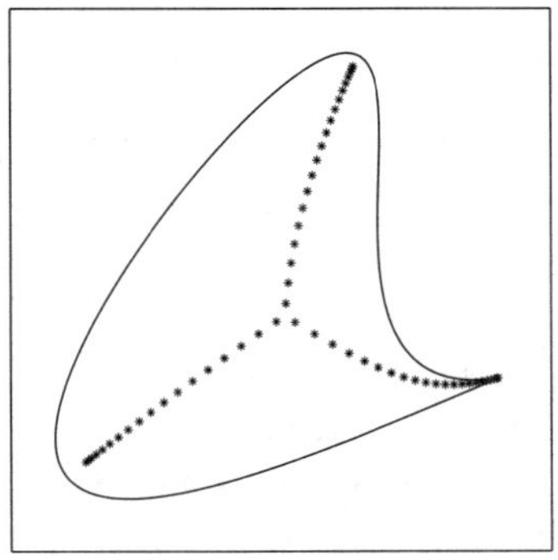

The pseudospectrum $\mathrm{sp}_\varepsilon T_n(b)$ measures the extent to which the spectrum of $T_n(b)$ may change by an arbitrary perturbation of norm at most ε. We now consider perturbations that have the same Toeplitz band structure as $T_n(b)$. In this way we can find the distance of $T_n(b)$ to the nearest singular matrix within the set of all matrices of the same banded structure as $T_n(b)$. Various condition numbers measure the sensitivity of properties of a matrix subject to perturbations. If the matrix has a certain structure, it is natural to require that the perturbations be of the same structure. This leads to the notion of structured condition numbers. In this chapter we study structured condition numbers for the Toeplitz structure. We give in particular a probabilistic argument which shows that in general we do not win anything by passing from unstructured condition numbers of banded Toeplitz matrices to Toeplitz-structured condition numbers.

13.1 Toeplitz Pseudospectra

Let $\mathcal{P}_{r,s}$ denote the set of all Laurent polynomials c of the form $c(t) = \sum_{j=-r}^{s} c_j t^j$ $(t \in \mathbf{T})$. For $b \in \mathcal{P}_{r,s}$ and $\varepsilon > 0$, we define the Toeplitz-structured pseudospectrum $\mathrm{sp}_\varepsilon^{\mathrm{Toep}[r,s]} T_n(b)$ by

$$\mathrm{sp}_\varepsilon^{\mathrm{Toep}[r,s]} T_n(b) = \bigcup_{\varphi \in \mathcal{P}_{r,s},\ \|\varphi\|_\infty \le \varepsilon} \mathrm{sp}\, T_n(b+\varphi).$$

Clearly, $\mathrm{sp}_\varepsilon^{\mathrm{Toep}[r,s]} T_n(b)$ is a subset of $\mathrm{sp}_\varepsilon T_n(b)$. We denote by $U_\sigma(\lambda_0)$ the open disk of radius σ centered at λ_0.

Lemma 13.1. *Let $c \in \mathcal{P}_{r,s}$. If $\lambda_0 \notin \Lambda(c)$, then there exist $n_0 \in \mathbf{N}$, $\sigma > 0$, $\delta > 0$ such that $U_\sigma(\lambda_0) \cap \mathrm{sp}\, T_n(c+\varphi) = \emptyset$ whenever $\varphi \in \mathcal{P}_{r,s}$, $\|\varphi\|_\infty \le \delta$, and $n \ge n_0$.*

Proof. From Theorem 11.3 we infer that there is a number $\varrho \in (0, \infty)$ such that λ_0 does not belong to $\mathrm{sp}\, T(c_\varrho)$. Hence, by Theorem 3.7, there exist $n_0 \in \mathbf{N}$ and $M \in (0, \infty)$ such that $\|T_n^{-1}(c_\varrho - \lambda_0)\|_2 \le M$ for all $n \ge n_0$. Put $\sigma = 1/(4M)$ and suppose that $\lambda \in U_\sigma(\lambda_0)$ and $\varphi \in \mathcal{P}_{r,s}$. We have

$$T_n((c+\varphi)_\varrho - \lambda) = T_n(c_\varrho - \lambda_0) + T_n(\varphi_\varrho) + (\lambda_0 - \lambda)P_n$$

and $\|T_n(c_\varrho - \lambda_0)\|_2 \ge 1/M > 1/(2M)$. Since $|\lambda_0 - \lambda| < 1/(4M)$ and $\|T_n(\varphi_\varrho)\|_2 \le \|\varphi_\varrho\|_\infty < 1/(4M)$ if only $\|\varphi\|_\infty \le \delta$ for some sufficiently small $\delta > 0$, it follows that $T_n((c+\varphi)_\varrho - \lambda)$ is invertible for all $\lambda \in U_\sigma(\lambda_0)$ and all $\varphi \in \mathcal{P}_{r,s}$ with $\|\varphi\|_\infty \le \delta$. As the invertibility of $T_n(c+\varphi-\lambda)$ is equivalent to the invertibility of $T_n((c+\varphi)_\varrho - \lambda)$ (notice that the two matrices are similar), we arrive at the assertion. □

Theorem 13.2. *If $b \in \mathcal{P}_{r,s}$ then*

$$\liminf_{n\to\infty} \mathrm{sp}_\varepsilon^{\mathrm{Toep}[r,s]} T_n(b) = \limsup_{n\to\infty} \mathrm{sp}_\varepsilon^{\mathrm{Toep}[r,s]} T_n(b) = \bigcup_{\varphi\in\mathcal{P}_{r,s},\ \|\varphi\|_\infty\le\varepsilon} \Lambda(b+\varphi).$$

Proof. Let $\lambda \in \limsup \mathrm{sp}_\varepsilon^{\mathrm{Toep}[r,s]} T_n(b)$. Then there are φ_{n_k} in $\mathcal{P}_{r,s}$ and λ_{n_k} in $\mathrm{sp}\, T_{n_k}(b+\varphi_{n_k})$ such that $\|\varphi_{n_k}\|_\infty \le \varepsilon$ and $\lambda_{n_k} \to \lambda$. As the unit ball of $\mathbf{C}^{r+s+1}$ is compact, the sequence $\{\varphi_{n_k}\}$ has a subsequence $\{\varphi_{n_{k_\ell}}\}$ converging to some $\varphi \in \mathcal{P}_{r,s}$ of ∞-norm at most ε. We claim that $\lambda \in \Lambda(b+\varphi)$. Indeed, if $\lambda \notin \Lambda(b+\varphi)$ then, by Lemma 13.1, $U_\sigma(\lambda) \cap \mathrm{sp}\, T_{n_{k_\ell}}(b+\varphi_{n_{k_\ell}}) = \emptyset$ for some $\sigma > 0$ and all sufficiently large n_{k_ℓ}, which is impossible because $\lambda_{n_{k_\ell}} \to \lambda$. Thus, we have proved that λ is in $\cup \Lambda(b+\varphi)$, the union over all $\varphi \in \mathcal{P}_{r,s}$ for which $\|\varphi\|_\infty \le \varepsilon$.

Now let $\lambda \in \Lambda(b+\varphi)$ for some $\varphi \in \mathcal{P}_{r,s}$ with $\|\varphi\|_\infty \le \varepsilon$. Then there are λ_n in $\mathrm{sp}\, T_n(b+\varphi)$ such that $\lambda_n \to \lambda$, which shows that λ is in $\liminf \mathrm{sp}_\varepsilon^{\mathrm{Toep}[r,s]} T_n(b)$. □

13.2 The Nearest Singular Matrix

Given an $n \times n$ matrix A_n, the distance $d(A_n)$ to the nearest singular matrix is defined as the infimum of the set of all $\varepsilon > 0$ for which there exists an $n \times n$ matrix K_n such that $\|K_n\|_2 \le \varepsilon$ and $0 \in \mathrm{sp}\,(A_n + K_n)$. If A_n is singular, then $d(A_n) = 0$. If A_n is invertible, we have

$$\begin{aligned} d(A_n) &= \inf\{\varepsilon > 0 : 0 \in \mathrm{sp}_\varepsilon A_n\} \\ &= \inf\{\varepsilon > 0 : \|A_n^{-1}\|_2 \ge 1/\varepsilon\} \\ &= \inf\{\varepsilon > 0 : \varepsilon \ge \|A_n^{-1}\|_2^{-1}\} = \|A_n^{-1}\|_2^{-1} = \sigma_1(A_n). \end{aligned}$$

Thus, in either case $d(A_n)$ equals the minimal singular value of A_n.

Analogously, if A is a bounded linear operator on ℓ^2, we define $d(A)$ as the infimum of all $\varepsilon > 0$ such that $0 \in \mathrm{sp}_\varepsilon A$. As above, $d(A) = 0$ if A is not invertible, while $d(A) = \|A^{-1}\|_2^{-1}$ in case A is invertible.

Now let b be a Laurent polynomial and suppose b has no zeros on the unit circle $\mathbf{T}$. If $\mathrm{wind}\, b = 0$, then (6.16) implies that

$$d(T_n(b)) = \sigma_1(T_n(b)) = \|T_n^{-1}(b)\|_2^{-1} \to \|T^{-1}(b)\|_2^{-1} = d(T(b)),$$

that is, $d(T_n(b))$ stays away from zero. However, if $\mathrm{wind}\, b \ne 0$, we deduce from Theorem 9.4 that $d(T_n(b)) = \sigma_1(T_n(b))$ converges to zero with at least exponential speed.

Things are less dramatic when restricting ourselves to structured perturbations. For $b \in \mathcal{P}_{r,s}$, we define

$$d^{\mathrm{Toep}[r,s]}(T_n(b)) = \inf\left\{\varepsilon > 0 : 0 \in \operatorname{sp} T_n(b+\varphi),\, \varphi \in \mathcal{P}_{r,s},\, \|\varphi\|_\infty \le \varepsilon\right\}.$$

In the notation of Section 13.1,

$$d^{\mathrm{Toep}[r,s]}(T_n(b)) = \inf\left\{\varepsilon > 0 : 0 \in \operatorname{sp}_\varepsilon^{\mathrm{Toep}[r,s]} T_n(b)\right\}.$$

Theorem 13.3. *If $b \in \mathcal{P}_{r,s}$, then*

$$\lim_{n\to\infty} d^{\mathrm{Toep}[r,s]}(T_n(b)) = \inf\left\{\varepsilon > 0 : 0 \in \bigcup_{\varphi\in\mathcal{P}_{r,s},\, \|\varphi\|_\infty \le \varepsilon} \Lambda(b+\varphi)\right\}, \tag{13.1}$$

and we always have

$$\operatorname{dist}(0, \operatorname{sp} T(b)) \le \lim_{n\to\infty} d^{\mathrm{Toep}[r,s]}(T_n(b)) \le \operatorname{dist}(0, \Lambda(b)). \tag{13.2}$$

Proof. Clearly, the infimum on the right of (13.1) is equal to

$$\inf\left\{\|f\|_\infty : f \in \mathcal{P}_{r,s},\, 0 \in \Lambda(b+f)\right\} =: \varrho.$$

Let $f \in \mathcal{P}_{r,s}$, $0 \in \Lambda(b+f)$, $\|f\|_\infty < \varrho + \varepsilon$. There are $\lambda_n \in \operatorname{sp} T_n(b+f)$ such that $\lambda_n \to 0$. As $0 \in \operatorname{sp} T_n(b+f-\lambda_n)$, we have $d^{\mathrm{Toep}[r,s]}(T_n(b)) \le \|f - \lambda_n\|_\infty$, whence $\limsup d^{\mathrm{Toep}[r,s]}(T_n(b)) \le \|f\|_\infty < \varrho + \varepsilon$. Since $\varepsilon > 0$ can be chosen arbitrarily, it follows that $\limsup d^{\mathrm{Toep}[r,s]}(T_n(b)) \le \varrho$. Let

$$\sigma := \liminf_{n\to\infty} d^{\mathrm{Toep}[r,s]}(T_n(b)).$$

Given any $\varepsilon > 0$, there exist $f_{n_k} \in \mathcal{P}_{r,s}$ such that $0 \in \operatorname{sp} T_{n_k}(b + f_{n_k})$ and $\|f_{n_k}\|_\infty \le \sigma + \varepsilon$. We may assume that the f_{n_k}'s converge in L^∞ to some $f \in \mathcal{P}_{r,s}$ satisfying $\|f\|_\infty \le \sigma + \varepsilon$ (if necessary, we can pass to a subsequence). Assume $0 \notin \Lambda(b+f)$. Then, by Lemma 13.1, $U_\sigma(0) \cap \operatorname{sp} T_{n_k}(b + f_{n_k}) = \emptyset$ for all sufficiently large n_k, which is impossible. Hence $0 \in \Lambda(b+f)$. We arrive at the conclusion that $\varrho \le \sigma + \varepsilon$, and as $\varepsilon > 0$ is an arbitrary number, it follows that $\varrho \le \sigma$. This completes the proof of equality (13.1).

Since $\Lambda(b+\delta) = \Lambda(b) + \delta$ for every complex number δ, we see that always $\lim d^{\mathrm{Toep}[r,s]}(T_n(b)) \le \operatorname{dist}(0, \Lambda(b))$, and as $\Lambda(b+\varphi) \subset \operatorname{sp} T(b+\varphi)$ and

$$\bigcup_{\|\varphi\|_\infty \le \varepsilon} \operatorname{sp} T(b+\varphi) = \operatorname{sp} T(b) + \varepsilon\overline{\mathbf{D}},$$

we obtain that $\operatorname{dist}(0, \operatorname{sp} T(b)) \le \lim d^{\mathrm{Toep}[r,s]}(T_n(b))$. This gives (13.2). □

The following two examples show that the estimates (13.2) are sharp.

Example 13.4. Let $b \in \mathcal{P}_{0,1}$ be given by $b(t) = b_0 + b_1 t$ with $|b_0| > |b_1| > 0$. Then

$$\operatorname{sp} T(b) = b_0 + b_1\overline{\mathbf{D}}, \quad \Lambda(b) = \{b_0\},$$
$$\operatorname{dist}(0, \operatorname{sp} T(b)) = |b_0| - |b_1|, \quad \operatorname{dist}(0, \Lambda(b)) = |b_0|,$$
$$\bigcup_{\varphi\in\mathcal{P}_{0,1},\, \|\varphi\|_\infty \le \varepsilon} \Lambda(b+\varphi) = b_0 + \varepsilon\overline{\mathbf{D}}.$$

Thus, by (13.1),

$$\operatorname{dist}(0, \operatorname{sp} T(b)) < \lim_{n\to\infty} d^{\text{Toep}[0,1]}(T_n(b)) = \operatorname{dist}(0, \Lambda(b)). \qquad \square$$

Example 13.5. Define $b \in \mathcal{P}_{1,1}$ by $b(t) = 4 + 2t + t^{-1}$. The range $b(\mathbf{T})$ is the ellipse $\{4 + 3\cos\theta + i \sin\theta : \theta \in [0, 2\pi)\}$. Hence $\operatorname{dist}(0, \operatorname{sp} T(b)) = 1$. From Theorem 2.4 we know that if $c(t) = c_0 + c_1 t + c_{-1}t^{-1}$, then $\Lambda(c)$ is the line segment between the foci of the ellipse $c(\mathbf{T})$,

$$\Lambda(c) = [c_0 - 2\sqrt{c_1 c_{-1}}, c_0 + 2\sqrt{c_1 c_{-1}}]. \tag{13.3}$$

It follows that in the case at hand $\Lambda(b) = [4 - 2\sqrt{2}, 4 + 2\sqrt{2}]$, whence $\operatorname{dist}(0, \Lambda(b)) = 4 - 2\sqrt{2} > 1$. Put $\varphi(t) = t^{-1}$. Then, by (13.3),

$$\Lambda(b + \varphi) = [4 - 2\sqrt{2 \cdot 2}, 4 + 2\sqrt{2 \cdot 2}] = [0, 8],$$

which together with Theorem 13.3 implies that $\lim d^{\text{Toep}[1,1]}(T_n(b)) \le \|\varphi\|_\infty = 1$. Consequently,

$$\operatorname{dist}(0, \operatorname{sp} T(b)) = \lim_{n\to\infty} d^{\text{Toep}[1,1]}(T_n(b)) < \operatorname{dist}(0, \Lambda(b)). \qquad \square$$

The following lemma is needed to prove Corollary 13.7. This corollary clarifies what happens with the distance of a large Toeplitz band matrix to the nearest singular matrix when the distance is measured within the Toeplitz band matrices of the same structure as the original matrix.

Lemma 13.6. *The function $b \mapsto \Lambda(b)$ is upper-semicontinuous on $\mathcal{P}_{r,s}$, that is, given any $\varepsilon > 0$, there is a $\delta > 0$ such that $\Lambda(b+\varphi) \subset \Lambda(b) + \varepsilon\overline{\mathbf{D}}$ whenever $\varphi \in \mathcal{P}_{r,s}$ and $\|\varphi\|_\infty \le \delta$.*

Proof. The case where b vanishes identically is trivial. We may therefore without loss of generality assume that $\varepsilon \in (0, \|b\|_\infty)$. Let K be the compact set

$$K = \{\lambda \in \mathbf{C} : |\lambda| \le 2\|b\|_\infty, \lambda \notin \Lambda(b) + \varepsilon\mathbf{D}\}.$$

By virtue of Lemma 13.1, for each $\lambda \in K$ there are $n \in \mathbf{N}$, $\sigma > 0$, $\delta > 0$ such that $U_\sigma(\lambda) \cap \operatorname{sp} T_k(b + \varphi) = \emptyset$ for all $\varphi \in \mathcal{P}_{r,s}$ with $\|\varphi\|_\infty \le \delta$ and all $k \ge n$. Since K is compact, we can find finitely many $\lambda_j \in K$, $n_j \in \mathbf{N}$, $\sigma_j > 0$, $\delta_j > 0$ $(j = 1, \dots, m)$ such that

$$K \subset \bigcup_{j=1}^{m} U_{\sigma_j}(\lambda_j), \quad U_{\sigma_j}(\lambda_j) \cap \operatorname{sp} T_k(b + \varphi) = \emptyset$$

for $\varphi \in \mathcal{P}_{r,s}$, $\|\varphi\|_\infty \le \delta_j$, $k \ge n_j$. Put $n_0 = \max n_j$ and $\delta = \min(\delta_1, \dots, \delta_m, \|b\|_\infty)$. If $\lambda \in K$, then $\lambda \in U_{\sigma_j}(\lambda_j)$ for some j and hence $T_k(b + \varphi - \lambda)$ is invertible for $\varphi \in \mathcal{P}_{r,s}$, $\|\varphi\|_\infty \le \delta$, $k \ge n_0$. If $\lambda > 2\|b\|_\infty$, then $T_k(b + \varphi - \lambda)$ is invertible for all k and all $\varphi \in \mathcal{P}_{r,s}$ with $\|\varphi\|_\infty \le \delta$ because

$$\|T_k(b + \varphi)\| \le \|b\|_\infty + \|\varphi\|_\infty \le 2\|b\|_\infty < |\lambda|.$$

Thus, we have shown that if $\lambda \notin \Lambda(b) + \varepsilon\mathbf{D}$, then $T_k(b + \varphi - \lambda)$ is invertible for $k \geq n_0$ and $\varphi \in \mathcal{P}_{r,s}$, $\|\varphi\|_\infty \leq \delta$. Consequently, $\operatorname{sp} T_k(b + \varphi) \subset \Lambda(b) + \varepsilon\overline{\mathbf{D}}$ for all $k \geq n_0$ and all $\varphi \in \mathcal{P}_{r,s}$ satisfying $\|\varphi\|_\infty \leq \delta$. This implies that $\Lambda(b + \varphi) \subset \Lambda(b) + \varepsilon\overline{\mathbf{D}}$ for $\varphi \in \mathcal{P}_{r,s}$ with $\|\varphi\|_\infty \leq \delta$. □

Corollary 13.7. *Let $b \in \mathcal{P}_{r,s}$. If $0 \in \Lambda(b)$, then $d^{\mathrm{Toep}[r,s]}(T_n(b)) \to 0$ as $n \to \infty$. If $0 \notin \Lambda(b)$, then there is an $\varepsilon > 0$ such that $d^{\mathrm{Toep}[r,s]}(T_n(b)) \geq \varepsilon$ for all sufficiently large n.*

Proof. If $0 \in \Lambda(b)$, then $d^{\mathrm{Toep}[r,s]}(T_n(b))$ goes to zero due to Theorem 13.3. If $0 \notin \Lambda(b)$, we can find an $\varepsilon > 0$ such that $0 \notin \Lambda(b) + \varepsilon\overline{\mathbf{D}}$. Theorem 13.3 and Lemma 13.6 so imply that $\lim d^{\mathrm{Toep}[r,s]}(T_n(b)) \geq \varepsilon$. □

13.3 Structured Normwise Condition Numbers

Let $\mathbf{K}$ stand for $\mathbf{R}$ or $\mathbf{C}$, let $M_n(\mathbf{K})$ be the collection of all $n \times n$ matrices with entries in $\mathbf{K}$, and let $\mathrm{Str}_n(\mathbf{K})$ denote the matrices in $M_n(\mathbf{K})$ which possess a certain prescribed structure. For example, $\mathrm{Str}_n(\mathbf{K})$ might be the set of all symmetric Toeplitz matrices with entries in $\mathbf{K}$. For an invertible matrix $A_n \in \mathrm{Str}_n(\mathbf{K})$, a vector $b \in \mathbf{K}^n$, and $x \in \mathbf{K}^n \setminus \{0\}$, one defines

$$\begin{aligned}\kappa_b^{\mathrm{Str}}(A_n, x) = \lim_{\varepsilon \to 0} \sup \Big\{ \frac{\|\delta x\|_2}{\varepsilon \|x\|_2} &: (A_n + \delta A_n)(x + \delta x) = A_n x + \delta b, \\ &\delta A_n \in \mathrm{Str}_n(\mathbf{K}),\ \|\delta A_n\|_2 \leq \varepsilon \|A_n\|_2 \\ &\delta b \in \mathbf{K}^n,\ \|\delta b\|_2 \leq \varepsilon \|b\|_2 \Big\}.\end{aligned}$$

Two natural choices are $b = 0$ (no perturbations to the right-hand side) and $b = A_n x$ (right-hand sides with the same relative error as in the matrix of the system). In the first case we speak of the *structured (normwise) condition number* and in the second case of the *full structured (normwise) condition number*, and we introduce the notations

$$\kappa^{\mathrm{Str}}(A_n, x) := \kappa_0^{\mathrm{Str}}(A_n, x), \quad \kappa_{\mathrm{full}}^{\mathrm{Str}}(A_n, x) := \kappa_{A_n x}^{\mathrm{Str}}(A_n, x).$$

In what follows we also need the number

$$\Phi^{\mathrm{Str}}(A_n, x) = \sup \big\{ \|A_n^{-1} \delta A_n x\|_2 : \delta A_n \in \mathrm{Str}_n(\mathbf{K}),\ \|\delta A_n\|_2 \leq 1 \big\}.$$

Proposition 13.8. *If $\mathrm{Str}_n(\mathbf{K})$ is invariant under multiplication by real numbers, then*

$$\frac{\Phi^{\mathrm{Str}}(A_n, x)}{\|x\|_2} \left(\|A_n\|_2 + \frac{\|b\|_2}{\|x\|_2} \right) \leq \kappa_b^{\mathrm{Str}}(A_n, x) \leq \|A_n^{-1}\|_2 \left(\|A_n\|_2 + \frac{\|b\|_2}{\|x\|_2} \right).$$

Proof. Let $\delta A_n \in \mathrm{Str}_n(\mathbf{K})$, $\delta b \in \mathbf{K}^n$ and suppose $\|\delta A_n\|_2 \leq \varepsilon \|A_n\|_2$, $\|\delta b\|_2 \leq \varepsilon \|b\|_2$. If $\varepsilon > 0$ is sufficiently small, then $\|A_n^{-1} \delta A_n\|_2 < 1$. In this case the equation $(A_n + \delta A_n)(x + \delta x) = A_n x + \delta b$ gives

$$\begin{aligned} x + \delta x &= (I + A_n^{-1} \delta A_n)^{-1}(x + A_n^{-1} \delta b) \\ &= (I - A_n^{-1} \delta A_n)(x + A_n^{-1} \delta b) + O(\varepsilon^2) \\ &= x + A_n^{-1} \delta b - A_n^{-1} \delta A_n x + O(\varepsilon^2). \end{aligned} \tag{13.4}$$

It follows that

$$\begin{aligned}\frac{\|\delta x\|_2}{\varepsilon\|x\|_2} &\le \frac{\|A_n^{-1}\|_2\|\delta b\|_2}{\varepsilon\|x\|_2} + \frac{\|A_n^{-1}\|_2\|\delta A_n\|_2\|x\|_2}{\varepsilon\|x\|_2} + O(\varepsilon)\\ &\le \frac{\|A_n^{-1}\|_2\|b\|_2}{\|x\|_2} + \|A_n^{-1}\|_2\|A_n\|_2 + O(\varepsilon),\end{aligned}$$

whence

$$\kappa_b^{\mathrm{Str}}(A_n, x) \le \frac{\|A_n^{-1}\|_2\|b\|_2}{\|x\|_2} + \|A_n^{-1}\|_2\|A_n\|_2,$$

which is the asserted estimate from above. To estimate $\kappa_b^{\mathrm{Str}}(A_n, x)$ from below, choose

$$\delta b = -\frac{\|b\|_2}{\|A_n\|_2\|x\|_2}\,\delta A_n x.$$

Clearly, $\|\delta b\|_2 \le \varepsilon\|b\|_2$. By (13.4),

$$\begin{aligned}\delta x &= A_n^{-1}\delta b - A_n^{-1}\delta A_n x + O(\varepsilon^2)\\ &= -\left(1 + \frac{\|b\|_2}{\|A_n\|_2\|x\|_2}\right) A_n^{-1}\delta A_n x + O(\varepsilon^2),\end{aligned}$$

which yields

$$\begin{aligned}\frac{\|\delta x\|_2}{\varepsilon\|x\|_2} &= \left(1 + \frac{\|b\|_2}{\|A_n\|_2\|x\|_2}\right)\frac{\|A_n^{-1}\delta A_n x\|_2}{\varepsilon\|x\|_2} + O(\varepsilon)\\ &= \left(1 + \frac{\|b\|_2}{\|A_n\|_2\|x\|_2}\right)\frac{\|A_n\|_2}{\|x\|_2}\left\|A_n^{-1}\left(\frac{1}{\varepsilon\|A_n\|_2}\,\delta A_n\right)x\right\|_2 + O(\varepsilon).\end{aligned}$$

Since $(1/(\varepsilon\|A_n\|_2))\delta A_n \in \mathrm{Str}_n(\mathbf{K})$, we obtain that

$$\begin{aligned}\kappa_b^{\mathrm{Str}}(A_n, x) &\ge \left(1 + \frac{\|b\|_2}{\|A_n\|_2\|x\|_2}\right)\frac{\|A_n\|_2}{\|x\|_2}\,\Phi^{\mathrm{Str}}(A_n, x)\\ &= \frac{\Phi^{\mathrm{Str}}(A_n, x)}{\|x\|_2}\left(\|A_n\|_2 + \frac{\|b\|_2}{\|x\|_2}\right),\end{aligned}$$

which proves the asserted lower estimate. □

From Proposition 13.8 we see in particular that

$$\kappa^{\mathrm{Str}}(A_n, x) \le \kappa_{\mathrm{full}}^{\mathrm{Str}}(A_n, x) \le 2\kappa(A_n) := \|A_n\|_2\|A_n^{-1}\|_2.$$

For $b = 0$, Proposition 13.8 can be sharpened.

Proposition 13.9. *If* $\mathrm{Str}_n(\mathbf{K})$ *is invariant under multiplication by real numbers, then*

$$\kappa^{\mathrm{Str}}(A_n, x) = \frac{\|A_n\|_2}{\|x\|_2}\,\Phi^{\mathrm{Str}}(A_n, x).$$

Proof. Formula (13.4) now reads $\delta x = -A_n^{-1}\delta A_n x + O(\varepsilon^2)$, and hence

$$\frac{\|\delta x\|_2}{\varepsilon\|x\|_2} = \frac{\|A_n^{-1}\,\delta A_n\,x\|_2}{\varepsilon\|x\|_2} + O(\varepsilon).$$

Since

$$\sup\left\{\frac{\|A_n^{-1}\,\delta A_n\,x\|_2}{\varepsilon\|x\|_2} : \delta A_n \in \mathrm{Str}_n(\mathbf{K}),\ \|\delta A_n\|_2 \le \varepsilon\|A_n\|_2\right\}$$
$$= \frac{\|A_n\|_2}{\|x\|_2}\,\sup\left\{\|A_n^{-1}\delta B_n x\|_2 : \delta B_n \in \mathrm{Str}_n(\mathbf{K}),\ \|\delta B_n\|_2 \le 1\right\},$$

we arrive at the assertion. □

In the case where $\mathrm{Str}_n(\mathbf{K}) = M_n(\mathbf{K})$, we write $\kappa_b(A_n, x)$, $\kappa(A_n, x)$, $\kappa_{\mathrm{full}}(A_n, x)$, and $\Phi(A_n, x)$ instead of $\kappa_b^{\mathrm{Str}}(A_n, x)$, $\kappa^{\mathrm{Str}}(A_n, x)$, $\kappa_{\mathrm{full}}^{\mathrm{Str}}(A_n, x)$, and $\Phi^{\mathrm{Str}}(A_n, x)$.

Proposition 13.10. *We have*

$$\Phi(A_n, x) = \|A_n^{-1}\|_2\,\|x\|_2, \tag{13.5}$$

$$\kappa(A_n, x) = \|A_n^{-1}\|_2\,\|A_n\|_2 = \kappa(A_n), \tag{13.6}$$

$$\kappa_{\mathrm{full}}(A_n, x) = \|A_n^{-1}\|_2\,\|A_n\|_2 + \|A_n^{-1}\|_2\frac{\|A_n x\|_2}{\|x\|_2}. \tag{13.7}$$

Proof. It is clear that $\Phi(A_n, x) \le \|A_n^{-1}\|_2\,\|x\|_2$. To prove that equality actually holds, choose a nonzero $y \in \mathbf{K}^n$ such that $\|A_n^{-1}y\|_2 = \|A_n^{-1}\|_2\,\|y\|_2$ and define $\delta A_n : \mathbf{K}^n \to \mathbf{K}^n$ by

$$\delta A_n\,\xi = \left(\xi, \frac{x}{\|x\|_2}\right)\frac{y}{\|y\|_2}.$$

Then $\|\delta A_n\|_2 = 1$ and

$$\|A_n^{-1}\,\delta A_n\,x\|_2 = \left\|A_n^{-1}\left(x, \frac{x}{\|x\|_2}\right)\frac{y}{\|y\|_2}\right\|_2 = \frac{\|x\|_2}{\|y\|_2}\|A_n^{-1}y\|_2 = \|A_n^{-1}\|_2\,\|x\|_2,$$

which completes the proof of (13.5). Combining Proposition 13.9 and (13.5) we get (13.6), while Proposition 13.8 in conjunction with (13.5) gives (13.7). □

Theorem 13.11. *Suppose* $\mathrm{Str}_n(\mathbf{K})$ *is invariant under multiplication by real numbers. If* $A_n \in \mathrm{Str}_n(\mathbf{K})$ *is an invertible matrix,* $x \in \mathbf{K}^n$ *is a nonzero vector, and*

$$\Phi^{\mathrm{Str}}(A_n, x) \ge \omega\|(A_n^{-1})^* x\|_2 \tag{13.8}$$

with some $\omega \in (0, \infty)$*, then*

$$\kappa_{\mathrm{full}}^{\mathrm{Str}}(A_n, x) \ge \sqrt{\frac{\omega}{2}\,\|A_n^{-1}\|_2\|A_n\|_2}\,. \tag{13.9}$$

Proof. Without loss of generality assume that $\|x\|_2 = 1$. From (13.4) we see that

$$\kappa_{\text{full}}^{\text{Str}}(A_n, x) = \lim_{\varepsilon \to 0} \sup \left\{ \frac{1}{\varepsilon} \|A_n^{-1} \delta A_n x - A_n^{-1} \delta b\|_2 \right\},$$

the supremum over all $\delta A_n \in \text{Str}_n(\mathbf{K})$ and $\delta b \in \mathbf{K}^n$ such that $(A_n + \delta A_n)(x + \delta x) = A_n x + \delta b$, $\|\delta A_n\|_2 \le \varepsilon \|A_n\|_2$, $\|\delta b\|_2 \le \varepsilon \|A_n x\|_2$. Since $\|-\delta b\|_2 = \|\delta b\|_2$, it follows that

$$\kappa_{\text{full}}^{\text{Str}}(A_n, x) = \lim_{\varepsilon \to 0} \sup \left\{ \frac{1}{\varepsilon} \max \left(\|A_n^{-1} \delta A_n x - A_n^{-1} \delta b\|_2, \|A_n^{-1} \delta A_n x + A_n^{-1} \delta b\|_2 \right) \right\},$$

the supremum over the same set as before. For arbitrary $u, v \in \mathbf{K}^n$ we have

$$\max \left(\|u + v\|_2, \|u - v|_2 \right) \ge \sqrt{\|u\|_2^2 + \|v\|_2^2} \ge \frac{1}{\sqrt{2}} (\|u\|_2 + \|v\|_2)$$

(note that if we are given a parallelogram with the sides a and b, then the length of its longest diagonal is $d^2 = a^2 + b^2 - 2ab \cos \varphi \ge a^2 + b^2$ because $\cos \varphi \le 0$). Consequently,

$$\kappa_{\text{full}}^{\text{Str}}(A_n, x) \ge \lim_{\varepsilon \to 0} \sup \left\{ \frac{1}{\sqrt{2}\, \varepsilon} \left(\|A_n^{-1} \delta A_n x\|_2 + \|A_n^{-1} \delta b\|_2 \right) \right\},$$

the supremum again taken over the same set as above, whence

$$\kappa_{\text{full}}^{\text{Str}}(A_n, x) \ge \frac{1}{\sqrt{2}} \left(\|A_n\|_2 \Phi^{\text{Str}}(A_n, x) + \|A_n x\|_2 \|A_n^{-1}\|_2 \right).$$

Thus, we are left to show that

$$\|A_n\|_2 \Phi^{\text{Str}}(A_n, x) + \|A_n x\|_2 \|A_n^{-1}\|_2 \ge \sqrt{\omega \|A_n^{-1}\|_2 \|A_n\|_2}.$$

This is certainly true if $\|A_n x\|_2 \ge \sqrt{\omega \|A_n\|_2 / \|A_n^{-1}\|_2}$. So assume that

$$\|A_n x\|_2 < \sqrt{\omega \|A_n\|_2 / \|A_n^{-1}\|_2}. \tag{13.10}$$

The product of the $1 \times n$ matrix $x^* A_n^{-1}$ and the $n \times 1$ matrix $A_n x$ is 1. Hence

$$1 \le \|x^* A_n^{-1}\|_2 \|A_n x\|_2 = \|(A_n^{-1})^* x\|_2 \|A_n x\|_2. \tag{13.11}$$

From (13.8), (13.10), and (13.11) we get

$$\begin{aligned} &\|A_n\|_2 \Phi^{\text{Str}}(A_n, x) \ge \omega \|A_n\|_2 \|(A_n^{-1})^* x\|_2 \\ &\ge \frac{\omega \|A_n\|_2}{\|A_n x\|_2} > \frac{\omega \|A_n\|_2}{\sqrt{\omega \, |A_n\|_2 / \|A_n^{-1}\|_2}} = \sqrt{\omega \|A_n\|_2 \|A_n^{-1}\|_2}. \end{aligned}$$

This completes the proof of (13.9). □

Things are very simple for circulant matrices. Let $\text{Circ}_n(\mathbf{K})$ stand for the collection of the matrices $C_n(b)$ defined in Section 2.1 with entries in $\mathbf{K}$. Suppose b has no zeros on

$\mathbf{T}$. Then $C_n(b)$ is invertible for all n. Note that $(C_n(b))^* = C_n(\overline{b})$. Proposition 2.1 implies that $\kappa(C_n(b)) \to \|b\|_\infty \|b^{-1}\|_\infty$ as $n \to \infty$.

Corollary 13.12. *If $C_n(b)$ is invertible, then*

$$\Phi^{\mathrm{Circ}}(C_n(b), x) = \|C_n^{-1}(b)x\|_2 = \|C_n^{-1}(\overline{b})x\|_2,$$

$$\kappa^{\mathrm{Circ}}(C_n(b), x) = \frac{\|C_n(b)\|_2}{\|x\|_2} \|C_n^{-1}(b)x\|_2,$$

$$\kappa_{\mathrm{full}}^{\mathrm{Circ}}(C_n(b), x) \geq \sqrt{\frac{1}{2}}\, \kappa(C_n(b)).$$

Proof. By Proposition 2.1, $C_n(b) = F_n^* D_n(b) F_n$ with a unitary matrix F_n and the diagonal matrix $D_n(b) := \operatorname{diag}\,(b(1), b(\omega_n), \ldots, b(\omega_n^{n-1}))$. Thus we can write $C_n^{-1}(b) = F_n^* D_n(b^{-1}) F_n$. By definition,

$$\Phi^{\mathrm{Circ}}(C_n(b), x) = \sup\{\|C_n^{-1}(b) C_n(g) x\|_2 : \|C_n(g)\|_2 \leq 1\},$$

and since $C_n^{-1}(b)$ is also a circulant matrix and circulant matrices commute, we see that $\Phi^{\mathrm{Circ}}(C_n(b), x) = \sup\{\|C_n(g) C_n^{-1}(b)x\|_2 : \|C_n(g)\|_2 \leq 1\} \leq \|C_n^{-1}(b)x\|_2$. As $I \in \mathrm{Circ}_n(\mathbf{K})$, we have $\Phi^{\mathrm{Circ}}(C_n(b), x) \geq \|C_n^{-1}(b)x\|_2$. Thus, $\Phi^{\mathrm{Circ}}(C_n(b), x) = \|C_n^{-1}(b)x\|_2$. Furthermore, $\|C_n^{-1}(\overline{b})x\|_2 = \|F_n^* D_n(\overline{b}^{-1}) F_n x\|_2 = \|D_n(\overline{b}^{-1}) F_n x\|_2$, and since $D_n(\overline{b}^{-1}) = S_n D_n(b^{-1})$ with a unitary diagonal matrix S_n, we get

$$\begin{aligned} &\|D_n(\overline{b}^{-1}) F_n x\|_2 = \|S_n D_n(b^{-1}) F_n x\|_2 \\ &= \|D_n(b^{-1}) F_n x\|_2 = \|F_n^* D_n(b^{-1}) F_n x\|_2 = \|C_n^{-1}(b)x\|_2. \end{aligned}$$

Now the formula for $\kappa^{\mathrm{Circ}}(C_n(b), x)$ follows from Proposition 13.9, while the estimate for $\kappa_{\mathrm{full}}^{\mathrm{Circ}}(C_n(b), x)$ results from Theorem 13.11. $\square$

Formula (9.33) and Corollary 13.12 show that if the Laurent polynomial b has no zeros on $\mathbf{T}$, then for a typical x,

$$\kappa^{\mathrm{Circ}}(C_n(b), x) \approx \frac{\|b\|_\infty}{\|x\|_2} \frac{\|b^{-1}\|_2}{\|b^{-1}\|_\infty} \|b^{-1}\|_\infty \|x\|_2 = \|b\|_\infty \|b^{-1}\|_2$$

for all sufficiently large n. This is a little better than $\kappa(C_n(b)) \approx \|b\|_\infty \|b^{-1}\|_\infty$, but the improvement is hardly significant.

13.4 Toeplitz Systems

Define $\mathrm{Toep}_n(\mathbf{K})$ as the set of all $n \times n$ Toeplitz matrices with entries in $\mathbf{K}$. In this section we estimate $\kappa_{\mathrm{full}}^{\mathrm{Toep}}(T_n(b), x)$ and show that these full structured condition numbers *always* have the same (good or bad) behavior as the usual condition numbers $\kappa(T_n(b))$ as $n \to \infty$.

Following the Highams [160] and Rump [236], we associate the $n \times (2n-1)$ matrix

$$\Psi_x := \Psi_x^{\mathrm{Toep}} := \begin{pmatrix} x_{n-1} & \ldots & x_1 & x_0 & & & \\ & x_{n-1} & \ldots & x_1 & x_0 & & \\ & & \ldots & \ldots & \ldots & \ldots & \\ & & & x_{n-1} & \ldots & x_1 & x_0 \end{pmatrix} \tag{13.12}$$

with a vector $x = (x_0, x_1, \dots, x_{n-1}) \in \mathbf{K}^n$. For every $n \times n$ Toeplitz matrix

$$T_n(g) = \begin{pmatrix} g_0 & g_{-1} & \dots & g_{-(n-1)} \\ g_1 & g_0 & \dots & g_{-(n-2)} \\ \dots & \dots & \dots & \dots \\ g_{n-1} & g_{n-2} & \dots & g_0 \end{pmatrix}$$

we have the equality

$$T_n(g)x = \Psi_x \begin{pmatrix} g_{-(n-1)} \\ g_{-(n-2)} \\ \dots \\ g_{n-2} \\ g_{n-1} \end{pmatrix} =: \Psi_x g, \tag{13.13}$$

which for $n = 3$ reads

$$\begin{pmatrix} g_0 & g_{-1} & g_{-2} \\ g_1 & g_0 & g_{-1} \\ g_2 & g_1 & g_0 \end{pmatrix} \begin{pmatrix} x_0 \\ x_1 \\ x_2 \end{pmatrix} = \begin{pmatrix} x_2 & x_1 & x_0 & 0 & 0 \\ 0 & x_2 & x_1 & x_0 & 0 \\ 0 & 0 & x_2 & x_1 & x_0 \end{pmatrix} \begin{pmatrix} g_{-2} \\ g_{-1} \\ g_0 \\ g_1 \\ g_2 \end{pmatrix},$$

and which can be readily verified for general n. We also define the map $\overline{W}_n$ by

$$\overline{W}_n : \mathbf{K}^n \to \mathbf{K}^n, \; (x_0, x_1, \dots, x_{n-1}) \mapsto (\overline{x}_{n-1}, \dots, \overline{x}_1, \overline{x}_0),$$

the bar denoting passage to the complex conjugate.

Lemma 13.13. *For each $x \in \mathbf{K}^n$ there exists a Toeplitz matrix $T_n(g) \in \mathrm{Toep}_n(\mathbf{K})$ such that $T_n(g)\overline{W}_n x = x$ and $\|T_n(g)\|_2 \le 1$.*

Proof. Put $y = \overline{W}_n x$. In accordance with (13.12),

$$\Psi_y = \begin{pmatrix} \overline{x}_0 & \overline{x}_1 & \dots & \overline{x}_{n-1} & & & \\ & \overline{x}_0 & \overline{x}_1 & \dots & \overline{x}_{n-1} & & \\ & & \dots & \dots & \dots & \dots & \\ & & & \overline{x}_0 & \overline{x}_1 & \dots & \overline{x}_{n-1} \end{pmatrix}.$$

We extend the $n \times (2n-1)$ matrix Ψ_y to a $(2n-1) \times (2n-1)$ circulant matrix C_y by adding $n-1$ more rows. For example, if $n = 3$,

$$\Psi_y = \begin{pmatrix} \overline{x}_0 & \overline{x}_1 & \overline{x}_2 & 0 & 0 \\ 0 & \overline{x}_0 & \overline{x}_1 & \overline{x}_2 & 0 \\ 0 & 0 & \overline{x}_0 & \overline{x}_1 & \overline{x}_2 \end{pmatrix}, \quad C_y = \begin{pmatrix} \overline{x}_0 & \overline{x}_1 & \overline{x}_2 & 0 & 0 \\ 0 & \overline{x}_0 & \overline{x}_1 & \overline{x}_2 & 0 \\ 0 & 0 & \overline{x}_0 & \overline{x}_1 & \overline{x}_2 \\ \overline{x}_2 & 0 & 0 & \overline{x}_0 & \overline{x}_1 \\ \overline{x}_1 & \overline{x}_2 & 0 & 0 & \overline{x}_0 \end{pmatrix}.$$

By Section 2.1, $C_y = U^*DU$ with the unitary matrix $U = F_{2n-1}$ and a diagonal matrix $D = \text{diag}\,(d_1, \ldots, d_{2n-1})$. Put $d_j^+ = 0$ if $d_j = 0$ and $d_j^+ = d_j^{-1}$ if $d_j \neq 0$. Then let $D^+ = \text{diag}\,(d_1^+, \ldots, d_{2n-1}^+)$ and $C = U^*D^+D^*U$. We have

$$C_yC = U^*DUU^*D^+D^*U = U^*DD^+D^*U = U^*D^*U = C_y^*.$$

Since Ψ_y consists of the first n rows of C_y, it follows that Ψ_yC is the matrix constituted by the first n rows of C_y^*: $\Psi_yC = P_nC_y^*$. Now define

$$g := \begin{pmatrix} g_{-(n-1)} \\ g_{-(n-2)} \\ \cdots \\ g_{n-2} \\ g_{n-1} \end{pmatrix} := C \begin{pmatrix} 1 \\ 0 \\ \cdots \\ 0 \\ 0 \end{pmatrix} =: Ce_1. \tag{13.14}$$

From (13.13), (13.14), and the equality $\Psi_yC = P_nC_y^*$ we obtain that $T_n(g)y = \Psi_yg = \Psi_yCe_1 = P_nC_y^*e_1 = x$. It remains to show that $\|T_n(g)\|_2 \leq 1$. The matrix C is obviously a circulant matrix, and by virtue of (13.14), g is the first column of C. This implies that $T_n(g)$ coincides with the lower-left $n \times n$ block of C. For example, if $n = 2$,

$$C = \begin{pmatrix} c_1 & c_2 & c_3 \\ c_3 & c_1 & c_2 \\ c_2 & c_3 & c_1 \end{pmatrix}, \quad g = \begin{pmatrix} g_{-1} \\ g_0 \\ g_1 \end{pmatrix} = \begin{pmatrix} c_1 \\ c_2 \\ c_3 \end{pmatrix},$$

$$T_2(g) = \begin{pmatrix} g_0 & g_{-1} \\ g_1 & g_0 \end{pmatrix} = \begin{pmatrix} c_2 & c_1 \\ c_3 & c_2 \end{pmatrix}.$$

Consequently, $\|T_n(g)\|_2 \leq \|C\|_2 = \|D^+D^*\|_2 = 1$. □

Theorem 13.14. *For every vector $x \in \mathbf{K}^n$,*

$$\kappa_{\text{full}}^{\text{Toep}}(T_n(b), x) \geq \sqrt{\frac{1}{2}\kappa(T_n(b))}\,.$$

Proof. By definition,

$$\Phi^{\text{Toep}}(T_n(b), x) = \sup\{\|T_n^{-1}(b)T_n(f)x\|_2 : \|T_n(f)\|_2 \leq 1\}.$$

Since $\overline{W}_nT_n(a)\overline{W}_n = T_n(\overline{a})$ and $\overline{W}_n$ is isometric, we get

$$\begin{aligned}\Phi^{\text{Toep}}(T_n(b), x) &= \sup\{\|\overline{W}_nT_n^{-1}(b)\overline{W}_n\overline{W}_nT_n(f)\overline{W}_n\overline{W}_nx\|_2 : \|T_n(f)\|_2 \leq 1\} \\ &= \sup\{\|T_n^{-1}(\overline{b})T_n(\overline{f})\overline{W}_nx\|_2 : \|T_n(\overline{f})\|_2 \leq 1\}.\end{aligned}$$

Lemma 13.13 tells us that there is an f such that $T_n(\overline{f})\overline{W}_nx = x$ and $\|T_n(\overline{f})\|_2 \leq 1$. Hence

$$\Phi^{\text{Toep}}(T_n(b), x) \geq \|T_n^{-1}(\overline{b})x\|_2 = \|(T_n^{-1}(b))^*x\|_2,$$

and Theorem 13.11 now gives the assertion. □

Thus, if $\kappa(T_n(b))$ increases at least exponentially or polynomially, then so does $\kappa_{\text{full}}^{\text{Toep}}(T_n(b), x)$. In other words, we do not win much when passing from unstructured condition numbers to full structured condition numbers.

13.5 Exact Right-Hand Sides

We now consider the structured condition numbers $\kappa^{\text{Toep}}(T_n(b), x)$. We begin with a simple auxiliary result.

Lemma 13.15. *Let $n \le k$, let A be an $n \times n$ matrix, and let B be an $n \times k$ matrix. Then $\|AB\|_2 \ge \|A\|_2\, \sigma_{\min}(B)$, where $\sigma_{\min}(B)$ denotes the smallest singular value of the matrix B.*

Proof. There is nothing to prove for $\sigma_{\min}(B) = 0$. So assume that $\sigma_{\min}(B) > 0$ and hence rank $B = n$. There is an x such that $\|A\|_2 = \|Ax\|_2$ and $\|x\|_2 = 1$. Since rank $B = n$, there exists a y with $By = x$. It follows that $1 = \|By\|_2 \ge \sigma_{\min}(B)\|y\|_2$, and this gives $\|AB\|_2 \ge \|ABy\|_2/\|y\|_2 = \|A\|_2/\|y\|_2 \ge \|A\|_2\sigma_{\min}(B)$. □

Theorem 13.16. *For every vector $x \in \mathbf{K}^n$,*

$$\frac{\kappa^{\text{Toep}}(T_n(b), x)}{\kappa(T_n(b), x)} \ge \frac{1}{\sqrt{n}} \frac{\sigma_{\min}(\Psi_x)}{\|x\|_2}.$$

Proof. We know from (13.13) that $T_n(f)x = \Psi_x f$. Since

$$\|T_n(f)\|_2 \le \|T_n(f)\|_F \le \sqrt{n}\, \|f\|_2,$$

we have

$$\begin{aligned}
\Phi^{\text{Toep}}(T_n(b), x) &= \sup\{\|T_n^{-1}(b)T_n(f)x\|_2 : \|T_n(f)\|_2 \le 1\} \\
&= \sup\{\|T_n^{-1}(b)\Psi_x f\|_2 : \|T_n(f)\|_2 \le 1\} \\
&\ge \sup\{\|T_n^{-1}(b)\Psi_x f\|_2 : \sqrt{n}\, \|f\|_2 \le 1\} \\
&= \frac{1}{\sqrt{n}} \sup\{\|T_n^{-1}(b)\Psi_x \sqrt{n}\, f\|_2 : \|\sqrt{n}\, f\|_2 \le 1\} \\
&= \frac{1}{\sqrt{n}} \|T_n^{-1}(b)\Psi_x\|_2 \ge \frac{1}{\sqrt{n}} \|T_n^{-1}(b)\|_2\, \sigma_{\min}(\Psi_x),
\end{aligned}$$

the last inequality resulting from Lemma 13.15. The assertion now follows from Propositions 13.9 and 13.10. □

Thus, the ratio $\kappa^{\text{Toep}}(T_n(b), x)/\kappa(T_n(b), x)$ can be estimated from below by the smallest singular value of the $n \times (2n-1)$ matrix Ψ_x. Rump [236] took 10^6 samples of $x \in \mathbf{R}^{100}$ with independent x_j that are either uniformly distributed random variables in $[-1, 1]$ or random variables with standard normal distribution. In either case, he observed that the mean of $\sigma_{\min}(\Psi_x)/\|x\|_2$ is 0.31 and that the standard deviation is 0.069. Thus, vectors x with small $\sigma_{\min}(\Psi_x)/\|x\|_2$ are rare. One such rare vector is composed of the coefficients of the polynomial

$$x(t) = (t+1)^{n-1}.$$

This vector was proposed by Georg Heinig, and Rump showed numerically that for this vector

$$\frac{\sigma_{\min}(\Psi_x)}{\|x\|_2} \simeq \left(\frac{2}{5}\right)^n$$

(see Figure 13.1). Notice, however, that a small value of $\sigma_{\min}(\Psi_x)/\|x\|_2$ does not yet imply that the quotient $\kappa^{\text{Toep}}(T_n(b),x)/\kappa(T_n(b),x)$ is also small, because Theorem 13.16 contains only a lower *estimate* (see also Example 13.19 below). Moreover, in general the determination of $\sigma_{\min}(\Psi_x)$ is difficult.

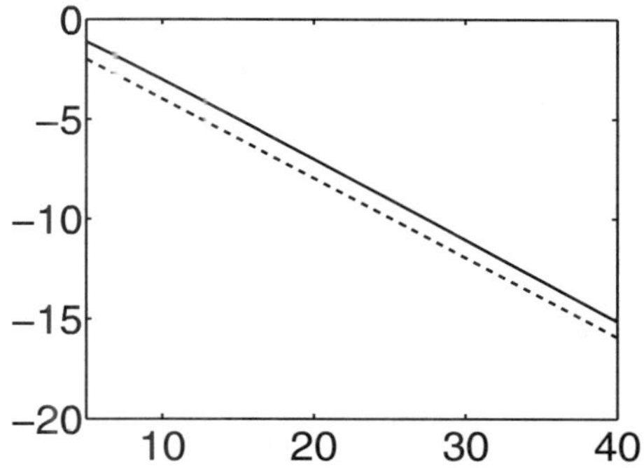

Figure 13.1. *The curve* $\log_{10}(\sigma_{\min}(\Psi_x)/\|x\|_2)$ *(solid) and the curve* $n\log_{10}(2/5)$ *(dashed) for* $5 \le n \le 40$.

Here is another estimate for $\kappa^{\text{Toep}}(T_n(b),x)$.

Theorem 13.17. *Let* $x = (x_0, x_1, \dots, x_{n-1}) \in \mathbf{K}^n$ *and suppose the polynomial* $x(t) = x_0 + x_1 t + \cdots + x_{n-1}t^{n-1}$ *has exactly* ℓ *zeros (counted with multiplicities) on the unit circle* $\mathbf{T}$. *Thus,*

$$x(t) = \prod_{j=1}^{\ell}(t-\mu_j)\,z(t), \quad \mu_j \in \mathbf{T}, \quad z(t) \neq 0 \text{ for } t \in \mathbf{T},$$

where $z(t)$ *is a polynomial of degree* $n-\ell-1$. *Put* $\min|z| = \min_{t\in\mathbf{T}}|z(t)|$. *If the matrix* $T_n(b)$ *is invertible, then*

$$\Phi^{\text{Toep}}(T_n(b),x) \ge \frac{\min|z|}{2^{\ell/2}n^{\ell+1/2}}\,\|T_n^{-1}(b)\|_2 \tag{13.15}$$

and hence

$$\frac{\kappa^{\text{Toep}}(T_n(b),x)}{\kappa(T_n(b),x)} \ge \frac{\min|z|}{2^{\ell/2}n^{\ell+1/2}}\,\frac{1}{\|x\|_2}. \tag{13.16}$$

Proof. To avoid involved notation, let us assume that $n=4$ and $\ell=2$. Thus,

$$x_0 + x_1 t + x_2 t^2 + x_3 t^3 = (t-\mu_1)(t-\mu_2)(z_0+z_1 t) \tag{13.17}$$

with $|\mu_1| = |\mu_2| = 1$ and $z_0 + z_1 t \neq 0$ for $t \in \mathbf{T}$. Factorization (13.17) is equivalent to the matrix equality

$$\begin{pmatrix} x_0 \\ x_1 \\ x_2 \\ x_3 \end{pmatrix} = \begin{pmatrix} -\mu_1 & 0 & 0 \\ 1 & -\mu_1 & 0 \\ 0 & 1 & -\mu_1 \\ 0 & 0 & 1 \end{pmatrix} \begin{pmatrix} -\mu_2 & 0 \\ 1 & -\mu_2 \\ 0 & 1 \end{pmatrix} \begin{pmatrix} z_0 \\ z_1 \end{pmatrix},$$

or, in other and self-evident notation,

$$x = T_{4,3}(t-\mu_1)T_{3,2}(t-\mu_2)z. \tag{13.18}$$

Choose $y \in \mathbf{K}^4$ so that $\|y\|_2 = 1$ and $\|T_4^{-1}(b)y\|_2 = \|T_4^{-1}(b)\|_2$.

We first assume that $\mathbf{K} = \mathbf{C}$. We begin with determining an $h \in \mathcal{P}_3$ such that

$$T_{4,2}(h)z = y, \tag{13.19}$$

that is,

$$\begin{pmatrix} h_0 & h_{-1} \\ h_1 & h_0 \\ h_2 & h_1 \\ h_3 & h_2 \end{pmatrix} \begin{pmatrix} z_0 \\ z_1 \end{pmatrix} = \begin{pmatrix} y_0 \\ y_1 \\ y_2 \\ y_3 \end{pmatrix}. \tag{13.20}$$

Equation (13.20) is certainly satisfied if

$$\begin{pmatrix} z_1 & z_0 & 0 & 0 & 0 \\ 0 & z_1 & z_0 & 0 & 0 \\ 0 & 0 & z_1 & z_0 & 0 \\ 0 & 0 & 0 & z_1 & z_0 \\ z_0 & 0 & 0 & 0 & z_1 \end{pmatrix} \begin{pmatrix} h_{-1} \\ h_0 \\ h_1 \\ h_2 \\ h_3 \end{pmatrix} = \begin{pmatrix} y_0 \\ y_1 \\ y_2 \\ y_3 \\ 0 \end{pmatrix},$$

or, equivalently, $C_5(z_1 + z_0 t^{-1})h = y$. Thus, we can take

$$h = C_5^{-1}(z_1 + z_0 t^{-1}) \begin{pmatrix} y \\ 0 \end{pmatrix},$$

whence

$$\|h\|_2 \le \frac{1}{\min_{t\in\mathbf{T}} |z_1 + z_0 t^{-1}|} \|y\|_2 = \frac{1}{\min |z|} \|y\|_2. \tag{13.21}$$

Next we seek a $g \in \mathcal{P}_3$ such that

$$T_{4,3}(g)T_{3,2}(t-\mu_2) = T_{4,2}(h), \tag{13.22}$$

which is the system

$$\begin{pmatrix} g_0 & g_{-1} & g_{-2} \\ g_1 & g_0 & g_{-1} \\ g_2 & g_1 & g_0 \\ g_3 & g_2 & g_1 \end{pmatrix} \begin{pmatrix} -\mu_2 & 0 \\ 1 & -\mu_2 \\ 0 & 1 \end{pmatrix} = \begin{pmatrix} h_0 & h_{-1} \\ h_1 & h_0 \\ h_2 & h_1 \\ h_3 & h_2 \end{pmatrix}.$$

The last system is satisfied as soon as $g_{-2} = 0$ and

$$\begin{pmatrix} -\mu_2 & 0 & 0 & 0 & 0 \\ 1 & -\mu_2 & 0 & 0 & 0 \\ 0 & 1 & -\mu_2 & 0 & 0 \\ 0 & 0 & 1 & -\mu_2 & 0 \\ 0 & 0 & 0 & 1 & -\mu_2 \end{pmatrix} \begin{pmatrix} g_{-1} \\ g_0 \\ g_1 \\ g_2 \\ g_3 \end{pmatrix} = \begin{pmatrix} h_{-1} \\ h_0 \\ h_1 \\ h_2 \\ h_3 \end{pmatrix}, \tag{13.23}$$

that is, $T_5(t-\mu_2)g = h$. The solution of (13.23) is

$$\begin{pmatrix} g_{-1} \\ g_0 \\ g_1 \\ g_2 \\ g_3 \end{pmatrix} = -\frac{1}{\mu_2} \begin{pmatrix} 1 & & & & \\ 1/\mu_2 & 1 & & & \\ 1/\mu_2^2 & 1/\mu_2 & 1 & & \\ 1/\mu_2^3 & 1/\mu_2^2 & 1/\mu_2 & 1 & \\ 1/\mu_2^4 & 1/\mu_2^3 & 1/\mu_2^2 & 1/\mu_2 & 1 \end{pmatrix} \begin{pmatrix} h_{-1} \\ h_0 \\ h_1 \\ h_2 \\ h_3 \end{pmatrix},$$

which gives $\|g\|_2^2 \le (5+4+3+2+1)\,\|h\|_2^2$ and thus

$$\|g\|_2 \le \sqrt{2}\cdot 4\,\|h\|_2 \tag{13.24}$$

(for general n the sum $5+4+3+2+1$ and the factor $\sqrt{2}\cdot 4$ become $(2n-3)+(2n-4)+\cdots+1$ and $\sqrt{2}\,n$, respectively). Finally, we want an $f \in \mathcal{P}_3$ such that

$$T_4(f)T_{4,3}(t-\mu_1) = T_{4,3}(g), \tag{13.25}$$

or, equivalently,

$$\begin{pmatrix} f_0 & f_{-1} & f_{-2} & f_{-3} \\ f_1 & f_0 & f_{-1} & f_{-2} \\ f_2 & f_1 & f_0 & f_{-1} \\ f_3 & f_2 & f_1 & f_0 \end{pmatrix} \begin{pmatrix} -\mu_1 & 0 & 0 \\ 1 & -\mu_1 & 0 \\ 0 & 1 & -\mu_1 \\ 0 & 0 & 1 \end{pmatrix} = \begin{pmatrix} g_0 & g_{-1} & 0 \\ g_1 & g_0 & g_{-1} \\ g_2 & g_1 & g_0 \\ g_3 & g_2 & g_1 \end{pmatrix}.$$

This is satisfied provided $f_{-2} = f_{-3} = 0$ and

$$\begin{pmatrix} -\mu_1 & 0 & 0 & 0 & 0 \\ 1 & -\mu_1 & 0 & 0 & 0 \\ 0 & 1 & -\mu_1 & 0 & 0 \\ 0 & 0 & 1 & -\mu_1 & 0 \\ 0 & 0 & 0 & 1 & -\mu_1 \end{pmatrix} \begin{pmatrix} f_{-1} \\ f_0 \\ f_1 \\ f_2 \\ f_3 \end{pmatrix} = \begin{pmatrix} g_{-1} \\ g_0 \\ g_1 \\ g_2 \\ g_3 \end{pmatrix}. \tag{13.26}$$

As above, we have

$$\|f\|_2 \le \sqrt{2}\cdot 4\,\|g\|_2. \tag{13.27}$$

From (13.18), (13.25), (13.22), (13.19) we get

$$\begin{aligned} T_4(f)x &= T_4(f)T_{4,3}(t-\mu_1)T_{3,2}(t-\mu_2)z \\ &= T_{4,3}(g)T_{3,2}(t-\mu_2)z \\ &= T_{4,2}(h)z \\ &= y \end{aligned} \tag{13.28}$$

and from (13.27), (13.24), (13.21) we obtain

$$\begin{aligned} \|T_4(f)\|_2 &\le \|T_4(f)\|_F \le \sqrt{4}\|f\|_2 \\ &\le \sqrt{4}(\sqrt{2}\cdot 4)(\sqrt{2}\cdot 4)\,\frac{1}{\min|z|}\,\|y\|_2 \\ &= 2^{\ell/2}n^{\ell+1/2}\,\frac{1}{\min|z|}\,\|y\|_2 \\ &= 2^{\ell/2}n^{\ell+1/2}\,\frac{1}{\min|z|} =: M. \end{aligned} \tag{13.29}$$

We have

$$\Phi^{\text{Toep}}(T_n(b), x) = \sup\left\{\|T_n^{-1}(b)T_n(\varphi)x\|_2 : \|T_n(\varphi)\|_2 \le 1\right\},$$

and since $\|T_n(f/M)\|_2 \le 1$ due to (13.29), it follows that

$$\Phi^{\text{Toep}}(T_n(b), x) \ge \|T_n^{-1}(b)T_n(f)x\|_2/M.$$

Equality (13.28) now implies that

$$\Phi^{\text{Toep}}(T_n(b), x) \ge \|T_n^{-1}(b)y\|_2/M = \|T_n^{-1}(b)\|_2/M.$$

This completes the proof of (13.15).

Estimate (13.16) follows from (13.15) and Propositions 13.9 and 13.10:

$$\frac{\kappa^{\text{Toep}}(T_n(b), x)}{\kappa(T_n(b), x)} = \frac{\|T_n(b)\|_2\Phi^{\text{Toep}}(T_n(b), x)}{\|x\|_2\|T_n(b)\|_2\|T_n^{-1}(b)\|_2} \ge \frac{1}{M}\,\frac{1}{\|x\|_2}.$$

This completes the proof in the case $\mathbf{K} = \mathbf{C}$.

Now assume $\mathbf{K} = \mathbf{R}$. Then the zeros of $x(t)$ are either real or form pairs $\mu, \overline{\mu}$. In the case of real zeros, we can proceed as before. So assume, for the sake of definiteness, that $\mu_1 = \mu$ and $\mu_2 = \overline{\mu}$ in (13.17). Then the two systems (13.23) and (13.26), which read $T_5(t-\mu)f = g$ and $T_5(t-\overline{\mu})g = h$, can be united in the single system $A_5 f = h$ with $A_5 = T_5(t-\overline{\mu})T_5(t-\mu) = T_5(t^2 - 2\,\text{Re}\,\mu\, t + 1)$. Since A_5 is an invertible real matrix, it follows that the solution f is also real. The rest is as above. □

In the following two examples we assume that b is a Laurent polynomial without zeros on $\mathbf{T}$ but with nonzero winding number. By Theorem 4.1 and Proposition 13.10, in this case the usual condition number increases at least exponentially,

$$\kappa(T_n(b), x) = \|T_n(b)\|_2\,\|T_n^{-1}(b)\|_2 \ge Ce^{\gamma n} \tag{13.30}$$

with constants $C > 0$ and $\gamma > 0$, where $\|T_n^{-1}(b)\|_2 := \infty$ if $T_n(b)$ is not invertible.

Example 13.18. Fix $x = (x_0, x_1, \ldots, x_{m-1}) \in \mathbf{K}^m$ and extend x by zeros before x_0 or after x_{m-1} to a vector in $\mathbf{K}^n$ $(n \ge m)$. Then ℓ and $\min|z|$ are constant, and the right-hand side of (13.16) goes to zero as $n^{-\ell-1/2}$. This in conjunction with (13.30) shows that $\kappa^{\text{Toep}}(T_n(b), x)$ grows exponentially. Since in the case at hand $\sigma_{\min}(\Psi_x)$ does not depend on n, the same conclusion can also be drawn from Theorem 13.16. □

Example 13.19. Suppose $x(t) = 1 + t^{n-1}$. Then $x(t)$ has $n-1$ zeros on $\mathbf{T}$, and (13.16) tells us that

$$\frac{\kappa^{\text{Toep}}(T_n(b), x)}{\kappa(T_n(b), x)} \ge \frac{1}{2^{(n-1)/2}n^{n-1/2}}\,\frac{1}{\sqrt{2}}.$$

This nourishes the hope that $\kappa^{\text{Toep}}(T_n(b), x)$ might grow essentially more slowly than $\kappa(T_n(b), x)$. However, this is not the case. To see this, we modify the argument in the proof of Theorem 13.17. Take $y \in \mathbf{K}^n$ so that $\|y\|_2 = 1$ and $\|T_n^{-1}(b)y\|_2 = \|T_n^{-1}(b)\|_2$. We seek a $g \in \mathcal{P}_{n-1}$ with coefficients in $\mathbf{K}$ such that $T_n(b)x = y$, that is,

$$\begin{aligned} g_0 + g_{-(n-1)} &= y_0 \\ g_1 + g_{-(n-2)} &= y_1 \\ &\dots \\ g_{n-1} + g_0 &= y_{n-1}. \end{aligned}$$

This is satisfied for

$$\begin{aligned} &g_0 = g_{-1} = \cdots = g_{-(n-2)} = 0, \\ &g_{-(n-1)} = y_0, \quad g_1 = y_1, \quad g_2 = y_2, \quad \dots, \quad g_{n-1} = y_{n-1}. \end{aligned}$$

It follows that $\|T_n(g)\|_2 \le \|T_n(g)\|_F \le \sqrt{n}\,\|g\|_2 = \sqrt{n}\,\|y\|_2 = \sqrt{n}$, whence

$$\Phi^{\mathrm{Toep}}(T_n(b), x) \ge \|T_n^{-1}(b)T_n(g/\sqrt{n})x\|_2 = \frac{1}{\sqrt{n}}\|T_n^{-1}(b)y\|_2 = \frac{1}{\sqrt{n}}\|T_n^{-1}(b)\|_2,$$

and thus

$$\frac{\kappa^{\mathrm{Toep}}(T_n(b), x)}{\kappa(T_n(b), x)} \ge \frac{1}{\sqrt{n}}\frac{1}{\|x\|_2} = \frac{1}{\sqrt{2n}}.$$

Consequently, $\kappa^{\mathrm{Toep}}(T_n(b), x)$ increases exponentially together with $\kappa(T_n(b), x)$.

Moral: Inequality (13.16) is a lower *estimate* that does a good job in many cases but that may be too crude in other cases. □

Here is a probabilistic argument which shows that if (13.30) holds, then we *at least in general* do not win anything by passing from unstructured condition numbers to structured condition numbers. Recall that a Rademacher variable is a random variable that assumes the value 1 with probability $1/2$ and the value -1 with probability $1/2$. A complex random variable is said to have a certain distribution if its real and imaginary parts are independent random variables with that distribution.

Theorem 13.20. *Let $x_0, x_1, \dots, x_{n-1} \in \mathbf{K}$ be independent standard normal or independent Rademacher variables and put $x = (x_0, x_1, \dots, x_{n-1})$. There exist universal constants $\varepsilon \in (0, \infty)$ and $n_0 \in \mathbf{N}$ such that*

$$P\left(\frac{\kappa^{\mathrm{Toep}}(T_n(b), x)}{\kappa(T_n(b), x)} \ge \frac{\varepsilon}{n^{3/2}}\right) > \frac{99}{100}$$

for all Laurent polynomials b with coefficients in $\mathbf{K}$ and all $n \ge n_0$.

Proof. We confine ourselves to the case $\mathbf{K} = \mathbf{C}$; the case $\mathbf{K} = \mathbf{R}$ is analogous (and a little simpler). Thus, let $u_0, v_0, \dots, u_{n-1}, v_{n-1}$ be independent standard normal or independent Rademacher variables. Put $x_0 = u_0 + iv_0, \dots, x_{n-1} = u_{n-1} + iv_{n-1}$ and let

$$\begin{aligned} &u(t) = u_0 + u_1 t + \cdots + u_{n-1}t^{n-1}, \quad v(t) = v_0 + v_1 t + \cdots + v_{n-1}t^{n-1}, \\ &x(t) = u(t) + iv(t). \end{aligned}$$

A deep recent result by Konyagin and Schlag [181] states that there is a universal constant $D \in (0, \infty)$ such that

$$\limsup_{n\to\infty} P\left(\min_{t\in\mathbf{T}} |u(t)| < \frac{\delta}{\sqrt{n}}\right) \le D\delta$$

for each $\delta > 0$ (they even proved this with the unit circle $\mathbf{T}$ replaced by the annulus $\{t \in \mathbf{C} : ||t| - 1| < \varepsilon/n^2\}$). Choose $\delta > 0$ so that $D\delta < 0.005$. Then there is an n_0 such that

$$P\left(\min_{t\in\mathbf{T}} |x(t)| < \frac{\delta}{\sqrt{n}}\right) \le P\left(\min_{t\in\mathbf{T}} |u(t)| < \frac{\delta}{\sqrt{n}}\right) < 0.005 \tag{13.31}$$

for all $n \ge n_0$. Theorem 13.17 tells us that

$$\frac{\kappa^{\mathrm{Toep}}(T_n(b), x)}{\kappa(T_n(b), x)} \ge \frac{1}{\|x\|_2} \min_{t\in\mathbf{T}} |x(t)| \frac{1}{\sqrt{n}}. \tag{13.32}$$

In the case of Rademacher variables we have $\|x\|_2 = \sqrt{2n}$, and (13.31) in conjunction with (13.32) gives

$$P\left(\frac{\kappa^{\mathrm{Toep}}(T_n(b), x)}{\kappa(T_n(b), x)} < \frac{\delta}{\sqrt{2}\,n^{3/2}}\right) \le P\left(\frac{1}{\|x\|_2} \min_{t\in\mathbf{T}} |x(t)| \frac{1}{\sqrt{n}} < \frac{\delta}{\sqrt{2}\,n^{3/2}}\right) < 0.005,$$

which yields the assertion with $\varepsilon = \delta/\sqrt{2}$. If u_j, v_j are independent standard normal variables, then $\|x\|_2^2 = \|u\|_2^2 + \|v\|_2^2$ is χ_{2n}^2-distributed. Thus the expected value of $\|x\|_2^2$ is $2n$, and from Chebychev's inequality we get

$$P\left(\frac{1}{\|x\|_2} \le \frac{1}{20\sqrt{n}}\right) = P(\|x\|_2^2 \ge 400n) \le \frac{2n}{400n} = 0.005. \tag{13.33}$$

Combining (13.31) and (13.33) we obtain

$$P\left(\min_{t\in\mathbf{T}} |x(t)| < \frac{\delta}{\sqrt{n}} \text{ or } \frac{1}{\|x\|_2} \le \frac{1}{20\sqrt{n}}\right) < 0.005 + 0.005 = 0.01,$$

and hence

$$P\left(\min_{t\in\mathbf{T}} |x(t)| \ge \frac{\delta}{\sqrt{n}} \text{ and } \frac{1}{\|x\|_2} > \frac{1}{20\sqrt{n}}\right) > 0.99.$$

Now (13.32) gives

$$P\left(\frac{\kappa^{\mathrm{Toep}}(T_n(b), x)}{\kappa(T_n(b), x)} \ge \frac{\delta}{20\,n^{3/2}}\right) \ge P\left(\frac{1}{\|x\|_2} \min_{t\in\mathbf{T}} |x(t)| \frac{1}{\sqrt{n}} \ge \frac{\delta}{20\,n^{3/2}}\right) > 0.99,$$

which is the assertion with $\varepsilon = \delta/20$. □

We conclude with an example of a Toeplitz-like structure for which there may occur indeed drastic differences between structured and unstructured condition numbers.

Example 13.21. Let $\mathrm{Str}_n(\mathbf{R}) = \mathrm{SymtridiagToep}_n(\mathbf{R})$ be the set of all $n \times n$ symmetric tridiagonal Toeplitz matrices with real entries. If $T_n(f)$ is a matrix in $\mathrm{SymtridiagToep}_n(\mathbf{R})$, then the equality $T_n(f)x = \Psi_x f$ is satisfied with

$$\Psi_x := \Psi_x^{\mathrm{SymtridiagToep}} := \begin{pmatrix} x_0 & x_1 \\ x_1 & x_0 + x_2 \\ x_2 & x_1 + x_3 \\ \dots & \dots \\ x_{n-2} & x_{n-3} + x_{n-1} \\ x_{n-1} & x_{n-2} \end{pmatrix}.$$

Since $\|T_n(f)\|_2$ is at least the ℓ^2 norm of its second column, we have $\|T_n(f)\|_2 \geq \|f\|_2$ and hence, for every $T_n(b) \in \mathrm{SymtridiagToep}_n(\mathbf{R})$,

$$\begin{aligned} &\Phi^{\mathrm{SymtridiagToep}}(T_n(b), x) \\ &= \sup\{\|T_n^{-1}(b)T_n(f)x\|_2 : T_n(f) \in \mathrm{SymtridiagToep}_n(\mathbf{R}), \|T_n(f)\|_2 \leq 1\} \\ &= \sup\{\|T_n^{-1}(b)\Psi_x f\|_2 : T_n(f) \in \mathrm{SymtridiagToep}_n(\mathbf{R}), \|T_n(f)\|_2 \leq 1\} \\ &\leq \sup\{\|T_n^{-1}(b)\Psi_x f\|_2 : T_n(f) \in \mathrm{SymtridiagToep}_n(\mathbf{R}), \|f\|_2 \leq 1\} \\ &\leq \|T_n^{-1}(b)\Psi_x\|_2. \end{aligned} \tag{13.34}$$

Now suppose that the natural number n is divisible by 6, that is, $n = 6m$. Put $z = (y, y, 0)$ and $x = (z, -z, z, -z, \dots, z, -z)$. Then Ψ_x has two equal columns, $\Psi_x = (x^\top \; x^\top)$. It is easily seen that $T_n(b)x = (b_0 + b_1)x$, whence

$$T_n^{-1}(b)\Psi_x = \frac{1}{b_0 + b_1}\,\Psi_x.$$

Consequently,

$$\|T_n^{-1}(b)\Psi_x\|_2 = \frac{1}{|b_0 + b_1|}\sqrt{2}\,\|x\|_2,$$

and from (13.34) we infer that

$$\Phi^{\mathrm{SymtridiagToep}}(T_n(b), x) \leq \frac{\sqrt{2}\,\|x\|_2}{|b_0 + b_1|},$$

which together with Proposition 13.9 gives

$$\kappa^{\mathrm{SymtridiagToep}}(T_n(b), x) \leq \|T_n(b)\|_2 \frac{\sqrt{2}}{|b_0 + b_1|} \leq \sqrt{2}\,\frac{|b_0| + 2|b_1|}{|b_0 + b_1|}.$$

Thus, $\kappa^{\mathrm{SymtridiagToep}}(T_n(b), x)$ remains bounded as $n = 6m \to \infty$. On the other hand, from Corollary 4.34 we know that if $|b_0| \leq 2|b_1|$, then

$$\kappa(T_n(b), x) = \|T_n(b)\|_2 \, \|T_n^{-1}(b)\|_2 \simeq n^2. \qquad \square$$

13.6 The Condition Number for Matrix Inversion

For an invertible matrix $A_n \in \mathrm{Str}_n(\mathbf{K})$, the *structured condition number for matrix inversion* is defined by

$$\kappa^{\mathrm{Str}}(A_n) = \lim_{\varepsilon\to 0} \sup \left\{ \frac{\|(A_n+\delta A_n)^{-1} - A_n^{-1}\|_2}{\varepsilon \|A_n^{-1}\|_2} : \delta A_n \in \mathrm{Str}_n(\mathbf{K}),\ \frac{\|\delta A_n\|_2}{\|A_n\|_2} \le \varepsilon \right\}.$$

The role played by $\Phi^{\mathrm{Str}}(A_n, x)$ in Section 13.3 is now figured by

$$\Omega^{\mathrm{Str}}(A_n) := \sup\{\|A_n^{-1}\delta A_n A_n^{-1}\|_2 : \delta A_n \in \mathrm{Str}_n(\mathbf{K}),\ \|\delta A_n\|_2 \le 1\}.$$

Proposition 13.22. *If* $\mathrm{Str}_n(\mathbf{K})$ *is invariant under multiplication by real scalars, then*

$$\kappa^{\mathrm{Str}}(A_n) = \frac{\|A_n\|_2}{\|A_n^{-1}\|_2}\,\Omega^{\mathrm{Str}}(A_n).$$

Proof. If $\varepsilon > 0$ is small enough, then $\|A_n^{-1}\|_2\|\delta A_n\|_2 < 1$ and hence

$$\begin{aligned}(A_n+\delta A_n)^{-1} &= [A_n(I + A_n^{-1}\delta A_n)]^{-1} = (I - A_n^{-1}\delta A_n + O(\varepsilon^2))A_n^{-1} \\ &= A_n^{-1} - A_n^{-1}\delta A_n A_n^{-1} + O(\varepsilon^2).\end{aligned}$$

It follows that

$$\begin{aligned}\frac{\|(A_n+\delta A_n)^{-1} - A_n^{-1}\|_2}{\varepsilon\|A_n^{-1}\|_2} &= \frac{\|A_n^{-1}\delta A_n A_n^{-1} + O(\varepsilon^2)\|_2}{\varepsilon\|A_n^{-1}\|_2} \\ &= \frac{\|A_n\|_2}{\|A_n^{-1}\|_2}\left\| A_n^{-1}\left(\frac{1}{\varepsilon\|A_n\|_2}\delta A_n\right)A_n^{-1}\right\|_2 + O(\varepsilon^2),\end{aligned}$$

which implies that

$$\kappa^{\mathrm{Str}}(A_n) = \frac{\|A_n\|_2}{\|A_n^{-1}\|_2}\sup\{\|A_n^{-1}\delta A_n A_n^{-1}\|_2 : \delta A_n \in \mathrm{Str}_n(\mathbf{K}),\ \|\delta A_n\|_2 \le 1\}. \qquad \square$$

As usual, we omit the superscript Str in case $\mathrm{Str}_n(\mathbf{K}) = M_n(\mathbf{K})$. The following observation shows that in the unstructured case we get back the usual condition number.

Proposition 13.23. *If* A_n *is invertible, then* $\kappa(A_n) = \|A_n\|_2\|A_n^{-1}\|_2$.

Proof. It is clear that $\Omega(A_n) \le \|A_n^{-1}\|_2^2$. To prove that $\Omega(A_n) \ge \|A_n^{-1}\|_2^2$, choose $x \in \mathbf{K}^n$ so that $\|x\|_2 = 1$ and $\|A_n^{-1}x\|_2 = \|A_n^{-1}\|_2$. Thus, $A_n^{-1}x = \|A_n^{-1}\|_2 y$ with $\|y\|_2 = 1$. We obtain

$$\begin{aligned}\Omega(A_n) &\ge \sup\{\|A_n^{-1}\delta A_n A_n^{-1}x\|_2 : \|\delta A_n\|_2 \le 1\} \\ &= \|A_n^{-1}\|_2 \sup\{\|A_n^{-1}\delta A_n y\|_2 : \|\delta A_n\|_2 \le 1\},\end{aligned}$$

and since δA_n defined by $\delta A_n z = (z, y)x$ obviously satisfies $\|\delta A_n\|_2 = 1$ and $\delta A_n y = x$, it results that $\Omega(A_n) \ge \|A_n^{-1}\|_2\|A_n^{-1}x\|_2 = \|A_n^{-1}\|_2^2$. The assertion is now immediate from Proposition 13.23. $\square$

The next result is the key for what follows.

Theorem 13.24 (Takagi). *Let $A_n \in M_n(\mathbf{K})$ be symmetric, $A_n^\top = A_n$, and let $\sigma_1 \le \cdots \le \sigma_n$ be the singular values of A_n. Then there exists a unitary matrix $V_n \in M_n(\mathbf{K})$ such that $A_n = V_n \operatorname{diag}(\sigma_1, \ldots, \sigma_n)\, V_n^\top$.*

A proof is in [166, Corollary 4.4.4], for example.

Now we are able to prove that for the structure $\mathrm{Toep}_n(\mathbf{K})$ there is actually no difference between the structured and the usual condition numbers for matrix inversion.

Theorem 13.25. *We have $\kappa^{\mathrm{Toep}}(T_n(b)) = \kappa(T_n(b))$.*

Proof. In view of Propositions 13.22 and 13.23 it remains to prove that

$$\Omega^{\mathrm{Toep}}(T_n(b)) \ge \|T_n^{-1}(b)\|_2^2.$$

Let $0 < \sigma_1 \le \cdots \le \sigma_n$ be the singular values on $T_n(b)$. Recall that W_n is the matrix with units on the anti-diagonal and zeros elsewhere. As this matrix is unitary, the singular values of $T_n(b)W_n$ coincide with those of $T_n(b)$. The (Hankel) matrix $T_n(b)W_n$ is symmetric, and hence Theorem 13.24 ensures the existence of a unitary matrix V_n such that $T_n(b)W_n = V_n \operatorname{diag}(\sigma_1, \ldots, \sigma_n)\, V_n^\top$. It follows that

$$W_n T_n^{-1}(b) = \overline{V}_n \operatorname{diag}\left(\frac{1}{\sigma_1}, \ldots, \frac{1}{\sigma_n}\right) V_n^*.$$

Put $x = V_n e_1$ and $y = \overline{V}_n e_1$. Then

$$W_n T_n^{-1}(b) x = \overline{V}_n \operatorname{diag}\left(\frac{1}{\sigma_1}, \ldots, \frac{1}{\sigma_n}\right) e_1 = \frac{1}{\sigma_1}\overline{V}_n e_1 = \frac{1}{\sigma_1} y$$

and, consequently,

$$\begin{aligned}\Omega^{\mathrm{Toep}}(T_n(b)) &= \sup\{\|T_n^{-1}(b)T_n(g)T_n^{-1}(b)\|_2 : \|T_n(g)\|_2 \le 1\}\\ &\ge \sup\{\|T_n^{-1}(b)T_n(g)T_n^{-1}(b)x\|_2 : \|T_n(g)\|_2 \le 1\}\\ &= \frac{1}{\sigma_1}\sup\{\|T_n^{-1}(b)T_n(g)W_n y\|_2 : \|T_n(g)\|_2 \le 1\}.\end{aligned}$$

By Lemma 13.13, there is a matrix $T_n(g)$ of norm at most 1 such that $T_n(g)W_n y = \overline{y}$. Thus,

$$\Omega^{\mathrm{Toep}}(T_n(b)) \ge \frac{1}{\sigma_1}\|T_n^{-1}(b)\overline{y}\|_2.$$

Since

$$\begin{aligned}T_n^{-1}(b)\overline{y} &= W_n W_n T_n^{-1}(b)\overline{y}\\ &= W_n \overline{V}_n \operatorname{diag}\left(\frac{1}{\sigma_1}, \ldots, \frac{1}{\sigma_n}\right) V_n^* V_n e_1 = W_n \overline{V}_n \frac{1}{\sigma_1} e_1,\end{aligned}$$

we finally obtain that

$$\Omega^{\mathrm{Toep}}(T_n(b)) \ge \frac{1}{\sigma_1}\|T_n^{-1}(b)\overline{y}\|_2 = \frac{1}{\sigma_1^2}\|W_n\overline{V}_n e_1\|_2 = \frac{1}{\sigma_1^2} = \|T_n^{-1}(b)\|_2^2. \qquad \square$$

13.7 Once More the Nearest Singular Matrix

In contrast to Section 13.1, we now measure the distance of a Toeplitz matrix to the nearest singular matrix within the set of *all* Toeplitz matrices. Thus, let

$$d^{\text{Toep}}(T_n(b)) = \inf\{\varepsilon > 0 : 0 \in \operatorname{sp} T_n(b+\varphi),\ \|T_n(\varphi)\|_2 \le \varepsilon\}.$$

The following theorem tells us that this distance is equal to the usual distance measured within the set of all $n \times n$ matrices.

Theorem 13.26. *We have* $d^{\text{Toep}}(T_n(b)) = \sigma_1(T_n(b))$.

Proof. Obviously, $d^{\text{Toep}}(T_n(b)) \ge d(T_n(b)) = \sigma_1(T_n(b))$ (recall Section 13.2). To prove the reverse inequality, we proceed as in the proof of Theorem 13.25. The matrix $T_n(b)W_n$ is symmetric and hence, by Theorem 13.24, $T_n(b)W_n = V_n S V_n^\top$, where V_n is unitary and $S = \operatorname{diag}(\sigma_1(T_n(b)), \dots, \sigma_n(T_n(b)))$. There is an x in $\mathbf{K}^n \setminus \{0\}$ such that $V_n^\top x = e_1$. With $\sigma_1 := \sigma_1(T_n(b))$ we therefore get

$$\begin{aligned} T_n(b)W_n x = V_n S V_n^\top x &= V_n S e_1 = V_n \sigma_1 e_1 = \sigma_1 V_n e_1 \\ &= \sigma_1 V_n \overline{e}_1 = \sigma_1 V_n \overline{V_n^\top x} = \sigma_1 V_n V_n^* \overline{x} = \sigma_1 \overline{x}. \end{aligned}$$

As $W_n x = \overline{W_n \overline{x}}$, we arrive at the equality $T_n(b)\overline{W_n \overline{x}} = \sigma_1 \overline{x}$. From Lemma 13.13 we infer the existence of a matrix $T_n(g)$ of norm at most 1 such that $T_n(g)\overline{W_n \overline{x}} = \overline{x}$. Thus, we have $(T_n(b) - \sigma_1 T_n(g))\overline{W_n \overline{x}} = 0$, which shows that $d^{\text{Toep}}(T_n(b))$ is not larger than $\sigma_1 \|T_n(g)\|_2 = \sigma_1$. □

Exercises

1. Let A_n and K_n be the $n \times n$ matrices

$$A_n = \begin{pmatrix} 1 & -1 & -1 & \dots & -1 \\ & 1 & -1 & \dots & -1 \\ & & 1 & \dots & -1 \\ & & & \ddots & \vdots \\ & & & & 1 \end{pmatrix}, \quad K_n = \begin{pmatrix} -\varepsilon & 0 & \dots & 0 \\ -\varepsilon & 0 & \dots & 0 \\ \vdots & \vdots & & \vdots \\ -\varepsilon & 0 & \dots & 0 \end{pmatrix}.$$

   Prove that $\det(A_n + K_n) = 1 - 2^{n-1}\varepsilon$ and hence $d(A_n) \le \sqrt{n}\,/\,2^{n-1}$. Show that, on the other hand, $d^{\text{Toep}[0,n]}(A_n) = 1$.

2. Let $b(t) = 2 + \alpha + t + t^{-1}$ where $\alpha > 0$ is small. Show that

$$\kappa(C_n(b)) = \frac{4+\alpha}{\alpha} \approx \frac{4}{\alpha}.$$

   Prove that if n is large and x is drawn from the unit sphere of $\mathbf{R}^n$ with the uniform distribution, then

$$\kappa^{\text{Circ}}(C_n(b), x) = \frac{(4+\alpha)^{1/4}(2+\alpha)^{1/2}}{\alpha^{3/4}} \approx \frac{2}{\alpha^{3/4}}$$

with probability near 1. Thus, for $\alpha = 0.01$,

$$\kappa(C_n(b)) \approx 400, \quad \kappa^{\mathrm{Circ}}(C_n(b), x) \approx 63,$$

and for $\alpha = 0.0001$,

$$\kappa(C_n(b)) \approx 40000, \quad \kappa^{\mathrm{Circ}}(C_n(b), x) \approx 2000.$$

How does this fit with the remark after the proof of Corollary 13.12?

3. Find the eigenvalues of the $2n \times 2n$ matrix

$$\begin{pmatrix} a & & & & b \\ & a & & b & \\ & & \cdots & & \\ & b & & a & \\ b & & & & a \end{pmatrix}.$$

4. Let

$$F(a,n) := T_n(|t-1|^2) + (a-2)E_{nn} := \begin{pmatrix} 2 & -1 & 0 & \dots & 0 & 0 \\ -1 & 2 & -1 & \dots & 0 & 0 \\ 0 & -1 & 2 & \dots & 0 & 0 \\ \vdots & \vdots & \vdots & \ddots & \vdots & \vdots \\ 0 & 0 & 0 & \dots & 2 & -1 \\ 0 & 0 & 0 & \dots & -1 & a \end{pmatrix}.$$

Prove that

$$\begin{aligned} F(1,n)^{-1} &= (\min(j,k))_{j,k=1}^n \\ F(2,n)^{-1} &= \left(\min(j,k) - \frac{jk}{n+1}\right)_{j,k=1}^n \\ F(3,n)^{-1} &= \left(\min(j,k) - \frac{2jk}{2n+1}\right)_{j,k=1}^n. \end{aligned}$$

Show that $\det F(a,n) = n(a-1)+1$.

5. (a) Let A and B be $k \times k$ matrices. Prove that the spectrum of the $nk \times nk$ block Toeplitz matrix

$$\begin{pmatrix} A & B & & & & \\ B & A & B & & & \\ & B & A & B & & \\ & & \ddots & \ddots & \ddots & \\ & & & B & A & B \\ & & & & B & A \end{pmatrix}$$

is $\bigcup_{j=1}^{n} \operatorname{sp}(A + 2B \cos \frac{\pi j}{n+1})$.

(b) Let C_n be the $nk \times nk$ matrix $(T_n(c_{ij}))_{i,j=1}^{k}$ with $c_{ij}(t) = a_{ij} + b_{ij}(t + t^{-1})$ ($a_{ij}, b_{ij} \in \mathbf{C}$). Put $A = (a_{ij})_{i,j=1}^{k}$ and $B = (b_{ij})_{i,j=1}^{k}$. Prove that

$$\operatorname{sp} C_n = \bigcup_{j=1}^{n} \operatorname{sp}\left(A + 2B \cos \frac{\pi j}{n+1}\right).$$

6. The *structured componentwise condition number* of a matrix $A_n \in \operatorname{Str}_n(\mathbf{R})$ at $x \in \mathbf{R}^n$ is defined by

$$\operatorname{cond}_{E,f}^{\operatorname{Str}}(A_n, x) = \lim_{\varepsilon \to 0} \sup \left\{ \frac{\|\delta x\|_\infty}{\varepsilon \|x\|_\infty} : (A_n + \delta A_n)(x + \delta x) = A_n x + \delta b, \right.$$
$$\delta A_n \in \operatorname{Str}_n(\mathbf{R}),\ |\delta A_n| \le \varepsilon |E_n|,$$
$$\left. \delta b \in \mathbf{R}^n,\ |\delta b| \le \varepsilon |f| \right\}.$$

Here $E_n \in \operatorname{Str}_n(\mathbf{R})$ is a given weight matrix and $f \in \mathbf{R}^n$ is a given weight vector. (The cases $E_n = A_n$, $f = b$ and $E_n = A_n$, $f = 0$ are of particular interest.) Furthermore, we use the absolute value and comparison of vectors and matrices componentwise.

(a) Prove that if $\operatorname{Str}_n(\mathbf{R}) = M_n(\mathbf{R})$, then

$$\operatorname{cond}_{E,f}^{\operatorname{Str}}(A_n, x) = \frac{\| \, |A_n^{-1}| \, |E_n| \, |x| + |A_n^{-1}| \, |f| \, \|_\infty}{\|x\|_\infty}.$$

(b) Prove that if $\operatorname{Str}_n(\mathbf{R}) = \operatorname{Toep}_n(\mathbf{R})$, then

$$\operatorname{cond}_{E,f}^{\operatorname{Str}}(A_n, x) = \frac{\| \, |A_n^{-1} \Psi_x| \, |p_{E_n}| + |A_n^{-1}| \, |f| \, \|_\infty}{\|x\|_\infty},$$

where Ψ_x is the matrix (13.12) and, for $E_n = T_n(h)$,

$$p_{E_n} := \begin{pmatrix} h_{-(n-1)} & h_{-(n-2)} & \dots & h_{n-1} & h_n \end{pmatrix}^{\top}$$

(so that $E_n x = \Psi_x p_{E_n}$ as in (13.13)).

(c) Prove that if $\operatorname{Str}_n(\mathbf{R}) = \operatorname{Circ}_n(\mathbf{R})$, then

$$\operatorname{cond}_{E,f}^{\operatorname{Str}}(A_n, x) = \frac{\| \, |E_n| \, |A_n^{-1} x| + |A_n^{-1}| \, |f| \, \|_\infty}{\|x\|_\infty}.$$

7. In our files, we found a copy of a transparency by Siegfried Rump showing the

following:

$$A = \begin{pmatrix} 2 & -1 & & \\ -1 & 2 & -1 & \\ & -1 & 2 & \ddots \\ & & \ddots & \ddots \end{pmatrix}, \quad x = \begin{pmatrix} 1 \\ -1 \\ 1 \\ -1 \\ \vdots \end{pmatrix},$$

$$\kappa(A) = \|A^{-1}\|\,\|A\| = 8.2 \cdot 10^6$$
$$\kappa^{\text{SymmToep}}(A, x) < 6.5 \cdot 10^4,$$
$$\text{cond}_{A,0}(A, x) = \frac{\|\,|A^{-1}|\,|A|\,|x|\,\|_\infty}{\|x\|_\infty} = 5.0 \cdot 10^5,$$
$$\text{cond}_{A,0}^{\text{SymmToep}}(A, x) = 1.$$

Which dimension does the matrix A have?

8. Let $A_n = T_n(1+\chi_1)+(-1)^{n+1}E_{1n}$, where E_{1n} is the matrix whose $1, n$ entry is 1 and whose other entries are all zero. Let $d_{\text{comp}}(A_n, |A_n|)$ be the infimum of all $\varepsilon > 0$ for which there exists a matrix $F_n \in M_n(\mathbf{R})$ such that $|F_n| \le \varepsilon |A_n|$ and $0 \in \text{sp}\,(A_n + F_n)$. Show that

$$\det A_n = 2, \quad d_{\text{comp}}(A_n, |A_n|) = 1, \quad \text{rad}\,(|A_n^{-1}|\,|A_n|) = n.$$

9. The full-Toeplitz structured pseudospectrum of a matrix $A_n \in M_n(\mathbf{C})$ is defined by

$$\text{sp}_\varepsilon^{\text{Toep}_n(\mathbf{C})}(A_n) = \bigcup_{E_n \in \text{Toep}_n(\mathbf{C}),\ \|E_n\|_2 \le \varepsilon} \text{sp}\,(A_n + E_n).$$

Prove that if $A_n \in \text{Toep}_n(\mathbf{C})$, then $\text{sp}_\varepsilon(A_n) = \text{sp}_\varepsilon^{\text{Toep}_n(\mathbf{C})}(A_n)$.

Notes

The problem of determining the distance of a structured matrix to the nearest singular matrix within the matrices of the same given structure and the problem of finding the componentwise (and not normwise) distance to the nearest singular matrix are studied in the work of Demmel [100], Gohberg and Koltracht [131], [132], D. J. Higham and N. J. Higham [160], [161], and Rump [231], [234], [235], [236], [237]. The results of Sections 13.1 and 13.2 are from our paper [57] with Kozak.

The structured condition numbers $\kappa_b^{\text{Str}}(A_n, x)$ and $\kappa^{\text{Str}}(A_n)$ as well as the distance $d^{\text{Str}}(A_n)$ were introduced by D. J. Higham and N. J. Higham [160], [161]. We learned of this topic from Siegfried Rump in 2001, and Sections 13.3 to 13.7 are based on Rump's paper [236]. In particular, all results and proofs of Sections 13.3, 13.6, and 13.7 are taken from [236]. The $T_n(g)x = \Psi_x g$ trick with the matrix (13.12) (and the analogue of this trick for other structures) was invented by D. J. Higham and N. J. Higham [160]. The marvelous Lemma 13.13 and Theorem 13.14 are Rump's [236]. Theorem 13.16 and Example 13.21 are also from [236]. Theorem 13.17 was established in [56]. Notice that it is this theorem that allows us to have recourse to the Konyagin and Schlag result [181] in an easy and luxurious way. We also remark that Theorems 13.14, 13.16, 13.17, 13.25, 13.26 are valid for general $n \times n$ Toeplitz matrices and are not limited to banded Toeplitz matrices.

Clearly, the study of Toeplitz band matrices within the structure $\text{Toep}_n(\mathbf{K})$ of all Toeplitz matrices is still too crude. Given $b \in \mathcal{P}_{r,s}$, the natural territory for the matrices $T_n(b)$ is the structure $\text{Toep}[r,s]_n(\mathbf{K})$ of all matrices $T_n(f)$ with $f \in \mathcal{P}_{r,s}$ and with the coefficients of f in $\mathbf{K}$. This strategy was pursued in Sections 13.1 and 13.2, and Corollary 13.7 is a really striking result: it says that outside $\Lambda(b)$, or at least at a certain distance from $\Lambda(b)$, no evil will happen. In contrast to this, the conclusion of Sections 13.4 to 13.7 is that the structured condition numbers and the distance to the nearest singular matrix in the structure $\text{Toep}_n(\mathbf{K})$ are never (or at least almost never) significantly better than their unstructured counterparts. We believe that for structured condition numbers things change when passing from $\text{Toep}_n(\mathbf{K})$ to $\text{Toep}[r,s]_n(\mathbf{K})$, that is, we conjecture that $\kappa_{\text{full}}^{\text{Toep}[r,s]}(T_n(b),x)$, $\kappa^{\text{Toep}[r,s]}(T_n(b),x)$, and $\kappa^{\text{Toep}[r,s]}(T_n(b))$ behave much better than $\kappa(T_n(b),x) = \kappa(T_n(b))$ and $\kappa(T_n(b))$. Example 13.21 and Corollary 13.7 show that there are good reasons for such a belief.

Exercise 1 is well known. We took it from [275]. We found Exercise 3 in [113]. Exercise 4 is a result of [122]. Part (a) of Exercise 6 is well known and due to Skeel [256] (see also [161]), part (b) is from [160] and [237], and part (c) is an observation of [237]. Exercise 8 is contained in [233]. The result of Exercise 9 was established in [140] and [238]. The last two papers have analogous results for plenty of other structures, too. Rump [238] also studies the case where $\text{Toep}_n(\mathbf{C})$ is replaced by $\text{Toep}_n(\mathbf{R})$. Theorem 8.2 of [62] implies that the analogue of the result of Exercise 9 for bounded Toeplitz operators on ℓ^2 is not true. This is another beautiful example of the fact that results being valid for all finite matrices need not extend to infinite matrices.

Further results: componentwise distance to the nearest singular matrix. For a matrix $C = (c_{jk})$, we denote by $|C|$ the matrix $(|c_{jk}|)$. An inequality of the form $C \le D$ is here (in contrast to the rest of the text) understood entrywise, that is, $C \le D$ means $c_{jk} \le d_{jk}$ for all j,k. Let $E_n \in M_n(\mathbf{R})$ be a given matrix with nonnegative entries. The componentwise distance $d_{\text{comp}}(a_n, E_n)$ of a matrix $A_n \in M_n(\mathbf{R})$ to the nearest singular matrix is defined as the infimum of all $\varepsilon > 0$ for which there exists a matrix $F_n \in M_n(\mathbf{R})$ such that $|F_n| \le \varepsilon E_n$ and $A_n + F_n$ is singular. One always has $d_{\text{comp}}(A_n, E_n) \ge 1/\text{rad}\,(|A_n^{-1}|\,E_n)$.

In 1992, N. J. Higham and J. Demmel raised the conjecture that there exists a $\gamma(n) < \infty$ such that

$$d_{\text{comp}}(A_n, |A_n|) \le \frac{\gamma(n)}{\text{rad}\,(|A_n^{-1}|\,|A_n|)}$$

(see [100]). This conjecture was confirmed by Rump [232], [233] (see also [234]), who even proved that for arbitrary weight matrices E_n the inequality

$$\frac{1}{\text{rad}\,(|A_n^{-1}|\,E_n)} \le d_{\text{comp}}(A_n, E_n) \le \frac{(3+2\sqrt{2})n}{\text{rad}\,(|A_n^{-1}|\,E_n)}$$

holds. Notice that Exercise 8 implies that $\gamma(n) \ge n$.

Chapter 14

Impurities

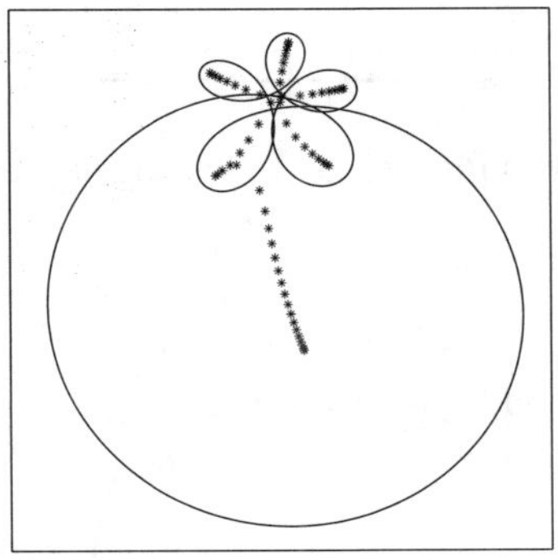

In this chapter we describe several phenomena that arise when randomly perturbing a relatively small number of entries of an infinite or a large finite Toeplitz matrix. We illustrate the appearance of localized eigenvectors by a very simple example. We also use the pseudospectral approach to explain the emergence of bubbles and antennae. It turns out that there are again big differences between the cases of infinite and large finite Toeplitz matrices. The chapter ends with some results on the important question whether structured pseudospectra can jump.

14.1 The Discrete Laplacian

Let $\sigma(t) = t + t^{-1}$ ($t \in \mathbf{T}$). The matrix $T_n(\sigma)$ is referred to as the *discrete Laplacian* and matrices of the form $T_n(\sigma) + V_n$ with a real diagonal matrix $V_n = \operatorname{diag}(v_1, \ldots, v_n)$ are called *discrete Hamiltonians*. We here consider the case where V_n has only one nonzero entry, that is, we study the *discrete Laplacian with a single impurity*. In this simple situation a fairly complete analysis is possible. Let E_{jj} denote the matrix whose jj entry is 1 and all other entries of which are zero. The spectrum of $T_n(\sigma)$ is a subset of $(-2, 2)$. Moreover, we know from Section 2.2 that $T_n(\sigma)$ has the eigenvalues

$$\lambda_j^{(n)} = 2\cos\frac{\pi j}{n+1} \quad (j = 1, \ldots, n)$$

and that

$$x^{(j)} = \left(\sin\frac{\pi j}{n+1}, \sin\frac{2\pi j}{n+1}, \ldots, \sin\frac{n\pi j}{n+1}\right)$$

is an eigenvector for $\lambda_j^{(n)}$. Notice that all eigenvectors are extended.

Proposition 14.1. *If $v > 1$ and $j \in \{2, \ldots, n-1\}$, then $T_n(\sigma) + vE_{jj}$ has an eigenvalue $\lambda > 2$ for each $n \geq 3$. The matrices $T_n(\sigma) + vE_{11}$ and $T_n(\sigma) + vE_{nn}$ have an eigenvalue $\lambda > 2$ whenever $v > 1 + 1/n$.*

Proof. Let $v > 1$. The eigenvalues of $T_3(\sigma) + vE_{22}$ are

$$\lambda_1 = 0, \quad \lambda_2 = \frac{v}{2} - \frac{\sqrt{v^2+8}}{2}, \quad \lambda_3 = \frac{v}{2} + \frac{\sqrt{v^2+8}}{2},$$

and $\lambda_3 > 2$ for $v > 1$. Since $T_3(\sigma) + vE_{22}$ is a submatrix of $T_n(\sigma) + vE_{jj}$ for every $n \geq 3$ and every $j \in \{2, \ldots, n-1\}$, we see that for these n and j,

$$\|T_n(\sigma) + vE_{jj}\|_2 \geq \|T_3(\sigma) + vE_{22}\|_2 = \lambda_3 > 2,$$

which implies that the maximal eigenvalue of $T_n(\sigma) + vE_{jj}$ is also greater than 2 (note that this conclusion can also be drawn from Theorem 9.19).

The determinant of $T_n(\sigma - \lambda) + vE_{11}$ is a polynomial in λ of degree n with the leading coefficient $(-1)^n$. Hence, this determinant has a zero in $(2, \infty)$ if its value at $\lambda = 2$ is $(-1)^n$ times a negative number. By formula (2.11),

$$\begin{aligned}
&\det(T_n(\sigma - 2) + vE_{11}) \\
&\quad = (v-2)\det T_{n-1}(\sigma - 2) - \det T_{n-2}(\sigma - 2) \\
&\quad = (v-2)(-1)^{n-1}n - (-1)^{n-2}(n-1) \\
&\quad = (-1)^n\Big(-(v-2)n - (n-1)\Big) = (-1)^n(-nv + n + 1)
\end{aligned}$$

and $-nv + n + 1 < 0$ whenever $v > 1 + 1/n$. The case where $T_n(\sigma) + vE_{nn}$ can be disposed of analogously, or it can be simply reduced to the case $T_n(\sigma) + vE_{11}$ by a similarity transformation. □

Proposition 14.2. *Consider the matrix* $T_n(\sigma) + \operatorname{diag}(v_1, \ldots, v_n)$*. If* $\lambda > 2$ *is an eigenvalue and* $x = (x_1, \ldots, x_n)$ *is an eigenvector for* λ*, then*

$$|x_1|^2 + \cdots + |x_n|^2 \leq \frac{1}{(\lambda - 2)^2}\left(|v_1 x_1|^2 + \cdots + |v_n x_n|^2\right).$$

Proof. Put $V = \operatorname{diag}(v_1, \ldots, v_n)$. We have $T_n(\sigma - \lambda)x = -Vx$, and taking into account that $\|T_n(\sigma)\|_2 \leq \|\sigma\|_\infty = 2 < \lambda$, we get

$$x = -T_n^{-1}(\sigma - \lambda)Vx = \frac{1}{\lambda}\sum_{k=0}^{\infty}\frac{1}{\lambda^k}T_n^k(\sigma)Vx,$$

whence

$$\|x\|_2 \leq \frac{1}{\lambda}\sum_{k=0}^{\infty}\left(\frac{2}{\lambda}\right)^k\|Vx\|_2 = \frac{1}{\lambda - 2}\|Vx\|_2. \qquad \square$$

Applying Proposition 14.2 to the matrix $T_n(\sigma) + vE_{jj}$, we obtain that

$$|x_1|^2 + \cdots + |x_n|^2 \leq \frac{|v|^2}{(\lambda - 2)^2}|x_j|^2. \tag{14.1}$$

This inequality shows that if n is large, then the vector x must be localized in some sense. For example, if $|v|^2/(\lambda-2)^2 = 25$ and $|x_j| = 1$, then

$$|x_1|^2 + \cdots + |x_{j-1}|^2 + |x_{j+1}|^2 + \cdots + |x_n|^2 \le 24,$$

which implies that at most 24 values of $|x_k|^2$ can be near 1, at most 48 values of $|x_k|^2$ can be near 1/2, etc. Clearly, the same localization phenomenon can be deduced from Proposition 14.2 in case diag $(v_1, \ldots, v_n)$ has only a finite number of nonzero entries and this number is small in comparison with n.

Put $D_k(\lambda) = \det T_k(\sigma - \lambda)$. From Theorem 2.6 we deduce that if the zeros

$$q_1 = -\frac{\lambda}{2} + \frac{\sqrt{\lambda^2-4}}{2}, \quad q_2 = -\frac{\lambda}{2} - \frac{\sqrt{\lambda^2-4}}{2}$$

of the polynomial $q^2 + \lambda q + 1$ are distinct, then

$$D_k(\lambda) = \frac{q_2^{k+1} - q_1^{k+1}}{q_2 - q_1}.$$

Let

$$p_1 = \frac{\lambda}{2} - \frac{\sqrt{\lambda^2-4}}{2}, \quad p_2 = \frac{\lambda}{2} + \frac{\sqrt{\lambda^2-4}}{2} \tag{14.2}$$

be the zeros of the polynomial $p^2 - \lambda p + 1$ and put

$$F_k(\lambda) = \frac{p_2^{k+1} - p_1^{k+1}}{p_2 - p_1}$$

for $p_1 \ne p_2$. Clearly, for $k \ge 1$,

$$p_1 = -q_1, \quad p_2 = -q_2, \quad F_k(\lambda) = (-1)^k D_k(\lambda).$$

Lemma 14.3. *Let $v \in \mathbf{R}$. A number $\lambda \in \mathbf{R}$ is an eigenvalue of $T_n(\sigma) + vE_{jj}$ if and only if*

$$(v-\lambda)F_{j-1}(\lambda)F_{n-j}(\lambda) + F_{j-2}(\lambda)F_{n-j}(\lambda) + F_{j-1}(\lambda)F_{n-j-1}(\lambda) = 0. \tag{14.3}$$

Proof. Expanding the determinant $\det(T_n(\sigma) + vE_{jj} - \lambda I)$ by the jth row, we get $(-1)^{n+1}$ times the left-hand side of (14.3). □

Proposition 14.4. *Let $v > 1$ and let $\lambda_n(v, 1)$ be the maximal eigenvalue of the matrix $T_n(\sigma) + vE_{11}$. Then, as $n \to \infty$,*

$$\lambda_n(v, 1) = v + \frac{1}{v} + o(1).$$

Proof. Proposition 14.1 implies that $\lambda_n := \lambda_n(v, 1) > 2$ for all sufficiently large n. Moreover, the argument of the first half of the proof of Proposition 14.1 shows that $\lambda_n \le \lambda_{n+1}$

for all n. Since $\lambda_n \le \|T_n(\sigma) + vE_{11}\|_2 \le 2 + v$, it follows that λ_n converges to some limit λ_∞. From Lemma 14.3 we infer that

$$v - \lambda_n = -\frac{F_{n-2}(\lambda_n)}{F_{n-1}(\lambda_n)} = -\frac{p_{2,n}^{n-1} - p_{1,n}^{n-1}}{p_{2,n}^{n} - p_{1,n}^{n}} \tag{14.4}$$

with

$$p_{1/2,n} = \frac{\lambda_n}{2} \mp \frac{\sqrt{\lambda_n^2 - 4}}{2} \to \frac{\lambda_\infty}{2} \mp \frac{\sqrt{\lambda_\infty^2 - 4}}{2} =: p_{1/2}.$$

Since $p_2 > 1 + \varepsilon$ and $0 < p_1 < 1 - \varepsilon$ with some $\varepsilon > 0$ for all sufficiently large n, we conclude that the right-hand side of (14.4) converges to $-1/p_2 = -p_1$. Thus,

$$v - \lambda_\infty = -\frac{\lambda_\infty}{2} + \frac{\sqrt{\lambda_\infty^2 - 4}}{2},$$

whence $\lambda_\infty = v + 1/v$. □

Proposition 14.5. *Let $j_n = [n/2]+1$. For $v > 1$, define $\lambda_n(v, j_n)$ as the maximal eigenvalue of $T_n(\sigma) + vE_{j_n,j_n}$. Then, as $n \to \infty$,*

$$\lambda_n(v, j_n) = \sqrt{v^2 + 4} + o(1).$$

Proof. The matrix $T_n(\sigma) + vE_{j_n,j_n}$ is a submatrix of $T_{n+1}(\sigma) + vE_{j_{n+1},j_{n+1}}$. Hence $\lambda_n := \lambda_n(v, j_n)$ converges monotonically to some limit $\lambda_\infty \in (2, 2 + v)$ as $n \to \infty$ (recall the proof of Proposition 14.4). From Lemma 14.3 we obtain

$$v - \lambda_n = -\frac{F_{j_n-2}(\lambda_n)}{F_{j_n-1}(\lambda_n)} - \frac{F_{n-j_n-1}(\lambda_n)}{F_{n-j_n}(\lambda_n)},$$

and as in the proof of Proposition 14.4 it follows that

$$-\frac{F_{j_n-2}(\lambda_n)}{F_{j_n-1}(\lambda_n)} - \frac{F_{n-j_n-1}(\lambda_n)}{F_{n-j_n}(\lambda_n)} \to -\frac{1}{p_2} - \frac{1}{p_2} = -2p_1.$$

Consequently, $v - \lambda_\infty = -\lambda_\infty + \sqrt{\lambda_\infty^2 - 4}$, which gives $\lambda_\infty = \sqrt{\lambda_\infty^2 + 4}$. □

The last two propositions provide us with two cases in which the constant $v/(\lambda - 2)$ appearing in (14.1) is seen to be no astronomic number. In the case of a single corner impurity we have

$$\frac{v}{\lambda_n(v, 1) - 2} \approx \frac{v}{v + 1/v - 2} \le 5 \ \text{ for all } \ v > 1.8,$$

while in the case of a single center impurity we even get

$$\frac{v}{\lambda_n(v, [n/2] + 1) - 2} \approx \frac{v}{\sqrt{v^2 + 4} - 2} = \sqrt{1 + \frac{4}{v^2}} + \frac{2}{v} < \sqrt{5} + 2 < 5 \ \text{ for all } \ v > 1.$$

Proposition 14.6. *Let $v \in \mathbf{R}$ and let $\lambda \in \mathbf{R} \setminus \{-2, 2\}$ be an eigenvalue of the matrix $T_n(\sigma) + vE_{jj}$. Then an eigenvector for λ is given by*

$$x = (\psi_1, \dots, \psi_j, \varphi_{n-j}, \dots, \varphi_1), \tag{14.5}$$

where

$$\psi_k = F_{n-j}(\lambda)F_{k-1}(\lambda) = \frac{p_2^{n-j+1} - p_1^{n-j+1}}{p_2 - p_1} \frac{p_2^k - p_1^k}{p_2 - p_1} \quad (1 \le k \le j),$$

$$\varphi_k = F_{j-1}(\lambda)F_{k-1}(\lambda) = \frac{p_2^j - p_1^j}{p_2 - p_1} \frac{p_2^k - p_1^k}{p_2 - p_1} \quad (1 \le k \le n-j),$$

$$p_{1/2} = \frac{\lambda}{2} \mp \frac{\sqrt{\lambda^2 - 4}}{2}.$$

Proof. We must verify that

$$\Big(T_n(\sigma - \lambda)x + vE_{jj}x\Big)_k = 0 \tag{14.6}$$

for $k \in \{1, \dots, n\}$. If $k \in \{1, \dots, j-1\}$, then the left-hand side of (14.6) is

$$\psi_{k-1} - \lambda\psi_k + \psi_{k+1} = F_{n-j}(\lambda)\Big(F_{k-2}(\lambda) - \lambda F_{k-1}(\lambda) + F_k(\lambda)\Big)$$

(note that $F_{-1}(\lambda) = 0$), and this is zero because p_1, p_2 are the zeros of the polynomial $p^2 - \lambda p + 1$. If $k = j + \ell$ with $\ell \in \{1, \dots, n-j\}$, then the left-hand side of (14.6) equals

$$x_{j+\ell-1} - \lambda x_{j+\ell} + x_{j+\ell+1} = \varphi_{n-j-\ell+2} - \lambda\varphi_{n-j+\ell+1} + \varphi_{n-j-\ell}$$

(notice that $x_j = \psi_j = F_{n-j}(\lambda)F_{j-1}(\lambda) = \varphi_{n-j+1}$), which is

$$F_{j-1}(\lambda)\Big(F_{n-j-\ell+1}(\lambda) - \lambda F_{n-j-\ell}(\lambda) + F_{n-j-\ell-1}(\lambda)\Big) = 0,$$

again because p_1, p_2 are the roots of the equation $p^2 - \lambda p + 1 = 0$. Finally, the jth equation of (14.6) is

$$\begin{aligned} 0 &= \psi_{j-1} + (v - \lambda)\psi_j + \varphi_{n-j} \\ &= F_{n-j}(\lambda)F_{j-2}(\lambda) + (v - \lambda)F_{n-j}(\lambda)F_{j-1}(\lambda) + F_{j-1}(\lambda)F_{n-j-1}(\lambda), \end{aligned}$$

which is equivalent to (14.3). □

The following corollary is illustrated in Figure 14.1.

Corollary 14.7. *Let $v \in \mathbf{R}$ and let $\lambda \in \mathbf{R}$ be an eigenvalue of $T_n(\sigma) + vE_{jj}$. If $|\lambda| < 2$, then the eigenvector* (14.5) *is extended, while if $|\lambda| > 2$, then the eigenvector* (14.5) *is localized around j.*

Proof. This follows from Proposition 14.6 along with the observation that $|p_1| = |p_2| = 1$ if $|\lambda| < 2$, $0 < p_1 < 1 < p_2$ if $\lambda > 2$, and $p_1 < -1 < p_2 < 0$ if $\lambda < -2$. □

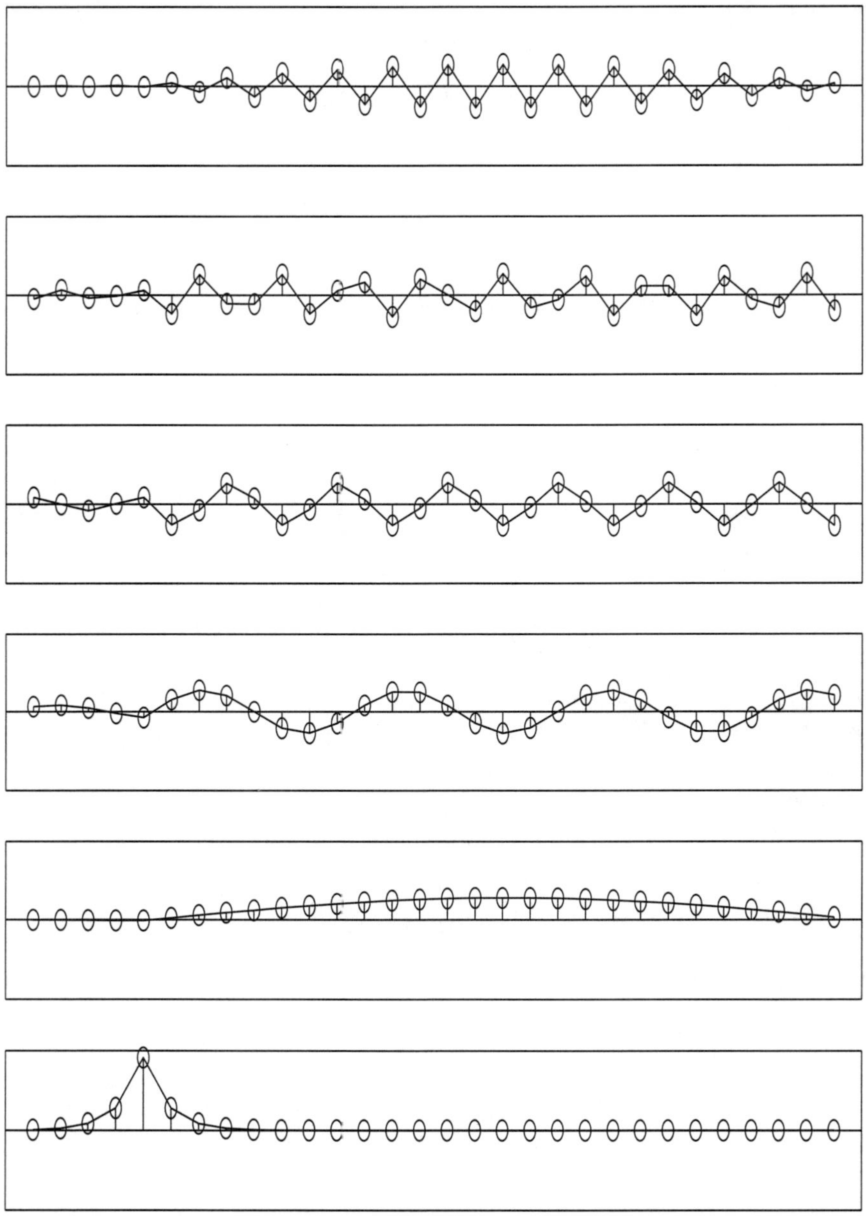

Figure 14.1. *The matrix $T_{30}(\sigma) + 3\,E_{5,5}$ has* 30 *real eigenvalues $\lambda_1 < \cdots < \lambda_{30}$. The maximal eigenvalue λ_{30} is* 3.6055*; the remaining eigenvalues are located in* $(-2, 2)$. *The pictures show eigenvectors to $\lambda_1 = -1.9858$, $\lambda_8 = -1.3385$, $\lambda_{15} = -0.0243$, $\lambda_{22} = 1.3102$, $\lambda_{29} = 1.9850$, and $\lambda_{30} = 3.6055$ (from the top to the bottom). The vertical axis is always from* -1 *to* 1.

Finally, let n be large, choose m small in comparison with n (say, $m = 10$), and let

$$1 = j_0 < j_1 < \cdots < j_m < j_{m+1} = n.$$

Consider the matrix

$$A_n = T_n(\sigma) + \sum_{\ell=1}^{m} v_\ell E_{j_\ell, j_\ell}$$

and suppose $\lambda \in \mathbf{R}\backslash\{-2, 2\}$ is an eigenvalue of A_n. Let x be an eigenvector for λ. Assume we have an ℓ for which $j_{\ell+1} - j_\ell$ is large (say, $n/5$). The values of x_k for $k \in \{j_\ell+1, \ldots, j_{\ell+1}-1\}$ are determined by the difference equation

$$x_{k-1} - \lambda x_k + x_{k+1} = 0$$

and the two boundary conditions

$$x_{j_\ell} - \lambda x_{j_\ell+1} + x_{j_\ell+2} = x_{j_{\ell+1}-2} - \lambda x_{j_{\ell+1}-1} + x_{j_{\ell+1}} = 0$$

(with $x_{-1} = x_{n+1} = 0$). This gives

$$x_k = c_1 p_1^k + c_2 p_2^k$$

with certain constants c_1, c_2 and

$$p_{1/2} = \lambda/2 \mp (1/2)\sqrt{\lambda^2 - 4}.$$

If $|\lambda| > 2$, then necessarily $0 < p_1 < 1 < p_2$ or $p_1 < -1 < p_2 < 0$. For the sake of definiteness, assume $0 < p_1 < 1 < p_2$. If $c_2 \neq 0$, then x_k is localized near $j_{\ell+1}$. If $c_2 = 0$ and $c_1 \neq 0$, then x_k is localized near j_ℓ. In case $c_1 = c_2 = 0$, x_k vanishes identically for $k \in \{j_\ell + 1, \ldots, j_{\ell+1} - 1\}$, which may also be viewed as being localized. If $|\lambda| < 2$, then $|p_1| = |p_2| = 1$ and hence x_k is extended unless $c_1 = c_2 = 0$.

14.2 An Uncertain Block

We now turn to general Toeplitz band matrices with a finite number of impurities. Let A be a bounded linear operator on $\ell^2(\mathbf{Z})$, $\ell^2(\mathbf{N})$, or $\mathbf{C}^n = \ell^2(\{1, \ldots, n\})$. Recall that P_m is the orthogonal projection onto the first m coordinates. For $\varepsilon > 0$, we set

$$\mathrm{sp}_\varepsilon^m A = \bigcup_{\|K\|_2 \le \varepsilon} \mathrm{sp}\,(A + P_m K P_m).$$

Thus, $\mathrm{sp}_\varepsilon^m A$ measures the extent to which $\mathrm{sp}\, A$ can increase by a perturbation (impurity, uncertainty) of norm at most ε localized in the upper-left $m \times m$ block of A.

In the notation of Section 7.1, $\mathrm{sp}_\varepsilon^m A$ is just $\mathrm{sp}_\varepsilon^{P_m, P_m} A$, and the results of Section 7.1 imply that

$$\mathrm{sp}_\varepsilon^m A = \bigcup_{\|K\|_2 \le \varepsilon,\ \mathrm{rank}\, K = 1} \mathrm{sp}\,(A + P_m K P_m)$$

and

$$\mathrm{sp}_\varepsilon^m A = \mathrm{sp}\, A \cup \left\{ \lambda \notin \mathrm{sp}\, A : \|P_m(A - \lambda I)^{-1} P_m\|_2 \geq \frac{1}{\varepsilon} \right\}. \tag{14.7}$$

If $b(t) = b_0$ is constant (equivalently, if $T(b)$ is diagonal), then, obviously,

$$\mathrm{sp}_\varepsilon^m T(b) \supset \mathrm{sp}_\varepsilon^1 T(b) = b_0 + \varepsilon \overline{\mathbf{D}}, \quad \mathrm{sp}\, T(b) = \{b_0\}.$$

Thus, $\mathrm{sp}_\varepsilon^m T(b)$ is strictly larger than $\mathrm{sp}\, T(b)$ for every $\varepsilon > 0$. As the following theorem shows, nondiagonal infinite Toeplitz band matrices behave differently.

Theorem 14.8. *If b is a nonconstant Laurent polynomial, then there exists a number $\varepsilon_1 = \varepsilon_1(b, m) > 0$ such that*

$$\mathrm{sp}_\varepsilon^m T(b) = \mathrm{sp}\, T(b) \text{ for all } \varepsilon \in (0, \varepsilon_1).$$

Proof. Due to (14.7), it suffices to show that

$$\sup_{\lambda \notin \mathrm{sp}\, T(b)} \|P_m T^{-1}(b - \lambda) P_m\|_2 < \infty.$$

If $\lambda \notin \mathrm{sp}\, T(b)$, we can write $b - \lambda = b_- b_+$ with

$$b_-(t) = \prod_{j=1}^r \left(1 - \frac{\delta_j}{t}\right), \quad b_+(t) = b_s \prod_{k=1}^s (t - \mu_k),$$

where $|\delta_j| < 1$, $|\mu_k| > 1$, $b_s \neq 0$, and hence

$$T^{-1}(b - \lambda) = T(b_+^{-1}) T(b_-^{-1})$$

(recall Section 1.4). Since

$$P_m T(b_+^{-1}) = P_m T(b_+^{-1}) P_m, \quad T(b_-^{-1}) P_m = P_m T(b_-^{-1}) P_m,$$

it follows that

$$P_m T^{-1}(b - \lambda) P_m = T_m(b_+^{-1}) T_m(b_-^{-1}).$$

We have

$$b_-^{-1}(t) = \left(1 + \frac{\delta_1}{t} + \frac{\delta_1^2}{t^2} + \cdots\right) \cdots \left(1 + \frac{\delta_r}{t} + \frac{\delta_r^2}{t^2} + \cdots\right) =: \sum_{n=0}^\infty \frac{c_n}{t^n}$$

with

$$|c_n| = \left| \sum_{\alpha_1 + \cdots + \alpha_r = n} \delta_1^{\alpha_1} \ldots \delta_r^{\alpha_r} \right| \leq (|\delta_1| + \cdots + |\delta_r|)^n \leq r^n.$$

Thus,

$$\|T_m(b_-^{-1})\|_2^2 \le m|c_0|^2 + (m-1)|c_1|^2 + \cdots + |c_{m-1}|^2$$
$$\le m - (m-1)r^2 + \cdots + r^{2m-2}.$$

On writing

$$b_+(t) = b_s \prod_{k=1}^{s}(-\mu_k) \prod_{k=1}^{s}\left(1 - \frac{t}{\mu_k}\right),$$

we obtain analogously that

$$\|T(b_+^{-1})\|_2^2 \le \frac{1}{|b_s|^2}\left(m + (m-1)s^2 + \cdots + s^{2m-2}\right).$$

In summary,

$$\|P_m T^{-1}(b-\lambda)P_m\|_2^2 \le \frac{1}{|b_s|^2}\left(\sum_{\ell=0}^{m-1}(m-\ell)r^{2\ell}\right)\left(\sum_{\ell=0}^{m-1}(m-\ell)s^{2\ell}\right), \tag{14.8}$$

which is the desired result. □

Let

$$b(t) = \sum_{j=-r}^{s} b_j t^j, \quad r \ge 1, \quad s \ge 1, \quad b_{-r}b_s \ne 0.$$

Inequality (14.8) implies that Theorem 14.8 is true with

$$\varepsilon_1 = |b_s|\left(\sum_{\ell=0}^{m-1}(m-\ell)r^{2\ell}\right)^{-1/2}\left(\sum_{\ell=0}^{m-1}(m-\ell)s^{2\ell}\right)^{-1/2}.$$

Considering the transpose of $T(b)$, we can slightly improve (and symmetrize) this estimate to

$$\varepsilon_1 \le \max(|b_{-r}|, |b_s|)\left(\sum_{\ell=0}^{m-1}(m-\ell)r^{2\ell}\right)^{-1/2}\left(\sum_{\ell=0}^{m-1}(m-\ell)s^{2\ell}\right)^{-1/2}. \tag{14.9}$$

We now turn to large finite Toeplitz band matrices. From Theorems 11.3 and 11.17 we know that

$$\lim_{n\to\infty} \operatorname{sp} T_n(b) = \Lambda(b) = \bigcap_{\varrho>0} \operatorname{sp} T(b_\varrho). \tag{14.10}$$

If $T(b)$ is triangular, then, obviously,

$$\operatorname{sp}_\varepsilon^m T_n(b) \supset \operatorname{sp}_\varepsilon^1 T_n(b) = b_0 + \varepsilon\overline{\mathbf{D}}$$

for each $\varepsilon > 0$ and each $n \geq 1$. This implies that the limit of $\mathrm{sp}_\varepsilon^m T_n(b)$ as $n \to \infty$ is strictly larger than $\Lambda(b) = \{b_0\}$ for each $\varepsilon > 0$. The following theorem shows that this cannot happen for nontriangular Toeplitz band matrices provided $\varepsilon > 0$ is sufficiently small.

Theorem 14.9. *Let b be a Laurent polynomial and suppose $T(b)$ is not a triangular matrix. Then there exists an $\varepsilon_2 = \varepsilon_2(b, m) > 0$ such that*

$$\lim_{n\to\infty} \mathrm{sp}_\varepsilon^m T_n(b) = \Lambda(b) \quad \text{for all} \quad \varepsilon \in (0, \varepsilon_2).$$

The proof of this theorem is based on three lemmas. Since $T(b)$ is supposed to be nontriangular, we can write

$$b(t) = \sum_{j=-r}^{s} b_j t^j, \quad r \geq 1, \quad s \geq 1, \quad b_{-r} \neq 0, \quad b_s \neq 0.$$

Lemma 14.10. *There exists a constant $\delta = \delta(b) > 1$ such that*

$$\Lambda(b) = \bigcap_{\varrho\in[1/\delta,\delta]} \mathrm{sp}\, T(b_\varrho).$$

Proof. We have

$$b_\varrho(t) = b_s \varrho^s t^s \left(1 + \frac{b_{s-1}}{b_s}\frac{1}{\varrho t} + \cdots + \frac{b_{-r}}{b_s}\frac{1}{\varrho^{r+r} t^{r+s}}\right).$$

Hence, if ϱ is large enough, then, for all $\lambda \in \mathrm{sp}\, T(b)$, $b_\varrho - \lambda$ has no zeros on $\mathbf{T}$ and $\mathrm{wind}\,(b_\varrho - \lambda) = s \neq 0$. This implies there is a $\varrho_1 \in (1, \infty)$ such that $\mathrm{sp}\, T(b) \subset \mathrm{sp}\, T(b_\varrho)$ for all $\varrho > \varrho_1$. Analogously, from the representation

$$b_\varrho(t) = b_{-r} \varrho^{-r} t^{-r} \left(1 + \frac{b_{-r+1}}{b_{-r}} \varrho t + \cdots + \frac{b_s}{b_{-r}} \varrho^{r+s} t^{r+s}\right)$$

we infer that there exists a $\varrho_2 \in (0, 1)$ such that $\mathrm{sp}\, T(b) \subset \mathrm{sp}\, T(b_\varrho)$ for all $\varrho < \varrho_2$. Letting $\delta := \max(\varrho_1, 1/\varrho_2)$ we get

$$\bigcap_{\varrho\notin[1/\delta,\delta]} \mathrm{sp}\, T(b_\varrho) \supset \mathrm{sp}\, T(b),$$

whence

$$\bigcap_{\varrho\in(0,\infty)} \mathrm{sp}\, T(b_\varrho) \supset \mathrm{sp}\, T(b) \cap \left[\bigcap_{\varrho\in[1/\delta,\delta]} \mathrm{sp}\, T(b_\varrho)\right] = \bigcap_{\varrho\in[1/\delta,\delta]} \mathrm{sp}\, T(b_\varrho). \qquad \square$$

Lemma 14.11. *For every $\varepsilon > 0$,*

$$\limsup_{n\to\infty} \mathrm{sp}_\varepsilon^m T_n(b) \subset \mathrm{sp}_\varepsilon^m T(b).$$

Proof. Pick $\lambda \in \mathbf{C} \setminus \mathrm{sp}_\varepsilon^m T(b)$. Then $\lambda \notin \mathrm{sp}\, T(b)$ and, by (14.7),

$$\| P_m T^{-1}(b-\lambda) P_m \|_2 < 1/\varepsilon.$$

It follows that there is an open neighborhood $U \subset \mathbf{C}$ of λ such that if $\mu \in U$, then $\mu \notin \mathrm{sp}\, T(b)$ and

$$\| P_m T^{-1}(b-\mu) P_m \|_2 < 1/\varepsilon.$$

From Corollary 3.8 we infer that $T_n^{-1}(b-\mu)$ converges strongly to $T^{-1}(b-\mu)$. Consequently,

$$\| P_m T_n^{-1}(b-\mu) P_m - P_m T^{-1}(b-\mu) P_m \|_2 \to 0 \text{ as } n \to \infty, \tag{14.11}$$

and it is straightforward to check that the convergence in (14.11) is uniform with respect to μ in compact subsets of U. Thus, there exist an open neighborhood $V \subset U$ of λ and a natural number n_0 such that

$$\| P_m T_n^{-1}(b-\mu) P_m \|_2 < 1/\varepsilon \text{ for all } \mu \in V \text{ and all } n \geq n_0.$$

This, in conjunction with (14.7), implies that $V \cap \mathrm{sp}_\varepsilon^m T_n(b) = \emptyset$ for all $n \geq n_0$, whence $\lambda \notin \limsup_{n\to\infty} \mathrm{sp}_\varepsilon^m T_n(b)$. □

Lemma 14.12. *Let δ be the constant given by Lemma* 14.10. *If $\varrho \in [1/\delta, \delta]$ and $\varepsilon > 0$, then*

$$\limsup_{n\to\infty} \mathrm{sp}_\varepsilon^m T_n(a) \subset \mathrm{sp}_\beta^m T(a_\varrho), \tag{14.12}$$

where $\beta = \varepsilon \max(\varrho^{m-1}, \varrho^{-m+1})$.

Proof. We have

$$T_n(b-\lambda) = D_n^{-1}(\varrho) T_n(b_\varrho - \lambda) D_n(\varrho),$$

where $D_n(\varrho)$ is the diagonal matrix diag $(\varrho, \varrho^2, \dots, \varrho^n)$. It follows that

$$P_m T_n^{-1}(b-\lambda) P_m = D_m^{-1}(\varrho) P_m T_n^{-1}(b_\varrho - \lambda) P_m D_m(\varrho),$$

whence

$$\| P_m T_n^{-1}(b-\lambda) P_m \|_2 \leq \kappa_m(\varrho) \| P_m T_n^{-1}(b_\varrho - \lambda) P_m \|_2,$$

where

$$\kappa_m(\varrho) := \| D_m^{-1}(\varrho) \|_2 \, \| D_m(\varrho) \|_2.$$

Since also $\mathrm{sp}\, T_n(b) = \mathrm{sp}\, T_n(b_\varrho)$, we conclude from (14.7) that

$$\begin{aligned}
&\mathrm{sp}_\varepsilon^m T_n(b) \\
&= \mathrm{sp}\, T_n(b) \cup \{\lambda \notin \mathrm{sp}\, T_n(b) : \| P_m T_n^{-1}(a-\lambda) P_m \|_2 \geq 1/\varepsilon\} \\
&= \mathrm{sp}\, T_n(b_\varrho) \cup \{\lambda \notin \mathrm{sp}\, T_n(b_\varrho) : \| P_m T_n^{-1}(b-\lambda) P_m \|_2 \geq 1/\varepsilon\} \\
&\subset \mathrm{sp}\, T_n(b_\varrho) \cup \{\lambda \notin \mathrm{sp}\, T_n(b_\varrho) : \kappa_m(\varrho) \| P_m T_n^{-1}(b_\varrho-\lambda) P_m \|_2 \geq 1/\varepsilon\} \\
&= \mathrm{sp}_{\varepsilon\kappa_m(\varrho)}^m T_n(b_\varrho),
\end{aligned}$$

and now Lemma 14.11 yields the inclusion

$$\limsup_{n\to\infty} \operatorname{sp}_\varepsilon^m T_n(b) \subset \operatorname{sp}_{\varepsilon\kappa_m(\varrho)}^m T(b_\varrho).$$

Because $\kappa_m(\varrho) = \varrho^{m-1}$ if $\varrho \geq 1$ and $\kappa_m(\varrho) = \varrho^{-m+1}$ if $\varrho < 1$, we are then able to arrive at the assertion. □

***Proof of Theorem* 14.9.** Denote the right-hand side of (14.9) by

$$C(r, s, m) \max(|b_{-r}|, |b_s|),$$

let δ be the constant from Lemma 14.10, and put

$$\varepsilon_2 := C(r, s, m)|b_s|/\delta^{s+m-1}.$$

We claim that Theorem 14.9 is true with this choice of ε_2.

Let $\varepsilon \in (0, \varepsilon_2)$. Lemma 14.12 shows that if $\varrho \in [1/\delta, \delta]$, then (14.12) holds with

$$\begin{aligned}
\beta &= \varepsilon \max(\varrho^{m-1}, \varrho^{-m+1}) \\
&< C(r, s, m)|b_s| \max(\varrho^{m-1}, \varrho^{-m+1})/\delta^{s+m-1} \\
&\leq C(r, s, m)|b_s|\varrho^s\delta^s \max(\varrho^{m-1}, \varrho^{-m+1})/\delta^{s+m-1} \quad \text{(since } 1 \leq \varrho\delta) \\
&= C(r, s, m)|b_s|\varrho^s \max(\varrho^{m-1}, \varrho^{-m+1})/\delta^{m-1} \\
&\leq C(r, s, m)|b_s|\varrho^s.
\end{aligned} \tag{14.13}$$

Applying Theorem 14.8 to the matrix $T(b_\varrho)$ and taking into account that $|b_s|$ does not exceed $\max(|b_{-r}|, |b_s|)$, we see that

$$\operatorname{sp}_\beta^m T(b_\varrho) = \operatorname{sp} T(b_\varrho) \quad \text{for} \quad \beta < C(r, s, m)|b_s|\varrho^s.$$

Thus, (14.12) and (14.13) give

$$\limsup_{n\to\infty} \operatorname{sp}_\varepsilon^m T_n(b) \subset \operatorname{sp} T(b_\varrho)$$

for every $\varrho \in [1/\delta, \delta]$. By Lemma 14.10, this implies that

$$\limsup_{n\to\infty} \operatorname{sp}_\varepsilon^m T_n(b) \subset \Lambda(b).$$

To establish the inclusion in the other direction, note that (14.10) implies

$$\Lambda(b) \subset \liminf_{n\to\infty} \operatorname{sp} T_n(b) \subset \liminf_{n\to\infty} \operatorname{sp}_\varepsilon^m T_n(b),$$

thus completing the proof of Theorem 14.9. □

In summary, Theorems 14.8 and 14.9 show that $\operatorname{sp}_\varepsilon^m T(b)$ and $\lim \operatorname{sp}_\varepsilon^m T_n(b)$ stabilize at constant sets *before* ε reaches zero and that, moreover, these two constant sets are in general *different*. Thus, except for some trivial cases, we always have

$$\lim_{n\to\infty} \operatorname{sp}_\varepsilon^m T_n(b) \subsetneqq \operatorname{sp}_\varepsilon^m T(b)$$

for sufficiently small ε, implying that the passage from the "finite volume case" to the "infinite volume case" is discontinuous.

14.3 Emergence of Antennae

Let A be a bounded operator on ℓ^2 over $\mathbf{Z}$, $\mathbf{N}$, or $\{1, \dots, n\}$. We denote by E_j the projection defined by

$$(E_j x)_k = \begin{cases} x_j & \text{for } k = j, \\ 0 & \text{for } k \neq j \end{cases}$$

and by E_{jk} the operator given by the matrix whose j, k entry is 1 and all other entries of which are zero. For a subset Ω of $\mathbf{C}$, we put

$$\operatorname{sp}_\Omega^{(j,k)} A = \bigcup_{\omega \in \Omega} \operatorname{sp}(A + \omega E_{jk}). \tag{14.14}$$

Thus, $\operatorname{sp}_\Omega^{(j,k)} A$ is the union of all possible spectra that may emerge as the result of a perturbation of A in the j, k site by a number randomly chosen in Ω. Notice that in the case where $\Omega = \varepsilon\overline{\mathbf{D}}$, the set (14.14) is just the set $\operatorname{sp}_\varepsilon^{E_j, E_k} A$ introduced in Section 7.1 (recall Theorem 7.2).

For a set $M \subset \mathbf{C}$, we define $-1/M$ as the set

$$-1/M := \{z \in \mathbf{C} : 1 + \mu z = 0 \text{ for some } \mu \in M\}.$$

Here is an analogue to (7.3).

Lemma 14.13. *If* $0 \in \Omega$, *then*

$$\operatorname{sp}_\Omega^{(j,k)} A = \operatorname{sp} A \cup \{\lambda \notin \operatorname{sp} A : [(A - \lambda I)^{-1}]_{kj} \in -1/\Omega\},$$

where $[(A - \lambda I)^{-1}]_{kj}$ *is the* k, j *entry of the resolvent* $(A - \lambda I)^{-1}$.

Proof. Since $0 \in \Omega$, we have $\operatorname{sp}_\Omega^{(j,k)} A = \operatorname{sp} A \cup X$ with some set $X \subset \mathbf{C} \setminus \operatorname{sp} A$. Fix $\lambda \in \mathbf{C} \setminus \operatorname{sp} A$. Obviously, λ belongs to X if and only if there is an $\omega \in \Omega$ such that $A - \lambda I + \omega E_{jk}$ is not invertible or, equivalently, such that $I + (A - \lambda I)^{-1}\omega E_{jk}$ is not invertible. As E_{jk} is a trace class operators, the operator

$$I + (A - \lambda I)^{-1}\omega E_{jk}$$

is not invertible if and only if

$$0 = \det(I + (A - \lambda I)^{-1}\omega E_{jk}) = 1 + \omega[(A - \lambda I)^{-1}]_{kj},$$

which proves the assertion. $\square$

Now let b be a Laurent polynomial. For $\lambda \notin \operatorname{sp} T(b)$, we denote the j, k entry of $T^{-1}(b - \lambda)$ by $d_{jk}(\lambda)$. Thus,

$$T^{-1}(b - \lambda) = (d_{jk}(\lambda))_{j,k=1}^{\infty} \text{ for } \lambda \in \mathbf{C} \setminus \operatorname{sp} T(b).$$

We also put

$$H_\Omega^{jk}(b) = \{\lambda \in \mathbf{C} \setminus \operatorname{sp} T(b) : d_{kj}(\lambda) \in -1/\Omega\}. \tag{14.15}$$

Theorem 14.14. *If* $0 \in \Omega$, *then*

$$\mathrm{sp}_\Omega^{(j,k)} T(b) = \mathrm{sp}\, T(b) \cup H_\Omega^{jk}(b).$$

Proof. This is immediate from Lemma 14.13 and the definition of $H_\Omega^{jk}(b)$. □

Example 14.15. Let $b(t) = t + \alpha^2 t^{-1}$ ($t \in \mathbf{T}$) with $\alpha \in [0, 1)$. The set $b(\mathbf{T})$ is the ellipse

$$\{(1+\alpha^2)\cos\theta + i(1-\alpha^2)\sin\theta : \theta \in [0, 2\pi)\}.$$

Let E_+ and E_- denote the interior and the exterior of the ellipse $b(\mathbf{T})$, respectively. Clearly,

$$\mathbf{C} \setminus \mathrm{sp}\, T(b) = E_-.$$

For $\varrho > 1$, define b_ϱ by $b_\varrho(t) = \varrho t + \alpha^2 \varrho^{-1} t^{-1}$. Then $b_\varrho(\mathbf{T})$ is the ellipse

$$\{(\varrho + \alpha^2\varrho^{-1})\cos\theta + i(\varrho - \alpha^2\varrho^{-1})\sin\theta : \theta \in [0, 2\pi)\}.$$

Since

$$\varrho + \alpha^2\varrho^{-1} > 1 + \alpha^2 \quad \text{and} \quad \varrho - \alpha^2\varrho^{-1} > 1 - \alpha^2 \quad \text{for} \quad \varrho > 1,$$

the ellipses $b_\varrho(\mathbf{T})$ are contained in E_- for every $\varrho > 1$, and each point of E_- lies on exactly one of these ellipses. In other words, each point $\lambda \in E_-$ can be uniquely written as

$$\lambda = \varrho e^{i\theta} + \alpha^2\varrho^{-1}e^{-i\theta} \text{ with } \varrho \in (1, \infty), \theta \in [0, 2\pi).$$

We have

$$b(t) - \lambda = t^{-1}(t^2 - \lambda t + \alpha^2) = t^{-1}(t - z_1)(t - z_2)$$

with

$$z_1 = \alpha^2\varrho^{-1}e^{-i\theta}, \quad z_2 = \varrho e^{i\theta}.$$

Since $|z_1| = \alpha^2\varrho^{-1} < 1$ and $|z_2| = \varrho > 1$, we get $T^{-1}(b - \lambda) = T(b_+^{-1})T(b_-^{-1})$ with

$$b_+(t) = t - z_2 = -z_2\left(1 - \frac{t}{z_2}\right), \quad b_-(t) = t^{-1}(t - z_1) = 1 - \frac{z_1}{t}.$$

Taking into account that

$$b_+^{-1}(t) = -\frac{1}{z_2}\left(1 + \frac{t}{z_2} + \frac{t^2}{z_2^2} + \cdots\right), \quad b_-^{-1}(t) = 1 + \frac{z_1}{t} + \frac{z_1^2}{t^2} + \cdots,$$

we arrive at the representation

$$T^{-1}(b - \lambda) = -\frac{1}{z_2}\begin{pmatrix} 1 & & & \\ 1/z_2 & 1 & & \\ 1/z_2^2 & 1/z_2 & 1 & \\ \dots & \dots & \dots & \dots \end{pmatrix}\begin{pmatrix} 1 & z_1 & z_1^2 & \dots \\ & 1 & z_1 & \dots \\ & & 1 & \dots \\ & & & \dots \end{pmatrix}.$$

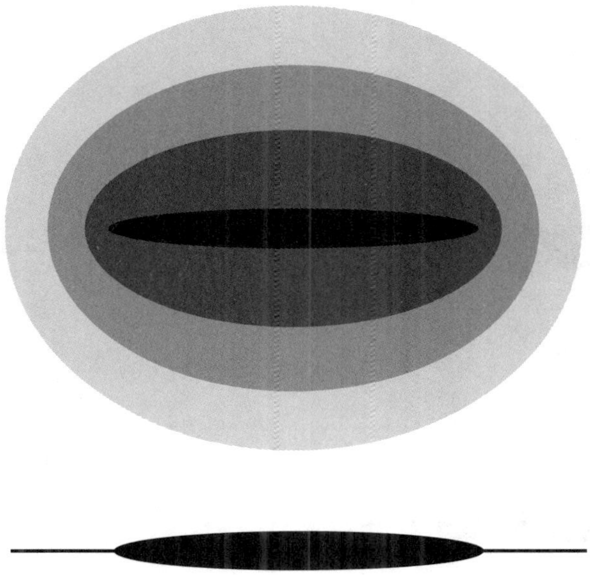

$\mathrm{sp}^{(1,1)}_{\varepsilon\overline{\mathbf{D}}} T(b)$

for $\varepsilon = 0, \frac{3}{2}, 2, \frac{5}{2}$

$\mathrm{sp}^{(1,1)}_{[-\varepsilon,\varepsilon]} T(b)$

for $\varepsilon = \frac{5}{2}$

Figure 14.2. *The sets* $\mathrm{sp}^{(1,1)}_{\varepsilon\overline{\mathbf{D}}} T(b)$ *and* $\mathrm{sp}^{(1,1)}_{[-\varepsilon,\varepsilon]} T(b)$ *for* $b(t) = t + \alpha^2 t^{-1}$ *with* $\alpha = \frac{9}{10}$.

In particular,

$$d_{11}(\lambda) = -\frac{1}{z_2} = -\frac{1}{\varrho e^{i\theta}}, \quad d_{12}(\lambda) = -\frac{z_1}{z_2} = -\frac{\alpha^2}{\varrho^2 e^{2i\theta}},$$

$$d_{21}(\lambda) = -\frac{1}{z_2^2} = -\frac{1}{\varrho^2 e^{2i\theta}}, \quad d_{22}(\lambda) = -\frac{1}{z_2}\left(\frac{z_1}{z_2} + 1\right) = -\frac{1}{\varrho e^{i\theta}}\left(\frac{\alpha^2}{\varrho^2 e^{2i\theta}} + 1\right).$$

These formulas and their analogues for general $d_{jk}(\lambda)$ can be used to compute the sets $H^{jk}_{\Omega}(b)$.

In the cases $\Omega = \varepsilon\overline{\mathbf{D}}$ and $\Omega = [-\varepsilon, \varepsilon]$, (14.15) reads

$$H^{jk}_{\varepsilon\overline{\mathbf{D}}}(b) = \left\{\lambda \in E_- : |d_{kj}(\lambda)| > \frac{1}{\varepsilon}\right\},$$

$$H^{jk}_{[-\varepsilon,\varepsilon]}(b) = \left\{\lambda \in E_- : d_{kj}(\lambda) \in \left(-\infty, -\frac{1}{\varepsilon}\right) \cup \left(\frac{1}{\varepsilon}, \infty\right)\right\}.$$

Figure 14.2 shows examples of $\mathrm{sp}^{(1,1)}_{\varepsilon\overline{\mathbf{D}}} T(b)$ and $\mathrm{sp}^{(1,1)}_{[-\varepsilon,\varepsilon]} T(b)$: An elliptic "halo" emerges in the former spectrum and two "wings" or "antennae" arise in the latter. In Figures 14.3 and 14.4, we illustrate $\mathrm{sp}^{(2,2)}_{\varepsilon\overline{\mathbf{D}}} T(b)$ and $\mathrm{sp}^{(2,2)}_{[-\varepsilon,\varepsilon]} T(b)$, and $\mathrm{sp}^{(3,3)}_{\varepsilon\overline{\mathbf{D}}} T(b)$ and $\mathrm{sp}^{(3,3)}_{[-\varepsilon,\varepsilon]} T(b)$ for the same values of ε. Notice that $\mathrm{sp}^{(2,2)}_{\varepsilon\overline{\mathbf{D}}} T(b)$ and $\mathrm{sp}^{(3,3)}_{\varepsilon\overline{\mathbf{D}}} T(b)$ contain holes that disappear for larger values of ε. Computing the sets $\mathrm{sp}^{(j,k)}_{[-\varepsilon,\varepsilon]} T(b)$ for all (j, k) with $\varepsilon = 5$, we arrive at Figure 14.5, which is examined in more detail in the three close-ups of Figure 14.7. Finally, Figure 14.6 is an attempt to obtain Figure 14.5 by replacing the infinite matrix $T(b)$ with $T_{30}(b)$; we will say more about this in the following section. □

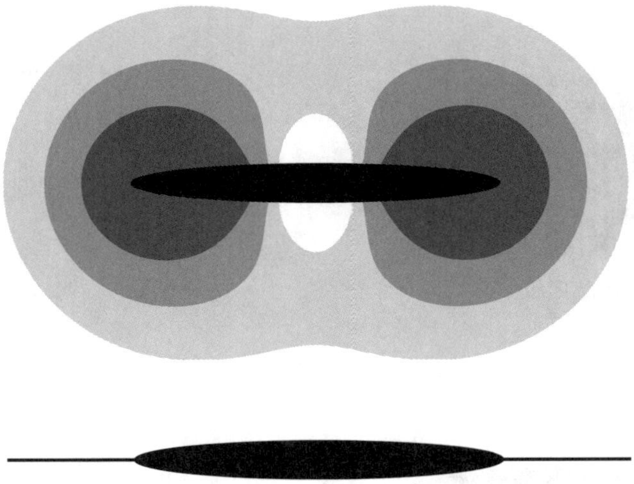

$\mathrm{sp}_{\varepsilon\overline{\mathbf{D}}}^{(2,2)}\, T(b)$

for $\varepsilon = 0, \frac{3}{2}, 2, \frac{5}{2}$

$\mathrm{sp}_{[-\varepsilon,\varepsilon]}^{(2,2)}\, T(b)$

for $\varepsilon = \frac{5}{2}$

Figure 14.3. *The sets* $\mathrm{sp}_{\varepsilon\overline{\mathbf{D}}}^{(2,2)} T(b)$ *and* $\mathrm{sp}_{[-\varepsilon,\varepsilon]}^{(2,2)} T(b)$ *for* $b(t) = t + \alpha^2 t^{-1}$ *with* $\alpha = \frac{9}{10}$.

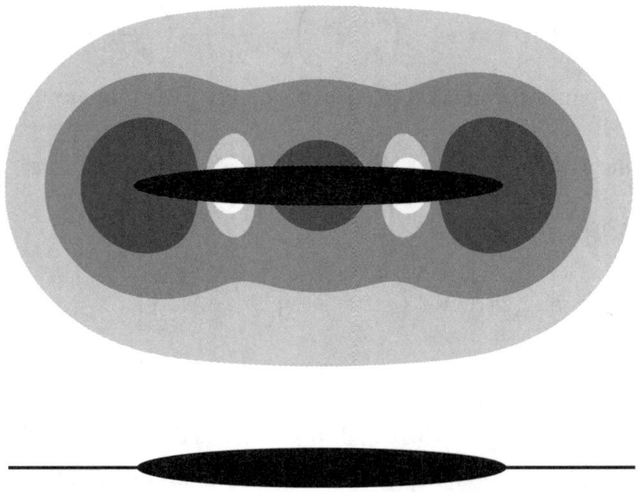

$\mathrm{sp}_{\varepsilon\overline{\mathbf{D}}}^{(3,3)}\, T(b)$

for $\varepsilon = \frac{3}{2}, 2, \frac{5}{2}$

$\mathrm{sp}_{[-\varepsilon,\varepsilon]}^{(3,3)}\, T(b)$

for $\varepsilon = \frac{5}{2}$

Figure 14.4. *The sets* $\mathrm{sp}_{\varepsilon\overline{\mathbf{D}}}^{(3,3)} T(b)$ *and* $\mathrm{sp}_{[-\varepsilon,\varepsilon]}^{(3,3)} T(b)$ *for* $b(t) = t + \alpha^2 t^{-1}$ *with* $\alpha = \frac{9}{10}$.

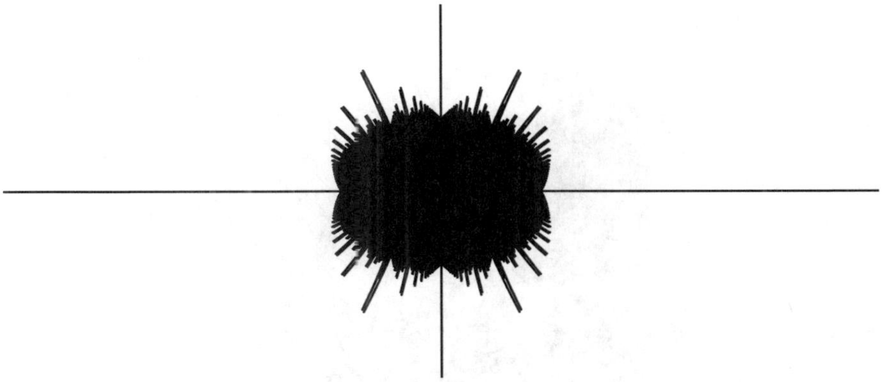

Figure 14.5. *The set* $\bigcup_{(j,k)\in\mathbf{N}\times\mathbf{N}} \mathrm{sp}^{(j\,k)}_{[-\varepsilon,\varepsilon]} T(b)$ *for* $b(t) = t + \alpha^2 t^{-1}$ *with* $\alpha = \frac{2}{5}$ *and* $\varepsilon = 5$.

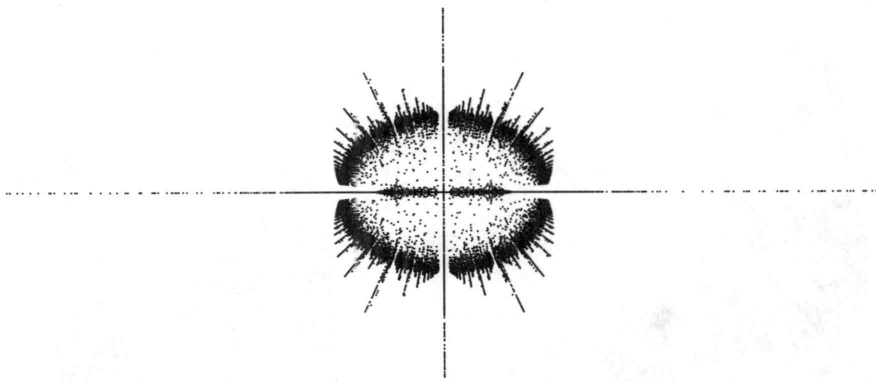

Figure 14.6. *Computed eigenvalues of* 5000 *Toeplitz matrices of dimension* 30 *with* $b(t) = t + \alpha^2 t^{-1}$ *for* $\alpha = \frac{2}{5}$, *each perturbed in a single random entry by a random number uniformly distributed in* $[-5, 5]$.

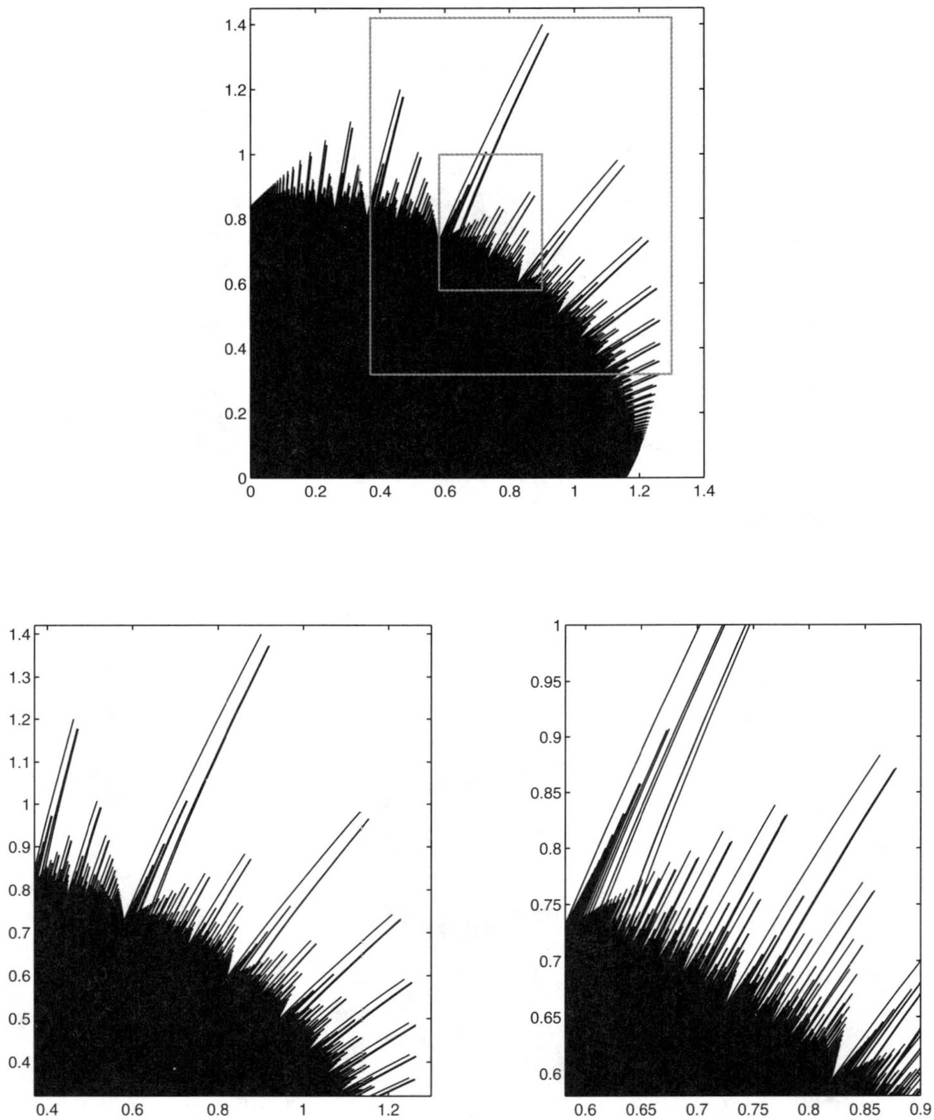

Figure 14.7. *Close-ups of the set shown in Figure* 14.5*; the gray boxes in the top image indicate the axes of the images below.*

14.4 Behind the Black Hole

We now study the behavior of $\operatorname{sp}_{\Omega}^{(j,k)} T_n(b)$ as n goes to infinity. Let b be a Laurent polynomial,

$$b(t) = \sum_{k=-r}^{s} b_k t^k, \quad b_{-r} b_s \neq 0, \tag{14.16}$$

and for $\varrho \in (0, \infty)$, define b_ϱ by $b_\varrho(t) = \sum_k b_k \varrho^k t^k$. The limiting set $\Lambda(b)$ given by (11.4) can be characterized as in Theorem 11.3. Thus, if $\lambda \in \mathbf{C} \setminus \Lambda(b)$, then there is a $\varrho > 0$ such that $T(b_\varrho - \lambda)$ is invertible.

Lemma 14.16. *The analytic functions $d_{jk} : \mathbf{C} \setminus \operatorname{sp} T(b) \to \mathbf{C}$ defined in Section 14.3 can be continued to analytic functions $d_{jk} : \mathbf{C} \setminus \Lambda(b) \to \mathbf{C}$ such that if $\lambda \in \mathbf{C} \setminus \Lambda(b)$, $\varrho > 0$, and $T(b_\varrho - \lambda)$ is invertible, then*

$$T^{-1}(b_\varrho - \lambda) = (\varrho^{j-k} d_{jk}(\lambda))_{j,k=1}^{\infty}. \tag{14.17}$$

Proof. Suppose first that $T(b)$ is not triangular; that is, let b be of the form (14.16) with $r \geq 1$ and $s \geq 1$. Pick $\lambda \in \mathbf{C} \setminus \Lambda(b)$ and choose any $\varrho > 0$ such that $T(b_\varrho - \lambda)$ is invertible. One can write

$$b(t) - \lambda = t^{-r} b_s \prod_{j=1}^{r+s} (t - z_j(\lambda)),$$

whence

$$b_\varrho(t) - \lambda = \varrho^{-r} t^{-r} b_s \prod_{j=1}^{r+s} (\varrho t - z_j(\lambda)).$$

Taking into account that $T(b_\varrho - \lambda)$ is invertible if and only if $b_\varrho - \lambda$ has no zeros on $\mathbf{T}$ and has winding number zero, it is not difficult to check that the invertibility of $T(b_\varrho - \lambda)$ is equivalent to the existence of a labelling of the zeros $z_j(\lambda)$ such that

$$|z_1(\lambda)| \leq \cdots \leq |z_r(\lambda)| < \varrho < |z_{r+1}(\lambda)| \leq \cdots \leq |z_{r+s}(\lambda)|. \tag{14.18}$$

Abbreviating $z_j(\lambda)$ to z_j, we have $b_\varrho(t) - \lambda = b_s b_-(t) b_+(t)$ with

$$b_-(t) = \prod_{j=1}^{r} \left(1 - \frac{z_j}{\varrho t}\right), \quad b_+(t) = \prod_{j=r+1}^{r+s} (\varrho t - z_j)$$

and thus

$$T^{-1}(b_\varrho - \lambda) = b_s^{-1} T(b_+^{-1}) T(b_-^{-1}). \tag{14.19}$$

Clearly,

$$b_-^{-1}(t) = \prod_{j=1}^{r} \left(1 + \frac{z_j}{\varrho t} + \frac{z_j^2}{\varrho^2 t^2} + \cdots\right) =: \sum_{n=0}^{\infty} b_n \varrho^{-n} t^{-n},$$

$$b_+^{-1}(t) = \frac{(-1)^s}{z_{r+1} \cdots z_{r+s}} \prod_{j=r+1}^{r+s} \left(1 + \frac{\varrho t}{z_j} + \frac{\varrho^2 t^2}{z_j^2} + \cdots\right) = \sum_{n=0}^{\infty} c_n \varrho^n t^n,$$

where $b_n = b_n(\lambda)$ and $c_n = c_n(\lambda)$ are independent of ϱ. Convergence of these series is a consequence of (14.18). Thus, by (14.19), $T^{-1}(b_\varrho - \lambda)$ equals

$$\frac{1}{b_s}\begin{pmatrix} c_0 & & & \\ c_1\varrho & c_0 & & \\ c_2\varrho^2 & c_1\varrho & c_0 & \\ \dots & \dots & \dots & \dots \end{pmatrix}\begin{pmatrix} b_0 & b_1/\varrho & b_2/\varrho^2 & \dots \\ & b_0 & b_1/\varrho & \dots \\ & & b_0 & \dots \\ & & & \dots \end{pmatrix},$$

which shows that

$$[T^{-1}(b_\varrho - \lambda)]_{j,k} = \varrho^{j-k} b_s^{-1}(c_{j-1}b_{k-1} + c_{j-2}b_{k-2} + \cdots).$$

This proves (14.17) with certain numbers $d_{jk}(\lambda)$ independent of ϱ.

If $T(b_\varrho - \lambda)$ is invertible, then so is $T(b_\varrho - \mu)$ for all μ in some open neighborhood of λ, and the entries of the inverse of $T(b_\varrho - \mu)$ are analytic functions of μ in this neighborhood. Thus, the functions d_{jk} are defined and analytic in $\mathbf{C} \setminus \Lambda(b)$. As outside $\operatorname{sp} T(b)$ these functions coincide with the functions d_{jk} introduced in Section 14.3, we arrive at the assertion of the lemma.

Finally, the proof is similar if $T(b)$ is triangular. □

The following two results will be used in the proof of Theorem 14.19, which is the main result of this section.

Lemma 14.17. *Fix a site (j, k) and let $\lambda \in \mathbf{C} \setminus \Lambda(b)$. If $\varrho > 0$ and $T(b_\varrho - \lambda)$ is invertible, then there exist an open neighborhood $U \subset \mathbf{C} \setminus \Lambda(b)$ of λ and a natural number n_0 such that $T(b_\varrho - \mu)$ is invertible for all $\mu \in U$, the matrices $T_n(b_\varrho - \mu)$ are invertible for all $\mu \in U$ and all $n \geq n_0$, and*

$$\left[T_n^{-1}(b_\varrho - \mu)\right]_{jk} \to \left[T^{-1}(b_\varrho - \mu)\right]_{jk} \quad \textit{as } n \to \infty$$

uniformly with respect to $\mu \in U$.

Proof. By Theorem 3.7, there exist an open neighborhood V of λ and a natural number m_0 such that $T(b_\varrho - \mu)$ is invertible for all $\mu \in V$,

$$M := \sup_{n \geq m_0} \sup_{\mu \in V} \|T_n^{-1}(b_\varrho - \mu)\| < \infty,$$

and

$$\left[T_n^{-1}(b_\varrho - \mu)\right]_{jk} \to \left[T^{-1}(b_\varrho - \mu)\right]_{jk} \quad \text{as } n \to \infty$$

for all $\mu \in V$. Hence, given any $\varepsilon > 0$, we can find a number $n_0 \geq m_0$ and an open neighborhood $U \subset V$ of λ such that

$$\left|\left[T_n^{-1}(b_\varrho - \lambda)\right]_{jk} - \left[T^{-1}(b_\varrho - \lambda)\right]_{jk}\right| < \varepsilon/3,$$

$$\left|\left[T^{-1}(b_\varrho - \mu)\right]_{jk} - \left[T^{-1}(b_\varrho - \lambda)\right]_{jk}\right| < \varepsilon/3,$$

and

$$\left|\left[T_n^{-1}(a_\varrho - \mu)\right]_{jk} - \left[T_n^{-1}(a_\varrho - \lambda)\right]_{jk}\right|$$
$$\le \|T_n^{-1}(a_\varrho - \mu) - T_n^{-1}(a_\varrho - \lambda)\|$$
$$\le |\mu - \lambda|\, \|T_n^{-1}(a_\varrho - \mu)\|\, \|T_n^{-1}(a_\varrho - \lambda)\| \le M^2|\mu - \lambda| < \varepsilon/3$$

for all $\mu \in U$ and all $n \ge n_0$. The assembly of these three $\varepsilon/3$ inequalities yields the assertion. □

Theorem 14.18 (Hurwitz). *Let $G \subset \mathbf{C}$ be an open set, let f be a function that is analytic in G and does not vanish identically, and let $\{f_n\}$ be a sequence of analytic functions in G that converges to f uniformly on compact subsets of G. If $f(\lambda) = 0$ for some $\lambda \in G$, then there is a sequence $\{\lambda_n\}$ of points $\lambda_n \in G$ such that $\lambda_n \to \lambda$ as $n \to \infty$ and $f_n(\lambda_n) = 0$ for all sufficiently large n.*

By virtue of Lemma 14.16 we can extend the sets $H_\Omega^{jk}(b) \subset \mathbf{C} \setminus \operatorname{sp} T(b)$ given by (14.15) to $\mathbf{C} \setminus \Lambda(b)$. Thus let henceforth

$$H_\Omega^{jk}(b) = \{\lambda \in \mathbf{C} \setminus \Lambda(b) : d_{kj}(\lambda) \in -1/\Omega\}.$$

Here is the analogue of Theorem 14.14 for large finite matrices. The technical assumptions made in the following result will be discussed later.

Theorem 14.19. *Let b be a Laurent polynomial and let Ω be a compact subset of $\mathbf{C}$ that contains the origin. If $d_{kj} : \mathbf{C} \setminus \Lambda(b) \to \mathbf{C}$ is identically zero or nowhere locally constant or assumes a constant value c that does not belong to $-1/\Omega$, then*

$$\lim_{n\to\infty} \operatorname{sp}_\Omega^{(j,k)} T_n(b) = \Lambda(b) \cup H_\Omega^{jk}(b). \tag{14.20}$$

Proof. Let G be a connected component of $\mathbf{C} \setminus \Lambda(b)$. We first prove that

$$\liminf_{n\to\infty} \operatorname{sp}_\Omega^{(j,k)} T_n(b) \cap \overline{G} \supset \left(\Lambda(b) \cup H_\Omega^{jk}(b)\right) \cap \overline{G}. \tag{14.21}$$

If $d_{kj}(\mu) = c \notin -1/\Omega$ or $d_{kj}(\mu) = 0$ for all $\mu \in \Omega$, then $H_\Omega^{jk}(b) = \emptyset$, and (14.21) is evident from Theorem 11.17. (Recall that $0 \in \Omega$.) Thus assume d_{kj} is not constant in G and $H_\Omega^{jk}(b)$ is not empty. Take λ in the right-hand side of (14.21). If λ is in the boundary ∂G of G, then λ is in $\Lambda(b)$ and hence in $\liminf \operatorname{sp}_\Omega^{(j,k)} T_n(b)$. Thus, let $\lambda \in G$. Since $\lambda \in H_\Omega^{jk}(b)$, there is an $\omega \in \Omega$ such that $1 + \omega d_{kj}(\lambda) = 0$. Choose $\varrho > 0$ so that $T(b_\varrho - \lambda)$ is invertible and let U and n_0 be as in Lemma 14.17. Due to Lemma 14.16,

$$f(\mu) := 1 + \omega d_{kj}(\mu) = 1 + \omega \varrho^{j-k} \left[T^{-1}(b_\varrho - \mu)\right]_{kj}$$

for $\mu \in U$. Lemma 14.17 shows that the functions f_n defined in U by

$$f_n(\mu) = 1 + \omega \varrho^{j-k} \left[T_n^{-1}(b_\varrho - \mu)\right]_{kj}$$

converge uniformly to f in U. Since f is not constant in U and is zero at $\lambda \in U$, Theorem 14.18 implies that there are $\lambda_n \in U$ such that $\lambda_n \to \lambda$ and $f_n(\lambda_n) = 0$. Let $D_\varrho = \operatorname{diag}(1, \varrho, \ldots, \varrho^{n-1})$. It can be readily verified that

$$T_n(b_\varrho - \mu) = D_\varrho T_n(b - \mu) D_\varrho^{-1}, \tag{14.22}$$

whence

$$\left[T_n^{-1}(b_\varrho - \mu)\right]_{kj} = \varrho^{k-j} \left[T_n^{-1}(b - \mu)\right]_{kj} \tag{14.23}$$

and thus

$$0 = f_n(\lambda_n) = 1 + \omega \left[T_n^{-1}(b - \lambda_n)\right]_{kj}.$$

From Lemma 14.13 we now deduce that $\lambda_n \in \operatorname{sp}_\Omega^{(j,k)} T_n(b)$, and since $\lambda_n \to \lambda$, it follows that λ is in the left-hand side of (14.21).

We now show that

$$\limsup_{n \to \infty} \operatorname{sp}_\Omega^{(j,k)} T_n(b) \cap \overline{G} \subset \Big(\Lambda(b) \cup H_\Omega^{jk}(b)\Big) \cap \overline{G}. \tag{14.24}$$

Pick λ in the left-hand side of (14.24). If $\lambda \in \partial G \subset \Lambda(b)$, then λ is obviously in the right-hand side of (14.24). We can therefore assume that $\lambda \in G$. By the definition of the partial limiting set, there are $\lambda_{n_\ell} \in \operatorname{sp}_\Omega^{(j,k)} T_{n_\ell}(b) \cap G$ such that $\lambda_{n_\ell} \to \lambda$. Choose $\varrho > 0$ so that $T(b_\varrho - \lambda)$ is invertible. By Lemma 14.17, the matrices $T_{n_\ell}(b_\varrho - \lambda_{n_\ell})$ are invertible whenever n_ℓ is sufficiently large, and from (14.22) it then follows that the matrices $T_n(b - \lambda_{n_\ell})$ are also invertible for all n_ℓ large enough. Hence, taking into account that $\lambda_{n_\ell} \in \operatorname{sp}_\Omega^{(j,k)} T_{n_\ell}(b)$ and using Lemma 14.13, we see that there are $\omega_{n_\ell} \in \Omega$ such that $1 + \omega_{n_\ell}[T_{n_\ell}^{-1}(b - \lambda_{n_\ell})]_{kj} = 0$. Due to (14.23), this implies that

$$1 + \omega_{n_\ell} \varrho^{j-k} \left[T_{n_\ell}^{-1}(b_\varrho - \lambda_{n_\ell})\right]_{kj} = 0. \tag{14.25}$$

Since Ω is compact, the sequence $\{\omega_{n_\ell}\}$ has a partial limit ω in Ω. Consequently, (14.25) and Lemma 14.17 give $1 + \omega \varrho^{j-k} [T^{-1}(a_\varrho - \lambda)]_{kj} = 0$, and Lemma 14.16 now yields the equality $1 + \omega d_{kj}(\lambda) = 0$. It results that $d_{kj}(\lambda) \in -1/\Omega$ and thus that $\lambda \in H_\Omega^{jk}(b)$.

From (14.21) and (14.24) we obtain that $\liminf \operatorname{sp}_\Omega^{(j,k)} T_n(b) \cap \overline{G}$ is equal to $\overline{G} \cap (\Lambda(b) \cup H_\Omega^{jk}(b))$. Considering the union over all components of $\mathbf{C} \setminus \Lambda(b)$ we arrive at (14.20). □

Example 14.20. Let $b(t) = t + \alpha^2 t^{-1}$ be as in Example 14.15. The range $b(\mathbf{T})$ is an ellipse, $\operatorname{sp} T(b)$ equals $b(\mathbf{T}) \cup E_+$, where E_+ denotes the set of points inside this ellipse, and $\Lambda(b) = [-2\alpha, 2\alpha]$ is the line segment between the foci of the ellipse. The expressions for $d_{kj}(\lambda)$ found in Example 14.15 for λ outside the ellipse E_+ extend by analyticity to all $\lambda \in \mathbf{C} \setminus \Lambda(b)$. Since $\mathbf{C} \setminus \Lambda(b)$ is connected, it is not difficult to show that the functions d_{kj} are nowhere locally constant (details are given in the next section). Thus, (14.20) is valid in the case at hand.

At the intersection of the jth row and kth column of Figure 14.8 we see the set $\Lambda(b) \cup H_\Omega^{jk}(b)$ and thus $\lim_{n \to \infty} \operatorname{sp}_\Omega^{(j,k)} T_n(b)$ for $\Omega = [-5, 5]$.

Figure 14.9 illustrates the effects of complex single-entry perturbations. In the jth row and the kth column of Figure 14.9 we see $\operatorname{sp}_{\overline{\mathbf{D}}}^{(j,k)} T_{25}(b)$ for $b(t) = t + \frac{1}{4} t^{-1}$.

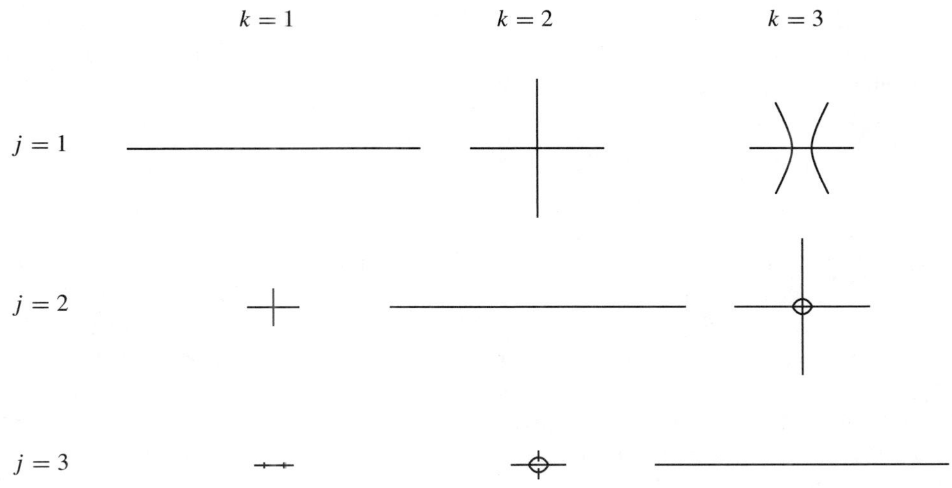

Figure 14.8. *The sets $\Lambda(b) \cup H_\Omega^{jk}(b)$ for $b(t) = t + \frac{1}{9}t^{-1}$, $\Omega = [-5, 5]$, and the nine possible choices of (j, k) with $j, k \in \{1, 2, 3\}$.*

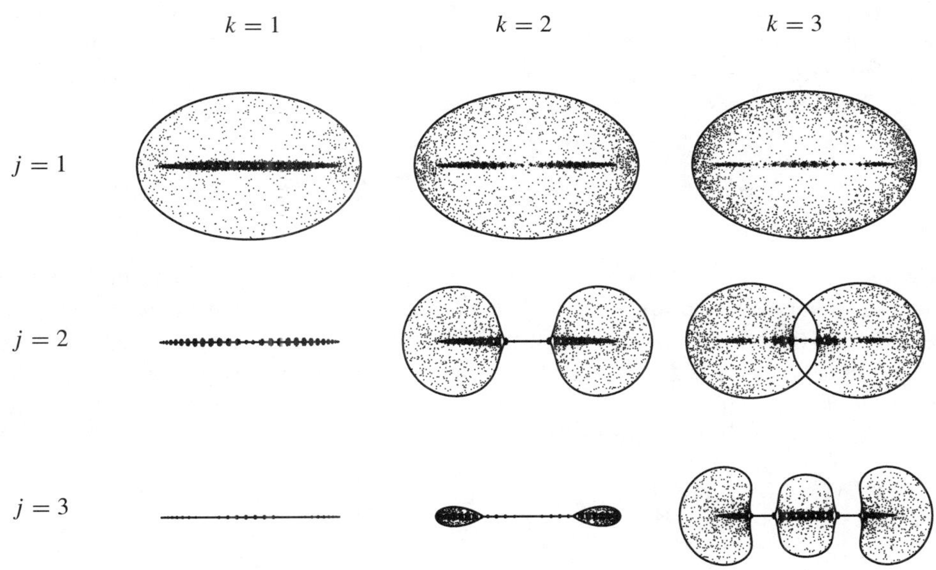

Figure 14.9. *Complex single-entry perturbations to $T(b)$ and $T_n(b)$ for $b(t) = t + \frac{1}{4}t^{-1}$ and $\Omega = \overline{\mathbf{D}}$, the closed unit disk.*

The plot in the jth row and kth column shows the superimposed eigenvalues of 1000 random perturbations to $T_{30}(b)$ in the (j,k) entry for $j,k \in \{1,2,3\}$. Each perturbation is a random number uniformly distributed in $\overline{\mathbf{D}}$. The boundaries of the regions

$$\lim_{n\to\infty} \mathrm{sp}_{\overline{\mathbf{D}}}^{(j,k)} T_n(b) = \Lambda(b) \cup H_{\overline{\mathbf{D}}}^{(j,k)}(b)$$

are drawn as solid curves. While the emergence of wings (or antennae) is typical for real-valued perturbations, one finds that complex perturbations usually lead to "bubbles." For example, we see two bubbles in $\mathrm{sp}_{\overline{\mathbf{D}}}^{(2,2)} T_{25}(b)$, which split into three bubbles in $\mathrm{sp}_{\overline{\mathbf{D}}}^{(3,3)} T_{30}(b)$.

Figures 14.10 and 14.11 illustrate the following experiment. We choose one of the entries of the upper $m \times m$ block of $T_n(b)$ randomly with probability $1/m^2$ and then perturb $T_n(b)$ in this entry by a random number uniformly distributed in $\Omega = [-5,5]$, plotting the n eigenvalues of the perturbed matrix. We repeat this N times and consider the superimposition of the Nn eigenvalues obtained. Equality (14.20) suggests that this superimposition should approximate

$$\bigcup_{1\le j,k\le m} \lim_{n\to\infty} \mathrm{sp}_{[-5,5]}^{(j,k)} T_n(b) = \bigcup_{1\le j,k\le m} \left(\Lambda(b) \cup H_{[-5,5]}^{(j,k)}(b)\right) \tag{14.26}$$

as $n \to \infty$ and $N \to \infty$. For $m = 3$ and $m = 5$, the sets (14.26) are shown in the middle pictures of Figures 14.10 and 14.11, respectively. Notice that, up to a change in scale, the middle picture of Figure 14.10 is nothing but the union of the nine pictures of Figure 14.8. The bottom pictures of Figures 14.10 and 14.11 depict the result of concrete numerical experiments with $N = 2000$ and $n = 20$. The agreement between the $n \to \infty$ theory and practice is striking even for modest values of n.

The top pictures of Figures 14.10 and 14.11 illustrate

$$\bigcup_{1\le j,k\le m} \mathrm{sp}_{[-5,5]}^{(j,k)} T(b) = \bigcup_{1\le j,k\le m} \left(\mathrm{sp}\, T(b) \cup H_{[-5,5]}^{(j,k)}(b)\right). \tag{14.27}$$

Obviously, these top pictures (infinite volume case) differ significantly from the middle pictures (finite volume case). Even more than that, in the finite volume case we discover a remarkable structure in the set (14.27). In the infinite volume case, this structure is hidden behind the black ellipse E_+, so we are only aware of the ends of certain arcs, resembling antennae sprouting from the ellipse.

The top picture of Figure 14.12 shows

$$\bigcup_{(j,k)\in\mathbf{N}\times\mathbf{N}} \mathrm{sp}_{[-5,5]}^{(j,k)} T(b) = \bigcup_{(j,k)\in\mathbf{N}\times\mathbf{N}} \left(\mathrm{sp}\, T(b) \cup H_{[-5,5]}^{(j,k)}(b)\right)$$

(infinite volume case), while in the middle picture we approximate

$$\bigcup_{(j,k)\in\mathbf{N}\times\mathbf{N}} \lim_{n\to\infty} \mathrm{sp}_{[-5,5]}^{(j,k)} T_n(b) = \bigcup_{(j,k)\in\mathbf{N}\times\mathbf{N}} \left(\Lambda(b) \cup H_{[-5,5]}^{(j,k)}(b)\right)$$

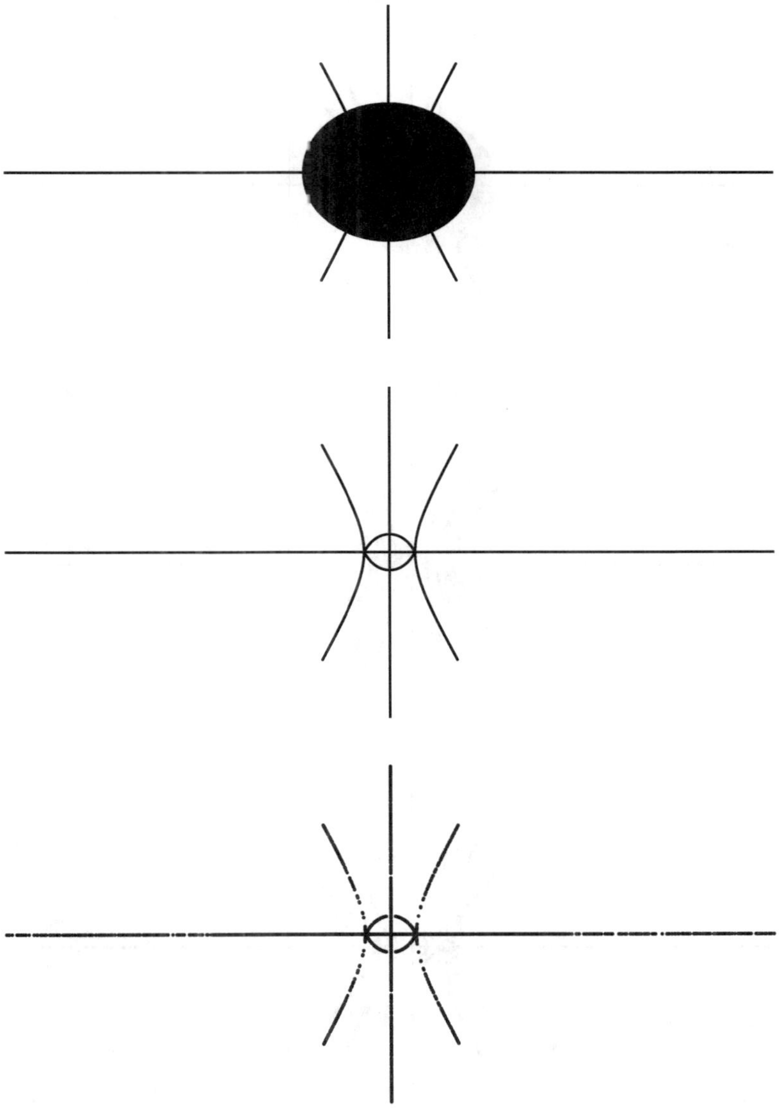

Figure 14.10. *Real single-entry perturbations to $T(b)$ and $T_n(b)$ for $b(t) = t + \frac{1}{9}t^{-1}$ and $\Omega = [-5, 5]$. The top picture shows the union of $\mathrm{sp}_\Omega^{(j,k)} T(b)$ over all (j, k) in the upper 3×3 block; the middle picture represents the union of $\lim_{n\to\infty} \mathrm{sp}_\Omega^{(j,k)} T_n(b)$ over the same (j, k). The bottom picture superimposes the eigenvalues of 2000 random single-entry perturbations of $T_{20}(b)$.*

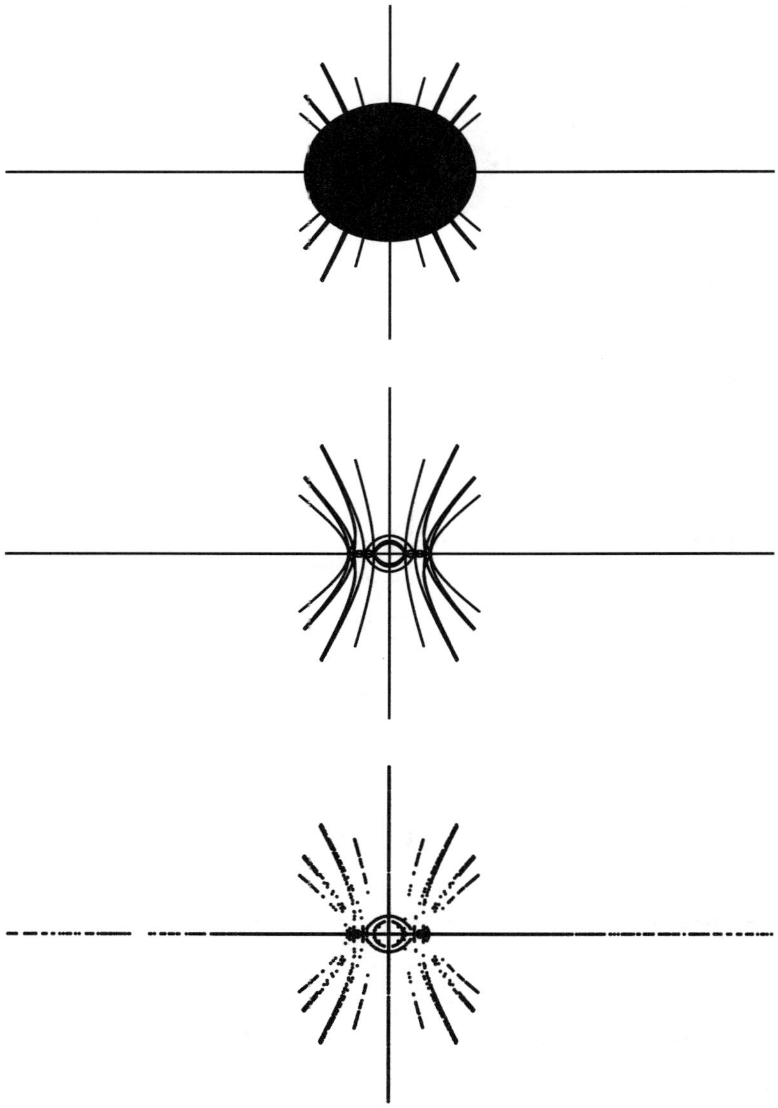

Figure 14.11. *This is the analogue of Figure* 14.10 *for real single-entry perturbations in the upper* 5×5 *block.*

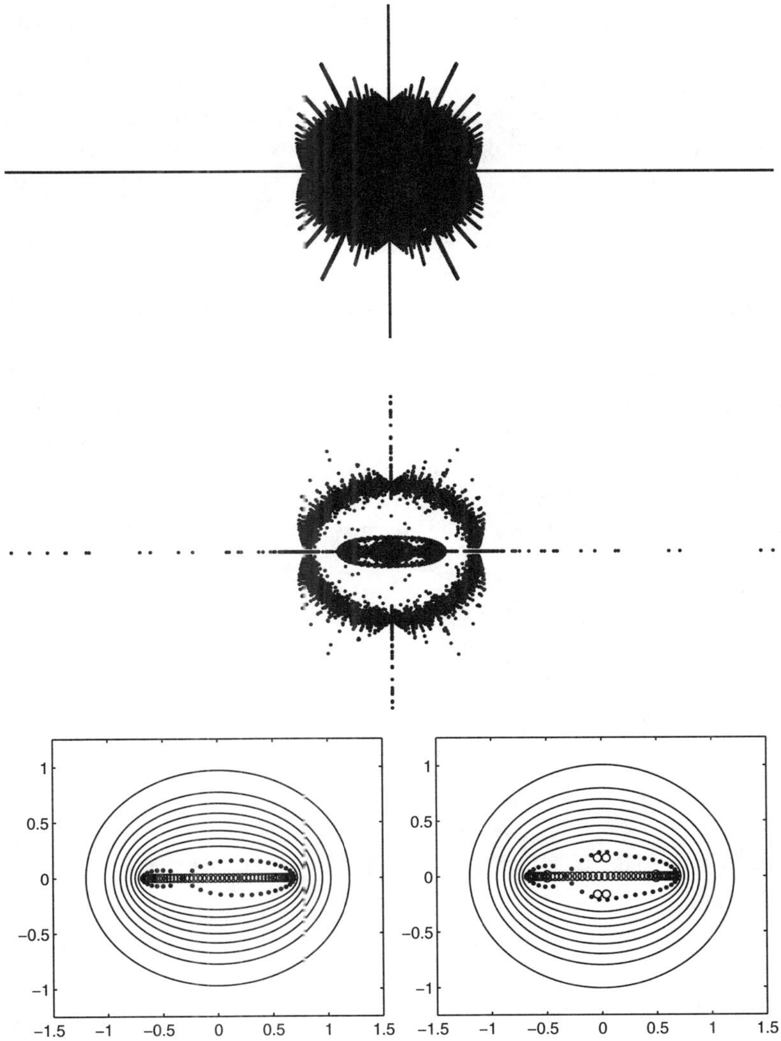

Figure 14.12. *Real single-entry perturbations to* $T(b)$ *and* $T_n(b)$ *for* $b(t) = t + \frac{1}{9}t^{-1}$ *and* $\Omega = [-5, 5]$. *The top picture is the union of* $\mathrm{sp}_{\Omega}^{(j,k)} T(b)$ *over* $(j, k) \in \mathbf{N} \times \mathbf{N}$. *The middle picture shows computed eigenvalues of* 1000 *single-entry perturbations to* $T_{50}(b)$, *where the perturbed entry is chosen by random anywhere in the matrix, and the perturbation itself is randomly chosen from the uniform distribution on* $[-5, 5]$. *The two bottom pictures intend to motivate the dark interior elliptic region in the middle picture. They are explained in the text.*

(finite volume case) by the union of sp $(T_{50}(a) + \omega E_{jk})$ for 1000 random choices of $j, k \in \{1, 2, \ldots, 50\}$ and $\omega \in [-5, 5]$. The eigenvalues of the finite Toeplitz matrices $T_n(a)$ are highly sensitive to perturbations even for modest dimensions. It is interesting that the single entry perturbations investigated here do not generally change the qualitative nature of that eigenvalue instability. This is revealed for a specific example by the pseudospectral plots at the bottom of Figure 14.12 (computed using [300]). The interior elliptical region of high eigenvalue concentration in the middle picture is an artifact of finite precision arithmetic; this is revealed by the two bottom pictures, which show the boundaries of the pseudospectra $\text{sp}_\varepsilon T_{50}(b)$ (left) and $\text{sp}_\varepsilon (T_{50}(b)+\omega E_{54})$ with $\omega = -3$ (right) for $\varepsilon = 10^{-1}, 10^{-3}, \ldots, 10^{-15}$. Dots $(\cdot)$ denote computed eigenvalues; circles $(\circ)$ show the true eigenvalue locations. This explains the dark interior ellipse in the center plot of Figure 14.12: Many of these computed eigenvalues are inaccurate due to rounding errors. Generic perturbations of norm 10^{-15} obscure the effects of our larger, single-entry perturbations. The true structure is more delicate, as emphasized by Figure 14.13, which zooms in on the middle image of Figure 14.11 for perturbations to the upper left 5×5 corner of $T_n(b)$. We compare the $n \to \infty$ structure to the eigenvalues of perturbations of $T_{10}(b)$ and $T_{50}(b)$. The convergence to the asymptotic limit is compelling, though from an applications perspective, any point in the interior of $b(\mathbf{T})$ will behave like an eigenvalue when n is large.

Finally, Figure 14.14 exhibits a Toeplitz matrix with six diagonals. The symbol b is given by

$$\begin{aligned} b(t) = {} & (1.5 - 1.2\,i)t^{-1} + (0.34 + 0.84\,i)t \\ & + (-0.46 - 0.1\,i)t^2 + (0.17 - 1.17\,i)t^3 + (-1 + 0.77\,i)t^4. \end{aligned}$$

Notice again the emergence of many wings, which make the set $\Lambda(b)$ (middle picture) become something reminiscent of a horse in cave paintings. $\square$

14.5 Can Structured Pseudospectra Jump?

Let H be a Hilbert space and let $A, B, C \in \mathcal{B}(H)$. In accordance with (7.2), we define the structured pseudospectrum

$$\text{sp}_\varepsilon^{B,C} A = \text{sp}\, A \cup \left\{\lambda \notin \text{sp}\, A : \|C(A - \lambda I)^{-1} B\| \geq 1/\varepsilon\right\}. \tag{14.28}$$

This section addresses the question whether $\text{sp}_\varepsilon^{B,C} A$ can jump as $\varepsilon \in (0, \infty)$ changes continuously. By virtue of (14.28), this question is equivalent to asking whether the norm $\|C(A - \lambda I)^{-1} B\|$ may be locally constant. If $H = \ell^2(\mathbf{N})$, we have

$$\text{sp}_{\varepsilon\overline{\mathbf{D}}}^{(j,k)} A = \text{sp}_\varepsilon^{E_j, E_k} A,$$

where E_j is the projection on the jth coordinate. Thus, the question whether

$$|d_{jk}(\lambda)| = |[T^{-1}(b_\varrho - \lambda)]_{jk}| = \|E_j T^{-1}(b_\varrho - \lambda) E_k\|$$

can locally be a nonzero constant, which is of importance in connection with Theorem 14.19, amounts to the problem whether $\text{sp}_\varepsilon^{E_j, E_k} T(b_\varrho)$ can jump.

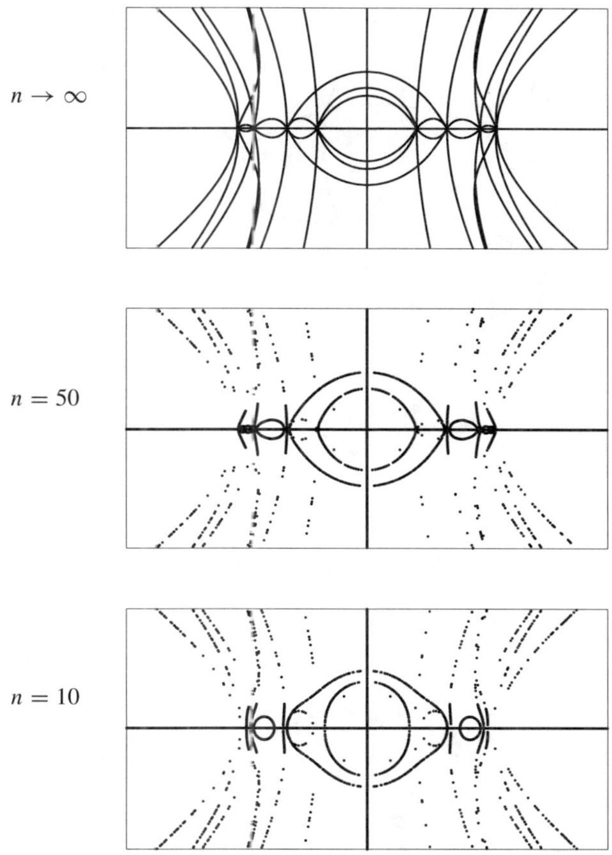

Figure 14.13. *Closer inspection of Figure* 14.11. *The top plot shows a portion of* $\lim_{n\to\infty} \mathrm{sp}_{\Omega}^{(j,k)} T_n(b)$ *over all* (j,k) *in the top* 5×5 *corner for* $\Omega = [-5, 5]$. *The middle image shows eigenvalues of* 10000 *random perturbations to a single entry in the top corner of* $T_{50}(b)$*; the bottom image shows the same for* $T_{10}(b)$. *(The accurate eigenvalues for* $N = 50$ *were obtained by reducing the nonnormality in the problem via a similarity transformation.)*

Proposition 14.21. *Let* A, B, C *be bounded linear operators on a Hilbert space* H *and suppose* B *or* C *is compact. Let* G_∞ *denote the unbounded component of* $\mathbf{C} \setminus \mathrm{sp}\, A$ *and let* G *be any component of* $\mathbf{C} \setminus \mathrm{sp}\, A$. *Suppose further that there exists a path* Γ *in the plane that connects* G *and* G_∞ *such that* $C(A - \lambda I)^{-1} B$ *can be analytically continued from* G *along* Γ *to some function* $f(\lambda)$ *defined in some open subset* V *of* G_∞ *and that* $f(\lambda) = C(A - \lambda I)^{-1} B$ *for* $\lambda \in V$. *Then* $\|C(A - \lambda I)^{-1} B\|$ *is either nowhere locally constant in* G *or identically zero in* G.

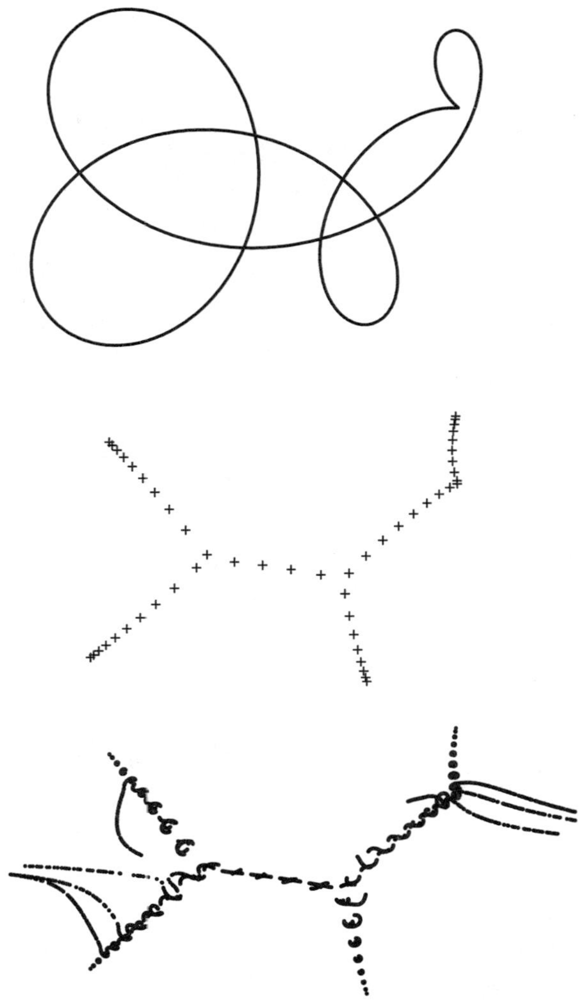

Figure 14.14. *Real single-entry perturbations to $T_n(b)$. The range $b(\mathbf{T})$ of the Laurent polynomial b is seen in the top picture. The middle picture shows* $\operatorname{sp} T_{50}(b)$ *and provides a very good idea of $\Lambda(b)$. The bottom picture depicts the superimposed eigenvalues of* 1000 *perturbations of $T_{50}(b)$ in a randomly chosen entry from* $(1, 1)$, $(2, 2)$, $(3, 3)$ *by a random number uniformly distributed in* $[-5, 5]$.

Of course, here "nowhere locally constant in G" means that there is no open subset of G on which the function is constant. Proposition 14.21 implies in particular that, provided B or C is compact, $\operatorname{sp}_\varepsilon^{B,C} A$ cannot jump if $\mathbf{C} \setminus \operatorname{sp} A$ is connected (which is certainly true for finite matrices A as well as for selfadjoint or compact operators A).

Proof. Abbreviate $C(A - \lambda I)^{-1}B$ to $f(\lambda)$. Pick $\lambda_0 \in G$. Since $f(\lambda_0)$ is compact, there exist $x_0, y_0 \in H$ of norm 1 such that $(x_0, f(\lambda_0)y_0) = \|f(\lambda_0)\|$. Put $h(\lambda) = (x_0, f(\lambda)y_0)$. The function h is analytic and hence either an open map or constant. In the former case, every neighborhood of λ_0 contains a $\lambda_1 \in G$ such that

$$\|f(\lambda_0)\| = |h(\lambda_0)| < |h(\lambda_1)| = |(x_0, f(\lambda_1)y_0)| \le \|f(\lambda_1)\|,$$

which shows that $\|f(\lambda)\|$ cannot be locally constant in G. In the latter case, analytic continuation of h along Γ to G_∞ and subsequently to infinity yields that the constant value assumed by h must be zero. □

Example 14.22. Let U be the forward shift on $\ell^2(\mathbf{Z})$, that is, $(Ux)_n = x_{n-1}$. Equivalently, U is the Laurent operator $L(b)$ induced by $b(t) = t$. Then $\operatorname{sp} U$ is the unit circle $\mathbf{T}$ and the central 4×4 block of the resolvent operator $(U - \lambda I)^{-1}$ is

$$\left(\begin{array}{c|ccc} 0 & 1 & \lambda & \lambda^2 \\ \hline 0 & 0 & 1 & \lambda \\ 0 & 0 & 0 & 1 \\ 0 & 0 & 0 & 0 \end{array}\right) \quad \text{and} \quad \left(\begin{array}{c|ccc} -1/\lambda & 0 & 0 & 0 \\ \hline -1/\lambda^2 & -1/\lambda & 0 & 0 \\ -1/\lambda^3 & -1/\lambda^2 & -1/\lambda & 0 \\ -1/\lambda^4 & -1/\lambda^3 & -1/\lambda^2 & -1/\lambda \end{array}\right)$$

for $|\lambda| < 1$ and $|\lambda| > 1$, respectively. Thus, $\|P_2(U - \lambda I)^{-1}P_2\| = 1$ for $|\lambda| < 1$. In this case the crucial hypothesis of Proposition 14.21 is not satisfied: although each entry of $(U - \lambda I)^{-1}$ can be analytically continued from $|\lambda| < 1$ to all of $\mathbf{C}$, the result of this continuation is different from the corresponding entry of $(U - \lambda I)^{-1}$ for $|\lambda| > 1$.

Using (14.28) and the explicit expressions for $(U - \lambda I)^{-1}$ displayed above, it is easy to compute $\operatorname{sp}_\varepsilon^{P_2,P_2} U$. Put $\varepsilon_0 = \sqrt{2/(3-\sqrt{5})} = 0.618\ldots$. There is a continuous and strictly monotonically increasing function $h \;\; [\varepsilon_0, \infty) \to [0, \infty)$ such that $h(\varepsilon_0) = 0$, $h(\infty) = \infty$, and

$$\operatorname{sp}_\varepsilon^{P_2,P_2} U = \begin{cases} \{\lambda \in \mathbf{C} : |\lambda| = 1\} & \text{for} \quad 0 < \varepsilon \le \varepsilon_0, \\ \{\lambda \in \mathbf{C} : 1 \le |\lambda| < 1 + h(\varepsilon)\} & \text{for} \quad \varepsilon_0 < \varepsilon \le 1, \\ \{\lambda \in \mathbf{C} : 0 \le |\lambda| < 1 + h(\varepsilon)\} & \text{for} \quad 1 < \varepsilon. \end{cases}$$

Clearly, $\operatorname{sp}_\varepsilon^{P_2,P_2} U$ jumps at $\varepsilon = 1$. □

Now let b be a Laurent polynomial of the form

$$b(t) = \sum_{k=-r}^{s} b_k t^k, \quad r \ge 0, \quad s \ge 0, \quad b_{-r} \ne 0, \quad b_s \ne 0. \tag{14.29}$$

Conjecture 14.23. *Let the Laurent polynomial b be of the form* (14.29) *and suppose that one of the operators B and C is compact. If G is any connected component of $\mathbf{C} \setminus \operatorname{sp} T(b)$, then $\|CT^{-1}(b - \lambda)B\|$ is either nowhere locally constant in G or identically zero in G.*

The rest of this section is devoted to the proof of the following result.

Theorem 14.24. *Suppose B and C are of the form $B = P_{m_1}\widetilde{B}$ and $C = \widetilde{C}P_{m_2}$ with bounded operators $\widetilde{B}$ and $\widetilde{C}$. Then Conjecture* 14.23 *is true in each of the following cases:*

(a) $\mathbf{C} \setminus \Lambda(b)$ *is connected;*

(b) $m_1 = m_2 = 1$;

(c) $T(b)$ *is Hessenberg, that is, r or s equals* 1;

(d) $r + s$ *is a prime number and r or s equals* 2;

(e) $r + s \le 5$ *or* $r + s = 7$.

Before proceeding to the proof, we cite a corollary of Theorem 14.24.

Corollary 14.25. *Let b be of the form* (14.29) *and let G be a connected component of $\mathbf{C} \setminus \Lambda(b)$. The function d_{11} is always nowhere locally constant in G. If $j \ge 1$ or $k \ge 1$, then the function d_{jk} is either identically zero in G or nowhere locally constant in G provided one of the following conditions is satisfied:*

(a) $\mathbf{C} \setminus \Lambda(b)$ *is connected;*

(b) $T(b)$ *is Hessenberg, that is, r or s equals* 1;

(c) $r + s$ *is a prime number and r or s equals* 2;

(d) $r + s \le 5$ *or* $r + s = 7$.

Proof. Pick $\lambda_0 \in G$. There is a $\varrho > 0$ such that $\lambda_0 \in \mathbf{C} \setminus \operatorname{sp} T(b_\varrho)$ and $d_{jk}(\lambda) = E_j T^{-1}(b_\varrho - \lambda)E_k$ for all λ in some open neighborhood of λ_0. It remains to use Theorem 14.24 with b replaced by b_ϱ, with $m_1 = j$ and $\widetilde{B} = E_j$, and with $m_2 = k$ and $\widetilde{C} = E_k$. □

We emphasize that $\mathbf{C} \setminus \Lambda(b)$ is connected in many cases. This set is in particular connected if $T(b)$ has at most three nonzero diagonals (see Example 11.18) or if $T(b)$ is triangular (in which case $\Lambda(b)$ is a singleton) or if $T(b)$ is Hermitian (which implies that $\Lambda(b)$ is a line segment of the real line). We do not know any b with $r + s \le 5$ for which $\mathbf{C} \setminus \Lambda(b)$ is disconnected. Theorem 11.20 delivers a b with $r = s = 3$ such that $\mathbf{C} \setminus \Lambda(b)$ is disconnected.

We now turn to the proof of Theorem 14.24. Let b be of the form (14.29) and put

$$n = r + s, \quad Q_n(z) = b_{-r} + b_{-r+1} + \cdots + b_s z^{r+s}. \tag{14.30}$$

For $\lambda \in \mathbf{C}$, let $z_1(\lambda), \ldots, z_n(\lambda)$ be the zeros of the polynomial $Q_n(z) - \lambda z^r$,

$$Q_n(z) - \lambda z^r = b_s(z - z_1(\lambda)) \cdots (z - z_n(\lambda)). \tag{14.31}$$

The operator $T(b - \lambda)$ is invertible if and only if r of the zeros $z_1(\lambda), \ldots, z_n(\lambda)$ have modulus less than 1 and the remaining s zeros are of modulus greater than 1. We denote the former zeros by $\delta_1(\lambda), \ldots, \delta_r(\lambda)$ and the latter zeros by $\mu_1(\lambda), \ldots, \mu_s(\lambda)$. We put

$$\mu(\lambda) = \mu_1(\lambda)\cdots\mu_s(\lambda), \quad u_0(\lambda) = 1, \quad v_0(\lambda) = 1,$$

$$u_m(\lambda) = \sum_{\substack{\alpha_1+\cdots+\alpha_s=m \\ \alpha_j\geq 0}} \mu_1(\lambda)^{-\alpha_1}\cdots\mu_s(\lambda)^{-\alpha_s} \quad (m \geq 1),$$

$$v_m(\lambda) = \sum_{\substack{\beta_1+\cdots+\beta_r=m \\ \beta_j\geq 0}} \delta_1(\lambda)^{\beta_1}\cdots\delta_r(\lambda)^{\beta_r} \quad (m \geq 1),$$

$$U(\lambda) = \begin{pmatrix} u_0(\lambda) & & & \\ u_1(\lambda) & u_0(\lambda) & & \\ u_2(\lambda) & u_1(\lambda) & u_0(\lambda) & \\ \dots & \dots & \dots & \dots \end{pmatrix},$$

$$V(\lambda) = \begin{pmatrix} v_0(\lambda) & v_1(\lambda) & v_2(\lambda) & \dots \\ & v_0(\lambda) & v_1(\lambda) & \dots \\ & & v_0(\lambda) & \dots \\ & & & \dots \end{pmatrix}.$$

Then

$$T^{-1}(b-\lambda) = \frac{(-1)^s}{b_s}\cdot\frac{1}{\mu(\lambda)}U(\lambda)V(\lambda). \tag{14.32}$$

From (14.32) we see that each entry of $T^{-1}(b-\lambda)$ is of the form

$$[T^{-1}(b-\lambda)]_{jk} = R_{jk}(\delta_1(\lambda),\dots,\delta_r(\lambda);\mu_1(\lambda),\dots,\mu_s(\lambda)), \tag{14.33}$$

where R_{jk} is a rational function of $n = r + s$ variables with coefficients in $\mathbf{Z}$. Moreover, R_{jk} is symmetric in the first r variables and in the last s variables.

Throughout what follows we let $f(\lambda) = CT^{-1}(b-\lambda)B$ and we assume that G is some bounded component of $\mathbf{C}\setminus \operatorname{sp} T(b)$.

We consider the Riemann surface of $Q_n(z) - \lambda z^r = 0$. The points $\lambda \in \mathbf{C}$ for which $Q_n(z) - \lambda z^r$ has a multiple zero are called the finite *branch points*. There exist at most n finite branch points. We denote them by $\lambda_1,\dots,\lambda_k$. The point at infinity is also a branch point. Thus, the set of all branch points is

$$W := \{\lambda_1,\dots,\lambda_k,\infty\}.$$

We join λ_1 to λ_2 by a cut S_1, λ_2 to λ_3 by a cut $S_2,\dots,$ and λ_k to ∞ by a cut S_k. Put

$$S = S_1 \cup \cdots \cup S_k \cup W.$$

We can draw $S_1,\dots,S_k$ so that $\mathbf{C}\setminus S$ is connected. The zeros in (14.31) can be chosen so that $z_1(\lambda),\dots,z_n(\lambda)$ are analytic functions in $\mathbf{C}\setminus S$. Take n copies $\Sigma_1,\dots,\Sigma_n$ of $\mathbf{C}\setminus S$ and think of z_j as a map of Σ_j to $\mathbf{C}$. We glue Σ_i and Σ_j along the cut S_ℓ whenever the function $z_i(\lambda)$ can be continued analytically to the function $z_j(\lambda)$ across S_ℓ. The resulting set Σ is the Riemann surface of $Q_n(z) - \lambda z^r = 0$, and Σ_j is referred to as the jth branch of Σ.

Each path in $\mathbf{C}\setminus W$ induces a permutation of the branches of Σ in a natural way. The set of all these permutations is a group $\mathbf{G}$, the *monodromy group* of $Q_n(z)-\lambda z^r = 0$. Let π_j

$(j = 1, \dots, k)$ be the permutation corresponding to a small counterclockwise-oriented circle around the branch point λ_j. Clearly, $\mathbf{G}$ contains $\pi_1, \dots, \pi_k$. We put $\pi_\infty = \pi_1 \dots \pi_k$. Thus, π_∞ is the permutation of the branches resulting from a large counterclockwise-oriented circle containing all finite branch points in its interior. The group $\mathbf{G}$ is generated by $\pi_1, \dots, \pi_k$.

Let $\lambda \in \mathbf{C} \setminus \operatorname{sp} T(b)$. We have $|z_j(\lambda)| < 1$ for exactly r values of j. We call the branches Σ_j corresponding to these values of j the *small branches at* λ. The s branches Σ_j for which $|z_j(\lambda)| > 1$ will be called the *large branches at* λ.

Proposition 14.26. *If there is a $\pi \in \mathbf{G}$ that permutes the set of the small branches at the points of G to the set of the small branches at the points of the unbounded component G_∞, then $\|f(\lambda)\|$ is either nowhere locally constant in G or identically zero in G.*

Proof. Let Γ be the path in $\mathbf{C} \setminus W$ that corresponds to the permutation π. From (14.32) and (14.33) we see that each entry of $P_{m_2} T^{-1}(b - \lambda) P_{m_1}$ and hence also $\widetilde{C} P_{m_2} T^{-1}(b - \lambda) P_{m_1} \widetilde{B}$ can be analytically continued along Γ. Since π permutes the small branches into themselves, the result of the analytic continuation coincides with the operator $\widetilde{C} P_{m_2} T^{-1}(b - \lambda) P_{m_1} \widetilde{B}$ for $\lambda \in G_\infty$. It remains to apply Proposition 14.21. □

From Chapter 11 we know that when labelling the zeros $z_1(\lambda), \dots, z_{r+s}(\lambda)$ of (14.31) so that

$$|z_1(\lambda)| \le |z_2(\lambda)| \le \dots \le |z_{r+s}(\lambda)|, \tag{14.34}$$

then

$$\Lambda(b) = \{\lambda \in \mathbf{C} : |z_r(\lambda)| = |z_{r+1}(\lambda)|\}. \tag{14.35}$$

***Proof of Theorem* 14.24(a).** Pick a point $\lambda_0 \in G$. Since $\mathbf{C} \setminus \Lambda(b)$ is connected, there is a path Γ in $\mathbf{C} \setminus (W \cup \Lambda(b))$ joining λ_0 to infinity. By (14.35), we have $|z_r(\lambda)| < |z_{r+1}(\lambda)|$ throughout this path. This means that when moving along this path we will eventually stay in the r small branches of Σ at infinity. In other words, there is a $\pi \in \mathbf{G}$ permuting the set of the small branches at λ_0 to the set of the small branches at infinity. The assertion is therefore immediate from Proposition 14.26. □

The group $\mathbf{G}$ is said to be *r-transitive* if for every two r-tuples $(\Sigma_{i_1}, \dots, \Sigma_{i_r})$ and $(\Sigma_{j_1}, \dots, \Sigma_{j_r})$ of r distinct branches of Σ there is a $\pi \in \mathbf{G}$ such that

$$\pi(\Sigma_{i_1}) = \Sigma_{j_1}, \dots, \pi(\Sigma_{i_r}) = \Sigma_{j_r}.$$

We call the group $\mathbf{G}$ *weakly r-transitive* if for every two sets $\{\Sigma_{i_1}, \dots, \Sigma_{i_r}\}$ and $\{\Sigma_{j_1}, \dots, \Sigma_{j_r}\}$ of r distinct branches of Σ there exists a $\pi \in \mathbf{G}$ such that

$$\{\pi(\Sigma_{i_1}), \dots, \pi(\Sigma_{i_r})\} = \{\Sigma_{j_1}, \dots, \Sigma_{j_r}\},$$

i.e., such that $\pi(\Sigma_{i_1}), \dots, \pi(\Sigma_{i_r})$ coincide with $\Sigma_{j_1}, \dots, \Sigma_{j_r}$ up to the arrangement. Clearly, weak 1-transitivity and 1-transitivity are equivalent.

Proposition 14.27. *If the monodromy group of $Q_n(z) - \lambda z^r = 0$ is weakly r-transitive, then $\|f(\lambda)\|$ is either nowhere locally constant in G or identically zero in G.*

Proof. Weak r-transitivity means that we can permute any prescribed set of r branches into any prescribed set of r branches. We can in particular permute the r small branches at the points of G into the r small branches at the points of G_∞. The assertion is therefore a direct consequence of Proposition 14.26. □

Proposition 14.28. *The monodromy group of the polynomial $Q_n(z) - \lambda z^r = 0$ is always 1-transitive.*

Proof. It is well known (see, e.g., [175, Section 4.14]) that $\mathbf{G}$ is 1-transitive if and only if $Q_n(z) - \lambda z^r$ is irreducible in $\mathbf{C}[z, \lambda]$. But the irreducibility of $Q_n(z) - \lambda z^r$ in $\mathbf{C}[z, \lambda]$ can be readily verified. □

***Proof of Theorem* 14.24(b).** Let λ be a point in $\mathbf{C} \setminus \operatorname{sp} T(b)$. We label the zeros $z_1(\lambda), \dots, z_{r+s}(\lambda)$ so that (14.34) holds. Thus, the small branches at λ are $\Sigma_1, \dots, \Sigma_r$ and the large branches at λ are $\Sigma_{r+1}, \dots, \Sigma_{r+s}$. By formula (14.32), the function $g(\lambda) := [T^{-1}(b-\lambda)]_{11}$ equals

$$(-1)^s b_s^{-1} \mu_1(\lambda)^{-1} \cdots \mu_s(\lambda)^{-1} = (-1)^s b_s^{-1} z_{r+1}(\lambda)^{-1} \cdots z_{r+s}(\lambda)^{-1}.$$

This shows that $g(\lambda) \neq 0$. Since the group $\mathbf{G}$ is 1-transitive (Proposition 14.28), there is a path in $\mathbf{C} \setminus W$ starting and terminating at λ such that the s large branches at λ are permuted into s branches $\Sigma_{i_1}, \dots, \Sigma_{i_s}$ containing at least one small branch. Consequently, after analytic continuation along this curve, $g(\lambda)$ becomes $z_{i_1}(\lambda)^{-1} \dots z_{i_s}(\lambda)^{-1}$, and because, obviously,

$$|z_{i_1}(\lambda) \cdots z_{i_s}(\lambda)| < |z_{r+1}(\lambda) \cdots z_{r+s}(\lambda)|,$$

it follows that $|g(\lambda)|$ cannot be locally constant. Since $f(\lambda) = \widetilde{C} P_1 g(\lambda) P_1 \widetilde{B} = g(\lambda) \widetilde{C} P_1 \widetilde{B}$, we arrive at the conclusion that $\|f(\lambda)\| = |g(\lambda)| \, \|\widetilde{C} P_1 \widetilde{B}\|$ is either identically zero or nowhere locally constant in Ω. □

***Proof of Theorem* 14.24(c).** Without loss of generality assume $r = 1$; the case $s = 1$ can be reduced to the case $r = 1$ by passage to adjoint operators. Combining Propositions 14.27 and 14.28 we arrive at the assertion. □

We now turn to the case $n = r + s = 4$. By what was already proved, we are left with the constellation $r = s = 2$. Everything would be fine if the monodromy group of $Q_4(z) - \lambda z^2 = 0$ were always weakly 2-transitive (Proposition 14.27). Unfortunately, this need not be the case. Indeed, let $Q_4(z)$be the polynomial $\mu z^2 + (z - \alpha)^2 (z - \beta)^2$, where $\alpha, \beta, \mu \in \mathbf{C}$, $\alpha \neq \beta$, and $\alpha\beta \neq 0$. We have exactly two finite branch points, λ_1 $(= \mu)$ and λ_2, and the monodromy group is generated by $\pi_1 = (12)(43)$ and $\pi_2 = (13)(24)$. Here and in what follows we identify $\Sigma_1, \Sigma_2, \dots$ with the numbers $1, 2, \dots$. Clearly, we cannot permute $\{1, 2\}$ into $\{1, 3\}$. Fortunately, in this case we can have recourse to Theorem 11.20, which shows that $\mathbf{C} \setminus \Lambda(b)$ is connected.

Let us return for a moment to the case of general n and r, s. The permutation π associated with a branch point λ can be written as a product of cycles. We say that λ is of the type $(L_1, L_2, \dots)$ if the cycle lengths of π are $L_1, L_2, \dots$. In the previous paragraph, we encountered two finite branch points of the type $(2, 2)$.

Proposition 14.29. *If all finite branch points of the Riemann surface of the polynomial*

$Q_n(z) - \lambda z^2 = 0$ are of the type $(2, 1, \dots, 1)$, then the monodromy group of $Q_n(z) - \lambda z^2 = 0$ is weakly 2-transitive.

Proof. It suffices to show that we can permute $\{1, 2\}$ to $\{1, m\}$ for arbitrary $m \neq 1$. Let $\lambda_1, \dots, \lambda_k$ be the finite branch points. Without loss of generality suppose that the permutations $(12), (13), \dots, (1\,p)$ are among $\pi_1, \dots, \pi_k$ and that the permutations $(1\,p+1), \dots, (1n)$ are not among $\pi_1, \dots, \pi_k$. It is easily seen that we can permute $\{1, 2\}$ to $\{1, m\}$ for every $m \in \{2, \dots, r\}$. Thus, let $m \geq p + 1$. Since **G** is 1-transitive (Proposition 14.28), there is a path on Σ joining Σ_1 to Σ_m. This path goes through the branches $x_1, x_2, \dots, x_\ell$ and we may assume that $x_j \neq 1$ and $x_j \neq m$ for all j and that $x_1 \in \{2, \dots, p\}$. As each branch point is of the type $(2, 1, \dots, 1)$, it follows that the path goes from the branch x_1 to the branch 1 and stays there. Thus, we can permute $\{1, x_1\}$ to $\{1, m\}$. □

Proposition 14.30. *If $r = s = 2$, then $\|f(\lambda)\|$ is either nowhere locally constant in G or identically zero in G.*

Proof. If there is a λ such that $Q_4(z) - \lambda z^2$ has two distinct zeros of multiplicity 2 or one zero of multiplicity 4, then the polynomial $Q_4(z)$ is of the form

$$\lambda z^2 + (z - \alpha)^2 (z - \beta)^2,$$

and the assertion follows from Theorem 11.20 and Theorem 14.24(a). Thus, we may restrict ourselves to the case where all of the (at most four) finite branch points are of the types $(3, 1)$ or $(2, 1, 1)$. It is easily seen that the branch point at infinity is of the type $(2, 2)$. Our goal is to show that **G** is weakly 2-transitive so that the present theorem follows from Proposition 14.27.

If all finite branch points have the type $(2, 1, 1)$, then Proposition 14.29 implies weak 2-transitivity. Thus, assume there is at least one finite branch point, λ_1, of the type $(3, 1)$ and that $\pi_1 = (123)$. Since **G** is 1-transitive, there must exist at least one more finite branch point λ_2.

We first consider the case where all finite branch points different from λ_1 are of the type $(2, 1, 1)$. As **G** is 1-transitive, branch 4 must be in the cycle of length 2 of one of these branch points, say λ_2. By symmetry, we may assume that the permutation π_2 is (14). But a group of permutations of $1, 2, 3, 4$ containing $\pi_1 = (123)$ and $\pi_2 = (14)$ is easily seen to be weakly 2-transitive.

We are left with the case where λ_2 is of the type $(3, 1)$. If branch 4 is in the cycle of length 3 of π_2, then **G** is readily checked to be weakly 2-transitive. So assume branch 4 is left fixed by π_2. Then there must exist a third finite branch point λ_3 of the type $(2, 1, 1)$ such that branch 4 is contained in a cycle of length 2. By direct inspection of the few possible cases one sees that the group **G** is always weakly 2-transitive. □

Lemma 14.31. *The order of the monodromy group* **G** *of $Q_n(z) - \lambda z^r = 0$ is divisible by n.*

Proof. For $1 \leq j \leq n$, let $\mathbf{G}_j = \{\pi \in \mathbf{G} : \pi(1) = j\}$. Clearly, $\mathbf{G} = \mathbf{G}_1 \cup \dots \cup \mathbf{G}_n$ and $\mathbf{G}_i \cap \mathbf{G}_j = \emptyset$ for $i \neq j$. Given two distinct numbers i, j in $\{1, \dots, n\}$, we can find a $\sigma \in \mathbf{G}$ such that $\sigma(i) = j$ (Proposition 14.28). The map $\mathbf{G}_i \to \mathbf{G}_j, \pi \mapsto \sigma\pi$ is obviously bijective.

This implies that all the sets $\mathbf{G}_j$ have the same number of elements, say ℓ. Consequently, the order of $\mathbf{G}$ is $n\ell$. $\square$

Lemma 14.32. *If n is a prime number, then the monodromy group* $\mathbf{G}$ *of the polynomial* $Q_n(z) - \lambda z^r = 0$ *contains an n-cycle, that is, after appropriately labelling the branches,* $(12\ldots n) \in \mathbf{G}$.

Proof. This follows from Lemma 14.31 and a well-known theorem by Cauchy, which says that if n is a prime and the order of a finite group is divisible by n, then the group contains an element of the order n. $\square$

Lemma 14.33 (well known). *If a subgroup* $\mathbf{G}$ *of the full symmetric group S_n contains an n-cycle and a transposition, then* $\mathbf{G} = S_n$. $\square$

***Proof of Theorem* 14.24(d).** If $r = 2$ or $s = 2$, then the branch point at infinity of the Riemann surface of $Q_n(z) - \lambda z^r = 0$ is of the type $(n-2, 2)$. Hence π_∞^{n-2} is a transposition (notice that $n-2$ is odd). Combining Proposition 14.27 with Lemmas 14.32 and 14.33, we arrive at the assertion. $\square$

***Proof of Theorem* 14.24(e).** Without loss of generality, assume $r \le s$. Parts (a) and (c) of Theorem 14.24 imply the desired result for $r = 0$ and $r = 1$. Proposition 14.30 disposes of the case $r = s = 2$, and Theorem 14.24(d) gives the assertion for $r = 2$, $s = 3$ and $r = 2$, $s = 5$. We are left with the case where $r = 3$ and $s = 4$. In this case we deduce from Lemma 14.32 that $\mathbf{G}$ contains a 7-cycle, say $\pi = (12\ldots 7)$. The branch point at infinity provides us with a permutation of the type $\sigma = (1xyz)(abc)$. By checking all possible cases, one sees that the group generated by π and σ is always weakly 3-transitive. It remains to make use of Proposition 14.27. $\square$

Remark. The arguments used above are standard in Galois theory, and in fact the monodromy group $\mathbf{G}$ of $Q_n(z) - \lambda z^r = 0$ is known to be isomorphic to the Galois group $\mathbf{G}_0$ of $Q_n(z) - \lambda z^r = 0$. Let $\mathcal{R} = \mathbf{C}(\lambda)$ be the field of the rational functions (of λ) with complex coefficients. We think of $Q_n(z) - \lambda z^r$ as an element of $\mathcal{R}[z]$. The splitting field of this polynomial is $\mathcal{R}(z_1(\lambda), \ldots, z_n(\lambda))$, where $z_1(\lambda), \ldots, z_n(\lambda)$ are given by (14.31). The Galois group $\mathbf{G}_0$ may be identified with the group of all permutations of $z_1(\lambda), \ldots, z_n(\lambda)$ that can be extended to an automorphism of the field $\mathcal{R}(z_1(\lambda), \ldots, z_n(\lambda))$ that leaves the elements of $\mathcal{R}$ fixed. When working with the monodromy group, we transform valid equalities into new equalities by analytic continuation, while from the algebraic point of view, valid equalities are transformed into new equalities by the action of the Galois group.

The polynomial $Q_n(z) - \lambda z^r$ can be shown to be the minimal polynomial of each of its zeros $z_1(\lambda), \ldots, z_n(\lambda)$. This implies that the dimension of $\mathcal{R}(z_1(\lambda))$ over $\mathcal{R}$ is n. Lemma 14.31 so amounts to saying that this dimension divides the order of $\mathbf{G}_0$, and this is one of the conclusions of the fundamental theorem of Galois theory.

Exercises

1. Let $U_n(\lambda)$ denote the Chebyshev polynomial of the second kind introduced in Section 1.9.

   (a) Show that the eigenvalues of $T_n(\sigma) + vE_{nn}$ are the zeros of the polynomial $U_n(\lambda/2) - vU_{n-1}(\lambda/2)$.

   (b) Let λ_j be an eigenvalue of $T_n(\sigma) + vE_{nn}$ and let $\theta_j \in \mathbf{C}$ be any number such that $\cos\theta_j = \lambda_j$. Prove that

   $$x_j = \left(\frac{2\sin(k\theta_j)}{\sqrt{2n+1-U_{2n}(\lambda_j/2)}} \right)_{k=1}^{n}$$

   is an eigenvector for λ_j and that $\|x_j\|_2 = 1$.

   (c) Use (b) to show that the eigenvector x_j is extended for $|\lambda_j| \le 2$ and localized for $|\lambda_j| > 2$.

2. Let $A_n = T_n(\sigma) + vI_{n,n}$, where v is uniformly distributed in $[-M, M]$ and $I_{n,n}$ is the $n \times n$ identity matrix. The density of the eigenvalues of A_n is defined as

   $$\varrho_n(\lambda) = E\left(\frac{1}{n} \sum_{j=1}^{n} \delta(\lambda - \lambda_j(A_n)) \right).$$

   By Theorem 2.4,

   $$\varrho_n(\lambda) = \frac{1}{2Mn} \int_{-M}^{M} \sum_{j=1}^{n} \delta\left(\lambda - v - 2\cos\frac{\pi j}{n+1}\right) dv.$$

   Clearly, $\varrho_n(\lambda) = 0$ for $\lambda \in \mathbf{C} \setminus \mathbf{R}$. Prove that for $\lambda \in \mathbf{R}$,

   $$\lim_{n\to\infty} \varrho_n(\lambda) = \frac{1}{2M\pi} \int_{\lambda-M}^{\lambda+M} \frac{\chi_{[-2,2]}(x)}{\sqrt{4-x^2}} \, dx,$$

   where $\chi_{[-2,2]}(x)$ is 1 for $-2 \le x \le 2$ and zero otherwise.

3. (a) Let $a \in W$ and $\lambda \in \mathbf{C} \setminus \operatorname{sp} T(a)$. Show that

   $$[T^{-1}(a-\lambda)]_{1,1} = 1/G(a-\lambda),$$

   where $G(b)$ is defined by (2.25), and deduce that

   $$\operatorname{sp}_{\varepsilon\overline{\mathbf{D}}}^{(1,1)} T(a) = \operatorname{sp} T(a) \cup \{\lambda \notin \operatorname{sp} T(a) : G(|a-\lambda|) \le \varepsilon\}.$$

   (b) Show that there is a dense subset $\mathcal{M}$ of W such that if $a \in \mathcal{M}$, then $\operatorname{sp}_{\varepsilon\overline{\mathbf{D}}}^{(1,1)} T(a)$ is strictly larger than $\operatorname{sp} T(a)$ for every $\varepsilon > 0$.

4. For $a \in W$, we denote by $L(a)$ the Laurent matrix $(a_{j-k})_{j,k=-\infty}^{\infty}$ (recall the notes to Chapter 1).

   (a) Show that $L(a)$ generates a bounded operator on $\ell^2(\mathbf{Z})$.

   (b) Show that $\operatorname{sp} L(a) = a(\mathbf{T})$.

   (c) Prove that $\operatorname{sp}^m_{\varepsilon\overline{\mathbf{D}}} L(a) = \operatorname{sp} L(a)$ for all sufficiently small $\varepsilon > 0$ if and only if

$$\max_{|k|\le m-1} \sup_{\lambda \notin a(\mathbf{T})} \left| \int_0^{2\pi} \frac{e^{ik\theta}}{a(e^{i\theta}) - \lambda}\, d\theta \right| < \infty.$$

   (d) Deduce that if $a \in \mathcal{P}$ is real valued on $\mathbf{T}$, then $\operatorname{sp}^m_{\varepsilon\overline{\mathbf{D}}} L(a)$ is strictly larger than $\operatorname{sp} L(a)$ for every $\varepsilon > 0$.

5. Show that the matrix

$$\begin{pmatrix} v_1 & e^{g_1} & & & e^{-g_n} \\ e^{-g_1} & v_2 & e^{g_2} & & \\ & \dots & \dots & \dots & \\ & & e^{-g_{n-2}} & v_{n-1} & e^{g_{n-1}} \\ e^{g_n} & & & e^{-g_{n-1}} & v_n \end{pmatrix}$$

   is similar to the matrix

$$\begin{pmatrix} v_1 & e^{g} & & & e^{-g} \\ e^{-g} & v_2 & e^{g} & & \\ & \dots & \dots & \dots & \\ & & e^{-g} & v_{n-1} & e^{g} \\ e^{g} & & & e^{-g} & v_n \end{pmatrix},$$

   where $g = (g_1 + \dots + g_n)/n$.

6. Show that the matrix

$$\begin{pmatrix} v_1 & \alpha^2 & & & \\ 1 & v_2 & \alpha^2 & & \\ & \dots & \dots & \dots & \\ & & 1 & v_{n-1} & \alpha^2 \\ & & & 1 & v_n \end{pmatrix} \quad (\alpha \ne 0)$$

   is similar to the matrix

$$\alpha \begin{pmatrix} v_1/\alpha & 1 & & & \\ 1 & v_2/\alpha & 1 & & \\ & \dots & \dots & \dots & \\ & & 1 & v_{n-1}/\alpha & 1 \\ & & & 1 & v_n/\alpha \end{pmatrix}.$$

7. Show that the matrix

$$\begin{pmatrix} v_1 & 1 & & & \\ & v_2 & 1 & & \\ & & \ddots & \ddots & \\ & & & v_{n-1} & 1 \\ & & & & v_n \end{pmatrix}$$

is similar to the matrix

$$\begin{pmatrix} v_1 & \beta & & & \\ & v_2 & \beta & & \\ & & \ddots & \ddots & \\ & & & v_{n-1} & \beta \\ \beta^{-n} & & & & v_n \end{pmatrix}.$$

8. Show that

$$\begin{pmatrix} a & -1 & \\ & b & -1 \\ & & c \end{pmatrix}^{-1} = \begin{pmatrix} a^{-1} & a^{-1}b^{-1} & a^{-1}b^{-1}c^{-1} \\ & b^{-1} & b^{-1}c^{-1} \\ & & c^{-1} \end{pmatrix}$$

and generalize the result from 3×3 to $n \times n$ matrices.

9. Let A be an $n \times n$ matrix and let u and v be column vectors of length n. Show that

$$\det(A + uv^\top) = \det A + v^\top \operatorname{adj}(A)\, u,$$

where $\operatorname{adj}(A)$ is the adjugate matrix of A.

Notes

The study of discrete Hamiltonians, either with deterministic or with random potential V_n, is a big business and goes beyond the scope of the present book. We therefore leave things with a few remarks.

Anderson [4] considered the selfadjoint matrix $T_n(\sigma)$ with random perturbations on the main diagonal and discovered numerically that the eigenvectors for eigenvalues outside the original spectrum in $[-2, 2]$ are localized. Figure 14.15 illustrates the phenomenon. The top picture shows the 50 eigenvalues of the matrix $T_{50}(\sigma) + \operatorname{diag}(v_j)_{j=1}^{50}$ for a single realization of 50 independent v_j's drawn from the uniform distribution on $[-1, 1]$. The (extended) eigenvector for the encircled eigenvalue in the middle is shown in the bottom left picture, and the (localized) eigenvector for the encircled boundary eigenvalue is seen in the bottom right picture.

The discrete Laplacian $T_n(\sigma)$ is selfadjoint, as it should be in quantum mechanics. The interest in randomly perturbed general tridiagonal Toeplitz matrices $T_n(t + \alpha^2 t^{-1})$ comes from so-called non-Hermitian quantum mechanics and was pioneered by Hatano and Nelson

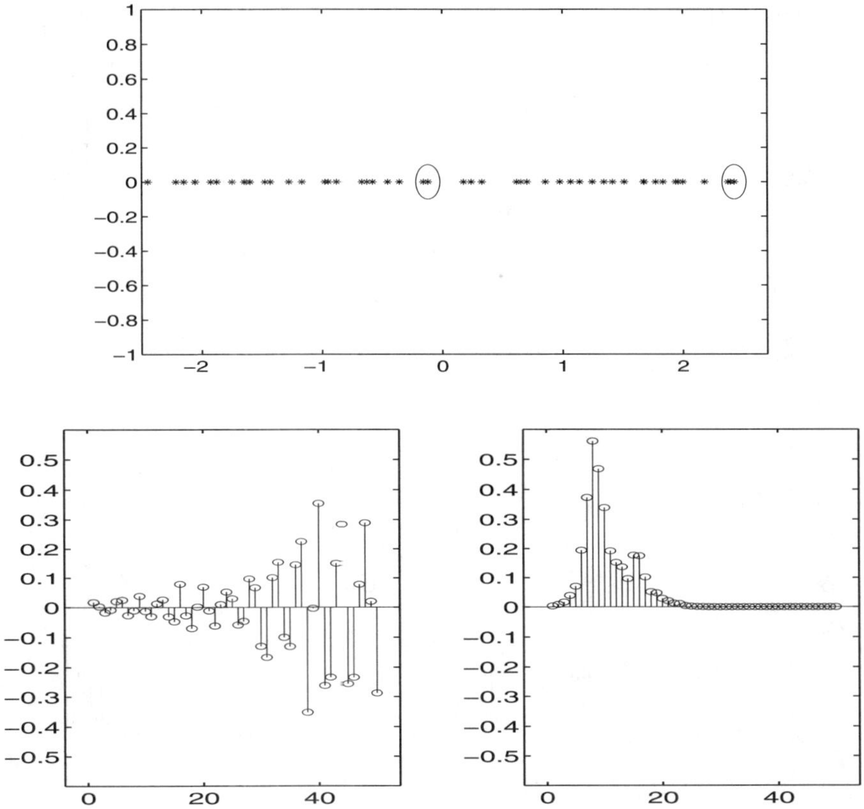

Figure 14.15. *The Anderson model.*

[152]. Their discovery is illustrated in Figure 14.16. To make the effect more visible, one replaces $T_n(t+\alpha^2 t^{-1})$ by $C_n(t+\alpha^2 t^{-1})$. In the top picture we plotted the 50 eigenvalues of $C_n(t+0.49t^{-1}) + \operatorname{diag}(v_j)_{j=1}^{50}$ for a single realization of 50 independent diagonal entries v_j from the uniform distribution on $[-1, 1]$. We see a bubble and two wings. The absolute value of the (extended) eigenvector for the encircled eigenvalue on the bubble is shown in the bottom left picture; the absolute value of the (localized) eigenvector for the encircled eigenvector on the wing is in the bottom right picture.

Profound theoretical investigations about this topic are due to Brézin and Zee [75], Brouwer, Silvestrov, and Beenakker [76], Davies [95], Feinberg and Zee [117], [118], Goldsheid and Khoruzhenko [135], [136], and Janik et al. [173], to cite only a few figures. Applications to population dynamics appear in [91], [191], and applications to small world networks can be found in [189], [259], for instance. An introduction to the ideas of Feinberg and Zee follows below.

In this chapter we confine ourselves to showing how structured pseudospectra can be employed in order to get interesting information about the possible spectra of banded (and not necessarily Hermitian) Toeplitz matrices with impurities. The idea of formulating

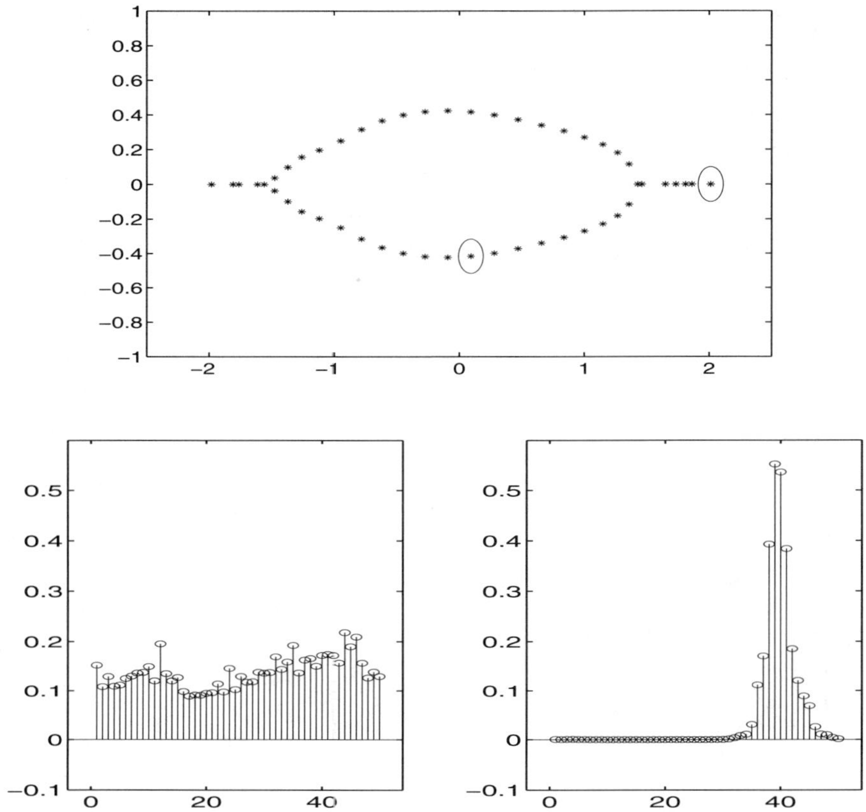

Figure 14.16. *The Hatano-Nelson model.*

problems on random Toeplitz matrices in the language of pseudospectral analysis goes back to Trefethen, Contedini, and Embree [274]. For the reader's convenience, we cite their main results below. The approach of [274] depends heavily on the circumstance that a simple explicit inverse for bidiagonal matrices is available (see Exercise 8). For matrices with more than two diagonals, things become significantly more intricate. Papers [41] to [44] are devoted to this more general situation and exhibit several phenomena that can be explained with the help and in the language of *structured* pseudospectra.

The purpose of Section 14.1 is to illustrate a few phenomena by a transparent example. Paper [189] is closely related to this section. The results of Section 14.2 were established in [43], and Section 14.3 is based on [42]. In Section 14.4 we follow [44]. All of Section 14.5 is from [50].

Exercise 3 is a result of [122]. Exercises 3 and 4 are from [42]. These two exercises reveal that Theorem 14.8 is not a consequence of some sort of general perturbation theory. Exercise 5 is taken from [91]. Exercise 7 is probably well known. We found it in [274]. Note that if $|\beta|$ is large, then the matrix with the β's differs from a purely bidiagonal matrix in an exponentially small term only—the spectral properties of $C_n(\chi)$ + random diagonal and $T_n(\chi_1)$ + random diagonal are nevertheless significantly different. Exercise 8 is the basis of the approach of [274], because it delivers explicit expressions for the entries of the inverse of the matrix $\lambda I - (T_n(\chi_1) + \mathrm{diag}\,(v_j))$. Exercise 9 is explicit in [233].

Further results: Arveson and C^*-algebras II. The following is intended as another illustration of the usefulness of C^*-algebras in numerical analysis on the one hand and as another instance of the emergence of Toeplitz plus diagonal matrices on the other. The results and ideas cited here are due to Arveson [7], [8], [9] (see also [21], [22], [149]).

The general problem is to find the spectrum of a selfadjoint operator A in $\mathcal{B}(\ell^2(\mathbf{Z}))$ by spectral approximation, say by considering the $2n \times 2n$ matrices

$$A_n = P_n A P_n | \mathrm{Im}\, P_n,$$

where now (in contrast to the rest of the book) P_n denotes the projection on $\ell^2(\mathbf{Z})$ defined by

$$P_n : \{\xi_k\}_{k=-\infty}^{\infty} \mapsto \{\dots, 0, \xi_{-n}, \dots, \xi_{-1}, \xi_0, \dots, \xi_{n-1}, 0, \dots\}.$$

Put $Q_n = I - P_n$. The Følner algebra $\mathcal{F}(\{P_n\})$ associated with the sequence $\{P_n\}_{n=1}^{\infty}$ is the C^*-subalgebra of $\mathcal{B}(\ell^2(\mathbf{Z}))$ that is constituted by all operators A for which

$$\lim_{n\to\infty} \frac{\mathrm{tr}\, P_n A^* Q_n A P_n}{n} = 0, \qquad \lim_{n\to\infty} \frac{\mathrm{tr}\, P_n A Q_n A^* P_n}{n} = 0.$$

It is easily seen that $\mathcal{F}(\{P_n\})$ contains all bounded Laurent operators and all bounded diagonal operators.

A tracial state of a unital C^*-algebra $\mathcal{A}$ is a bounded linear functional $\tau : \mathcal{A} \to \mathbf{C}$ satisfying $\tau(a^*a) \ge 0$ and $\tau(ab) = \tau(ba)$ for all $a, b \in \mathcal{A}$ and $\tau(e) = 1$ for the unit e of $\mathcal{A}$. A unital C^*-algebra is said to have a unique tracial state if the set of its tracial states is a singleton.

Let $\mathcal{A}$ have a unique tracial state τ. Then for each selfadjoint $a \in \mathcal{A}$ the map $C_0(\mathbf{R}) \to \mathbf{C}$, $\varphi \mapsto \tau(\varphi(a))$ is a positive linear functional. (Here $C_0(\mathbf{R})$ stands for the compactly supported continuous functions on $\mathbf{R}$.) By the Riesz-Markov theorem, there is a probability measure μ_a on $\mathbf{R}$ such that

$$\tau(\varphi(a)) = \int_{-\infty}^{\infty} \varphi(x) d\mu_a(x) \quad \text{for all} \quad \varphi \in C_0(\mathbf{R}).$$

This measure μ_a is called the spectral distribution of a.

Arveson's theorem. *Let $\mathcal{A}$ be a unital C^*-subalgebra of the Følner algebra $\mathcal{F}(\{P_n\})$ and suppose $\mathcal{A}$ has a unique tracial state τ. Let $A \in \mathcal{A}$ be a selfadjoint operator, let $\lambda_j(A_n)$ $(j = 1, \dots, 2n)$ be the eigenvalues of A_n, and let μ_A be the spectral distribution of A. Then*

$$\lim_{n\to\infty} \frac{1}{2n} \sum_{j=1}^{2n} \varphi(\lambda_j(A_n)) = \int_{-\infty}^{\infty} \varphi(x) f \mu_A(x) \ \textit{for every}\ \varphi \in C_0(\mathbf{R}). \tag{14.36}$$

An irrational rotation C^*-algebra is a C^*-algebra that is generated by two unitary elements u and v which satisfy $uv = e^{2\pi i \theta} vu$ with some irrational number $\theta \in \mathbf{R}$. It turns out that all such algebras with the same θ are isomorphic. To be more precise, if $\mathcal{A}^{(j)}$ $(j = 1, 2)$ are C^*-algebras that are generated by unitary elements u_j, v_j satisfying $u_j v_j = e^{2\pi i\theta} v_j u_j$, then the map $u_1 \mapsto u_2$, $v_1 \mapsto v_2$ extends to a C^*-algebra isomorphism of $\mathcal{A}^{(1)}$ onto $\mathcal{A}^{(2)}$.

A simple example of an irrational rotation C^*-algebra is the C^*-subalgebra of $\mathcal{B}(\ell^2(\mathbf{Z}))$ that is generated by the two unitary operators

$$(U\xi)_n = \xi_{n-1}, \quad (V\xi)_n = e^{-2\pi in\theta}\xi_n. \tag{14.37}$$

We denote this C^*-algebra by $\mathcal{A}_\theta$. Since, obviously, U and V belong to the Følner algebra $\mathcal{F}(\{P_n\})$, the entire C^*-algebra $\mathcal{A}_\theta$ is contained in $\mathcal{F}(\{P_n\})$. One can also show that $\mathcal{A}_\theta$ has a unique tracial state.

Let us now turn to quantum mechanics. The one-dimensional Hamiltonian

$$(Hf)(x) = -\frac{1}{2} f''(x) + \psi(x) f(x)$$

is discretized by

$$H_\delta = \frac{1}{2} P_\delta^2 + \psi(Q_\delta),$$

where δ, the numerical step size, is a small positive rational number and P_δ, Q_δ (which should not be confused with the projections P_n, Q_n introduced above) are the selfadjoint operators defined on $L^2(\mathbf{R})$ by

$$(P_\delta f)(x) = \frac{f(x+\delta) - f(x-\delta)}{2i\delta}, \quad (Q_\delta f)(x) = \frac{\sin(\delta x)}{\delta} f(x).$$

With the unitary operators U and V given on $L^2(\mathbf{R})$ by

$$(U_\delta f)(x) = f(x + 2\delta), \quad (V_\delta f)(x) = e^{i\delta x} f(x),$$

we have

$$H_\delta = -\frac{1}{8\delta^2}\left[U_\delta + U_\delta^* - 8\delta^2 \psi\left(\frac{1}{2i\delta}(V_\delta - V_\delta^*)\right)\right] + \frac{1}{4\delta^2} I. \tag{14.38}$$

It is easily seen that $U_\delta V_\delta = e^{2i\delta^2} V_\delta U_\delta = e^{2\pi i(\delta^2/\pi)} V_\delta U_\delta$. By what was said above, the C^*-algebra generated by U_δ and V_δ is isomorphic to $\mathcal{A}_{\delta^2/\pi}$ (recall that δ is rational and that hence δ^2/π is irrational). We may therefore identify the operator in the brackets of (14.38) with the operator

$$A = U + U^* - 8\delta^2 \psi\left(\frac{1}{2i\delta}(V - V^*)\right) \in \mathcal{A}_\theta \subset \mathcal{B}(\ell^2(\mathbf{Z})), \tag{14.39}$$

where U and V are given by (14.37) with $\theta = \delta^2/\pi$. Clearly, we can write (14.39) in the form

$$A = L(\chi_{-1} + \chi_1) + \operatorname{diag}(v_j)_{j=-\infty}^{\infty} \in \mathcal{A}_\theta \subset \mathcal{B}(\ell^2(\mathbf{Z})) \tag{14.40}$$

with $v_j = -8\delta^2 \psi((-1/\delta)\sin(2j\delta^2))$. Application of Arveson's theorem now yields that (14.36) is true with

$$A_n = \begin{pmatrix} v_{-n} & 1 & & & & \\ 1 & v_{-n+1} & 1 & & & \\ & \dots & \dots & \dots & & \\ & & & 1 & v_{n-2} & 1 \\ & & & & 1 & v_{n-1} \end{pmatrix}$$

$$= T_{2n}(\chi_{-1} + \chi_1) + \operatorname{diag}(v_j)_{j=-n}^{n-1}$$

and thus sets at least a theoretical foundation for the spectral approximation of A and hence H_δ.

Further results: Feinberg and Zee. The following material is taken from [117] and [118]. Let A_n be a random $n \times n$ matrix. The *eigenvalue density* of A_n is

$$\varrho(\lambda) = E\left(\frac{1}{n}\sum_{j=1}^{n} \delta(\lambda - \lambda_j(A_n))\right),$$

where E denotes the expected value (= mean) and δ is the Dirac delta function. The key formulas for computing $\varrho(\lambda)$ are

$$\varrho(\lambda) = \frac{1}{\pi}\,\partial^* E\left(\frac{1}{n}\operatorname{tr}[(\lambda I - A_n I)^{-1}]\right), \tag{14.41}$$

$$= \frac{1}{\pi}\,\partial\,\partial^* E\left(\frac{1}{n}\log\det(\lambda I - A_n)(\overline{\lambda} I - A_n^*)\right), \tag{14.42}$$

where

$$\partial^* = \frac{\partial}{\partial\overline{\lambda}} = \frac{1}{2}\left(\frac{\partial}{\partial x} + i\frac{\partial}{\partial y}\right), \quad \partial = \frac{\partial}{\partial\lambda} = \frac{1}{2}\left(\frac{\partial}{\partial x} - i\frac{\partial}{\partial y}\right).$$

The function in the parentheses on the right of (14.41) is called Green's function and denoted by $G(\lambda)$. Thus,

$$\varrho(\lambda) = \frac{1}{\pi}\,\partial^* G(\lambda), \quad G(\lambda) = E\left(\frac{1}{n}\operatorname{tr}[(\lambda I - A_n I)^{-1}]\right).$$

Let now $A_n = T_n(b) + vE_{11}$ where $b \in \mathcal{P}$ is fixed and v is drawn from some probability distribution. If $|\lambda|$ is large, we have

$$(\lambda I - A_n)^{-1} = (C_n(\lambda - b) - vE_{11})^{-1} = (I - C_n^{-1}(\lambda - b)vE_{11})^{-1}C_n^{-1}(\lambda - b)$$

$$= C_n^{-1}(\lambda - b) + \sum_{k=1}^{\infty} v^k([C_n^{-1}(\lambda - b)]_{11})^{k-1} C_n^{-1}(\lambda - b)E_{11}C_n^{-1}(\lambda - b)$$

and hence

$$\frac{1}{n}\operatorname{tr}[(\lambda I - A_n)^{-1}]$$

$$= \frac{1}{n}\operatorname{tr} C_n^{-1}(\lambda - b) + \sum_{k=1}^{\infty} v^k([C_n^{-1}(\lambda - b)]_{11})^{k-1}\operatorname{tr} C_n^{-1}(\lambda - b)E_{11}C_n^{-1}(\lambda - b).$$

We know from Proposition 2.1 that

$$C_n^{-1}(\lambda - b) = U_n^* \operatorname{diag} \frac{1}{\lambda - b(\omega_n^j)} U_n, \quad C_n^{-2}(\lambda - b) = U_n^* \operatorname{diag} \frac{1}{(\lambda - b(\omega_n^j))^2} U_n,$$

where $\omega_n = e^{2\pi i/n}$. Put

$$G_0(\lambda) = \frac{1}{n} \operatorname{tr} C_n^{-1}(\lambda - b) = \frac{1}{n} \sum_{j=0}^{n-1} \frac{1}{\lambda - b(\omega_n^j)}.$$

Then

$$\operatorname{tr} E_{11} C_n^{-2}(\lambda - b) E_{11} = [C_n^{-2}(\lambda - b)]_{11} = \frac{1}{n} \sum_{j=0}^{n-1} \frac{1}{(\lambda - b(\omega_n^j))^2} = -\frac{\partial G_0(\lambda)}{\partial \lambda}$$

and consequently,

$$G(\lambda) = G_0(\lambda) - \frac{1}{n} \frac{\partial G_0(\lambda)}{\partial \lambda} E\left(\frac{v}{1 - G_0(\lambda) v}\right). \tag{14.43}$$

As both sides of (14.43) are analytic functions of λ, equality (14.43) is actually true for all λ outside $\operatorname{sp} C_n(b) = \{b(1), b(\omega_n), \dots, b(\omega_n^{n-1})\}$.

Now take $b(t) = \chi_1(t) := t$. For large n,

$$G_0(\lambda) = \frac{1}{n} \sum_{j=0}^{n-1} \frac{1}{\lambda - \omega_n^j} \approx \frac{1}{2\pi} \int_0^{2\pi} \frac{d\theta}{\lambda - e^{i\theta}} = \begin{cases} 0 & \text{for } |\lambda| < 1, \\ 1/\lambda & \text{for } |\lambda| > 1, \end{cases}$$

and thus, by (14.43),

$$G(\lambda) \approx \begin{cases} 0 & \text{for } |\lambda| < 1, \\ \frac{1}{\lambda} + \frac{1}{n\lambda} E\left(\frac{v}{\lambda - v}\right) & \text{for } |\lambda| > 1. \end{cases}$$

With the Heaviside function $H(x) = 0$ for $x < 0$ and $H(x) = 1$ for $x > 1$, we can write

$$G(\lambda) \approx H(\lambda\bar{\lambda} - 1)\left[\frac{1}{\lambda} + \frac{1}{n\lambda} E\left(\frac{v}{\lambda - v}\right)\right]. \tag{14.44}$$

As a first concrete example, assume v takes the two values r and $-r$ with equal probability,

$$P(v = r) = P(v = -r) = \frac{1}{2}. \tag{14.45}$$

Then (14.44) becomes

$$\begin{aligned} G(\lambda) &\approx H(\lambda\bar{\lambda} - 1)\left[\frac{1}{\lambda} + \frac{1}{2n\lambda}\left(\frac{r}{\lambda - r} - \frac{r}{\lambda + r}\right)\right] \\ &= H(\lambda\bar{\lambda} - 1)\left[\left(1 - \frac{1}{n}\right)\frac{1}{\lambda} + \frac{1}{2n}\left(\frac{1}{\lambda - r} + \frac{1}{\lambda + r}\right)\right]. \end{aligned}$$

Taking into account that $H'(x) = \delta(x)$ and $\partial^*(1/\lambda) = \pi\delta(\lambda)$ we get

$$\begin{aligned}\pi\varrho(\lambda) &= \partial^* G(\lambda)\\ &\approx \delta(\lambda\bar{\lambda}-1)\left[1-\frac{1}{n}+\frac{\lambda}{2n}\left(\frac{r}{\lambda-r}+\frac{r}{\lambda+r}\right)\right]\\ &+ H(\lambda\bar{\lambda}-1)\left[\left(1-\frac{1}{n}\right)\pi\delta(\lambda)+\frac{1}{2n}\pi\delta(\lambda-r)+\frac{1}{2n}\pi\delta(\lambda+r)\right].\end{aligned}$$

We have $\delta(\lambda) = 0$ for $|\lambda| > 1$, and the term

$$\delta(\lambda\bar{\lambda}-1)\,\frac{\lambda}{2n}\left(\frac{r}{\lambda-r}+\frac{r}{\lambda+r}\right)$$

can be neglected for large n. In summary

$$\varrho(\lambda) \approx \frac{1}{\pi}\left(1-\frac{1}{n}\right)\delta(|\lambda|^2-1) + H(|\lambda|^2-1)\left[\frac{1}{2n}\delta(\lambda-r)+\frac{1}{2n}\delta(\lambda+r)\right].$$

Thus, in the $n \to \infty$ limit the situation is as follows. If $r \in (0, 1)$, then the eigenvalues are distributed on $\mathbf{T}$, and if $r \in (1, \infty)$, then $n-2$ eigenvalues are distributed on $\mathbf{T}$, while two eigenvalues are located at $-r$ and r with the density $1/(2n)$. Notice that the coefficient $1/\pi$ before $\delta(|\lambda|^2-1)$ is correct, since if Ω is any region containing $\mathbf{T}$, then

$$\begin{aligned}\int_\Omega \frac{1}{\pi}\,\delta(|\lambda|^2-1)\,dA(\lambda) &= \int_0^{2\pi}\int_{1-\varepsilon}^{1+\varepsilon}\frac{1}{\pi}\,\delta(r^2-1)\,r\,dr\,d\theta\\ = 2\int_{1-\varepsilon}^{1+\varepsilon}\delta(r^2-1)\,r\,dr &= 2\int_{(1-\varepsilon)^2}^{(1+\varepsilon)^2}\delta(s-1)\,\frac{1}{2}\,ds = 1.\end{aligned}$$

We remark that the special distribution (14.45) can also be treated in a more elementary manner. Namely, the eigenvalues of $C_n(\chi_1) + vE_{11}$ are the roots of the polynomial $\lambda^n - v\lambda^{n-1} - 1$. If $|v| > 1$ then, in the $n \to \infty$ limit, one of these roots is at v with the density $1/n$ and the remaining roots are distributed on $\mathbf{T}$. If $|v| \le 1$ then, in the $n \to \infty$ limit, all zeros are distributed on $\mathbf{T}$.

We now take v from the uniform distribution on $[-M, M]$. Proceeding as above, one can show that then

$$G(\lambda) \approx H(\lambda\bar{\lambda}-1)\left[\left(1-\frac{1}{n}\right)\frac{1}{\lambda}-\frac{1}{2nM}\log\frac{\lambda-M}{\lambda+M}\right]$$

and

$$\varrho(\lambda) \approx \frac{1}{\pi}\left(1-\frac{1}{n}\right)\delta(|\lambda|^2-1)+\frac{H(|\lambda|^2-1)}{2nM}\delta(\operatorname{Im}\lambda)H(M-|\operatorname{Re}\lambda|).$$

We see in particular that if $M > 1$, then asymptotically the eigenvalues lie on $\mathbf{T}$ and two wings $[-M, -1]$ and $[1, M]$. Figure 14.17 shows the results of two numerical realizations.

If $x = (x_1, \ldots, x_n)$ is an eigenvector for $C_n(\chi_1) + vE_{11}$, then $x_{j+1} = \lambda x_j$ for $j = 2, \ldots, n-1$. Thus, if $|\lambda| = 1$ then x is extended, whereas for $\lambda = \pm r$ with $r > 1$ the vector x is localized at the beginning.

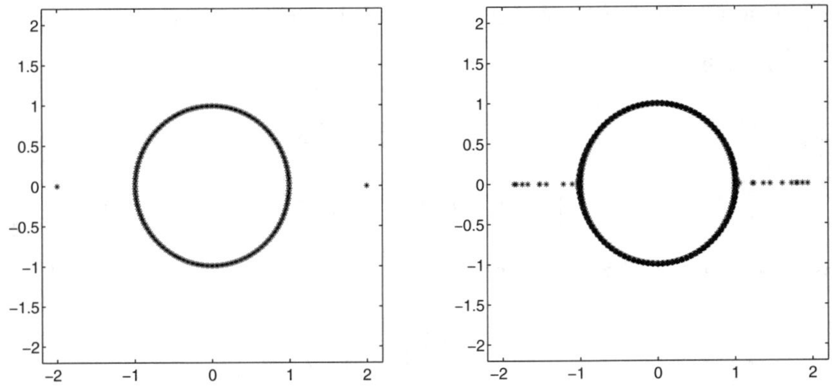

Figure 14.17. *The superposition of the eigenvalues of* $C_{100}(\chi_1) + vE_{11}$ *for* 50 *random choices of* v*. In the left picture,* v *takes the values* -2 *and* 2 *with equal probability. In the right picture,* v *is drawn from the uniform distribution on the segment* $[-2, 2]$.

Feinberg and Zee also consider $A_n = C_n(\chi_1) + \text{diag}\,(v_1, \ldots, v_n)$, where the v_j are independent and $P(v_j = r) = P(v_j = -r) = 1/2$ for all j. In this case one can employ (14.42). In the case at hand,

$$\det A_n = \left(\prod_{k=1}^{n} (\lambda - v_j) \right) - 1$$

and hence

$$E[\log \det (\lambda I - A_n)] = \frac{1}{2^n} \sum_{j=0}^{n} \binom{n}{j} \log \left[(\lambda - r)^j (\lambda + r)^{n-j} - 1 \right].$$

In the limit $n \to \infty$, the binomial coefficients are sharply peaked around $j \approx n/2$, and thus

$$E[\log \det (\lambda I - A_n)] \approx \log \left[(\lambda - r)^{n/2} (\lambda + r)^{n/2} - 1 \right]. \tag{14.46}$$

A similar approximation is true for $E[\log \det (\overline{\lambda} I - A_n^*)]$. By (14.42), the support of the density $\varrho(\lambda)$ is the singularities of the right-hand side of (14.46). These singularities occur at $(\lambda^2 - r^2)^{n/2} = 1$, that is, at

$$\lambda_j = \pm\sqrt{r^2 + e^{4\pi i j/n}} \quad (j = 0, 1, \ldots, n/2). \tag{14.47}$$

It follows that the original circle spectrum is distorted by the specific randomness considered into the curve $\lambda^2 = r^2 + e^{i\theta}$ $(0 \le \theta < 2\pi)$. Figure 14.18 shows some examples.

Further results: Trefethen, Contedini, and Embree. All of the following is from [274]. This paper is devoted to matrices of the form $T_n(\chi_1)$ + random diagonal and to $C_n(\chi_1)$ + random diagonal, and it exhibits a Hatano-Nelson bubble and an associated localization-delocalization phenomenon in the context of pseudospectra and resolvents instead of eigenvalues and eigenvectors.

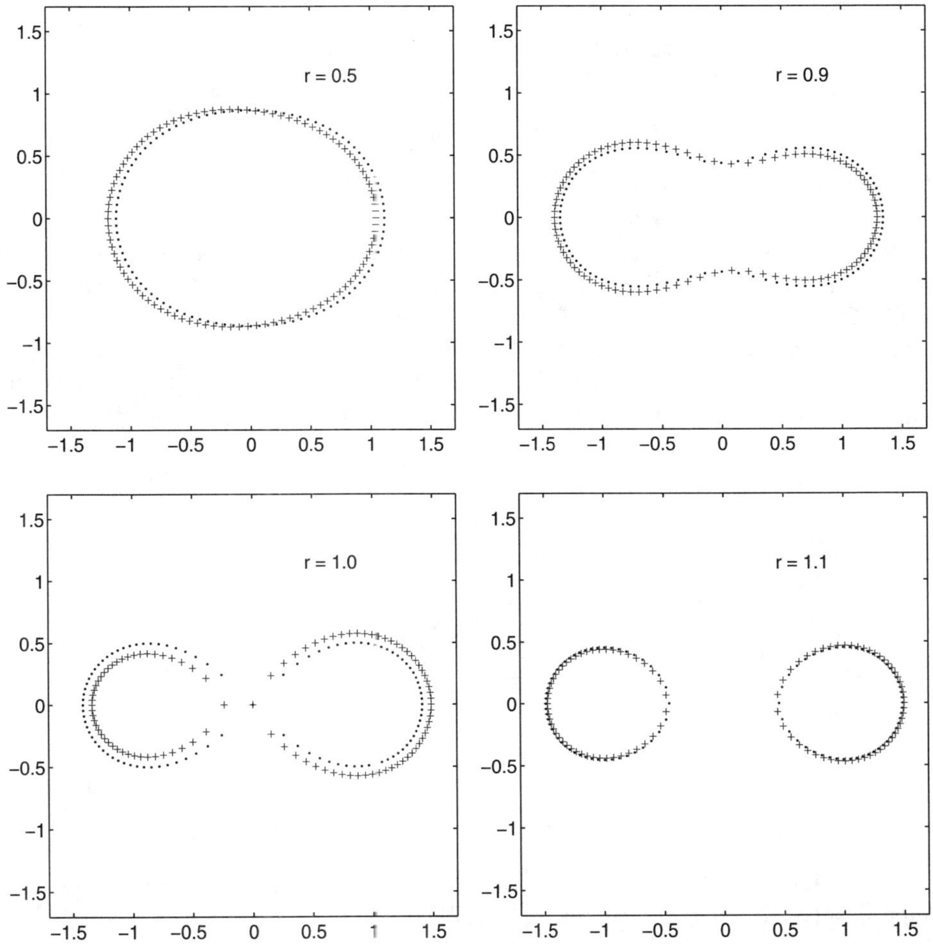

Figure 14.18. *Each picture shows the* 100 *eigenvalues (marked by* $+$*) of the matrix* $C_{100}(\chi_{-1}+\chi_1)+\mathrm{diag}\,(v_j)$ *with independent* v_j *taking the values* $-r$ *and* r *with equal probability. The values of* r *are* 0.5, 0.9, 1.0, 1.1. *The dots are the points* (14.47) *for* $n = 100$.

Let V be a random variable taking values dense in a compact subset $\mathrm{supp}\,V$ of the complex plane. For $\lambda \in \mathbf{C}$, we define

$$d_{\min}(\lambda) = \min_{v\in\mathrm{supp}\,V} |\lambda - v|, \quad d_{\max}(\lambda) = \max_{v\in\mathrm{supp}\,V} |\lambda - v|,$$
$$d_{\mathrm{mean}}(\lambda) = \exp E(\log|\lambda - V|).$$

We then have $0 \le d_{\min}(\lambda) \le d_{\mathrm{mean}}(\lambda) \le d_{\max}(\lambda) < \infty$ for every $\lambda \in \mathbf{C}$. Let

$$\Omega_{\rm I} = \{\lambda \in \mathbf{C} : d_{\max}(\lambda) < 1\},$$
$$\Omega_{\rm II} = \{\lambda \in \mathbf{C} : d_{\rm mean}(\lambda) < 1 \le d_{\max}(\lambda)\},$$
$$\Omega_{\rm III} = \{\lambda \in \mathbf{C} : d_{\min}(\lambda) \le 1 \le d_{\rm mean}(\lambda)\},$$
$$\Omega_{\rm IV} = \{\lambda \in \mathbf{C} : 1 < d_{\min}(\lambda)\}.$$

Obviously, these four sets are disjoint, and their union is $\mathbf{C}$.

Here are two concrete examples. Suppose first that $\operatorname{supp} V = \{-1, 1\}$ and that $P(V = -1) = P(V = 1) = 1/2$. Then $\Omega_{\rm I}$ is empty, $\Omega_{\rm II}$ is the closed set bounded by the lemniscate $|\lambda^2 - 1|^2 = 1$, $\Omega_{\rm III}$ is the two closed disks $\{-1, 1\} + \overline{\mathbf{D}}$ minus $\Omega_{\rm II}$, and $\Omega_{\rm IV}$ is $\mathbf{C} \setminus \Omega_{\rm III}$ (see the left picture of Figure 14.19). If V is uniformly distributed on $[-2, 2]$, then $\Omega_{\rm I}$ is again empty, $\Omega_{\rm II}$ is the bubble bounded by two symmetric arcs in the upper and lower half-planes meeting at $\lambda_0 \in (1, 2)$ and $-\lambda_0 \in (-2, -1)$, $\Omega_{\rm III}$ is $[-2, 2] + \overline{\mathbf{D}}$ minus $\Omega_{\rm III}$, and $\Omega_{\rm IV}$ is the complement of $\Omega_{\rm III}$ (see the right picture of Figure 14.19).

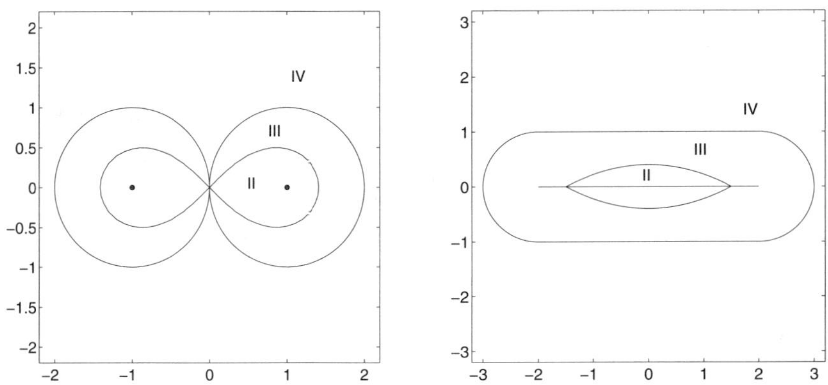

Figure 14.19. *The sets $\Omega_{\rm II}$, $\Omega_{\rm III}$, $\Omega_{\rm IV}$ for the case of the uniform distribution on* $\operatorname{supp} V = \{-1, 1\}$ *(left) and* $\operatorname{supp} V = [-2, 2]$ *(right).*

Now put

$$A_n = T_n(\chi_1) + \operatorname{diag}(v_1, \ldots, v_n), \quad A = T(\chi_1) + \operatorname{diag}(v_j)_{j=1}^{\infty},$$
$$B_n = C_n(\chi_1) + \operatorname{diag}(v_1, \ldots, v_n), \quad B = L(\chi_1) + \operatorname{diag}(v_j)_{j=1}^{\infty},$$

where $L(c) := (c_{j-k})_{j,k=-\infty}^{\infty}$ and the v_j are independent random variables with the same distribution as V. We say some result holds almost surely if it is true with probability 1.

Theorem on A_n. (a) *If $\lambda \in \Omega_{\rm I}$, then $\|(\lambda I - A_n)^{-1}\|_2 \ge (1/d_{\max}(\lambda))^n$ (guaranteed exponential growth).*

(b) *If $\lambda \in \Omega_{\rm I} \cup \Omega_{\rm II}$, then $\|(\lambda I - A_n)^{-1}\|_2 \to \infty$ as $n \to \infty$ almost surely. If, in addition, $\lambda \notin \operatorname{supp} V$, then $\|(\lambda I - A_n)^{-1}\|_2^{1/n} \to 1/d_{\rm mean}(\lambda)$ as $n \to \infty$ almost surely (almost sure exponential growth).*

(c) *If $\lambda \in \Omega_{\rm III}$, then $\|(\lambda I - A_n)^{-1}\|_2 \to \infty$ as $n \to \infty$ almost surely. If, in addition, $\lambda \notin$ $\operatorname{supp} V$, then $\|(\lambda I - A_n)^{-1}\|_2^{1/n} \to 1$ as $n \to \infty$ almost surely (almost sure subexponential growth).*

(d) *If* $\lambda \in \Omega_{\rm IV}$, *then* $\|(\lambda I - A_n)^{-1}\|_2 < 1/(d_{\min}(\lambda) - 1)$ *(guaranteed boundedness).*

The spectrum $\operatorname{sp} A_n$ *converges in the Hausdorff metric to* $\operatorname{supp} V$ *as* $n \to \infty$ *almost surely.*

Theorem on $\boldsymbol{A}$**.** *If* $\lambda \in \Omega_{\rm I}$, *then* $\|(\lambda I - A)^{-1}\|_2 = \infty$. *If* $\lambda \in \Omega_{\rm II} \cup \Omega_{\rm III}$, *then* $\|(\lambda I - A)^{-1}\|_2 = \infty$ *almost surely. If* $\lambda \in \Omega_{\rm IV}$, *then* $\|(\lambda I - A)^{-1}\|_2 \le 1/(d_{\min}(\lambda) - 1)$ *and this inequality is an equality almost surely. The spectrum* $\operatorname{sp} A$ *equals* $\Omega_{\rm I} \cup \Omega_{\rm II} \cup \Omega_{\rm III}$ *almost surely.*

We define

$$S_{\rm bubble} = \{\lambda \in \mathbf{C} : d_{\rm mean}(\lambda) = 1\}, \quad S_{\rm wings} = \{\lambda \in \operatorname{supp} V : d_{\rm mean}(\lambda) > 1\}.$$

Theorem on $\boldsymbol{B_n}$**.** (a) *If* $\lambda \in \Omega_{\rm I}$, *then* $\|(\lambda I - B_n)^{-1}\|_2 \le 1/(1 - d_{\max}(\lambda))$.

(b) *If* $\lambda \in \Omega_{\rm II}$, *then* $\|(\lambda I - B_n)^{-1}\|_2 \to \infty$ *as* $n \to \infty$ *almost surely and the quantity* $\|(\lambda I - B_n)^{-1}\|_2^{1/n}$ *goes to* 1 *as* $n \to \infty$ *almost surely.*

(c) *If* $\lambda \in \Omega_{\rm III}$, *then* $\|(\lambda I - B_n)^{-1}\|_2 \to \infty$ *as* $n \to \infty$ *almost surely. If in addition* $\lambda \notin S_{\rm bubble} \cup S_{\rm wings}$, *then* $\|(\lambda I - B_n)^{-1}\|_2^{1/n} \to 1$ *as* $n \to \infty$ *almost surely.*

(d) *If* $\lambda \in \Omega_{\rm IV}$, *then* $\|(\lambda I - B_n)^{-1}\|_2 \le 1/(d_{\min}(\lambda) - 1)$.

If $S_{\rm bubble}$ *consists only of curves disjoint from* $\operatorname{supp} V$ *except at isolated points, then* $\operatorname{sp} B_n$ *converges in the Hausdorff metric to* $S_{\rm bubble} \cup S_{\rm wings}$ *as* $n \to \infty$ *almost surely.*

Theorem on $\boldsymbol{B}$**.** *If* $\lambda \in \Omega_{\rm I}$ *then* $\|(\lambda I - B)^{-1}\|_2 \le 1/(1 - d_{\max}(\lambda))$, *and this inequality is an equality almost surely. If* $\lambda \in \Omega_{\rm II} \cup \Omega_{\rm III}$, *then* $\|(\lambda I - B)^{-1}\|_2 = \infty$ *almost surely. If* $\lambda \in \Omega_{\rm IV}$ *then* $\|(\lambda I - B)^{-1}\|_2 \le 1/(d_{\min}(\lambda) - 1)$, *and this inequality is an equality almost surely. The spectrum* $\operatorname{sp} B$ *coincides with* $\Omega_{\rm II} \cup \Omega_{\rm III}$ *almost surely.*

Bibliography

[1] V. M. ADAMYAN, *Asymptotic properties for positive and Toeplitz matrices*, Oper. Theory Adv. Appl., 43 (1990), pp. 17–38.

[2] E. L. ALLGOWER, *Exact inverses of certain band matrices*, Numer. Math., 21 (1973), pp. 279–284.

[3] G. AMMAR AND P. GADER, *A variant of the Gohberg–Semencul formula involving circulant matrices*, SIAM J. Matrix Anal. Appl., 12 (1991), pp. 534–540.

[4] P. W. ANDERSON, *Absence of diffusion in certain random lattices*, Phys. Rev. (Second Series), 109 (1958), pp. 1492–1505.

[5] A. L. ANDREW, *Eigenvectors of certain matrices*, Linear Algebra Appl., 7 (1973), pp. 151–162.

[6] P. M. ANSELONE AND I. H. SLOAN, *Spectral approximations for Wiener-Hopf operators II*, J. Integral Equations Appl., 4 (1992), pp. 465–489.

[7] W. ARVESON, *Noncommutative spheres and numerical quantum mechanics*, in Operator Algebras, Mathematical Physics, and Low-Dimensional Topology (Istanbul, 1991), Res. Notes Math., Vol. 5, A. K. Peters, Wellesley, MA, 1993, pp. 1–10.

[8] W. ARVESON, *The role of C^*-algebras in infinite-dimensional numerical linear algebra*, Contemp. Math., 167 (1994), pp. 114–129.

[9] W. ARVESON, *C^*-algebras and numerical linear algebra*, J. Funct. Anal., 122 (1994), pp. 333–360.

[10] F. AVRAM, *On bilinear forms in Gaussian random variables and Toeplitz matrices*, Probab. Theory Related Fields, 79 (1988), pp. 37–45.

[11] O. AXELSSON, *Iterative Solution Methods*, Cambridge University Press, Cambridge, UK, 1996.

[12] M. BAKONYI AND D. TIMOTIN, *On an extension problem for polynomials*, Bull. London Math. Soc., 33 (2001), pp. 599–605.

[13] E. BASOR, *Review of "Invertibility and Asymptotics of Toeplitz Matrices"*, Linear Algebra Appl., 68 (1985), pp. 275–278.

[14] E. Basor and T. Ehrhardt, *Asymptotic formulas for determinants of a sum of finite Toeplitz and Hankel matrices*, Math. Nachr., 228 (2001), pp. 5–45.

[15] E. Basor and K. E. Morrison, *The Fisher-Hartwig conjecture and Toeplitz eigenvalues*, Linear Algebra Appl., 202 (1994), pp. 129–142.

[16] E. Basor and H. Widom, *On a Toeplitz determinant identity of Borodin and Okounkov*, Integral Equations Operator Theory, 37 (2000), pp. 397–401.

[17] G. Baxter, *Polynomials defined by a difference system*, J. Math. Anal. Appl., 2 (1961), pp. 223–263.

[18] G. Baxter, *A norm inequality for a finite-section Wiener-Hopf equation*, Illinois J. Math., 7 (1963), pp. 97–103.

[19] G. Baxter and P. Schmidt, *Determinants of a certain class of non-Hermitian Toeplitz matrices*, Math. Scand., 9 (1961), pp. 122–128.

[20] R. M. Beam and R. F. Warming, *The asymptotic spectra of banded Toeplitz and quasi-Toeplitz matrices*, SIAM J. Sci. Comput., 14 (1993), pp. 971–1006.

[21] E. Bédos, *On filtrations for C^*-algebras*, Houston J. Math., 20 (1994), pp. 63–74.

[22] E. Bédos, *On Følner nets, Szegö's theorem and other eigenvalue distribution theorems*, Exposition. Math., 15 (1997), pp. 193–228.

[23] F. Di Benedetto, G. Fiorentino, and S. Serra, *CG preconditioning for Toeplitz matrices*, J. Comput. Math. Appl., 25 (1993), pp. 35–45.

[24] L. Berg, *Über eine Identität von W. F. Trench zwischen der Toeplitzschen und einer verallgemeinerten Vandermondeschen Determinante*, Z. Angew. Math. Mech., 66 (1986), pp. 314–315.

[25] L. Berg, *Lineare Gleichungssysteme mit Bandstruktur und ihr asymptotisches Verhalten*, Deutscher Verlag der Wissenschaften, Berlin, 1986.

[26] A. Berman and R. J. Plemmons, *Nonnegative Matrices in the Mathematical Sciences*, Classics Appl. Math. 9, SIAM, Philadelphia, 1994.

[27] R. Bhatia, *Matrix Analysis*, Springer-Verlag, New York, 1997.

[28] D. Bini and A. Böttcher, *Polynomial factorization through Toeplitz matrix computations*, Linear Algebra Appl., 366 (2003), pp. 25–37.

[29] P. du Bois-Reymond, *Untersuchungen über die Convergenz und Divergenz der Fourierschen Darstellungsformeln*, Abh. d. Math.-Phys. Classe d. Königl. Bayerischen Akad. d. Wiss., 12 (1876), pp. 1–13.

[30] F. F. Bonsall and J. Duncan, *Numerical Ranges of Operators on Normed Spaces and Elements of Normed Algebras*, Cambridge University Press, Cambridge, UK, 1971.

[31] A. BORODIN AND A. OKOUNKOV, *A Fredholm determinant formula for Toeplitz determinants*, Integral Equations Operator Theory, 37 (2000), pp. 386–396.

[32] A. BÖTTCHER, *Status Report on Rationally Generated Block Toeplitz and Wiener-Hopf Determinants*, unpublished manuscript, 34 pages, 1989 (available from the author on request).

[33] A. BÖTTCHER, *Truncated Toeplitz operators on the polydisk*, Monatshefte f. Math., 110 (1990), pp. 23–32.

[34] A. BÖTTCHER, *Pseudospectra and singular values of large convolution operators*, J. Integral Equations Appl., 6 (1994), pp. 267–301.

[35] A. BÖTTCHER, *On the approximation numbers of large Toeplitz matrices*, Documenta Mathematica, 2 (1997), pp. 1–29.

[36] A. BÖTTCHER, *C^*-algebras in numerical analysis*, Irish Math. Soc. Bull., 45 (2000), pp. 57–133.

[37] A. BÖTTCHER, *One more proof of the Borodin-Okounkov formula for Toeplitz determinants*, Integral Equations Operator Theory, 41 (2001), pp. 123–125.

[38] A. BÖTTCHER, *On the determinant formulas by Borodin, Okounkov, Baik, Deift, and Rains*, Oper. Theory Adv. Appl., 135 (2002), pp. 91–99.

[39] A. BÖTTCHER, *Transient behavior of powers and exponentials of large Toeplitz matrices*, Electron. Trans. Numer. Anal., 18 (2004), pp. 1–41.

[40] A. BÖTTCHER, *The constants in the asymptotic formulas by Rambour and Seghier for inverses of Toeplitz matrices*, Integral Equations Operator Theory, 50 (2004), pp. 43–55.

[41] A. BÖTTCHER, M. EMBREE, AND M. LINDNER, *Spectral approximation of banded Laurent matrices with localized random perturbations*, Integral Equations Operator Theory, 42 (2002), pp. 142–165.

[42] A. BÖTTCHER, M. EMBREE, AND V. I. SOKOLOV, *Infinite Toeplitz and Laurent matrices with localized impurities*, Linear Algebra Appl., 343/344 (2002), pp. 101–118.

[43] A. BÖTTCHER, M. EMBREE, AND V. I. SOKOLOV, *On large Toeplitz band matrices with an uncertain block*, Linear Algebra Appl., 366 (2003), pp. 87–97.

[44] A. BÖTTCHER, M. EMBREE, AND V. I. SOKOLOV, *The spectra of large Toeplitz band matrices with a randomly perturbed entry*, Math. Comp., 72 (2003), pp. 1329–1348.

[45] A. BÖTTCHER, M. EMBREE, AND L. N. TREFETHEN, *Piecewise continuous Toeplitz matrices and operators: Slow approach to infinity*, SIAM J. Matrix Anal. Appl., 24 (2002) pp. 484–489.

[46] A. BÖTTCHER AND S. GRUDSKY, *On the condition numbers of large semi-definite Toeplitz matrices*, Linear Algebra Appl., 279 (1998), pp. 285–301.

[47] A. Böttcher and S. Grudsky, *Toeplitz band matrices with exponentially growing condition numbers*, Electron. J. Linear Algebra, 5 (1999), pp. 104–125.

[48] A. Böttcher and S. Grudsky, *Toeplitz Matrices, Asymptotic Linear Algebra, and Functional Analysis*, Hindustan Book Agency, New Delhi, 2000, and Birkhäuser Verlag, Basel, 2000.

[49] A. Böttcher and S. Grudsky, *Condition numbers of large Toeplitz-like matrices*, Contemp. Math., 280 (2001), pp. 273–299.

[50] A. Böttcher and S. Grudsky, *Can spectral value sets of Toeplitz band matrices jump?*, Linear Algebra Appl., 351/352 (2002), pp. 99–116.

[51] A. Böttcher and S. Grudsky, *Asymptotic spectra of dense Toeplitz matrices are unstable*, Numer. Algorithms, 33 (2003), pp. 105–112.

[52] A. Böttcher and S. Grudsky, *The norm of the product of a large matrix and a random vector*, Electronic J. Probability, 8 (2003), Paper 7, pp. 1–29.

[53] A. Böttcher and S. Grudsky, *Fejér means and norms of large Toeplitz matrices*, Acta Sci. Math. (Szeged), 69 (2003), pp. 889–900.

[54] A. Böttcher and S. Grudsky, *Toeplitz matrices with slowly growing pseudospectra*, in Factorization, Singular Integral Operators, and Related Topics, S. Samko, A. Lebre, and A. F. dos Santos, eds., Kluwer Academic Publishers, Dordrecht, The Netherlands, 2003, pp. 43–54.

[55] A. Böttcher and S. Grudsky, *Asymptotically good pseudomodes for Toeplitz matrices and Wiener-Hopf operators*, Oper. Theory Adv. Appl., 147 (2004), pp. 175–188.

[56] A. Böttcher and S. Grudsky, *Structured condition numbers of large Toeplitz matrices are rarely better than usual condition numbers*, Numer. Linear Algebra Appl., 12 (2005), pp. 95–102.

[57] A. Böttcher, S. Grudsky, and A. Kozak, *On the distance of a large Toeplitz band matrix to the nearest singular matrix*, Oper. Theory Adv. Appl., 135 (2002), pp. 101–106.

[58] A. Böttcher, S. Grudsky, A. Kozak, and B. Silbermann, *Norms of large Toeplitz band matrices*, SIAM J. Matrix Anal. Appl., 21 (1999), pp. 547–561.

[59] A. Böttcher, S. Grudsky, A. Kozak, and B. Silbermann, *Convergence speed estimates for the norms of the inverses of large truncated Toeplitz matrices*, Calcolo, 36 (1999), pp. 103–122.

[60] A. Böttcher, S. Grudsky, and E. Ramírez de Arellano, *Approximating inverses of Toeplitz matrices by circulant matrices*, Methods Appl. Anal., 11 (2004), pp. 211–220.

[61] A. BÖTTCHER, S. GRUDSKY, AND E. RAMÍREZ DE ARELLANO, *On the asymptotic behavior of the eigenvectors of large banded Toeplitz matrices*, Math. Nachr., to appear.

[62] A. BÖTTCHER, S. GRUDSKY, AND B. SILBERMANN, *Norms of inverses, spectra, and pseudospectra of large truncated Wiener-Hopf operators and Toeplitz matrices*, New York J. Math., 3 (1997), pp. 1–31.

[63] A. BÖTTCHER AND B. SILBERMANN, *Notes on the asymptotic behavior of block Toeplitz matrices and determinants*, Math. Nachr., 98 (1980), pp. 183–210.

[64] A. BÖTTCHER AND B. SILBERMANN, *The asymptotic behavior of Toeplitz determinants for generating functions with zeros of integral orders*, Math. Nachr., 102 (1981), pp. 79–105.

[65] A. BÖTTCHER AND B. SILBERMANN, *Über das Reduktionsverfahren für diskrete Wiener-Hopf-Gleichungen mit unstetigem Symbol*, Z. Anal. Anwendungen, 1, no. 2 (1982), pp. 1–5.

[66] A. BÖTTCHER AND B. SILBERMANN, *The finite section method for Toeplitz operators on the quarter-plane with piecewise continuous symbols*, Math. Nachr., 110 (1983), pp. 279–291.

[67] A. BÖTTCHER AND B. SILBERMANN, *Invertibility and Asymptotics of Toeplitz Matrices*, Akademie-Verlag, Berlin, 1983.

[68] A. BÖTTCHER AND B. SILBERMANN, *Toeplitz matrices and determinants with Fisher-Hartwig symbols*, J. Funct. Anal., 63 (1985), pp. 178–214.

[69] A. BÖTTCHER AND B. SILBERMANN, *Toeplitz operators and determinants generated by symbols with one Fisher-Hartwig singularity*, Math. Nachr., 127 (1986), pp. 95–123.

[70] A. BÖTTCHER AND B. SILBERMANN, *Analysis of Toeplitz Operators*, Akademie-Verlag, Berlin, 1989 and Springer-Verlag, Berlin, 1990.

[71] A. BÖTTCHER AND B. SILBERMANN, *Introduction to Large Truncated Toeplitz Matrices*, Springer-Verlag, New York, 1999.

[72] A. BÖTTCHER AND H. WIDOM, *Two remarks on spectral approximations for Wiener-Hopf operators*, J. Integral Equations Appl., 6 (1994), pp. 31–36.

[73] A. BÖTTCHER AND H. WIDOM, *Two elementary derivations of the pure Fisher-Hartwig determinant*, Integral Equations Operator Theory, to appear.

[74] A. BÖTTCHER AND H. WIDOM, *From Toeplitz eigenvalues through Green's kernels to higher-order Wirtinger-Sobolev inequalities*, Oper. Theory Adv. Appl., to appear.

[75] E. BRÉZIN AND A. ZEE, *Non-Hermitean delocalization: Multiple scattering and bounds*, Nuclear Phys. B, 509 (1998), pp. 599–614.

[76] P. W. Brouwer, P. G. Silvestrov, and C. W. J. Beenakker, *Theory of directed localization in one dimension*, Phys. Rev. B, 56 (1997), pp. R4333–R4335.

[77] A. Brown and P. Halmos, *Algebraic properties of Toeplitz operators*, J. Reine Angew. Math., 213 (1963/1964), pp. 89–102.

[78] E. S. Brown and I. M. Spitkovsky, *On matrices with elliptical numerical range*, Linear Multilinear Algebra, 52 (2004), pp. 177–193.

[79] D. Bump and P. Diaconis, *Toeplitz minors*, J. Combin. Theory Ser. A, 97 (2002), pp. 252–271.

[80] J. Burke and A. Greenbaum, *Some equivalent characterizations of the polynomial numerical hull of degree k*, Oxford University Computing Laboratory Report, no. 04/29 (2004).

[81] A. Calderón, F. Spitzer, and H. Widom, *Inversion of Toeplitz matrices*, Illinois J. Math., 3 (1959), pp. 490–498.

[82] A. Cantoni and F. Butler, *Eigenvalues and eigenvectors of symmetric centrosymmetric matrices*, Linear Algebra Appl., 13 (1976), pp. 275–288.

[83] R. H. Chan, *Toeplitz preconditioners for Toeplitz systems with nonnegative generating functions*, IMA J. Numer. Anal., 11 (1991), pp. 333–345.

[84] R. H. Chan, X.-Q. Jin, and M.-C. Yeung, *The circulant operator in the Banach algebra of matrices*, Linear Algebra Appl., 149 (1991), pp. 41–53.

[85] R. H. Chan and M. K. Ng, *Conjugate gradient methods for Toeplitz systems*, SIAM Rev., 38 (1996), pp. 427–482.

[86] R. H. Chan, M. K. Ng, and A. M. Yip, *A survey of preconditioners for ill-conditioned Toeplitz systems*, Contemp. Math., 281 (2001), pp. 175–191.

[87] R. H. Chan and G. Strang, *Toeplitz equations by conjugate gradients with circulant preconditioner*, SIAM J. Sci. Statist. Comput., 10 (1989), pp. 104–119.

[88] R. H. Chan and M.-C. Yeung, *Circulant preconditioners constructed from kernels*, SIAM J. Numer. Anal., 29 (1992), pp. 1093–1103.

[89] T. Chan, *An optimal circulant preconditioner for Toeplitz systems*, SIAM J. Sci. Statist. Comput., 9 (1988), pp. 766–771.

[90] R. Courant, K. Friedrichs, and H. Lewy, *Über die partiellen Differenzengleichungen der mathematischen Physik*, Math. Ann., 100 (1928), pp. 32–74.

[91] K. A. Dahmen, D. R. Nelson, and N. M. Shnerb, *Life and death near a windy oasis*, J. Math. Biol., 41 (2000), pp. 1–23.

[92] J. W. Daniel, *The conjugate gradient method for linear and nonlinear operator equations*, SIAM J. Numer. Anal., 4 (1967), pp. 10–26.

[93] S. DASGUPTA AND A. GUPTA, *An elementary proof of a theorem of Johnson and Lindenstrauss*, Random Structures & Algorithms, 22 (2003), pp. 60–65.

[94] E. B. DAVIES, *Spectral enclosures and complex resonances for general self-adjoint operators*, LMS J. Comput. Math., 1 (1998), pp. 42–74.

[95] E. B. DAVIES, *Spectral properties of random non-self-adjoint matrices and operators*, Proc. Roy. Soc. London Ser. A, 457 (2001), pp. 191–206.

[96] E. B. DAVIES, *Semigroup growth bounds*, J. Oper. Theory, to appear.

[97] K. M. DAY, *Toeplitz matrices generated by the Laurent series expansion of an arbitrary rational function*, Trans. Amer. Math. Soc., 206 (1975), pp. 224–245.

[98] K. M. DAY, *Measures associated with Toeplitz matrices generated by the Laurent expansion of rational functions*, Trans. Amer. Math. Soc., 209 (1975), pp. 175–183.

[99] P. DELSARTE AND Y. GENIN, *Spectral properties of finite Toeplitz matrices*, in Mathematical Theory of Networks and Systems (Beer Sheva, 1983), Lecture Notes in Control and Inform. Sci. 58, Springer-Verlag, London, 1984, pp. 194–213.

[100] J. DEMMEL, *The componentwise distance to the nearest singular matrix*, SIAM J. Matrix Anal. Appl., 13 (1992), pp. 10–19.

[101] P. DEUFLHARD AND A. HOHMANN, *Numerische Mathematik. Eine algorithmisch orientierte Einführung*, de Gruyter, Berlin, 1991.

[102] A. DEVINATZ, *The strong Szegö limit theorem*, Illinois J. Math., 11 (1967), pp. 160–175.

[103] R. G. DOUGLAS, *Banach Algebra Techniques in Operator Theory*, 2nd edition, Springer-Verlag, New York, 1998.

[104] R. G. DOUGLAS AND R. HOWE, *On the C^*-algebra of Toeplitz operators on the quarter-plane*, Trans. Amer. Math. Soc., 158 (1971), pp. 203–217.

[105] R. V. DUDUCHAVA, *On discrete Wiener-Hopf equations*, Trudy Tbilis. Mat. Inst., 50 (1975), pp. 42–59 (in Russian).

[106] T. EHRHARDT, *A generalization of Pincus' formula and Toeplitz operator determinants*, Arch. Math. (Basel), 80 (2003), pp. 302–309.

[107] M. EIERMANN, *Fields of values and iterative methods*, Linear Algebra Appl., 180 (1993), pp. 167–197.

[108] R. L. ELLIS AND I. GOHBERG, *Orthogonal Systems and Convolution Operators*, Birkhäuser Verlag, Basel, 2003.

[109] L. ELSNER AND S. FRIEDLAND, *The limit of the spectral radius of block Toeplitz matrices with nonnegative entries*, Integral Equations Operator Theory, 36 (2000), pp. 193–200.

[110] M. Embree and L. N. Trefethen, *Pseudospectra Gateway*, Web site: http://www.comlab.ox.ac.uk/pseudospectra.

[111] M. Embree and L. N. Trefethen, *Generalizing eigenvalue theorems to pseudospectra theorems*, SIAM J. Sci. Comput., 23 (2001), pp. 583–590.

[112] V. Faber, A. Greenbaum, and D. E. Marshall, *The polynomial numerical hulls of Jordan blocks and related matrices*, Linear Algebra Appl., 374 (2003), pp. 231–246.

[113] D. K. Faddeev and I. S. Sominskiĭ, *A Collection of Exercises in Higher Algebra*, 10th edition, Nauka, Moscow, 1972 (in Russian).

[114] D. R. Farenick, M. Krupnik, N. Krupnik, and W. Y. Lee, *Normal Toeplitz matrices*, SIAM J. Matrix Anal. Appl., 17 (1996), pp. 1037–1043.

[115] D. Fasino, *Spectral properties of Toeplitz-plus-Hankel matrices*, Calcolo, 33 (1996), pp. 87–98.

[116] D. Fasino and P. Tilli, *Spectral clustering properties of block multilevel Hankel matrices*, Linear Algebra Appl., 306 (2000), pp. 155–163.

[117] J. Feinberg and A. Zee, *Non-Hermitian localization and delocalization*, Phys. Rev. E, 59 (1999), pp. 6433–6443.

[118] J. Feinberg and A. Zee, *Spectral curves of non-hermitian hamiltonians*, Nuc. Phys. B, 552 (1999), pp. 599–623.

[119] L. Fejér, *Untersuchungen über Fouriersche Reihen*, Math. Ann., 58 (1904), pp. 501–569.

[120] G. M. Fichtenholz, *Differential- und Integralrechnung III*, fünfte Auflage, Deutscher Verlag d. Wisss., Berlin, 1972.

[121] B. Fischer, *Polynomial Based Iteration Methods for Symmetric Linear Systems*, John Wiley & Sons, Ltd., Chichester, UK, and B. G. Teubner, Stuttgart, 1996.

[122] J. Fortiana and C. M. Cuadras, *A family of matrices, the discretized Brownian bridge, and distance-based regression*, Linear Algebra Appl., 264 (1997), pp. 173–188.

[123] F. D. Gakhov, *On Riemann's boundary value problem*, Matem. Sbornik, 2 (44) (1937), pp. 673–683 (in Russian).

[124] E. Gallestey, D. Hinrichsen, and A. J. Pritchard, *Spectral value sets of infinite-dimensional systems*, in Open Problems in Mathematical Systems and Control Theory, Comm. Control Engrg. Ser., Springer-Verlag, London, 1999, pp. 109–113.

[125] E. Gallestey, D. Hinrichsen, and A. J. Pritchard, *Spectral value sets of closed linear operators*, Proc. Roy. Soc. Lond. Ser. A, 456 (2000), pp. 1397–1418.

[126] V. I. Gel'fgat, *A normality criterion for Toeplitz matrices*, Comput. Math. Math. Phys., 35 (1995), pp. 1147–1150.

[127] J. S. Geronimo and K. M. Case, *Scattering theory and polynomials orthogonal on the unit circle*, J. Math. Phys., 20 (1979), pp. 299–310.

[128] I. Gohberg, *On an application of the theory of normed rings to singular integral equations*, Uspekhi Mat. Nauk, 7 (1952), pp. 149–156 (in Russian).

[129] I. Gohberg, *On the number of solutions of homogeneous singular integral equations with continuous coefficients*, Dokl. Akad. Nauk SSSR, 122 (1958), pp. 327–330 (in Russian).

[130] I. Gohberg and I. A. Feldman, *Convolution Equations and Projection Methods for Their Solution*, AMS, Providence, RI, 1974.

[131] I. Gohberg and I. Koltracht, *Mixed, componentwise, and structured condition numbers*, SIAM J. Matrix Anal. Appl., 14 (1993), pp. 688–704.

[132] I. Gohberg and I. Koltracht, *Structured condition numbers for linear matrix structures*, in Linear Algebra for Signal Processing (Minneapolis, 1992), IMA Vol. Math. Appl. 69, Springer-Verlag, New York, 1995, pp. 17–26.

[133] I. Gohberg and M. G. Krein, *Introduction to the Theory of Linear Non-Selfadjoint Operators in Hilbert Space*, AMS, Providence, RI, 1969.

[134] I. Gohberg and A. A. Sementsul, *The inversion of finite Toeplitz matrices and their continual analogues*, Matem. Issled., 7 (1972), pp. 201–223 (in Russian).

[135] I. Ya. Goldsheid and B. A. Khoruzhenko, *Distribution of eigenvalues in non-Hermitian Anderson models*, Phys. Rev. Lett., 80 (1998), pp. 2897–2900.

[136] I. Ya. Goldsheid and B. A. Khoruzhenko, *Eigenvalue curves of asymmetric tridiagonal random matrices*, Electronic J. Probability, 5 (2000), Paper 16, pp. 1–28.

[137] G. H. Golub and C. F. Van Loan, *Matrix Computations*, 3rd edition, Johns Hopkins University Press, Baltimore, MD, 1996.

[138] G. M. Goluzin, *Some estimates for bounded functions*, Matem. Sbornik, N. S., 26 (68) (1950), pp. 7–18 (in Russian).

[139] G. M. Goluzin, *Geometric Theory of Functions of a Complex Variable*, AMS, Providence, RI, 1969.

[140] S. Graillat, *A note on structured pseudospectra*, J. Comput. Appl. Math., to appear.

[141] A. Greenbaum, *Generalizations of the field of values useful in the study of polynomial functions of a matrix*, Linear Algebra Appl., 347 (2002), pp. 233–249.

[142] A. Greenbaum, *Card shuffling and the polynomial numerical hull of degree k*, SIAM J. Sci. Comput., 25 (2003), pp. 408–416.

[143] A. Greenbaum, Personal communication, August 2003.

[144] A. GREENBAUM AND L. N. TREFETHEN, *Do the Pseudospectra of a Matrix Determine Its Behavior?*, Technical Report TR 93-1371, Comp. Sci. Dept., Cornell University, Ithaca, NY, August 1993.

[145] U. GRENANDER AND G. SZEGÖ, *Toeplitz Forms and Their Applications*, University of California Press, Berkeley, CA, 1958.

[146] S. GRUDSKY AND A. V. KOZAK, *On the convergence speed of the norms of inverses of truncated Toeplitz operators*, in Integro-Differential Equations and Their Applications, Rostov State Univ. Press, Rostov-on-Don, 1995, pp. 45–55 (in Russian).

[147] C. GU AND L. PATTON, *Commutation relations for Toeplitz and Hankel matrices*, SIAM J. Matrix Anal. Appl., 24 (2003), pp. 728–746.

[148] K. E. GUSTAFSON AND D. K. M. RAO, *Numerical Range. The Field of Values of Linear Operators and Matrices*, Springer-Verlag, New York, 1997.

[149] R. HAGEN, S. ROCH, AND B. SILBERMANN, *C^*-Algebras and Numerical Analysis*, Marcel Dekker, New York, 2001.

[150] P. HALMOS, *A Hilbert Space Problem Book*, D. van Nostrand, Princeton, 1967.

[151] M. HANKE AND J. G. NAGY, *Toeplitz approximate inverse preconditioner for banded Toeplitz matrices*, Numer. Algorithms, 7 (1994), pp. 183–199.

[152] N. HATANO AND D. R. NELSON, *Vortex pinning and non-Hermitian quantum mechanics*, Phys. Rev. B, 56 (1997), pp. 8651–8673.

[153] F. HAUSDORFF, *Set Theory*, Chelsea, New York, 1957.

[154] G. HEINIG AND F. HELLINGER, *The finite section method for Moore-Penrose inversion of Toeplitz operators*, Integral Equations Operator Theory, 19 (1994), pp. 419–446.

[155] G. HEINIG AND K. ROST, *Algebraic Methods for Toeplitz-Like Matrices and Operators*, Akademie-Verlag, Berlin, 1984 and Birkhäuser Verlag, Basel, 1984.

[156] G. HEINIG AND K. ROST, *DFT representations of Toeplitz-plus-Hankel Bezoutians with application to fast matrix-vector multiplication*, Linear Algebra Appl., 284 (1998), pp. 157–175.

[157] G. HEINIG AND K. ROST, *Introduction to Structured Matrices*, book to appear.

[158] J. W. HELTON AND R. E. HOWE, *Integral operators: Commutators, traces, index, and homology*, in Proceedings of a Conference on Operator Theory, Lecture Notes in Math. 345, Springer-Verlag, Berlin, 1973, pp. 141–209.

[159] D. HERTZ, *On the extreme eigenvalues of Toeplitz and real Hankel interval matrices*, Multidimens. Systems Signal Process., 4 (1993), pp. 83–90.

[160] D. J. HIGHAM AND N. J. HIGHAM, *Backward error and condition of structured linear systems*, SIAM J. Matrix Anal. Appl., 13 (1992), pp. 162–175.

[161] N. J. HIGHAM, *Accuracy and Stability of Numerical Algorithms*, SIAM, Philadelphia, 1996.

[162] D. HINRICHSEN AND B. KELB, *Spectral value sets: A graphical tool for robustness analysis*, Systems Control Lett , 21 (1993), pp. 127–136.

[163] D. HINRICHSEN AND A. J. PRITCHARD, *Real and complex stability radii: A survey*, in Control of Uncertain Systems, Progr. Systems Control Theory 6, Birkhäuser Verlag, Boston, 1990, pp. 119–162.

[164] I. I. HIRSCHMAN, JR., *On a formula of Kac and Achiezer*, J. Math. Mech., 16 (1966), pp. 167–196.

[165] I. I. HIRSCHMAN, JR., *The spectra of certain Toeplitz matrices*, Illinois J. Math., 11 (1967), pp. 145–159.

[166] R. A. HORN AND C. R. JOHNSON, *Matrix Analysis*, Cambridge University Press, Cambridge, UK, 1985.

[167] R. A. HORN AND C. R. JOHNSON, *Topics in Matrix Analysis*, Cambridge University Press, Cambridge, UK, 1991.

[168] Z. HURÁK, A. BÖTTCHER, AND M. ŠEBEK, *Minimum distance to the range of a banded lower triangular Toeplitz operator in ℓ^1 and application in ℓ^1-optimal control*, to appear.

[169] KH. D. IKRAMOV, *On a desciption of normal Toeplitz matrices*, Comput. Math. Math. Phys., 34 (1994), pp. 399–404.

[170] KH. D. IKRAMOV, *Classification of normal Toeplitz matrices with real elements*, Math. Notes, 57 (1995), pp. 463–469.

[171] KH. D. IKRAMOV AND V. N. CHUGUNOV, *A criterion for the normality of a complex Toeplitz matrix*, Comput. Math. Math. Phys., 36 (1996), pp. 131–137.

[172] T. ITO, *Every normal Toeplitz matrix is either of type* I *or of type* II, SIAM J. Matrix Anal. Appl., 17 (1996), pp. 998–1006.

[173] R. A. JANIK, M. A. NOWAK, G. PAPP, AND I. ZAHED, *Localization transitions from free random variables*, Acta Physica Polonica B, 30 (1999), pp. 45–58.

[174] K. JOHANSSON, *On random matrices from the compact classical groups*, Annals of Math., 145 (1997), pp. 519–545.

[175] G. A. JONES AND D. SINGERMAN, *Complex Functions - An Algebraic and Geometric Viewpoint*, Cambridge University Press, Cambridge, UK, 1987.

[176] M. KAC, W. L. MURDOCK, AND G. SZEGÖ, *On the eigenvalues of certain Hermitian forms*, J. Rational Mech. Anal., 2 (1953), pp. 767–800.

[177] T. KAILATH AND A. H. SAYED, EDS., *Fast Reliable Algorithms for Matrices with Structure*, SIAM, Philadelphia, 1999.

[178] H. Kesten, *On a theorem of Spitzer and Stone and random walks with absorbing barriers*, Illinois J. Math., 5 (1961), pp. 246–266.

[179] H. Kesten, *Random walks with absorbing barriers and Toeplitz forms*, Illinois J. Math., 5 (1961), pp. 267–290.

[180] E. M. Klein, *The numerical range of a Toeplitz operator*, Proc. Amer. Math. Soc., 35 (1972), pp. 101–103.

[181] S. V. Konyagin and W. Schlag, *Lower bounds for the absolute value of random polynomials on a neighborhood of the unit circle*, Trans. Amer. Math. Soc., 351 (1999), pp. 4963–4980.

[182] P. A. Kozhukhar, *Linear operators*, Matem. Issled., 54 (1980), pp. 50–55 (in Russian).

[183] P. A. Kozhukhar, *The absence of eigenvalues in a perturbed discrete Wiener-Hopf operator*, Izv. Akad. Nauk Moldav. SSR Mat., 1990/3 (1990), pp. 26–35 (in Russian).

[184] M. G. Krein, *Integral equations on the half-line with a kernel depending on the difference of the arguments*, Uspekhi Mat. Nauk, 13 (1958), pp. 3–120 (in Russian).

[185] H. Landau, *On Szegö's eigenvalue distribution theorem and non-Hermitian kernels*, J. Analyse Math., 28 (1975), pp. 335–357.

[186] H. Landau, *Loss in unstable resonators*, J. Opt. Soc. Amer., 66 (1976), pp. 525–529.

[187] H. Landau, *The notion of approximate eigenvalues applied to an integral equation of laser theory*, Quart. Appl. Math., April 1977, pp. 165–171.

[188] L. Lerer and M. Tismenetsky, *Generalized Bezoutians and the inversion problem for block matrices, I, General scheme*, Integral Equations Operator Theory, 9 (1986), pp. 790–819.

[189] X. Liu, G. Strang, and S. Ott, *Localized eigenvectors from widely spaced matrix modifications*, SIAM J. Discrete Math., 16 (2003), pp. 479–498.

[190] J. Nagy, R. Plemmons, and T. Torgersen, *Iterative image restoring using approximate inverse preconditioning*, IEEE Trans. Image Processing, 15 (1996), pp. 1151–1162.

[191] D. R. Nelson and N. M. Shnerb, *Non-Hermitian localization and population biology*, Phys. Rev. E, 58 (1998), pp. 1383–1403.

[192] O. Nevanlinna, *Convergence of Iterations for Linear Equations*, Birkhäuser Verlag, Basel, 1993.

[193] O. Nevanlinna, *Hessenberg matrices in Krylov subspaces and the computation of the spectrum*, Numer. Funct. Anal. Optim., 16 (1995), pp. 443–473.

[194] L. N. Nikolskaya and Yu. B. Farforovskaya, *Toeplitz and Hankel matrices as Hadamard-Schur multipliers*, Algebra i Analiz, 15 (2003), pp. 141–160 (in Russian).

[195] N. K. NIKOLSKI, *Treatise on the Shift Operator*, Springer-Verlag, Berlin, 1986.

[196] N. K. NIKOLSKI, *Operators, Functions, and Systems: An Easy Reading. Vol. 1. Hardy, Hankel, and Toeplitz*, AMS, Providence, RI, 2002.

[197] S. V. PARTER, *On the extreme eigenvalues of truncated Toeplitz matrices*, Bull. Amer. Math. Soc., 67 (1961), pp. 191–196.

[198] S. V. PARTER, *Extreme eigenvalues of Toeplitz forms and applications to elliptic difference equations*, Trans. Amer. Math. Soc., 99 (1961), pp. 153–192.

[199] S. V. PARTER, *On the extreme eigenvalues of Toeplitz matrices*, Trans. Amer. Math. Soc., 100 (1961), pp. 263–275.

[200] S. V. PARTER, *On the distribution of the singular values of Toeplitz matrices*, Linear Algebra Appl., 80 (1986), pp. 115–130.

[201] J. R. PARTINGTON, *An Introduction to Hankel Operators*, Cambridge University Press, Cambridge, UK, 1988.

[202] V. V. PELLER, *Smooth Hankel operators and their applications (ideals S_p, Besov classes, random processes)*, Dokl. Akad. Nauk SSSR, 252 (1980), pp. 43–48 (in Russian).

[203] V. V. PELLER, *Hankel operators of class S_p and their applications (rational approximation, Gaussian processes, the problem of majorization of operators)*, Mat. Sbornik, 113 (1980), pp. 538–581 (in Russian).

[204] V. V. PELLER, *Hankel Operators and Their Applications*, Springer-Verlag, New York, 2003.

[205] J. PLEMELJ, *Ein Ergänzungssatz zur Cauchy'schen Integraldarstellung analytischer Funktionen, Randwerte betreffend*, Monatshefte Math. Phys., 19 (1908), pp. 205–210.

[206] J. D. PINCUS, *On the Trace of Commutators in the Algebra of Operators Generated by an Operator with Trace Class Self-Commutator*, unpublished manuscript, 1972.

[207] N. I. POLSKI, *Projection methods for solving linear equations*, Uspekhi Mat. Nauk, 18 (1963), pp. 179–180 (in Russian).

[208] G. PÓLYA AND G. SZEGÖ, *Problems and Theorems in Analysis, Vols. I and II*, Springer-Verlag, Berlin, 1998.

[209] A. POMP, *Zur Konvergenz des Reduktionsverfahrens für Wiener-Hopfsche Gleichungen, Teil I: Ein allgemeines Operatorenschema*, Preprint P-MATH-03/81, Akad. Wiss. DDR, Inst. f. Math., Berlin, 1981.

[210] A. POMP, *Zur Konvergenz des Reduktionsverfahrens für Wiener-Hopfsche Gleichungen, Teil II: Anwendungen auf diskrete Wiener-Hopfsche Gleichungen und Fehlerabschätzungen*, Preprint P-MATH-05/81, Akad. Wiss. DDR, Inst. f. Math., Berlin, 1981.

[211] D. Potts, *Schnelle Polynomialtransformationen und Vorkonditionierer für Toeplitz-Matrizen*, Shaker Verlag, Aachen, 1998.

[212] D. Potts and G. Steidl, *Preconditioners for ill-conditioned Toeplitz matrices*, BIT, 39 (1999), pp. 513–533.

[213] S. C. Power, *Hankel Operators on Hilbert Space*, Pitman, Boston, London, 1982.

[214] P. Rambour, J.-M. Rinkel, and A. Seghier, *Inverse asymptotique de la matrice de Toeplitz et noyau de Green*, C. R. Acad. Sci. Paris, 331 (2000), pp. 857–860.

[215] P. Rambour and A. Seghier, *Exact and asymptotic inverse of the Toeplitz matrix with polynomial singular symbol*, C. R. Acad. Sci. Paris, 335 (2002), pp. 705–710; erratum in C. R. Acad. Sci. Paris, 336 (2003), pp. 399–400.

[216] P. Rambour and A. Seghier, *Formulas for the inverses of Toeplitz matrices with polynomially singular symbols*, Integral Equations Operator Theory, 50 (2004), pp. 83–114.

[217] M. Reed and B. Simon, *Methods of Modern Mathematical Physics, Vol. I*, Academic Press, New York, 1972.

[218] E. Reich, *On non-Hermitian Toeplitz matrices*, Math. Scand., 10 (1962), pp. 145–152.

[219] L. Reichel and L. N. Trefethen, *Eigenvalues and pseudo-eigenvalues of Toeplitz matrices*, Linear Algebra Appl., 162/164 (1992), pp. 153–185.

[220] S. Roch, *Numerical ranges of large Toeplitz matrices*, Linear Algebra Appl., 282 (1998), pp. 185–198.

[221] S. Roch, *Pseudospectra of operator polynomials*, Oper. Theory Adv. Appl., 124 (2001), pp. 545–558.

[222] S. Roch and B. Silbermann, *Limiting sets of eigenvalues and singular values of Toeplitz matrices*, Asymptotic Anal., 8 (1994), pp. 293–309.

[223] S. Roch and B. Silbermann, *C^*-algebra techniques in numerical analysis*, J. Oper. Theory, 35 (1996), pp. 241–280.

[224] S. Roch and B. Silbermann, *Index calculus for approximation methods and singular value decomposition*, J. Math. Anal. Appl., 225 (1998), pp. 401–426.

[225] S. Roch and B. Silbermann, *A note on singular values of Cauchy-Toeplitz matrices*, Linear Algebra Appl., 275/276 (1998), pp. 531–536.

[226] M. Rosenblum, *The absolute continuity of Toeplitz's matrices*, Pacific J. Math., 10 (1960), pp. 987–996.

[227] M. Rosenblum, *Self-adjoint Toeplitz operators and associated orthonormal functions*, Proc. Amer. Math. Soc., 13 (1962), pp. 590–595.

[228] M. Rosenblum, *A concrete spectral theory for self-adjoint Toeplitz operators*, Amer. J. Math., 87 (1965), pp. 709–718.

[229] M. Rosenblum and J. Rovnyak, *Hardy Classes and Operator Theory*, Oxford University Press, New York, 1985.

[230] W. Rudin, *Real and Complex Analysis*, 3rd edition, McGraw-Hill, New York, 1987.

[231] S. M. Rump, *Estimation of the sensitivity of linear and nonlinear algebraic problems*, Linear Algebra Appl., 153 (1991), pp. 1–34.

[232] S. M. Rump, *Almost sharp bounds for the componentwise distance to the nearest singular matrix*, Linear and Multilinear Algebra, 42 (1997), pp. 93–107.

[233] S. M. Rump, *Bounds for the componentwise distance to the nearest singular matrix*, SIAM J. Matrix. Anal. Appl., 18 (1997), pp. 83–103.

[234] S. M. Rump, *Structured perturbations and symmetric matrices*, Linear Algebra Appl., 278 (1998), pp. 121–132.

[235] S. M. Rump, *Ill-conditioned matrices are componentwise near to singularity*, SIAM Rev., 41 (1999), pp. 102–112.

[236] S. M. Rump, *Structured perturbations part* I: *Normwise distances*, SIAM J. Matrix Anal. Appl., 25 (2003), pp. 1–30.

[237] S. M. Rump, *Structured perturbations part* II: *Componentwise distances*, SIAM J. Matrix Anal. Appl., 25 (2003), pp. 31–56.

[238] S. M. Rump, *Eigenvalues, pseudospectrum and structured perturbations*, Linear Algebra Appl., to appear.

[239] D. E. Rutherford, *Some continuant determinants arising in physics and chemistry*, Proc. Royal Soc. Edin., 62 A (1947), pp. 229–236.

[240] D. E. Rutherford, *Some continuant determinants arising in physics and chemistry, II*, Proc. Royal Soc. Edin., 63 A (1952), pp. 232–241.

[241] A. L. Sakhnovich, *Szegö limits for infinite Toeplitz matrices determined by the Taylor series of two rational functions*, Linear Algebra Appl., 343/344 (2002), pp. 291–302.

[242] A. L. Sakhnovich and I. M. Spitkovsky, *Block-Toeplitz matrices and associated properties of a Gaussian model on the half axis*, Teoret. Mat. Fiz., 63 (1985), pp. 154–160 (in Russian).

[243] P. Schmidt and F. Spitzer, *The Toeplitz matrices of an arbitrary Laurent polynomial*, Math. Scand., 8 (1960) pp. 15–38.

[244] I. Schur and G. Szegö, *Über die Abschnitte einer im Einheitskreise beschränkten Potenzreihe*, Sitzungsberichte Preuss. Akad. Wiss. Berlin, 1925, pp. 545–560.

[245] D. SeLegue, *A C^*-algebraic extension of the Szegö trace formula*, talk given at the GPOTS, Arizona State University, Tempe, May 22, 1996.

[246] S. Serra, *Preconditioning strategies for Hermitian Toeplitz systems with nondefinite generating functions*, SIAM J. Matrix Anal. Appl., 17 (1996), pp. 1007–1019.

[247] S. Serra Capizzano, *On the extreme spectral properties of Toeplitz matrices generated by L^1 functions with several minima/maxima*, BIT, 36 (1996), pp. 135–142.

[248] S. Serra Capizzano, *On the extreme eigenvalues of Hermitian (block) Toeplitz matrices*, Linear Algebra Appl., 270 (1998), pp. 109–129.

[249] S. Serra, *How to choose the best iterative strategy for symmetric Toeplitz systems*, SIAM J. Numer. Anal., 36 (1999), pp. 1078–1103.

[250] S. Serra Capizzano, *Spectral behavior of matrix sequences and discretized boundary value problems*, Linear Algebra Appl., 337 (2001), pp. 37–78.

[251] S. Serra Capizzano and P. Tilli, *Extreme singular values and eigenvalues of non-Hermitian block Toeplitz matrices*, J. Comput. Appl. Math., 108 (1999), pp. 113–130.

[252] E. Shargorodsky, *Geometry of higher order relative spectra and projection methods*, J. Oper. Theory, 44 (2000), pp. 43–62.

[253] B. Silbermann, *Lokale Theorie des Reduktionsverfahrens für Toeplitzoperatoren*, Math. Nachr., 104 (1981), pp. 137–146.

[254] B. Simon, *Notes on infinite determinants of Hilbert space operators*, Advances in Math., 24 (1977), pp. 244–273.

[255] B. Simon, *Orthogonal Polynomials on the Unit Circle, Part 1*, AMS, Providence, RI, 2005.

[256] R. D. Skeel, *Scaling for numerical stability in Gaussian elimination*, J. Assoc. Comput. Mach., 26 (1979), pp. 494–526.

[257] F. Spitzer and C. J. Stone, *A class of Toeplitz forms and their application to probability theory*, Illinois J. Math., 4 (1960), pp. 253–277.

[258] G. Strang, *A proposal for Toeplitz matrix computations*, Stud. Appl. Math., 74 (1986), pp. 171–176.

[259] G. Strang, *From the SIAM President*, SIAM News, April 2000 and May 2000.

[260] T. Strohmer, *Four short stories about Toeplitz matrix calculations*, Linear Algebra Appl., 343/344 (2002), pp. 321–344.

[261] F.-W. Sun, Y. Jiang, and J. S. Baras, *On the convergence of the inverses of Toeplitz matrices and its applications*, IEEE Trans. Inform. Theory, 49 (2003), pp. 180–190.

[262] G. Szegö, *Ein Grenzwertsatz über die Toeplitzschen Determinanten einer reellen positiven Funktion*, Math. Ann., 76 (1915), pp. 490–503.

[263] G. SZEGÖ, *On certain Hermitian forms associated with the Fourier series of a positive function*, in Festschrift Marcel Riesz, Lund, 1952, pp. 222–238.

[264] P. TILLI, *Singular values and eigenvalues of non-Hermitian block Toeplitz matrices*, Linear Algebra Appl., 272 (1998), pp. 59–89.

[265] P. TILLI, *Some results on complex Toeplitz eigenvalues*, J. Math. Anal. Appl., 239 (1999), pp. 390–401.

[266] M. TISMENETSKY, *Determinant of block-Toeplitz band matrices*, Linear Algebra Appl., 85 (1987), pp. 165–184.

[267] O. TOEPLITZ, *Zur Theorie der quadratischen und bilinearen Formen von unendlichvielen Veränderlichen*, Math. Ann., 70 (1911), pp. 351–376.

[268] O. TOEPLITZ, *Das algebraische Analogon zu einem Satze von Fejér*, Math. Z., 2 (1918), pp. 187–197.

[269] L. N. TREFETHEN, *Approximation theory and numerical linear algebra*, in Algorithms for Approximation II, J. C. Mason and M. G. Cox, eds., Chapman and Hall, London, 1990, pp. 336–360.

[270] L. N. TREFETHEN, *Pseudospectra of matrices*, in Numerical Analysis 1991, D. F. Griffiths and G. A. Watson, eds., Longman Sci. Tech, Harlow, Essex, UK, 1992, pp. 234–266.

[271] L. N. TREFETHEN, *Computation of pseudospectra*, Acta Numerica, 8 (1999), pp. 247–295.

[272] L. N. TREFETHEN, Personal communication, February 2003.

[273] L. N. TREFETHEN AND D. BAU, III, *Numerical Linear Algebra*, SIAM, Philadelphia, 1997.

[274] L. N. TREFETHEN, M. CONTEDINI, AND M. EMBREE, *Spectra, pseudospectra, and localization for random bidiagonal matrices*, Comm. Pure Appl. Math., 54 (2001), pp. 594–623.

[275] L. N. TREFETHEN AND M. EMBREE, *Spectra and Pseudospectra: The Behavior of Nonnormal Matrices and Operators*, Princeton University Press, Princeton, 2005.

[276] W. F. TRENCH, *An algorithm for the inversion of finite Toeplitz matrices*, J. Soc. Indust. Appl. Math., 12 (1964), pp. 515–522.

[277] W. F. TRENCH, *Inversion of Toeplitz band matrices*, Math. Comp., 28 (1974), pp. 1089–1095.

[278] W. F. TRENCH, *On the eigenvalue problem for Toeplitz band matrices*, Linear Algebra Appl., 64 (1985), pp. 199–214.

[279] W. F. TRENCH, *Asymptotic distribution of the spectra of a class of generalized Kac-Murdock-Szegö matrices*, Linear Algebra Appl., 294 (1999), pp. 181–192; erratum in Linear Algebra Appl., 320 (2000), p. 213.

[280] W. F. TRENCH, *Asymptotic distribution of the even and odd spectra of real symmetric Toeplitz matrices*, Linear Algebra Appl., 302/303 (1999), pp. 155–162.

[281] W. F. TRENCH, *Spectral distribution of generalized Kac-Murdock-Szegö matrices*, Linear Algebra Appl., 347 (2002), pp. 251–273.

[282] E. E. TYRTYSHNIKOV, *Influence of matrix operations on the distribution of eigenvalues and singular values of Toeplitz matrices*, Linear Algebra Appl., 207 (1994), pp. 225–249.

[283] E. E. TYRTYSHNIKOV, *Circulant preconditioners with unbounded inverses*, Linear Algebra Appl., 216 (1995), pp. 1–24.

[284] E. E. TYRTYSHNIKOV, *A unifying approach to some old and new theorems on distribution and clustering*, Linear Algebra Appl., 232 (1996), pp. 1–43.

[285] E. E. TYRTYSHNIKOV AND N. L. ZAMARASHKIN, *Toeplitz eigenvalues for Radon measures*, Linear Algebra Appl., 343/344 (2002), pp. 345–354.

[286] J. L. ULLMAN, *A problem of Schmidt and Spitzer*, Bull. Amer. Math. Soc., 73 (1967), pp. 883–885.

[287] V. S. VLADIMIROV AND I. V. VOLOVICH, *A model of statistical physics*, Teoret. Mat. Fiz., 54 (1983), pp. 8–22 (in Russian).

[288] R. VREUGDENHIL, *The resolution of the identity for selfadjoint Toeplitz operators with rational matrix symbol*, Integral Equations Operator Theory, 20 (1994), pp. 449–490.

[289] E. WEGERT AND L. N. TREFETHEN, *From the Buffon needle problem to the Kreiss matrix theorem*, Amer. Math. Monthly, 101 (1994), pp. 132–139.

[290] H. WIDOM, *On the eigenvalues of certain Hermitian operators*, Trans. Amer. Math. Soc., 88 (1958), pp. 491–522.

[291] H. WIDOM, *Extreme eigenvalues of translation kernels*, Trans. Amer. Math. Soc., 100 (1961), pp. 252–262.

[292] H. WIDOM, *Extreme eigenvalues of N-dimensional convolution operators*, Trans. Amer. Math. Soc., 106 (1963), pp. 391–414.

[293] H. WIDOM, *Toeplitz determinants with singular generating functions*, Amer. J. Math., 95 (1973), pp. 333–383.

[294] H. WIDOM, *Asymptotic behavior of block Toeplitz matrices and determinants, II*, Advances in Math., 21 (1976), pp. 1–29.

[295] H. WIDOM, *On the singular values of Toeplitz matrices*, Z. Anal. Anwendungen, 8 (1989), pp. 221–229.

[296] H. WIDOM, *Eigenvalue distribution of nonselfadjoint Toeplitz matrices and the asymptotics of Toeplitz determinants in the case of nonvanishing index*, Oper. Theory Adv. Appl., 48 (1990), pp. 387–421.

[297] H. WIDOM, *Eigenvalue distribution for nonselfadjoint Toeplitz matrices*, Oper. Theory Adv. Appl., 71 (1994), pp. 1–8.

[298] J. H. WILKINSON, *The Algebraic Eigenvalue Problem*, Clarendon Press, Oxford, UK, 1965.

[299] T. WRIGHT, *EigTool software package*, Web site: http://www.comlab.ox.ac.uk/pseudospectra/eigtool.

[300] T. WRIGHT, *Algorithms and Software for Pseudospectra*, Thesis, University of Oxford, Oxford, UK, 2002.

[301] N. L. ZAMARASHKIN AND E. E. TYRTYSHNIKOV, *Distribution of the eigenvalues and singular numbers of Toeplitz matrices under weakened requirements on the generating function*, Sb. Math., 188 (1997), pp. 1191–1201.

[302] N. L. ZAMARASHKIN AND E. E. TYRTYSHNIKOV, *On the distribution of the eigenvectors of Toeplitz matrices under weakened requirements for the generating function*, Russian Math. Surveys, 52 (1997), pp. 1333–1334.

[303] P. ZIZLER, R. A. ZUIDWIJK, K. F. TAYLOR, AND S. ARIMOTO, *A finer aspect of eigenvalue distribution of selfadjoint band Toeplitz matrices*, SIAM J. Matrix Anal. Appl., 24 (2002), pp. 59–67.

[304] A. ZYGMUND, *Trigonometric Series, Vol. I*, Cambridge University Press, Cambridge, UK, 1988.

Index